APPLICATIONS OF HOLOGRAPHY
AND OPTICAL DATA PROCESSING

APPLICATIONS OF HOLOGRAPHY AND OPTICAL DATA PROCESSING

Proceedings of the International Conference

Jerusalem, August 23 - 26, 1976

Editors

E. MAROM

School of Engineering, Tel Aviv University, Israel

A. A. FRIESEM

Department of Electronics, The Weizmann Institute of Science, Israel

and

E. WIENER-AVNEAR

Department of Physics, Ben Gurion University of Negev, Israel

PERGAMON PRESS

OXFORD · NEW YORK · TORONTO · SYDNEY · PARIS · FRANKFURT

U.K.	Pergamon Press Ltd., Headington Hill Hall, Oxford OX3 0BW, England
U.S.A.	Pergamon Press Inc., Maxwell House, Fairview Park, Elmsford, New York 10523, U.S.A.
CANADA	Pergamon of Canada Ltd., 75 The East Mall, Toronto, Ontario, Canada
AUSTRALIA	Pergamon Press (Aust.) Pty. Ltd., 19a Boundary Street, Rushcutters Bay, N.S.W. 2011, Australia
FRANCE	Pergamon Press SARL, 24 Rue des Ecoles, 75240 Paris, Cedex 05, France
WEST GERMANY	Pergamon Press GmbH, 6242 Kronberg-Taunus, Pferdstrasse 1, Frankfurt-am-Main, West Germany

Library of Congress Cataloging in Publication Data

International Conference on Applications of Holography and Optical Data processing, Jersusalem, Israel, 1976.
Proceedings of the International Conference on Applications of Holography and Optical Data Processing, August 23-26, 1976, Jerusalem, Israel.

Includes bibliographical references.
1. Holography—Congresses. 2. Optical data processing—Congresses. I. Marom, E. II. Friesem, A. A. III. Wiener-Avnear, E. IV. Title: Proceedings of the International Conference on Applications of Holography and Optical Data Processing ...
TA1542.I58 1976   621.36   77-1864
ISBN 0-08-021625-0

Printed in Great Britain by William Clowes & Sons, Limited London, Beccles and Colchester

CONTENTS

v

SESSION 17. MEASURING METHODS II

Chairman: J. SHAMIR

SESSION 18. NON-OPTICAL HOLOGRAPHY

Chairman: J. POLITCH

LIST OF PARTICIPANTS

ABITBOL, C. I. Institut d'Optique, Paris, France
ABRAMSON, N. Royal Institute of Technology, Stockholm, Sweden
AGUILAR, M. Facultad de Ciencas, Valencia, Spain
ALEXANDER, G. Rafael, Haifa, Israel
ALMEIDA, S. P. Virginia Polytechnic Institute, Virginia, U.S.A.
ALMI, L. U. El-Op, Rehovot, Israel
APFELDORFER, H. C. National Physical Laboratory, Hebrew University, Jerusalem
   Israel
ASSA, A. Technion, Haifa, Israel
ATKINSON, J. T. Liverpool Polytechnic, Liverpool, England
BAIRD, K. M. National Research Council, Ontario, Canada
BALLY, G. von Hals-Nasen-Ohren-Klinik der Universität, Munster, Germany
BAR-CHAIM, N. Tel Aviv University, Tel Aviv, Israel
BAR-DAVID, I. Technion, Haifa, Israel
BARRETT, H. H. University of Arizona, Arizona, U.S.A.
BEDARIDA, F. Universita di Genova, Genova, Italy
BEN-YEHUDA, R. 22 Herzog, Tel Aviv, Israel
BEN-YOSEF, N. Jerusalem College of Technology, Jerusalem, Israel
BENTON, S. A. Polaroid Corporation, Mass., U.S.A.
BETHDIN, E. M.B.T., Yahud, Israel
BETSER, A. Technion, Haifa, Israel
BIEDERMANN, K. The Royal Institute of Technology, Stockholm, Sweden
BIRIN, G. National Physical Laboratory, Hebrew University, Jerusalem, Israel
BJELKHAGEN, H. Royal Institute of Technology, Stockholm, Sweden
BODENHEIMER, H. Jerusalem College of Technology, Jerusalem, Israel
BOUCLE, R. A.F.E., Paris, France
BRUNSTEIN, M. Tel Aviv University, Tel Aviv, Israel
BRYNGDAHL, O. Université de Paris-Sud, Orsay, France
CHAUVALLON, A. Maubourg, Paris, France
CHAVEL, P. Université de Paris-Sud, Orsay, France
CLAIR, J. J. Université de Paris VI, Paris, France
COZENS, J. R. Imperial College, London, England
DAN, M. Rafael, Haifa, Israel
DAN, S. Rafael, Haifa, Israel
DANDLIKER, R. Brown Boveri Research Center, Baden, Switzerland
DAS, P. Rensselaer, Polytechnic Institute, New York, U.S.A.
DIL, J. Phillips Laboratories, Netherlands
DOLAN, G. Hebrew University, Jerusalem, Israel
DOMBROWSKI, R.
DORFAN, M. El-Op, Rehovot, Israel
DZIALOWSKI, Y. Université de Paris VI, Paris, France
EFRON, U. M.M.G., Yavne, Israel
EICHMANN, G. City College of New York, New York, U.S.A.
EK, L. Royal Institute of Technology, Stockholm, Sweden
EREZ, R. Technion, Haifa, Israel
ESITI, E. Italy

FENIGSTEIN, I.   3 Habeluyem Street, Ramat Gan, Israel
FINKLER, R.   London, S.W.F., England
FISHMAN, M.   El-Op, Rehovot, Israel
FLEURET, J.   Ecole Nationale Superieure des Telecommunications, Paris, France
FOSSATI-BELLANI, V.   C.I.S.E. Milano, Italy
FOURNIER, J. M.   Laboratoire de Physique, Besanon, France
FRANKENA, H. J.   Delft University of Technology, Delft, Netherlands
FREJLICH, J.   Université de Paris VI, Paris, France
FRIESEM, A. A.   The Weizmann Institute, Rehovot, Israel
GANIEL, U.   The Weizmann Institute, Rehovot, Israel
GASSEND, M. L. A.   University of Manitoba, Winnipeg, Canada
GILL, D.   Ben-Gurion University, Beer-Sheva, Israel
GLASER, I.   The Weizmann Institute, Rehovot, Israel
GOLDBERG, J.   National Measurement Laboratory, Chippendale, Australia
GOLDSCHMIDT, S.   Hebrew University, Jerusalem, Israel
GREENFIELD, E.   PO Box 16042, Jerusalem, Israel
GRIMLAND, J.   PO Box 23004, Tel-Aviv, Israel
GRIMLAND, S.   3 Yehuda Halevi Street, Herzliya, Israel
GRINBERG, J.   Hughes Research Laboratories, California, U.S.A.
GRONEMANN, U.   Sci-Tex Corporation, PO Box 330, Herzliya, Israel
HAMALAINEN, R.   University of Joensuv, Joensuv, Finland
HAUSLER, G.   Universität Erlangen-Nurnberg, Erlangen, Germany
HEGEDUS, Z. S.   National Measurement Laboratory, Chippendale, Australia
HENRY, M.   Institut d'Optique, Paris, France
HOCHBERG, A.   M.M.G., Yavne, Israel
INBAR, D.   Elscint, Tel-Aviv, Israel
INEICHEN, B.   Brown Boveri Research Center, Baden, Switzerland
JAERISCH, W.   IBM Deutschland, Sindelfinger, Germany
JIFAL, S.   Rafael, Haifa, Israel
JOHANSSON, S.   Royal Institute of Technology, Stockholm, Sweden
JOSHUA, H.   Rafael, Haifa, Israel
KALISKY, A.   PO Box 2197, Jerusalem, Israel
KARSIKOV, N.   12 Shahevich Street, Gedera, Israel
KASHER, I.   12 Samuel Hanagid Street, Herzliya, Israel
KATO, M.   Metsushita Electric Industrial, Osaka, Japan
KATZ, J.   25 Eilat Street, Givatayim, Israel
KATZIR, Y.   The Weizmann Institute, Rehovot, Israel
KAUFMAN, K.   Kulso, Haifa, Israel
KERMISCH, D.   Xerox Corporation, New York, U.S.A.
KESSLER, D.   Tel-Aviv University, Tel Aviv, Israel
KONFORTI, N.   10 Tzeelim Street, Holon, Israel
KOPEIKA, N. S.   Ben-Gurion University, Beer-Sheva, Israel
KOPILOVITZ, B.   614 9th Street, Petach-Tikva, Israel
KORMAN, A.   Tel-Aviv University, Tel Aviv, Israel
KOZIMA, K.   Kyoto Technical University, Kyoto, Japan
KRIENS, R.   Institute of Applied Physics, Delft, Netherlands
KURZ, H.   Phillips, Forschungslaboratorium, Hamburg, Germany
LABRUNIE, G.   Leti/Ceng, Grenoble, France
LANDAU, J.   Technion, Haifa, Israel
LANGE, E.   National Physical Laboratory, Jerusalem, Israel
LAVI, A.   Ministry of Commerce and Industry, Jerusalem, Israel
LEITH, E. N.   University of Michigan, Ann Arbor, Michigan, U.S.A.
LEVEBY, K.   Royal Institute of Technology, Stockholm, Sweden
LEVI, L.   Jerusalem College of Technology, Jerusalem, Israel
LEVI, U.   The Weizmann Institute, Rehovot, Israel
LEVINSON, J.   M.M.G., Yavne

LEVITIN, L.   Tel Aviv University, Tel Aviv, Israel
LIPSON, S. G.   The Technion, Haifa, Israel
LOHMANN, A.   Universitat Erlange-Nurnberg, Germany
LOWENTHAL, S.   Université de Paris-Sud, Orsay, Paris, France
LUPO, A.   Tel Aviv University, Tel Aviv, Israel
MAIDANIK, G.   David W. Taylor Naval Ship Research and Development Center,
   Maryland, U.S.A.
MAKOSCH, G.   IBM Deutschland, Sindelfinger, Germany
MANOR, E.   Moshav Tal-Shahar, Israel
MARECHAL, A.   Université de Paris-Sud, Orsay, France
MAROM, E.   Tel Aviv University, Tel Aviv, Israel
MATSUDA, K.   Mechanical Engineering Laboratory, Tokyo, Japan
MAY, M.   Université de Paris VI, Paris, France
MICHERON, F.   Thomson-C.S.F., Orsay, France
MIYAKE, K. P.   Kyoiku University, Tokyo, Japan
MOSHE, G.   Rafael, Haifa, Israel
MOSSEL, M.   National Physical Laboratory, Jerusalem, Israel
NAGLER, M.   Rehovot Instruments Ltd., Rehovot, Israel
NESHER, Y.   Herzliya, Israel
NEUGARTEN, M.   Merkas Klita, Sc. Hameginim, Haifa, Israel
NILSSON, L. E.   Royal Institute of Technology, Stockholm, Sweden
ÖSTLUND, L. A.   Royal Institute of Technology, Stockholm, Sweden
PAGE, M.   College Militaire de Saint Jean, Quebec, Canada
PEER, J.   25 Charlton Ave., Hamilton, Canada
PERI, D.   The Weizmann Institute, Rehovot, Israel
PETKOVSEK, J.   724 Cesta 24, Junjija, Yugoslavia
PIPMAN, J.   Technion, Haifa, Israel
POLITCH, J.   Technion, Haifa, Israel
POVERENE, A.   Peran St. 18/6, Jerusalem, Israel
RAM, G.   PO Box 5258, Haifa, Israel
RATERINK, H. J.   Institute of Applied Physics, Delft, Netherlands
RAVNOY, Z.   Weizmann Institute, Rehovot, Israel
RECHAVI, E.   Tel Aviv University, Tel Aviv, Israel
REICH, S.   The Weizmann Institute, Rehovot, Israel
RIBAK, E.   The Weizmann Institute, Rehovot, Israel
ROBLIN, G.   Laboratoire Associé au CNRS, Paris, France
RONEN, J.   Hachochit 24, Ramat-Hasharon, Israel
ROUSSEAU, M.   Faculté de Sciences de Rouen, Paris, France
RUDOLPH, D.   Universitäts Sternwarte Göttingen, W. Germany
SANCHEZ, F.   Université de Paris Sud, Orsay, France
SEKLER, J.   Tel Aviv University, Tel Aviv, Israel
SELVI, W.   Rafael, Haifa, Israel
SHAMIR, J.   The Technion, Haifa, Israel
SHAW, R.   R.M.I.T., Melbourne, Australia
SHKLANSKY, D.   Haifa, Israel
SPITZ, E.   Thomson-CSF, Orsay, France
STEEL, W.   National Measurement Laboratory, Chippendale, Australia
STEIN, R.   Tel Hashomer Hospital, Israel
SURGET, J.   Office National de Recherches Aerospatiales, Chatillon, France
TAMIR, I.   Polytechnic Institute of New York, Brooklyn, New York, U.S.A.
TANOS, E.   Research Institute of Electric Industry, Budapest, Hungary
TREVES, D.   The Weizmann Institute, Rehovot, Israel
TRIBILLON, D.   B.I.S., 26 Blv. Victor, Paris, Armees, France
TRICOLES, G.   General Dynamics Electronics Division, San Diego, U.S.A.
TSCHUDI, T.   Institute für Angew. Physik, Bern, Switzerland
TSUJIUCHI, J.   Tokyo Institute of Technology, Midori-ku, Yokohama, Japan

VAN DER MEULEN, J. H. J. Netherlands Ship Basin, Wageningen, Netherlands
VIENOT, A. Ch. Université de Besançon, Besançon, France
VIRDI, S. S. Université de Paris Sud, Orsay, France
WALKUP, J. F. Texas Technical University, Lubbock, Texas, U.S.A.
WEIL, R. B. Technion, Haifa, Israel
WEINBERGER, Z. Jerusalem College of Technology, Jerusalem, Israel
WIENER-AVNEAR, E. Ben Gurion University, Beer-Sheva, Israel
WLOCH, G. R. 73 Eaglewood Road, London, England
YAHALOM, R. Hashora 6, Ramat-Gan, Israel
ZEFIRO, L. Universita di Genova, Genova, Italy
ZIEKMAN, R. S.K.F. European Research Center, Netherlands

EDITOR'S PREFACE

To the scores of travellers from every corner of the earth whom the ancient city of Jerusalem has welcomed and inspired over the centuries, can now be added another group - the participants in the August 1976 International Conference on Applications of Holography and Optical Data Processing.

As evidenced by the conference proceedings, collected in this volume, a wide range of topics, representing research on the forefront of applied and theoretical optics, was covered in the four day meeting. Included were holographic applications in medicine, interferometry, display, storage and retrieval, as well as image processing, measuring techniques, spectroscopy, and, of course, recording materials and devices.

Agreement on the high points of this conference would be hard to find for the meeting was characterized by a high level of scientific presentations and the active participation of many eminent scientists in the field of optics. However, as one participant wryly remarked while the conference excursion bus, headed down from Jerusalem to the Dead Sea (395 meter below sea level) -- "Well, there is no question as to what is going to be the low point of this conference".

It is also probably safe to say that the general concensus among the participants was that this August meeting served to reaffirm the vitality and viability of the research area to which we are all committed. If some people were questioning themselves about the future of Optical Data Processing, the participants at this conference reaffirmed their belief in its promising prospects, when coherent and incoherent techniques, complemented by electronic support will find practical applications. Indeed it has left us with renewed anticipation of upcoming gatherings and significant advances in this field.

To keep up the momentum generated by the conference, the editors decided to publish the proceedings at the earliest possible date. This goal has been achieved at the occasional expense of editing and polishing of language and style, since, with the exception of minor modifications, the material in this publication has been reproduced directly from copies supplied by the authors. However, the high scientific standards that governed the selection of papers for the conference were maintained and for this we owe our gratitude to the referees who screened the material for publication.

We would also like to extend our thanks to all those organizations who lent their support to the conference, and in particularly, to the International Commission of Optics, which was the prime sponsor.

In the convivial conference atmosphere where old acquaintances were renewed and new ones made, we were constantly reminded of the absence of the mentor of our field, Professor Dennis Gabor, prevented by illness from joining us in Jerusalem. It is to him that the editors would like to dedicate this volume.

E. Marom
A.A. Friesem
E. Wiener-Avnear

Jerusalem, August 1976

OPENING SESSION

Chairman: A. A. FRIESEM

RECENT TRENDS IN HOLOGRAPHY
AND OPTICAL DATA PROCESSING

E. N. Leith

*Environmental Research Institute of Michigan and The University
of Michigan, Ann Arbor, Michigan, U.S.A.*

Holography and optical data processing became, in the 1960's, areas of
intensive research. This activity still continues strong—over a decade
later. Indeed, each year sees fresh, new results, showing that these fields,
despite the many years of their existence, remain young. As we survey the
research results of recent years, we find much that is new and original. As
we look at the program of this meeting, we find significant, original work
in abundance. Certainly, all the major areas of these two disciplines are
well represented. In this introductory talk, however, I will confine the
number of topics to a relatively few, based in part upon my own interests.

One of the most promising applications of holography has been displays.
Since 1964, when high quality, 3-D holographic imagery was demonstrated,
displays became an obvious application, although this early promise never
seemed to be fulfilled, for certain rather obvious reasons, including the
expense of lasers and the speckly character of the image.

If one had in earlier years posed the question, what advances in holography
were needed in order to bring this promising application to fruition, he
would likely have received such replies as, eliminate the laser (or other
coherent source) from the viewing process, and eliminate the laser from the
process of recording information about the object. If only the remarkable
results of holography could be obtained with ordinary light instead of with
laser light, the success of display holography would be assured.

Indeed, these objectives have been achieved. The result has been in the
past two years, a tremendous surge in display holography, which has now been
brought to the stage of economic viability.

Of course, the Denisyuk-Lippmann holograms have been available for years.
Recent experimental procedures have vastly improved this process, primarily
by increasing the diffraction efficiency without lowering the image quality.
Since this type of hologram utilizes only a narrow wavelength band of the
incident light, high diffraction efficiency within this band is essential.

A completely different approach was given by Stephen Benton, in which the
entire visible spectrum is used in viewing the hologram, with the result
that the hologram can be exceedingly bright, even when the source is of very
moderate intensity [1]. This procedure involves an interesting tradeoff,
whereby this broad-spectrum capability is obtained in exchange for the loss
of vertical parallax. This is an acceptable tradeoff, and observers
unskilled in optics generally do not notice the missing vertical parallax.

But what about eliminating the coherence requirement in recording information about the object? A recent development by Lloyd Cross in effect accomplishes this. The key is to photograph the object in a conventional way, making hundreds (or thousands) of ordinary pictures, all from different positions. This array of pictures contains essentially the same information as a hologram; one has only to synthesize them into a hologram. Such techniques are by no means new, dating back to Pole (1966). Cross has built upon this earlier work, incorporated new ideas, and engineered it all into a highly workable system. Such holograms can be made of any object that can be photographed, and the resulting hologram is amenable to the Benton method of white light readout.

Of course, one does not entirely eliminate the coherence requirement in making the hologram, since the actual process of synthesizing the hologram from the many conventional pictures does use coherent light; however, what is significant is that the data acquisition portion of the process, wherein the object is photographed, is done with ordinary photography, using incoherent light. The actual hologram making process, although done with coherent light, now no longer imposes a limitation on the type of object that can be holographed.

These new developments have carried display holography to the threshold of commercial viability. For some applications, holography is now from the commercial viewpoint clearly the best form of 3-D imagery.

Real time optical processing is another area that has recently reached a stage of development that at last borders on the practical. For many years, workers in optical processing have sought a real time device that would be a substitute for photographic film in incoherent to coherent light transducers. Now, various types of devices, such as PROM, liquid crystals, and thermoplastic have reached a stage of development where they can serve practical purposes.

Some of the real time devices have been in existence for many years, although in a state of development inadequate for application. Such was the case with thermoplastic recorders, which have been available since the 1950's. Although they serve useful functions involving incoherent illumination, they had always been inadequate for coherent readout, primarily because they had never been of sufficiently high optical quality. Now, ERIM researchers have overcome this difficulty and have achieved optically flat thermoplastic coated surfaces of the order of a square inch [2].

Another area which continues to yield exciting, although primarily academic results is that of diffraction from volume gratings. The theory of diffraction from thick structures is old, having long been studied in the context of x-ray diffraction from crystals and the diffraction of ultrasonic waves from columns of water or quartz, in which travelling compressional waves have been induced.

Recorded interference fringes in thick phase materials have been used to diffract light waves. Such work, in the context of holography, dates from the early 1960's, and the list of researchers who have contributed to this area is long. Yet, even today we find new and interesting physics emerging from studies of this sort. Among the most recent contributors are S. Case and R. Alferness, who have studied diffraction from multiple gratings. In their several papers are to be found many interesting results, relating primarily to diffraction efficiencies possible under various circumstances [3-5]. For example, they find that in the Bragg diffraction process, incident light energy coupled into one grating can be transferred in its entirety to the other grating as the light propagates through the composite structure. Again, Alferness finds, for diffraction at the second order, a condition analogous to Kogelnik's condition for achieving 100% diffraction efficiency in the first order. But he finds in addition, an auxiliary condition, independent of thickness or index modulation, but dependent on spatial frequency of the recorded fringes, which must also be met in order to achieve 100% diffraction efficiency in the second order.

Thick phase diffraction gratings may not be competitive with conventional gratings for conventional uses, but as Kogelnik first showed, they can serve such interesting purposes as couplers to optical waveguides. Recent work by Case and co-workers has revealed a remarkable versatility of such gratings, as non-linear couplers, multimode couplers, mode converters, and the like [6].

The preponderance of papers in the area of hologram interferometry shows this area to remain as the principle application of holography, a position it has maintained for the past 11 years. This technique, along with the newer technique of laser speckle interferometry, has resulted in a wide variety of useful optical testing techniques.

It is gratifying to see that holography and optical processing, which have already produced high amounts of good research, continues to be productive areas, as though their potential were limitless. It is even more gratifying to see, more and more, useful devices emerging from this research.

REFERENCES

1. S. Benton, Hologram Reconstruction with Extended Incoherent Source, J. Opt. Soc. Am. 59, 1545 (1969).

2. G. Currie, I. Cindrich, C. Leonard, The ERIM TOPR in Optical Data Processing, 20 th Annual SPIE Technical Symposium, 1976.

3. R. Alferness, Analysis of Optical Propagation in Thick Holographic Gratings, Appl. Phys. 7, 29 (1975).

4. S. Case, Coupled Wave Theory for Multiply Exposed Thick Holographic Gratings, J. Opt. Soc. Am. 65, 724 (1975).

5. R. Alferness and S. Case, Coupling in Doubly Exposed, Thick Holographic Gratings, J. Opt. Soc. Am. 65, 730 (1975).

6. S. Case and M. Han, Multimode Holographic Waveguide Coupler, Opt. Comm. 15, 306 (1975).

SESSION 1.

IMAGE PROCESSING I

Chairman: J. Ch. VIENOT

HYBRID IMAGE PROCESSING

G. Häusler, A. Lohmann

Physikalisches Institut der Universität Erlangen-Nürnberg,
Erwin-Rommel-Str.1, 8520 Erlangen, Germany

ABSTRACT

In the past, optical image processing has been performed
primarily with coherent light. We want to promote the use of
incoherent light, since it alleviates some of the transducer
problem normally encountered when coupling optical with electronic
subsystems. This coupling (hybrid processing) may include either
digital electronics or analog electronics, such as commercial
TV components. We shall discuss several such hybrid systems.

I. INTRODUCTION

If we review the field of different image processing methods on
the basis of properties like flexibility, real time papability,
etc., it appears that the qualities and the drawbacks are spread
out uniformly (fig.1).

ASPECTS OF IMAGE PROCESSING METHODS

	Flexibility	Parallel cap.	2-dimensional?	Real time cap.	Noise cap.	Non linear cap.	Stability
Coherent	0	+	+	0	-	-	-
Incoherent	-	+	+	-	+	-	+
Analog electr.	-	-	-	+	0	+	+
Digital electr.	+	-	+	-	+	+	+
Hybrid	0	+	+	+	0	+	+

Fig.1 Comparison of image processing methods

It thus suggests itself to combine the virtues of optical processing, electronic analog processing and digital processing,
hoping at the same time, that the drawbacks cancel. By presenting
a few examples, we shall demonstrate that this should indeed be
possible.

At the same time we want to emphasize that hybrid processing is
more than just a happy union of two or three technologies: in our
opinion it is the only survival chance of optics in the business
of picture processing.
Perhaps this statement about the future of optical processing is
too pessimistic. However, claims of the fantastic Fourier transform and filtering capabilities of optical systems have been
greatly oversold, to the extent of eroding the credibility of the
optics community.
Furthermore, digital multiprocessors are vigorously developed
today. Hence, the need to use optical means for handling the
large data quantities in pictures is constantly diminishing. Most
systems architects will favor the trend to all-digital picture
processing, because it leads to unified modular hardware which
can be guided by means of highly developed software.
The advocates of all-digital processing seem to enjoy pointing at
the weaknesses of coherent optical processing:
coherent noise, lack of transducers (real time),
critical positioning requirements and program inflexibility.
We now want to point out that these weaknesses are predominantly
tied to coherent optical processing. When using incoherent
optical processors, in conjunction with TV electronics and/or a
limited amount of digital processing, one can do a great deal
indeed without suffering from the weaknesses of coherent optics.
For a description of some incoherent methods see e.g., Rogers (1).

A quasi monochromatic, incoherent spatial filtering device can do
almost everything a coherent device can do. But the coherent
noise is avoided. There is no need for a transducer from
electronic to coherent optical signals or from optical-incoherent
signals to coherent optical signals. Instead, even the self-
luminous pattern on a TV monitor may serve as the data input
device.
Another advantage of incoherent over coherent spatial filtering
is the generous positioning tolerance of the spatial filter.
The only drawback of incoherent systems is the restriction to
non-negative signals (intensities). However, it is possible to
handle one bipolar signal as two non-negative signals in two
coordinated incoherent channels. An alternative possibility is
to put the bipolar signal onto a bias in order to make it non-
negative. Spatial and temporal modulation procedures can be
helpful, making incoherent processing both versatile and
compatible with real time electronic devices, such as TV cameras
and monitors. In this context it is worth mentioning the
attractive price/performance of standardised TV technology.

A very common criticism of optical processing methods is the
lack of programming flexibility. A typical comment:
"Only linear and space invariant operations can be performed
optically." This criticism becomes less and less valid, however,
as Goodman pointed out in a recent summary article on non linear
and space variant optical processing methods (2). Most such
methods are not sufficiently perfected to be incorporated into
operating picture processing systems. Yet even today there is
a feasible approach for increasing substantially the program
flexibility of optical processors. This is by combining the

optical processors with digital or electronic analog systems into
suitable hybrid systems. Such an approach is sensible if the bulk
of the data processing labor is linear in nature, and only
occasionally a step is encountered that cannot be performed
properly by optical means. If these occasional steps are of the
non linear memoryless type (e.g., hard clipping), they may be
easily performed by a TV system. On the other hand, e.g., the
quotient of two complex functions is something that can be done
much better digitally.

Another example is the production of a computer generated spatial
filter for code translation (3) or for character recognition,
based on the principal component concept (4). This latter example
requires the inversion of a large matrix - obviously a job for a
digital computer.

In our view, a hybrid processor is similar to a digital computer
with micro programming capabilities. The optical spatial filtering
setup corresponds to the general operational part of the digital
processor, and the computer generated spatial filter plays the
same role as the "micro program card" does.

II. EXAMPLES OF HYBRID PROCESSING

In the following we present several examples of hybrid processing
systems, with emphasis on incoherent processing and on the use
of commercial TV systems.

In the first example we consider synthesis of arbitrary OTF's
using TV techniques for subtraction (5). As a second example we
describe the increase of depth of focus in real time, without
loss of resolution, using hybrid optical-electronic techniques
(6), (7).

1. Real Time OTF Synthesis

As we noted in the introduction, incoherent illumination has
many advantages, among them good image quality and real time
capability.

However, only positive-real point spread functions (or
equivalently autocorrelation-type transfer functions) can be
realized directly (8). We cannot, for example, directly perform
incoherent highpass filtering, a definite drawback since
highpass filtering is frequently a very important operation in
image processing. But it is well known that arbitrary OTF's can
be synthesized by splitting the imaging system into two channels
and subracting one output from the other. In using such an
approach three major questions arise:

1.) How should the corresponding pupil functions be realized?
2.) What should be done with a bipolar output signal to make
     it suitable for display?
3.) What method should be employed for the between channel
     subtraction?

With respect to the first question, off-axis reference wave
holograms can be used (9). For our own work, however, we prefer
to work with non-holographic computer generated transparencies.
In this context, ROACH-type complex pupil plane filters (10)
could be used to produce an arbitrary point spread function.
From an operational standpoint, however, it is more convenient,

to work with filter transparencies that are both real and
positive. Since the resulting OTF is real and symmetric, the
associated PSF is also real and symmetric.
In many situations - character recognition or code translation,
for example - this restriction is not serious; we simply work
with a correspondingly symmetrized version of the input object.
Generally the evaluation of pupil functions leads to both
positive and negative values. However, Hauck (11) has shown that
subsequent positive valued pupil functions can always be derived
by proper biasing of the bipolar pupil functions. Associated
effects of light efficiency and output contrast have yet to be
studied.
With respect to the second question - what should be done with
a bipolar output? - There is no clear best answer. The problem
is reasonably common in two channel operation; for certain
input objects the difference between the two output intensity
distributions may in some regions be negative. Since the
difference signal to be displayed is an intensity distribution
it is necessary either to bias or to rectify the bipolar signal.
Either operation has its drawbacks. Biasing decreases contrast,
while rectifying is a non-unique non-reversible non-linearity.

The third question - how to effect the subtraction - is the most
challenging operation. Much attention has been devoted to the
problem of image subtraction, as evidenced by Ebersole's recent
survey (12). It is common to all optical image subtraction
methods that incoherent real time operation cannot be achieved.
That is why in addition to interferometric and photographic
techniques, a variety of video techniques exist: synchronous
two camera scans, for example, or sequential scans with intermediate storage and subsequent subtraction of the video waveforms. Dashiell, Lohmann and Michaelson (13) have proposed a
single camera device for subtracting two mutually incoherent
images in real time. The single camera nature of the system
eliminates problems associated with distortion and adjustment.
Because this device is cheap and very simple in its operation,
we have investigated its potential in OTF synthesis.

The system we have worked with is shown in figure 2.

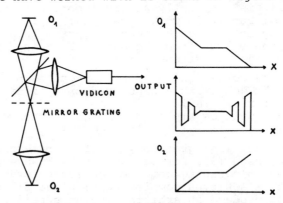

Fig. 2 Principle of real time TV image subtraction

Channels O_1 and O_2 are spatially interlaced by a reflective ronchi grating. The output signal is modulated only where the two input patterns are different. We separate the desired modulated term by highpass filtering of the video signal as shown in fig. 3.

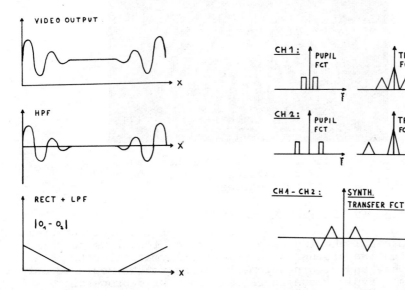

Fig. 3 Processing of video output

Fig. 4 Synthesizing a transfer function with contrast reversal and bandpass characteristics

As an example for the application of this system to OTF synthesis, we show the realization of combined bandpass filtering and a specific contrast reversal. Such an operation cannot be achieved with a conventional incoherent imaging system.
Figure 4 shows the pupil functions in the two channels and the resulting OTF's. The difference OTF is shown at the bottom. This OTF has been applied to a radial test pattern, with the result shown in fig. 5. Contrast reversal and bandpass characteristics are obvious.

It should be mentioned here that the subtraction method, like many other incoherent methods, suffers from resolution and contrast limitations. A 66% loss of resolution and roughly 50% loss of dynamic range occurs. The loss of dynamic range is probably the more serious problem. With low contrast objects, TV systems with high SNR or long time averaging capability must be used.

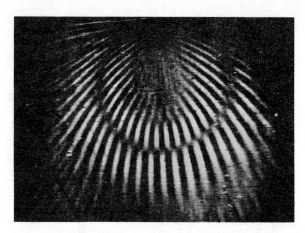

Fig. 5   Application of the synthesized
          transfer function of fig. 4 on a
          radial test pattern.

2. Real Time Increase of Depth of Focus

We proceed now to our second example of hybrid incoherent optical
analog processing: the increase of depth of focus. We concern
ourselves with the virtues of a priori information, with noise
considerations, and with problems relating to the positive output
requirements - typical considerations in hybrid processing.
As suggested by fig. 6a and Fig. 7a, because of limited depth of
focus, the imagery of a three dimensional object is highly space
variant; it cannot be deblurred by conventional deconvolution
techniques. Stated differently, the point spread function is
different for different object planes. A priori knowledge of the
nature of the image degradation is important in the development
of a technique for increasing the depth of focus without
sacrificing resolution.
In what can be viewed as a pre detection operation, we first
modify the image to obtain the same transfer function for all
object planes. This is achieved by moving the object - (e.g. that,
whose conventional microscopic image is shown in fig. 7a) - along
the optical axis through a distance somewhat larger than the
object thickness and integrating the image in time. The resultant
integrated transfer function is shown in fig. 6b.

The attenuation of the high frequency components is then
compensated by post detection filtering, see fig. 6c. Note that
there are no nulls in the integrated OTF to complicate the
compensation operation.

The operations described can be implemented using purely optical
systems. Real time operation, however, which is of course highly
desirable for practical use, requires a hybrid optical electronic
system. We describe briefly two alternative systems we have
investigated (6), (7).

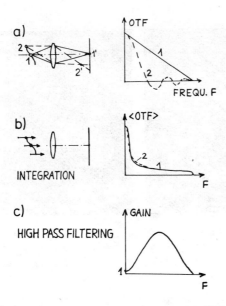

Fig. 6   Principle of focus depth increase

In the first system, the object was vibrated rapidly and a vidicon used to obtain the "integrated image".(The temporal integration characteristics of the vidivon photocathode were used to advantage here.)
The video output signal was then highpass filtered to increase the high frequency content in the horizontal direction. (We either attempted one dimensional and two dimensional filtering of the video signal; even one dimensional filtering yields good results, as is expected by the example of differential interference contrast.)

For an N fold increase in depth of focus, the maximum gain of the highpass filter must be approximately N. For large N, the low frequency bias effectively saturates the detector, thus limiting the achievable increase. The results of fig. 7b suggest that at least a 20-fold increase of depth can be achieved by this method.

In a more recent improvement on this first method we interchange the order of the two processing steps: High pass filtering is followed by image integration. Were the overall system truly linear, the interchange of two such linear operations would have no effect. Consider the result with a radial test pattern, positioned with its plane oblique to the optical axis (fig. 8). The out of focus image structure contains no appreciable high frequency content. As a consequence, the displayed   high pass iltering (which, recall, must be performed electronically, not

 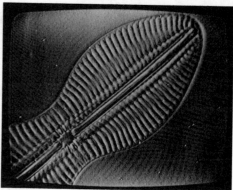

Fig. 7a Conventional diatomic micrograph. (N.A. 1.32 object thickness ± 8µ)

Fig. 7b Diatomic of fig. 7a with increased depth of focus.

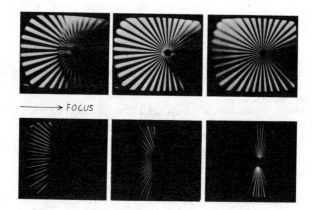

Fig. 8 Principle of "depth slicing".
above: different planes of focus of a radial test pattern, oblique to the optical axis.
below: corresponding high pass filtered images.

incoherent-optically), results in a bipolar output. Since our vidicon storage tube can only store positive signals, the bipolar signals must be rectified. The result is a non unique image. Except in the case of more or less pathological cases, this restriction is not especially severe, as may be confirmed by fig. 9, and considering the advantages of the method. For now defocussed image information does not consume available dynamic range, and - in principle, at least - there is no limit on the increase in depth of focus.

Fig. 9 Depth increase by hybrid optical
        electronic "depth slicing" and
        subsequent integration.

III. CONCLUSIONS

Hybrid image processing may be viewed as dividing problems into
subproblems, which can then be solved by using optical, digital
electronic, or analog electronic means - whichever is most
appropriate.
By approaching problems in that manner, we are often able to
process pictures in <u>real time</u>, with <u>incoherent</u> illumination and
with considerable <u>flexibility</u>, - all <u>important</u> considerations in
the context of practical applications.

Two examples have been presented which demonstrate the virtues of
these concepts, at the same time pointing out typical difficulties
related to two-channel operation and noise.

Though much work has yet to be done in investigating e.g., the
feasibility of synthesizing arbitrary OTF's (dynamic range-,
bandwidth limitations), we are of the hope,-confirmed by our
second example,- that various methods working until now optically
only "in principle", may be brought by hybrid methods out of the
journals into the factories.

REFERENCES:

(1) G. L. Rogers, Non coherent optical processing,
        Opt. and Laser Tech. 7, 153 (1975).
(2) J. W. Goodman, Operations achievable with optical data
        processing,
        Conference on Electrooptics Systems Design, Anaheim 1975

(3)   A. Lohmann, D. P. Paris, H. W. Werlich, A computer generated
      filter applied to code translation,
          Appl Opt. 6, 1139 (1967).

(4)   B. Braunecker, A. Lohmann, Character recognition by digital
      holography,
          Opt. Comm. 11, 141 (1974).

(5)   G. Häusler, B. Schnell, Inkohärente Echtzeitsynthese optischer
      Übertragungsfunktionen,
          Tagung der Deutschen Gesellschaft für Angewandte Optik,
          Nürnberg, Germany (1976), paper in preparation.

(6)   G. Häusler, A method to increase the depth of focus by two
      step image processing,
          Opt. Comm. 6, 38 (1972).

(7)   W. Girbig, G. Häusler, Ein optoelektronisches Verfahren zur
      Erweiterung der Schärfentiefe,
          Tagung der Deutschen Gesellschaft für Angewandte Optik,
          Nürnberg, Germany (1976), paper in preparation.

(8)   W. Lukosz, Properties of linear low pass filters for non-
      negative signals,
          JOSA 52, 827 (1962).

(9)   P. Chavel, S. Lowenthal, A method of incoherent optical-
      image processing using synthetic holograms,
          JOSA 66, 14 (1976).

(10) D. C. Chu, J. R. Fienup, J. W. Goodman, Multiemission on-
      axis computer generated hologram,
          Appl. Opt. 12, 1386 (1973).

(11) R. Hauck, Optische Zeichenerkennung mit inkohärenter
      Principal Component Filterung,
          Diplomarbeit am Physikalischen Institut der Universität
          Erlangen - Nürnberg, (1976).

(12) J. F. Ebersole, Optical image subtraction,
          Opt. Eng. 14, 436 (1975).

(13) S. R. Dashiell, A. Lohmann, J. D. Michealson, Real time
      incoherent optical-electronic image subtraction,
          Opt. Comm. 8, 105 (1973).

HYBRID OPTICAL AND ELECTRONIC IMAGE PROCESSING

Dorian Kermisch

*Webster Research Center, Xerox Corporation, Webster,
New York 14580, U.S.A.*

ABSTRACT

A method that combines efficiently and in real-time optical and electronic
processing is described and several ways to implement it are discussed. The
method is of interest in special purpose image processors when it is desirable to simplify and speed up the overall process. An example of its use for
the visualization of phase objects is given.

I. INTRODUCTION

We discuss here some methods that combine in one image processing system both
optical and electronic processing. These methods are of interest in the case
of special purpose, especially real-time, image processors that have to process an input image in a given way, and the process requires some steps that
must be done electronically. The objective is to reduce the overall complexity of the processor, and by the performance of optical pre-processing of the
input image, to reduce the number and complexity of the operations required
from the electronic stage; thus increasing the speed of the process.

In a sense, any special purpose electronic image processor, digital or analog, is a hybrid processor. It includes an optical stage that converts the
input, two-dimensional optical image into a one-dimensional electronic signal. This stage is usually a scanner that scans the input image line by line.
In the next two sections we discuss two scanning systems that can also perform optical image processing in a scanning mode. The electronic signal generated by these scanners corresponds to an optically processed image. The
scanners can emulate any conventional optical image processing system.

In general, electronic processing is both fast and simple if it requires only
operations along the lines of scan. It is slower, and more difficult to implement if it requires operations across the scan lines, like spatial frequency filtering for example. Such operations, that involve an area of the
image, if linear, are easily implemented by optical processors. Moreover,
electronic processors cannot operate on the complex amplitude of the light,
like it is done by coherent optical processors. Also, electronic processing
cannot remove aliasing effects introduced by the sampling process inherent in
any scanning method. Aliasing can be avoided only in the optical stage by
removing the high spatial frequencies from the input image.

If the process that has to be implemented can be divided into a linear spatial frequency filtering process (this includes differentiation, integration,
correlation etc.), followed by a nonlinear process, we can build an optical

19

scanner that will process the image at the speed of the scan. By doing that, we reduce appreciably the requirements on the electronic processor. In Section IV we give an example of such an application.

II. DOUBLE APERTURE SCANNING

It can be shown that the electronic signal produced by a microdensitometer type scanner represents approximately an image resulting from the convolution of the original image with an impulse response defined by the transmittance of the scanning aperture. We can shape the form of the impulse response by introducing a mask in front of the aperture. However, the impulse response is always positive, and it can perform only a very limited number of operations.

To create an impulse response that is both positive and negative, double aperture scanning is used (1). The light from each aperture is detected by a separate photoconductor, and the electric signals from one detector is subtracted from that of the other detector. Such a system is shown in Fig. 1, where the scanning spot is imaged by use of a beamsplitter onto two separate photodetectors. In front of each detector is a mask. The transmittance of one is proportional to the positive part of the desired impulse response, and zero in the areas where the impulse response is zero or negative. The transmittance of the other mask is zero where the desired impulse response is positive, and proportional to the magnitude of the impulse response where it is negative. The signal from one detector is subtracted from the other signal.

DOUBLE APERTURE SCANNER

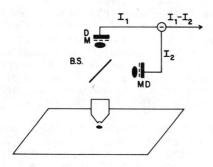

Fig. 1. Double aperture scanner

A double aperture scanner operates on the irradiance of the image. It works as well with diffusely or specularly transmitting (or reflecting) images. It can perform all the operations achievable by convolution with a real impulse function, that is, sharp cut-off filtering, differentiation, integration, etc.

The method has some drawbacks: It can operate only on real images, it cannot operate on complex amplitude or phase images. The generation of the masks is difficult and both the masks and the images of the scanning aperture have to be registered well. The method can be easily implemented only in so-called moving aperture scanners in which either the aperture or the image are translated mechanically; these scanners are slow.

III. LASER SCANNING

In a previous publication (2) we described a laser scanning system that can
perform the same image processing as conventional coherent, or partially coherent optical systems. We showed that for every conventional optical imaging
system we can find an equivalent scanning system with the same image processing properties. Since then, we found that the same equivalence principle is
put to use in electron microscopy. Zeitler and Tomson (3) have shown that the
imaging properties of the Scanning Transmission Electron Microscope (STEM) can
be made equivalent to those of a Conventional Electron Microscope.

In Fig. 2 we show an example of such a scanning system. The focused laser
beam, at P, is collimated by the lens L_1 is modified by the complex filter H,
is focused on a transparency, t, recollimated by the lens L_3 and detected by
the photodetector D, whose electric output is proportional to the light irradiance incident on it. To scan the transparency, t is translated along its own plane.

Fig. 2. Laser scanning system

We will prove that the electronic output of D, which we call the image of the scanning system, is proportional to the irradiance that will be produced at P, if we replace the detector by a spatially incoherent light source, of the same wave-
length as the laser. The source must have the same area and position as the detector. We note that the optical system images t in the plane of P. Because of the spatial frequency fil-
ter H, it produces a processed image of t, which may be coherently or partially coherently processed, depending on the relative size of H and D. If the
source in the conventional imaging system (or the detector in the equivalent
scanning system), is very small with respect to H, the process is coherent.
If the size of the source is very large with respect to H the process is incoherent. Otherwise, the process is partially coherent.

As suggested by Zeitler and Tomson (3) the proof can be based on Helmholtz's reciprocity theorem. It states that if a point source at P produces a certain light amplitude at a point Q; the same source at Q will produce the same amplitude at P. Consider the point source at P in Fig. 2. At each point x on the detector it produces a light amplitude e(x). The detector sums the irradiance of all light incident upon it, hence, its output is proportional to $|e(x)|^2$ integrated over the area of the detector.

We now replace the detector with a spatially incoherent source of uniform irradiance. At each point x on the detector we have a source that produces an amplitude e(x) at P. (We neglect a proportionality factor.) Because the source is spatially incoherent, the total irradiance at P is the sum of all $|e(x)|^2$, that is, the same integral as before.

We conclude that the scanner produces the same image as its equivalent conventional system. Moreover, the scanner can produce images that are not realizable with conventional systems. If we split the detector D, in Fig. 2, into two separate detectors, and subtract the signal of one detector from that of the other, we produce the same image as if we have a conventional system with two sources, one that emits positive light and one that emits negative light! We make use of this property in the example shown in the next section.

Unlike the double aperture scanner, this scanner operates on the amplitude of light and it can process phase images. Depending on the size of the detector it can perform coherent, partially coherent or incoherent image processing. However, in order to perform coherent optical processing, it requires that the input image be specularly transmitting (or reflecting). In the form shown in Fig. 2, its scanning speed is limited by the speed of translation of the transparency.

The configuration of the laser scanner can be changed to that shown in Fig. 3. The collimated laser beam is reflected by the mirror M and if focused by lens L_2. The mirror oscillates in both x and y directions. Its center of rotation is on the optical axis of L_2, in its focal plane. As the mirror rotates, the focused beam scans the transparency. The lens L_3 images the light in the focal plane of L_2 onto the plane of the detector D. If we shape the complex amplitude of the beam in the focal plane of L_2 so that it is proportional to the amplitude transmittance of H in Fig. 2, in the absence of lens aberrations, the images of the

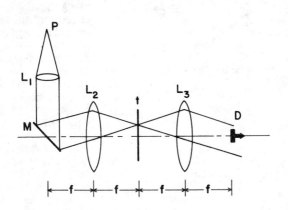

Fig. 3.   Laser scanning system
          for fast scanning rates

two systems are identical. The amplitude of the beam can be controlled by shaping the amplitude reflectance of the mirror M, or by other optical elements not shown in the figure. The scanner shown in Fig. 3 puts stronger demands on the lenses of the system, but enables us to scan the image at much higher speeds. The speed depends on the properties of the deflector, that can be either a galvo-mirror or an acousto-optic deflector.

III. PHASE VISUALIZATION

As an example of a hybrid optical-electronic processor we present a system for the visualization of large phase variation objects. The complex amplitude of a collimated beam of light, after passing through such an object is given by

$$t = e^{ik\phi(x,y)} \tag{1}$$

where k is the wavenumber and ϕ represents the optical path variation that characterizes the object. We want to display ϕ as a conventional image.

J. B. DeVelis and L. C. Martin (4) have proposed, and R. A. Sprague and B. J. Thompson (5) have implemented a method based on coherent optical processing, for displaying ϕ. The method consists of performing first a one-dimensional differentiation of the complex amplitude with an operator

$$(1 - ia\frac{\partial}{\partial x})e^{ik\phi} = (1 + ak\frac{\partial\phi}{\partial x})e^{ik\phi} \tag{2}$$

This operation can be achieved with a frequency filter consisting of a linear amplitude transmittance wedge.

The recorded image is proportional to

$$(1 + ak\frac{\partial\phi}{\partial x})^2 . \tag{3}$$

By careful photographic development we obtain $(1 + ak \frac{\partial\phi}{\partial x})$, we subtract 1 and integrate in the x direction. Assuming that the whole process starts in an area where $\phi(x,y) = o$, the integral reconstructs $\phi(x,y)$.

We considered first to implement this process in real-time by performing the optical differentiation in a scanning mode, followed by electronic processing. The system worked, but it had the following drawbacks. To make the process coherent we had to use a small detector, we detected only a small part of the total laser beam. The frequency filter reduced even more the irradiance of the detected

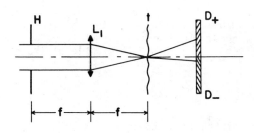

Fig. 4. Scanner for phase visualization

light. The dynamic range of the electronic processor had to be large, it had
to accommodate the square term of Eq. (3), but then used for integration only
its square root. Any slight absorptance of the phase object introduced errors
in the process.

We decided to make use of the scanner property that is the equivalent of imaging with negative light sources, and built the system (6) shown in schematic
form in Fig. 4. The filter H consisted of a rectangular aperture, and we used
a detector split at the optical axis into two parts D_+ and D_-. The detectors
were large enough to detect all the light passing through the system. The
transparent phase object was translated along its own plane. If we denote by,
respectively, I_+ and I_- the outputs of the two detectors, it can be shown (6)
that

$$\frac{\partial \phi}{\partial x} \; \alpha \; \frac{I_+ - I_-}{I_+ + I_-} \; , \tag{4}$$

where α denotes proportionality, and $\frac{\partial \phi}{\partial x}$ denotes the phase derivative at the
point coincident with the optical axis.

This system utilized all the light of the laser beam, was insensitive to any
slight absorptance of the phase object, and reduced the dynamic range required
from the electronic processor. The subtraction, division and integration were
performed electronically.

To achieve faster scanning rates, we used a modified scanning system. Its
schematic diagram was essentially the same as that shown in Fig. 3, where the
mirror M had a rectangular reflecting area, and the detector D was large and
split at the optical axis into two separate detectors. When we denote by I_+
and I_- the electronic outputs of the two detectors, Eq. (4) represents the
phase derivative at the point of incidence of the focused laser beam. With
this system we achieve faster scanning, but we must use more expensive lenses
(for the same resolution, and amount of lens aberrations as with the previous
system).

IV. CONCLUSIONS

We have discussed a method that combines efficiently, and in real-time optical
and electronic processing of images, and showed several ways to implement it.
This method is of interest in special purpose image processors where we desire
to simplify and speed up the overall process. We showed an example of its
use.

REFERENCES

(1) See for example, R. V. Shack, Proc. of the 14th Annual Symposium of SPIE,
     Aug. 11-14, 1964, p. 393.

(2) Dorian Kermisch, "Partially coherent processing by laser scanning",
     J. Opt. Soc. Am. 65, 887-891 (1975).

(3) E. Zeitler and M.G.R. Tomson, "Scanning transmission electron microscopy
     I & II", Optik 31, 258-280 and 359-366 (1970).

(4) J. B. DeVelis and G. O. Reynolds, Theory and Applications of Holography, Addison-Wesley, Reading, Massachusetts, 1967, p. 161.

(5) R. A. Sprague and B. J. Thompson, "Quantitative Visualization of Variation Phase Objects", Appl. Opt. 11, 1469-1479 (1972).

(6) Dorian Kermisch, "Visualization of Large Variation Phase Objects", Proc. of SPIE, Image Processing 74, 126-129 (1976).

IMAGE RECONSTRUCTION FROM A PARTIAL FOURIER TRANSFORM

S. G. Lipson

Physics Department, Technion-Israel Institute of Technology, Haifa

ABSTRACT

We describe an image scanning method which samples the Fourier Transform of the image at points lying on a circle in Fourier space. It works by transmitting the image through a rotating grating. Reconstructions of point and continuous images are illustrated. A system using two rotating gratings is proposed to improve the performance for continuous images.

INTRODUCTION

We were recently involved in an industrial development project to design an instrument in which certain numerical information had to be extracted from optical observations of an object. All stages of the work could be carried out satisfactorily in the laboratory, but the project eventually failed because of the practical problem of converting an optical image into numerical data cheaply, by scanning or otherwise.

In this paper we shall discuss a method which appears to work satisfactorily under some circumstances, and gives a mechanically simple way of converting images into numerical data. The method is not original; it was invented some years ago (1) by Mertz for X-ray astronomical imaging, but it seems to have remained almost neglected by all other fields of interest — sufficiently so that we rediscovered it independently. The method in its simplest form consists of transmitting the image through a periodic grating which rotates in its own plane. The total transmitted light, I, is collected in a single photodetector, and is measured as a function of the angle of rotation of the grating, ϕ. The resulting function $I(\phi)$ is a sample of the Fourier Transform of the image, and from this sample it is possible under some circumstances to reconstruct the image fairly well. The discussion which follows will give some examples of reconstructions, and will show some of the basic limitations of the method. Later we shall consider methods of improving the performance, particularly for continuous-tone images, which should make the method more generally useful.

Let us emphasize here the advantages of this approach. Firstly, it provides a mechanically simple method of scanning images, although the output requires computation to convert it to a useful form. Secondly the method is applicable to incoherent images at any wavelength and needs but a single detector, so that it avoids problems of detector arrays; for example, we are at present considering its application to microwave imaging. Thirdly, one might expect an improvement in signal-to-noise compared with straightforward scanning when the image is very weak and the signals are detector-noise dominated; this has yet to be demonstrated.

EXPERIMENTAL

Most of this article is concerned with the theory of the method, and is illustrated by several examples. Data for the examples was collected in a very primitive manner. A photographically produced Ronchi grating (3 ℓ/mm) was rotated in the image plane at about 15 min per revolution on an y-t recorder. light fell on a photomultiplier whose output was recorded on y-t recorder. Later, the data was read from the recordings, punched on to cards, and processed by the computer. No attempt whatsoever was made to work with invisible wavelengths, or at a realistic speed, since these are not relevant to a study of the mathematical capabilities of the method.

MATHEMATICAL PRINCIPLE

Suppose the grating has a spatial frequency ρ_o , and is situated at an angle ϕ . Then its transmittance $t(r,\theta)$ can be expressed in polar coordinates by

$$t(r,\theta) = \frac{1}{2} \{ 1 + \cos[\rho_o r\cos(\theta-\phi) + \Delta]\} \tag{1}$$

where Δ is a phase which can be determined to suit our needs. If the image incident on the detector is $f(r,\theta)$, the total transmitted light will be

$$I(\phi) = \int_o^{2\pi}\int_o^R f(r,\theta)t(r,\theta)rdrd\theta,$$

$$= \frac{1}{2}\int_o^{2\pi}\int_o^R f(r,\theta)\{1 + \cos[\rho_o r\cos(\theta-\phi) + \Delta]\}rdrd\theta \tag{2}$$

This is equal to a constant plus the Fourier Transform of $f(r,\theta)$ evaluated at the spatial frequency $(\rho_o\cos \phi, \rho_o \sin \phi)$. In other words the rotating grating samples the Fourier Transform $F(\rho,\phi)$ of $f(r,\theta)$ on a circle in the Fourier plane, having radius ρ_o . The imaging achieved by retransforming the measured $I(\phi)$ appropriately is therefore exactly analogous to coherent imaging in a system having an anular aperture in the Fourier plane.

Before continuing, we shall mention two important points:

a) Since $f(r,\theta)$ is a real function, we can use the fact that $F(\rho,\phi) = F*(\rho,\phi+\pi)$ to obtain both the real and imaginary parts of F , To do this we choose $\Delta = \pi/4$ in which case

$$F_R(\rho_o,\phi) = \sqrt{2}(I(\phi) + I(\phi+\pi) - I_o) \tag{3}$$

$$F_I(\rho_o, \phi) = \sqrt{2}(I(\phi) - I(\phi + \pi)) \tag{4}$$

where F_R and F_I are the real and imaginary parts of F.

b) In contrast to coherent imaging the reconstruction of the image is done mathematically, and so various possibilities of apodization are possible, which would be very difficult to carry out in a coherent imaging system.

Once $F(\rho_o, \phi)$ is calculated from $I(\phi)$ using (3) and (4) it is in principle simple to reconstruct the image. We can calculate the transform of $F(\rho_o, \phi)$ directly

$$f_o(r, \theta) = \int_o^{2\pi} F(\rho_o, \phi) \exp\{i\rho_o r \cos(\theta - \phi)\} d\phi \tag{5}$$

which is a simple one-dimensional integral, since we have no data on $F(\rho, \phi)$ at values of ρ other than ρ_o.

This reconstruction is not very good; it is equivalent to uncorrected coherent imaging through the anulus. The point spread function is obtained by putting $F(\rho_o, \phi) = 1$, and one obtains the result

$$p_o(r, \theta) = J_o(\rho_o r) \tag{6}$$

which is an intolerably bad point spread function. Much has been written elsewhere on the characteristics of imaging through an anular aperture (2) and we shall not go into further details here. It is possible to use a complex apodization method invented by Wild (3) to improve the point spread function. One calculates

$$f_n(r, \theta) = \int_o^{2\pi} F(\rho_o, \phi) \exp i\{n(\theta - \phi) + \rho_o r \cos(\theta - \phi)\} d\phi \tag{7}$$

in place of (5). This integral has a point spread function

$$p_n(r, \theta) = J_n(\rho_o r) e^{in\pi} \tag{8}$$

and Wild showed that the series

$$f_w(r, \theta) = \Sigma_n t_n |f_n(r, \theta)|^2 \tag{9}$$

for certain values of $t_n (0 < n < 10)$ gives a very much improved point spread function. Because of the computational complexity of evaluating (7) for eleven values of n, we have also tried a simplified form of Wild's series with $t_o = 1$, $t_2 = -1$ and all other t's zero. This gives a point spread function

$$p^{\dagger}(r, \theta) = J_o^2(\rho_o r) - J_2^2(\rho_o r) \tag{10}$$

Using experimentally produced data for single points, we illustrate in fig. 1 contour maps of the intensity of the point spread functions (6), (9) and (10).

EXTENDED IMAGES

The method works well for point images, which has been the secret of its success in astronomy. Continuous images are a much greater problem. The Fourier Transform of a point, or a collection of well separated points, extends throughout the whole Fourier plaee, and so the circle of radius ρ_o samples it tolerably well. A continuous image of dimensions greater than ρ_o^{-1} has a transform which has already fallen to a negligible value at the radius of the circle, and the sampling is very poor. To overcome this problem we can superimpose the image on a screen (generally square lattice; sometimes hexagonal) with spatial frequency ρ_o. This multiplication by a periodic function convolves the transform $F(\rho,\phi)$ with a lattice of δ-functions, the lowest order ones lying on the circle. Thus the circle now samples the very low frequencies of the transform very well. We have tried this approach and have obtained tolerable results with simple geometrical shapes (Fig. 2). However, the sampling at higher spatial frequencies is still not too good, and is very anisotropic; one can see from Fig. 3 how dependent the sampling is on the relative orientations of the grid and the image boundaries.

A method which we expect to give considerable improvement and will shortly be tried consists of rotating both the grating and the screen. The result should then be independent of the orientation of the image boundaries. Of course the data will now be a function of two variables (the angles of the screen and the grating) and so a much larger amount of input data $I(\phi_1,\phi_2)$ will be involved. In fact, one can no longer distinguish between the roles of screen and grating, and in fact the whole problem can be analysed as one of imaging through two crossed rotating gratings of equal spatial frequency. The scanning in the Fourier plane is no longer a circle, but becomes complex route of which two examples are shown in Fig. 4 (for different relative speeds of rotation). We expect considerably improved performance using this type of scheme.

PHYSICAL CONSIDERATIONS

By using the single rotating grating we transform a two-dimensional image into a one-dimensional array of data. For measurement and data storage this is a distinct advantage, but leads to a scarcity of information when a complicated image is being processed. In general, if we measure $I(\phi)$ at $2N$ points in $0 < \phi < 2\pi$ (obtaining N complex values of $F(\rho_o,\phi)$) we can reconstruct a region containing N^2/π resolution elements.   It is impossible that these points should be independent, but as Gabor (4) has pointed out, the interdependence may be quite small, and a reasonable reconstruction can be obtained if there are far fewer than N bright points in the image. This is rather restricting. Thus it seems reasonable to expect that the two-grating method (measuring $4N^2$ data) will lead to considerable improvement.

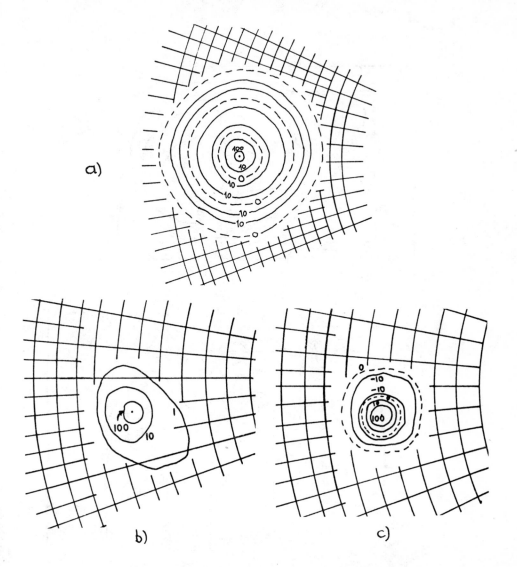

Figure 1: Contour maps of the point spread functions-
(a) for unapodized data, (b) using Wild's series,
(c) using the simplified series (eq. 10). All
drawings use the same input data (N = 71) and the
grids show the calculation network.

Figure 2: Two examples of the reconstructions of extended
images. The outline of the black-and-white originals is
shown broken, and contours of the reconstruction intensity
as full lines.   The screen period is indicated around the
edges of the examples.

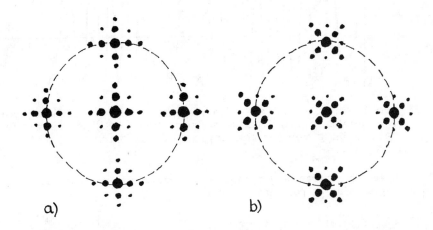

Figure 3:   Schematic drawings of the Fourier plane when sampling
an extended image superimposed on a periodic screen.   The
sampling circle is shown broken.   In (a) the screen is
parallel to the edges of the object;   in (b) it is at 45°.

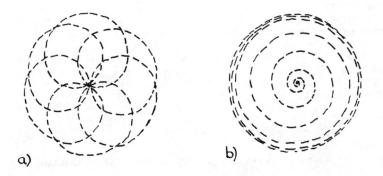

Figure 4:   Sampling curves in the Fourier plane for two superimposed rotating grids.   In (a) the rates of rotation are
very different (1:6);   in (b) they are similar (11:12),
and only the first half of the sampling curve is shown, the
second half being a mirror image.

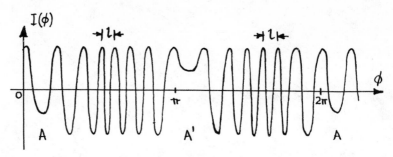

Figure 5:   The warble. The radius of the generating point,
r, is given by the maximum frequency $(1/l)$ whereas the
angle θ is given by the angles of the turning points
A and A'.

One can see the operational principle by considering the data corresponding
to a single point at position (a, θ_o)

$$F(\rho_o, \phi) = \exp i(\rho_o a \cos(\phi - \theta_o)) \tag{11}$$

The form of the real part of this function (a "warble" (1)) is shown in
Fig. 5. It contains enough information to deduce θ_o and a to theoretical
accuracy. However, two such functions with different a and θ_o are not
quite orthogonal, so that a combination of functions cannot be uniquely
decomposed into pure warbles.

SUMMARY

We have described an image-scanning method which samples the Fourier Transform of an image at points lying on a circle in the Fourier plane. Reconstruction of the image from the sampled data is good only if the image
consists of a few well-resolved points. Continuous images can be sampled by
this method if a periodic screen is superimposed. Results using the screen
are sometimes good, but leave a lot to be desired. A system using two rotating gratings is proposed, which should solve the problems arising with
continuous images.

ACKNOWLEDGEMENT

We are indebted to L. Mertz for a very informative discussion.

REFERENCES

1. L. Mertz, Proc. Conf. on Modern Optics, (Brooklyn Polytechnic Press)
    p. 787, 1967.

2. G. Toraldo di Francia, Jour. Opt. Soc. Am. 59, 799 (1969).

3. J.P. Wild, Proc. Roy. Soc. A286, 499 (1965).

4. D. Gabor, Rep. Prog. Phys. 32, 395 (1969).

CONTRIBUTION TO NUMERICAL HOLOGRAPHY: THREE DIFFERENT APPLICATIONS

Jacques Fleuret

Laboratoire Image, Ecole Nationale Supérieure des Télécommunications,
46 rue Barrault, Paris 13, France

ABSTRACT

Three improvements to numerical holography are presented.
- "Brain-Computed" hologram : a Fourier hologram of a binary object is easily implemented, without using any Fourier transform algorithm.
- Constant speed smear deblurring : images are restored by coherent Wiener filtering and by a very simple non-coherent correlation method.
- Pattern recognition : a unique optimized filter is described for binary detection and transcoding of characters.

DESCRIPTION OF THREE APPLICATIONS

Several improvements to numerical holography are presented, related to three different applications.

APPLICATION TO MEMORIES :A"BRAIN-COMPUTED" HOLOGRAM

Synthetic holograms usually involve computations which are complicated and time-consuming. B.C.H. method leads to extremely simple computations, allowing to generate the Fourier hologram of a binary object, without using any Fourier transform algorithm. Figure 1 gives a monodimensionnal description of hologram synthesis. A binary grating of period d has a ray spectrum of period 1/d. Consider another grating having a slightly different period :

$$d_m = d + m \frac{d}{p} \qquad\qquad \begin{array}{l} m, \ p \ \text{integers} \geqslant 0 \\ m << p \end{array}$$

Where p is fixed and m represents grating number. In restitution space, corresponding spectral points have positions :

$$\frac{n}{d_m} \ne\hspace{-0.6em}\ne \frac{n}{d} \ (1 - \frac{m}{p}) \tag{1}$$

Where n (integer $\gtrless$ 0) represents diffraction order. So, we get in spectrum approximately regularly spaced points for choice of periods d_m. And there is a correspondance between each point or ray in spectrum and each binary grating.

Therefore, image of a binary object will be restituted in first order of spectrum if hologram consists of <u>juxtaposition of binary gratings having different periods</u>. And hologram will be generated by filling or not filling binary cells whose position is predetermined.

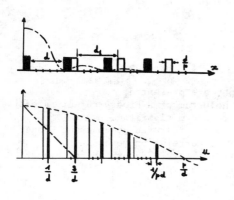 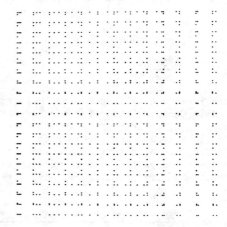

Fig. 1.   up :   Hologram synthesis by juxtaposition of
(left)   binary gratings.

                 down :Corresponding ray spectrum.(Impulse
                        response and transfer function of non-
                        coherent light filtering are in dotted
                        lines).

Fig. 2.   Real 2-dimensionnal binary hologram.
(right)

A great advantage of method is that <u>computations are drastically reduced</u>. Indeed, the only computations needed will be cyclic increments for writing indexes indicating coordinates of cells in each grating.

But method has some limitations. Spectrum slightly decreases, due to cell width in hologram. Number of points in image is limited, due to approximation used in (1) in order to get equispaced points. Noise results from random cell superpositions from different gratings.

Main defect results from various grating periods : filling rate is not uniform. Thus, hologram efficiency is not homogeneous. For low frequency points,
grating period is large and hologram efficiency for these points will be reduced, due to a small filling rate in hologram. In fact, brightness increases
approximately linearly when image point considered is going away from center.
So as to compensate this defect, hologram has been registered using a very
simple filtering process realized in non-coherent light. It simply consists
in photographing hologram using an adequate diaphragm. For a rectangular filter whose impulse response is $\sin^2 (\pi x/d) / \pi^2 x^2$, corresponding transfer
function (i.e. diaphragm autocorrelation) linearly decreases and cancels for
$u = 2/d$. This decrease rectifies efficiency inhomogeneity. Furthermore, as
I showed in (1), this kind of filter cuts out inconvenient high diffraction
orders, thus increasing hologram efficiency and improving image quality.

Figure 2 shows a real hologram, generated point by point by a printer. It
consists of a main 4 points pattern which is nothing else than object itself.
These 4 points are regularly repeated in both direction but with different period for each point.

Hologram is restituted as a Fourier hologram. Figure 3a shows restitution of
hologram, filtered as explained above. Final restitution of object is shown
Fig. 3b . The binary image obtained is highly acceptable.

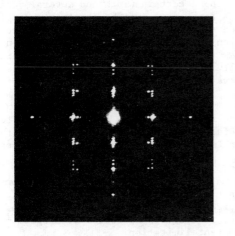
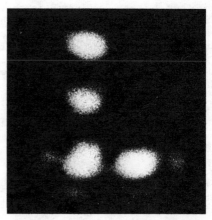

Fig. 3a. Filtered hologram restitution.

Fig. 3b. One diffraction order.

Nevertheless, for larger objects, number of samples needed is very high. It can be shown that, for Airy criterium, number of hologram cells is proportionnal to number of points in object <u>squared</u>. This condemns point by point hologram generation. In fact, a binary grating with a large number of cells should be generated once and then used with a zoom system allowing discrete size modification of this grating.

According to these principles, a very rapid automatic system for binary objects hologram generation could be implemented.

APPLICATION TO IMAGE RESTORATION

Linearly degraded images can be restored analogically by means of computed filters. Results presented here concern constant speed smear deblurring with special care to image quality (speckle noise). Two restoration methods are described : 1) coherent optimum Wiener filtering   2) non-coherent correlation method.

Coherent Optimum Wiener Filtering

The use of coherent Wiener computed filters is now classical (Ref 2,3). Nevertheless, Wiener filtering optimizes S/N ratio for a mean square error criterium, which is not a good criterium when image quality is looked for.

Due to linearity, it is sufficient to study restoration of a slit whose width equals smear-width. Theoretical study and numerical simulation shows that restoration consists of a central spike having a certain width and several equispaced other spikes. Conclusion is that a compromise must be chosen between central spike width and heights of inconvenient spikes. This is what can be called an <u>optimum Wiener filter</u>. Based on these principles, filter has been generated by two methods : double filter (amplitude + phase) and holographic amplitude filter - for this kind of filter (Ref 1) phase coding is suppressed by use of a high frequency reference wave and restoration is obtained in first diffraction order.

When applied to restoration of a smeared thin line - i.e. a slit, both filters give good results. An improvement of a factor 5 has been obtained in line thinness. But spurious lines are still visible : they will produce echoes in restored image.

When applied to restoration of a real image, problem of speckle noise becomes important. Figure 4 shows results for a constant speed smeared image. It is clear that image is restored. But restored image is noisy. And, in spite of careful choice of photographic emulsion, in spite of optical system optimization, <u>noise still remains a very important limitation</u>.

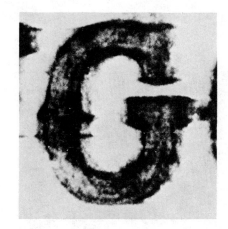

Fig. 4a.   Image degraded by a constant speed smear.

Fig. 4b.   Image restored in coherent optics by a double
            filter.

Non-coherent Correlation Method (N.C.C.)

This new method uses a very simple non-coherent optical system (Fig. 5).
Degraded object will be placed in plane (1), a computed mask will be placed
in plane (2) and restored image obtained in plane (3). Plane (1) is illuminated by a white light source through a broad diffusor. Then, it is easy to
show that intensities just after the three planes are related by a correlation relationship (neglecting scaling factors) - Cf. Fig. 5 . So, restoration will be obtained by generation of a computed mask representing impulse
response of filter. It has been shown (Ref 4) that, for a constant speed smear,
filter impulse response can be approximated by successive positive and negative Dirac impulses. A mask, whose transmittance profile approximates this
ideal model has been generated photographically. Density profile of this mask
is shown Fig. 6 . Distance between spike couples equals smear-width. Of course, mask intensity transmittance must be positive and a constant bias term
has been introduced. Figure 6 also shows restoration of a slit, obtained by
microdensitometry of plane (3).Practical result totally agree with theoritical correlation and a good restoration is obtained. Of course, due to non-
coherent Optics processing, method needs a further step of image substraction,
so as to suppress constant bias term.

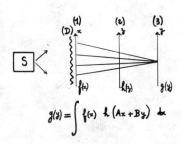 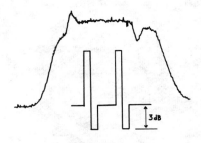

$$g(z) = \int f(x)\, \ell\,(Ax + By)\, dx$$

Fig. 5. Optical system for non-coherent correlation.

Fig. 6. up : Restoration of a slit by N.C.C.
(microdensitometry profile).

down : Density profile of impulse response mask.

Nevertheless, this experiment demonstrates that such a very simple method can be used for image restoration in white light, without any speckling problem.

APPLICATION TO PATTERN RECOGNITION :
FILTRE OPTIMISE UNIQUE DE TRANSCODAGE F.O.U.T.

This application is an example of what can be done in optimization and synthesis of filter by numerical holography. Figure 7 illustrates the use of a unique optimized filter for binary detection of one character out of 32.

By this method, a letter is identified only by one experiment by means of a filter whose impulse response is a multiplexing of all characters u_n (x, y) in font :

$$p\ (x,\ y) = \sum_{n=1}^{N} \sum_{k=1}^{K}\ a\ (n,\ k)\ u_n^{*}\ (-x + kx_o,\ -y)$$

Filter will be optimized by forcing the answer for any letter to be binary. Consequently, weighting coefficients a (n, k) will be determined by following equation system :

$$\sum_{n=1}^{N} a (n, k_1) \; C_{n_o n} (0, 0) = b (n_o, k_1)$$

Where b (n_o, k_1) is n_o th letter code-word. Matrix to be inverted is character correlation matrix.

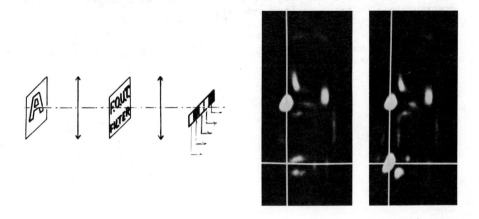

Fig.   7.   Character recognition by coherent F.O.U.T. filtering.

Fig.   8.   F.O.U.T. analogical recognition of "P" and "R".
left : correlation P   (R-P)
right :correlation R   (R-P)

Method has been tested with a very simple alphabet consisting of 2 letters "R" and "P". These letters are non independent - "P" is included in "R". Thus, matched filtering method would give very unfavorable detection rate. For this simple case, filter must be practically matched to difference between "R" and "P", that is tail of "R". This filter has been made analogically and applied to "P" and "R" recognition. Patterns obtained in output plane of a double diffraction optical system are shown Fig.   8 . An improvement of a factor 7 is obtained for detection rate, when compared with classical matched filtering.

I thank Mr. H. Maitre for his fruitful collaboration.

REFERENCES

(1) J. Fleuret, Calcul d'hologrammes à niveaux continus de gris. Application
     à la synthèse d'une mémoire,
     Nouv. Rev. Optique. 5, 4, 219 (1974).

(2) C. W. Helstrom, Image restoration by the method of least squares,
     J. Opt. Soc. Am. 57, 3, 297 (1967).

(3) J. L. Horner, Optical restoration of images blurred by atmospheric tur-
     bulence using optimum filter theory,
     Appl. Opt. 9, 1, 167 (1970).

(4) D. P. Jablonowski, S. H. Lee, Restoration of degraded images by composite
     gratings in a coherent optical processor,
     Appl. Opt. 12, 7, 1703 (1973).

(5) J. Fleuret, H. Maitre, Optimization of a binary transcoding single filter
     used for character recognition,
     Opt. Comm. 17, 1, 64 (1976).

SESSION 2.

IMAGE PROCESSING II

Chairman: E. MAROM

NOISE PROBLEMS IN OPTICAL
IMAGE PROCESSING

Serge Lowenthal and Pierre Chavel

Institut d'Optique, Université de Paris-Sud, 91405 Orsay, France

ABSTRACT

Noise problems in image processing by optical filtering are examined. A comparison between coherent and incoherent illumination is made.

1 - INTRODUCTION

It is often stated that image processing by spatial filtering in incoherent rather in coherent light improves the signal to noise ratio. In this paper, we develop some considerations on the comparison between the two methods. To this end we shall

1. examine first an approximate way (by linearization) to compare coherent and incoherent systems, which are not directly comparable because they belong to different spaces : amplitude space on one hand, intensity space on the other ;

2. show that, due to the film grain noise, a transmittance has not always an intrinsic definition. Therefore, a measurable or observable transmittance will be defined and its variations as a function of the degree of coherence in the object will be examined ;

3. show that, as a consequence of 2., the image contrast can strongly vary according to the type of illumination that is used ;

4. show that there is a definitive gain in signal to noise ratio when using incoherent instead of coherent illumination. This gain, nevertheless, varies according to the type of noise involved. Therefore, we shall evaluate the influence on the processed image of the different types of noise that we shall classify, for the sake of convenience, although somewhat arbitrarily, in three categories :

   - input noise,
   - optical system noise,
   - output detector noise.

2 - HOW TO COMPARE COHERENT AND INCOHERENT SYSTEMS

A. Comparison of Noise Free Images

In an optical system (Fig. 1), the detected output is always an intensity. For incoherent illumination, input and output are quantities of the same nature (powers) and the in-out relationship is linear. For coherent or partially coherent illumination, input and output are not of the same kind, being respectively amplitudes and intensities, and the in-out relationship is no longer linear.

In order to compare images obtained with different types of illumination, we have to linearize the in-out relationships by assuming, for instance, weak object modulation, i.e. low contrast. For the following noise comparison, we observe here that noise has its strongest effect on low contrast objects.

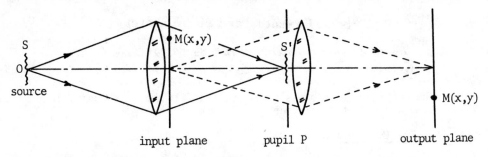

Fig. 1. Optical set-up

The light source is imaged on the pupil P, whose area is A_p.
- Coherent illumination : the source is a point O.
- Partial coherent illumination : the imaged source S' has a finite area A_s.
- Incoherent illumination : limit for $A_s \to \infty$.

Quasimonochromatism is assumed in all cases.
M(x,y) is a point in the input or in the output plane.

Input in absence of noise. We use the same object for all cases. We define it as an amplitude transmittance :

$$f(M) = \tau + m(M) \tag{1}$$

where τ is the space average transmittance and m(M), as mentioned above, is a real-valued weak modulation : $|m(M)| \ll \tau$ for all M.

Output for coherent illumination. From the object-image relation $I(M) = |f(M) * h(M)|^2$, we obtain the image illuminance :

$$I(M) \simeq \tau^2 + 2\tau\, m(M) * h(M) \tag{2}$$

where h(M) is the coherent impulse response normalized as $\int h(M)d(M) = 1$.

Output for incoherent illumination. The image illuminance is obtained in a similar way. We only have to replace the coherent impulse response h(M) by the intensity response in Eq. (2). But, incoherent processing is only interesting if we can perform operations similar to those of the coherent system, i.e. if the impulse response h(M) is the same in both cases.

Although complex impulse responses can be generated using subtraction, squaring, and addition processes, we shall restrict our discussion to the case of a real bipolar (positive and negative) impulse response h(M), obtained by subtraction {1}.

Then, for a true subtraction (for instance electronic), the image illuminance is again given by Eq. (2). For an optical subtraction, that implies an additional squaring step, the final image illuminance is proportional to :

$$|I(M)|^2 \propto \tau^2 + 4\tau \, m(M) * h(M) \tag{3}$$

Output for partially coherent illumination. In order to obtain a linear relation between the object Eq. (1) and the image illuminance I(M), we have to introduce an additional assumption, namely that the modulation m(M) only contains low frequencies. Then, we can write :

$$I(M) \simeq \tau^2 + 2\tau \, m(M) \tag{4}$$

This approximation is rather crude for imaging studies, but it will only be
used in § 3, where, for noise effect examination, it is valid.

B. Comparison of Noisy Images

Under the assumption of small modulations, the relationships involved in different types of imaging have been linearized. Therefore, the useful deterministic image due to the modulation alone is the same in both coherent and incoherent cases. Now, since the power in the modulation is small, the related
noise is also small and for a comparison of signal to noise from coherent to
incoherent, it may be dropped in the comparison ratio :

$$K = \frac{\sigma_{inc}/<I_{inc}>}{\sigma_{coh}/<I_{coh}>} \tag{5}$$

where σ^2_{inc} and σ^2_{coh} are the variances of the image intensities, and <I> the
statistical average of the image illuminance.
In fact, this comparison under the assumption of weak modulations means that
we compare the noise due to the d.c. components. Therefore, Eq. (5) is nothing
else than the ratio of the minimum perceptible image contrasts.

3 - WHAT IS A TRANSMITTANCE ?

The transmittance of a photographic film often has no real significance in itself. Its measured value depends on the operating conditions, i.e. on the
apertures of the systems of observation and illumination. The reason lies in
the film grain noise. Changing the operating conditions changes the transmitted d.c. term due to the noise. Therefore, the observable transmittance as
defined in § 3.C. and the image contrast change too. The calculations for the
influence of noise will be made in the next section. We shall here only make
some comments on the nature of the different kinds of transmittances involved
in optical processing.

A. Amplitude Transmittance

This quantity, defined as the ratio of the incoming and outgoing fields, is
not always a constant. When multiple scattering occurs in a noisy emulsion, it
may depend on the incoming field, for instance on the spatial location of an
illuminating point source. In this case, the amplitude transmittance has,
strictly speaking, no meaning. Nevertheless, it appears experimentally that
the average noise spectrum is shift-invariant with a tilt of the incoming wave
and that is all we need in the next section for noise calculations.

B. Intensity Transmittance

This quantity has no physical significance in itself since the measured transmittance depends greatly on the way it is measured, i.e. the relative aperture
of the illuminating and the imaging systems. This effect, the CALLIER effect,

has been investigated by VERNIER for a binary model of photographic emulsion {2}, and we will show how it acts on the contrast of the images, independently of the noise model.

C. Observable Transmittance

Now, since the intensity transmittance of an emulsion has no intrinsic meaning, we have to use a transmittance t that is observable or measurable, i.e.
- we assume an optical system working under any degree of coherence (even totally coherent), then,
- we observe (or calculate) the image illuminance without and with some object present,
- and define the transmittance as :

$$t(M) = \frac{\text{image illuminance with the object present}}{\text{image illuminance without the object}} \,. \tag{6}$$

t also may be called the instrumental transmittance, because its value depends on the observation and illumination system. It is nothing else than the normalized image intensity. For example, t would be represented by Eqs. (2) or (4) if no film grain noise were present.

4 - INFLUENCE OF FILM GRAIN NOISE ON THE OBSERVABLE TRANSMITTANCE

In order to calculate the influence of the film grain noise on the observable transmittance Eq. (6), let us suppose that, for a given optical system, the illumination changes in a continuous fashion from coherent to incoherent, i.e. for a varying degree of coherence. We do not restrict the discussion only to the coherent and incoherent limits, because the so-called incoherent illumination is always in practice more or less partially coherent.

A. Assumptions

We will restrict our discussion to a simple example with the following assumptions :

i) The same object is examined in both coherent and partially coherent illumination. Its deterministic amplitude transmittance is (see Eq. 1) :

$$f(M) = \tau + m(M)$$

The corresponding noisy transmittance always can be written as :

$$\tau' = \tau + m(M) + X(M) \qquad \text{with } <X> = 0 \tag{7}$$

where X is the noise fluctuation and < > indicates statistical average.

ii) The modulation m(M) is weak. Consequently, a) X(M) in Eq. (7) is a nearly stationary process, b) since m(M) contains a small power, its contribution to the noise is negligible and m can be discarded in case of noise calculations.

iii) The modulation m(M) only contains low spatial frequencies.

iv) Ordinary imaging is examined, i.e. the same uniform pupil (with no filter) is used in all cases of illumination.

B. Method for Noise Calculation in Partially Coherent Illumination

For noise calculations , we can neglect the weak modulation m in the object Eq. (7), as stated in assumption ii). The instrumental transmittance t is then calculated according to the following steps :

1) Determine the complex image amplitude $A(M,S)$ and illuminance $I(M,S) = |A|^2$ due to one point S of the source.
2) Write the illuminance due to the whole source of area A_s as :

$$I_{p.c.}(M) = \int_{A_s} I(M,S) \, dS$$

where S is a point on the source.
3) Average statistically $I_{p.c.}$; this step involves no difficulties, since only second order moments have to be calculated. For coherent imaging, averaging takes place directly after step 1).
4) Normalize to obtain the transmittance.

C. Results

Now, the effect of the film grain noise is only the addition of a d.c. term n to the transmittance Eq. (4) since, by assumption (ii), the modulation only adds a negligible amount of noise. Therefore, the average instrumental transmittance can be written :

$$<t_i> = \tau^2 + 2\tau \, m(M) + n_i \tag{8}$$

where the index i may denote partially coherent (p.c.), coherent (coh) or incoherent (inc).
We then obtain :

Result for coherent illumination. The noise contribution in Eq. (8), calculated according to section B. is :

$$n_{coh} = \int \Phi_x(M) \, \Phi_h(M) \, dM \tag{9}$$

where $\Phi_x(M)$ is the (statistical) autocorrelation function of the object fluctuation $X(M)$ (see Eq. 7), and $\Phi_h(M) = h(M) * h^*(-M)$.
A complete study of n_{coh} involves the knowledge of the amplitude autocorrelation function $\Phi_x(M)$. Nevertheless, an interesting and common case can be treated in a straightforward manner under the assumption of microscopic correlation, i.e. $\Phi_x(M)$ narrow compared with $\Phi_h(M)$. Eq. (9) can then be simplified as :

$$n_{coh} = \frac{\sigma_x^2}{N} \tag{10}$$

where $\sigma_x^2 = \Phi_x(0)$ is the variance of the object amplitude and

$$N = \frac{A_h}{A_c} = \frac{\text{resolution cell area}}{\text{noise correlation area}} \tag{11}$$

Therefore, the transmittance Eq. (8) is, for coherent illumination :

$$<t_{coh}> = \tau^2 + \frac{\sigma_x^2}{N} + 2\tau \, m(M) \tag{12}$$

Result for partially coherent illumination. In a similar way, we obtain for the noise term in Eq. (8) :

$$n_{p.c.} = \frac{1}{I_o} \int \Phi_x(M) \, h(M) \, \mu(M) \, dM \tag{13}$$

where $\Phi_x(M)$ is again the autocorrelation function of X, $h(M)$ is the coherent impulse response, and $\mu(M)$ the complex degree of coherence (Fourier transform, normalized, of the source luminance).

I_O is a normalizing factor : $I_O = \int h(M) \mu(M) dM$.

In order to interpret Eq.(13), we can rewrite it, using Parseval's theorem, as :

$$n_{p.c.} = \frac{1}{I_O} \int \tilde{\Phi}_x(\Omega) \left[\tilde{h}(\Omega) * \tilde{\mu}^*(-\Omega) \right] d\Omega \tag{14}$$

where $\tilde{}$ means Fourier transform, and where Ω is a point in the Fourier space. Then, keeping for instance the source size constant, and increasing the size of the pupil from zero, it is easy to show that the noise term $n_{p.c.}$ goes through a minimum for: source area(aperture) = pupil area. This fact is confirmed by theoretical and experimental investigations on Callier effect {2} {3}. Various limiting forms of Eq. (14) can be derived. We give two examples :

1) Case of microscopic correlation. The noise spectrum $\tilde{\Phi}_x(\Omega)$ is broad compared with both the source and pupils areas A_s and A_p and $A_p < A_s$. Then, the noise term Eq. (14) simplifies to :

$$n_{p.c.} = \sigma_x^2 / N' \tag{15}$$

where $N' = \dfrac{A_{coh}}{A_{corr}}$ is the ratio of the coherence to the correlation area.

Thus, the transmittance Eq. (8) can be written as :

$$<t_{p.c.}> = \tau^2 + \frac{\sigma_x^2}{N'} + 2\tau m \tag{16}$$

2) Case of totally incoherent illumination. We again obtain a simple expression :

$$<t_{inc}> = \tau^2 + \sigma_x^2 + 2\tau m \tag{17}$$

D. Comments

The effect of the film grain noise is the addition of a d.c. term n_i to the average observable transmittance. This term n is a function of the amplitude autocorrelation function of the film and of the apertures of the illuminating and observation systems. Its value may be stated explicitly for various cases. However, the interesting fact is that when, for a given pupil aperture, the complex degree of coherence in the object varies continuously from 1 (coherence) to 0 (incoherence), then n_i, i.e. the overall transmittance Eq.(8) first decreases, goes through a minimum when pupil = source aperture, then increases again up to the incoherent limit. Thus, the transmittance can vary strongly according the type of illumination, over a factor 10 in certain cases. This affects obviously the image contrast, as we shall examine in the next section.

5 - INPUT FILM GRAIN NOISE AND IMAGE CONTRAST

A way to analyze how the film grain noise acts on the transmittance when the type of illumination is changed, is to examine the corresponding image contrast changes.

A. Definition of Image Contrast

The average observable transmittance has been found (Eq. 8) as :

$$<t_i> = \tau^2 + 2\tau \, m(M) + n_i$$

The image contrast is defined as the ratio of the modulation to the total d.c. term :

$$C_i = \frac{2\tau m}{\tau^2 + n_i}$$

In general, we are not interested in the absolute value of the image contrasts but by the comparison of coherent to partially coherent. To this end, we will use the ratio :

$$\frac{C_{p.c.}}{C_{coh}} = \frac{\tau^2 + n_{coh}}{\tau^2 + n_{p.c.}} \tag{18}$$

B. Contrast Comparison for Coherent/Partially-Coherent Illumination

The noise terms $n_{p.c.}$ and n_{coh} are now replaced by their values according to Eqs. (13) and (9) in the expression Eq. (18) for the image contrast ratio. Now, when for a fixed pupil size, the illumination varies continuously from coherent to incoherent, this ratio starts from the value 1, then increases to a maximum obtained for : pupil = source aperture, then decreases till the incoherent limit, which, under the assumption of microscopic correlation, is :

$$\frac{C_{inc}}{C_{coh}} = \frac{\tau^2 + \sigma_x^2/N}{\tau^2 + \sigma_x^2} \tag{19}$$

Two examples will illustrate these contrast variations which can be very strong.

1) Let us consider a system where : pupil = source aperture. When the aperture varies from a small to a large value, then the image contrast may vary by a factor of 10 (from experimental data in Refs. {2} and {3}.
2) For a binary emulsion model, Eq. (19) transforms into :

$$\frac{C_{inc}}{C_{coh}} = \tau + (1-\tau)/N \tag{20}$$

which, for a large value of N (see Eq. 11) assumes just the value τ, i.e. the average object amplitude transmittance.

C. Comments

The influence on the image contrast of the d.c. term due to the film grain noise can simply be explained in the following way.
The optical system has a pupil of area A_p. The illuminating source is imaged in the pupil as an area A_S. Let A_S increase, starting from a small value (coherent illumination). Then :

- the signal power, nearly concentrated in the area A_S, increases like A_S until $A_S = A_p$, and then remains constant for $A_S > A_p$;
- the deterministic d.c. background varies in the same way ;
- the d.c. background due to the noise also increases with A_S, but more slowly than the signal power : an off-axis point of the source yields in the pupil an off-axis noise spectrum, and a part of the power smaller than for the on-axis point is transmitted through the pupil.
When A_S becomes larger than A_p, then the transmitted noise power is still increasing (whereas the signal power remains constant), through the tails of the noise spectra due to the outer points of the source.
- As a consequence, the image contrast (ratio of the signal to the total d.c. term) first increases with A_S up to a maximum for $A_S = A_p$, then decreases

for $A_s > A_p$ down to the incoherent limit.

6 - NOISE FLUCTUATIONS IN IMAGE PROCESSING

In coherent processing, the final step is square detection of complex amplitudes. When additive noise is present, then the detected intensity I is thus an interference pattern between the noise and the useful deterministic image. Through this interference phenomena, the noise fluctuation may be greatly amplified.
This has been explicitely formulated by J. GOODMAN {5} in the case of circular gaussian statistics for the noise.
A general investigation on noise in coherent processing can be found in Ref.6. In incoherent processing, several independent (or uncorrelated) noise samples add in power, without interference, and this involves a gain in signal to noise ratio. We shall try to evaluate this gain for :

- input noise,
- optical set-up noise,
- detector noise.

A - Input Noise

This type of noise consists of input phase noise, film grain noise and of what has been called by J. GOODMAN input impulse noise.
Phase noise, which yields laser speckle in coherent illumination, is suppressed with an ideal incoherent illumination, we shall therefore examine the two other cases.

a) Impulse input noise. Impulse input noise is due to small defects in the input emulsion. TICHENOR and GOODMAN {7} have assumed that such defects behave as delta functions and therefore, produce in the image plane replicas of the impulse response h(M).
Let us consider one single defect that we locate, for convenience, at M = 0. In the coherent case, the image intensity is then obtained from Ref. 7, as:

$$I = |r(M)|^2 - 2\varepsilon a \ r(M) \ h(M) \tag{21}$$

where r(M) is the processed image amplitude, ε is proportional to the area of the defect, a, the object amplitude at the neighbourhood of the defect. Thus, this type of noise is proportional to the area ε of the defect D, the signal amplitude a, at the location of D and depends on the deblurred image amplitude r(M).
In the case of small modulations m(M) (see Eq. (2)),Eq. (21) can be transformed into :

$$I = \tau^2 + n_{coh} + 2\tau \ r(M) - 2\varepsilon\tau^2 \ h(M) \tag{22}$$

where n_{coh} is the Callier term Eq. (9) and r(M) = m * h. Thus, the noise term is directly proportional to the average object amplitude transmittance squared τ^2, and the particle area ε.
In the incoherent case, with the same impulse response h(M) obtained by a true subtraction, we obtain :

$$I = \tau^2 + n_{inc} + 2\tau \ r(M) - \varepsilon\tau^2 \ h(M) \tag{23}$$

Comments : The gain incoherent to coherent is only by a factor of 2 for this type of noise.

b) Film grain input noise. We again take one example, namely pure imaging.

Our aim is to compare image fluctuations, in coherent and incoherent illumination, due to film grain noise.

Assumptions. We use assumptions i), ii) and iv) of §4 A. But, in order to determine variances of intensities, we have to introduce additional assumptions since fourth-order moments of amplitudes have to be calculated :

v) Since we do not know the object amplitude statistics, we assume that the noise is broadband compared to the bandwidth of the optical system (microscopic correlation). Then, by the Central limit theorem, the random image amplitude due to one point of the source is nearly gaussian circular (in general with non zero mean).

vi) The object amplitude transmittance is a pure transmittance, i.e. no multiple diffusions occur. This assumption is necessary for the incoherent case only.

For the coherent case the variance can be easily calculated through the properties of gaussian random variables.

For the incoherent case, we use first the three steps of calculation already outlined in §4B. Then we determine the variance by

$$\sigma^2_{inc} = \iint \{<I(M\,S_1)\,I(M\,S_2)> - <I(M\,S_1)><I(M\,S_2)>\}\,dS_1\,dS_2 \qquad (24)$$

From this calculation, we extract two limiting cases.

The average object transmittance is zero : $\tau = 0$. This case indeed does not correspond to a true photographic emulsion, but, for instance, to strongly diffusing objects (phase or/and amplitude). Then, the ratio (Eq. (5))of the variance is

$$K = \frac{\sigma_{inc}}{\sigma_{coh}} = 0 \qquad (25)$$

This means that in the coherent case we have pure speckle, whereas in coherent case, we have no noise. Physically this result is obvious : 1) for pure phase objects, the phase disappears in incoherent illumination, 2) for pure amplitude objects or mixed objects, coherent illumination will again give speckle however small the grain size (correlation radius) may be. But in incoherent illumination fluctuations only appear if the grains are resolved or nearly so, and that is not the case with our assumption v).

The average object transmittance $\tau \neq 0$, but the average image illuminance $<I_N>$ due to the noise alone is small compared with the average of the total image illuminance $<I_{inc}>$.

This case corresponds to a normal emulsion. The noise ratio (Eq. (5)) can then be expressed as

$$K = \frac{\sigma_{inc}/<I_{inc}>}{\sigma_{coh}/<I_{coh}>} = \frac{1}{3}\frac{\{\alpha^2+4\}^{1/2}}{\alpha^2+1} \qquad (26)$$

where $\alpha = \sigma_x/\tau$ is the ratio (for amplitudes) of the object standard deviation to the average object transmittance.

Thus, if we have $\alpha \gg 1$, then $K = 1/3\alpha$ and incoherent illumination yields lower noise than coherent illumination. If $\alpha \ll 1$, then few gain is obtained.

Comments

It may be noted that, in fact, for many types of emulsions, the assumption of no multiple diffusion does not hold. Therefore, noise samples due to the neighbouring points of the source are more decorrelated than in the present study and the gain of incoherent to coherent may be significant for all cases.

B - Optical set-up Noise

This kind of noise, due to scattering, reflexions, etc... by optical components, is one of the most serious causes of image degradation in coherent processing. Calculation of its effects confirms the following physical reasoning, where we shall use some simplifying assumptions

i) all the noise scatterers are concentrated in the pupil of the optical system ;

ii) the same instrument with a uniform pupil (pure imaging) is used for coherent and incoherent imaging ;

iii) the power in the input Fourier spectrum due to one point S of the source is nearly concentrated in one Fourier resolution cell located on the image S' of S (following GOODMAN in Ref. {7}).

Now, suppose the incoherent source produces N independent superposed images, then the incoherent to coherent gain is

$$K = \frac{\sigma_{inc}/<I_{inc}>}{\sigma_{coh}/<I_{coh}>} = N^{-1/2}$$

In order to determine N, we note that according to assumption iii), one elementary source domain covers one Fourier resolution area A_R. Furthermore, two different elementary source domains are focused on two different parts of the pupil and therefore, produce independent noise contributions. Therefore, denoting by A_p the pupil area, we have ;

$$N = \frac{A_p}{A_R} = \text{number of degrees of freedom of the optical system.}$$

Thus, the gain in signal to noise is significant in this case since N is usually a very large number.

C - Detector Noise

We restrict the discussion to the photographic detector. A quantitative evaluation is complicated in the present state of knowledge. But, we may note that BIEDERMANN {8} made measurements on scattered noise flux on different types of emulsions for holography. He found that plates exposed to laser light scatter more strongly than plates exposed to incoherent or partial coherent illumination. This fact is due to the light diffusion within the medium during the recording. In coherent illumination, superposition of scattered and unscattered light yields a random interference pattern where the fluctuations are amplified. This pattern then creates the latent image.

7 - CONCLUSION

What is better, image processing with coherent or with incoherent illumination ? The answer to this question depends on several factors such as the mathematical operations that can be performed in each case, the simplicity of the set-ups, the real time possibilities etc... and the noise.

Incoherent processing is restricted to real non negative inputs. However that is just the nature of most of the images that we want to process. With respect to the mathematical operations that can be performed we may conclude that the impulse responses are allowed to be complex in both cases, this as long as some artifices are incorporated. The set-up is probably somewhat simpler in the incoherent case since no liquid gate has to be used and only one optical element, the imaging lens, has to be of good quality ; however if a subtraction step is used, in order to simulate negative illuminances, then a certain amount of complication is introduced. The computer generated filters used for the processing are equally complicated (or simple) although cases can be found in incoherent processing when they are simpler. Real time processing is also simpler in the incoherent case as long as a good incoherent to coherent image converters will not be available.

And now to the question concerning the influence of the noise which is the crucial point. One of our intentions with the present paper is to show that not only fluctuations but also a d.c; component appear in the image, due to the film grain noise. This component lowers the image contrast which may vary by a factor even exceeding 10, according to the type of illumination and emulsion.

Our aim is also to show that incoherent systems are definitively better than coherent ones with respect to the noise fluctuations.The improvements vary according to the type of noise involved, but we will come back just to one example, the set-up noise due to scatterings or spurious reflexions (even with coated lenses). Assume that in the coherent system we have 0.1% of noise power. Then, even with such a small noise, the total image fluctuation is $2\sigma = 9\%$. For an incoherent system with only $N = 10^6$ degrees of freedom, this fluctuation is divided by $N = 1000$.

Therefore our personal opinion is : use incoherent illumination whenever it is possible.

8 - REFERENCES

{1} P. CHAVEL and S. LOWENTHAL, Journ. Opt. Soc. Am. 66 (1976), 14-23.

{2} P. VERNIER, Theses, Gauthiers-Villars, Editeur-Imprimeur, Paris 1958.

{3} R. GORISCH, Zeitsch. Wiss. Phot. 48 (1953), 86-102.

{4} H.C. SCHMITT, Jr. and J.H. ALTMAN, Appl. Opt. 9 (1970), 871-874.

{5} J.W. GOODMAN, Journ.Opt. Soc. Am., (1967), 493-502.

{6} J.W. GOODMAN, in Proceedings of the USSR symposium on optical information processing, (1975), Plenum Press, ed.

{7} D.A. TICHENOR and J.W. GOODMAN, International optical computing conference Washington D.C., april 23-25 (1975), digest of papers, IEEE Catalog n° 75 CHO941-5C, 82-85.

{8} K. BIEDERMANN, Optik 31 (1970), 1-23.

OPTICAL VIDEO RECORDING—
WHY AND HOW

M. E. Spitz*

Thomson-CSF - Laboratoire Central de Recherches. B.P. 10-91401-Orsay-France

INTRODUCTION

To day most of the video signals are recorded magnetically. The magnetic recording medium (mostly tapes) is inexpensive and is manufactured in vast quantities, the recording technique is simple, reversible and permits a given record to be erased and rerecorded repeatedly.

All of these properties are not found in optical recording but the following advantages make us believe that optical video recording will be important in the future.

1. The storage density of video recording is of the order of 10^8 bits/cm^2 in optics (one standart TV frame is stored in the area of less than 1 mm^2) compared to less that 10^7 bits/cm^2 (TV frame in 15 mm^2) in magnetics.

2. As a consequence of this storage density, optical recording allows us to store a reasonable program (for example 30 min) on a small surface, typically a 30 cm diameter disc. And if we store the signal on a spiral in such a way that each circumference of the disc contains one complete video frame then all kinds of important possibilities appear.

3. Instead of following the spiral we jump back once per rotation we get a still picture, by jumping more or less rapidly we can go through the picture like through a book.

4. Access to each picture (and they are of the order of 50000) can be very rapid.

5. There is no contact between the reading head and the disc.

6. The disc can be copied very rapidly and in parallel by pressing (like audio records)or by photographic means.

At present time these features are unrealistic in magnetic recording.

GENERALITIES ON OPTICAL READOUT - SOME ORDERS OF MAGNITUDE -

At first sight optical readout of videosignal stored on a disc seems to be a difficult and critical task due to what is paradoxically one of its most interesting features : no mechanical contact between the readout "stylus" and the information carrying surface of the disc. Since the readout "stylus" used here is a light beam, it cannot be simply mechanically guided by a groove, the way

---

Much of the work presented here has been carried out by a group of researchers headed by M. Le Carvennec and M. Tinet.

it is accomplished, for example with conventional audio records.

This paper intends to demonstrate that the servoing of the optical readout
stylus on the track is not a critical process. Various simple, reliable, low
cost solutions are possible and give the opportunity to take benefit of the
main features of optical readout : no wear, possibility of tricks. Work done
at Thomson concerning optical readout has dealt mostly with simplification of
solutions without being prejudicial to performance. New developments concerning this topics will be emphasized. Results obtained from prototypes built
according to some of these solutions will be illustrated by some data concerning picture quality.

Technical problems encountered in the realization of a videodisc storage system are mostly due to the necessarily high density of information. Several
solutions have been proposed and investigated to solve the problems of writing and reading this information (1, 2, 3, 4, 5, 6). We chose the solution
where the videosignal is stored in the form of "binary" microinformations
arranged along a spiral (with a typical pitch of 1.6 µm (fig. 1). The coding
system is a frequency modulated signal leading to a modulation of the duty factor of the "binary" microinformations (modulation of the zero crossing of the
electrical signal).

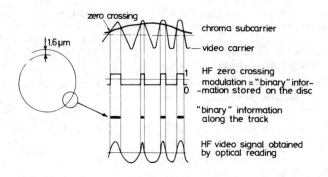

- Fig. 1 - Coding of information on optical videodisc.

Optical readout of those microinformations is accomplished by means of a focused laser beam (of the order of 1 µm). Transmitted or reflected light is modulated by these microinformations and detected with photocells. The original
electrical signal is then restored (fig.2, 3). Practically several ways of
storage can be used : phase relief modulation, absorption modulation, reflection modulation. One interesting advantage of phase relief modulation (fig.4)
is that it is well adapted to known replication processes. It renders possible
replication by embossing a thermoplastic material with a metallic stamper obtained from the master. This process is similar to the one used in audio domain.

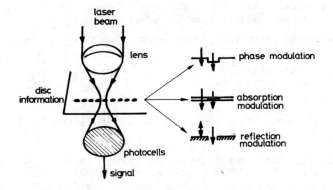

- Fig. 2 - Transmissive Readout.

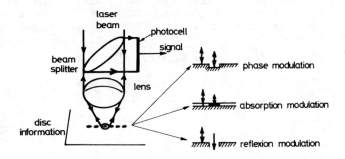

- Fig. 3 - Reflective Readout.

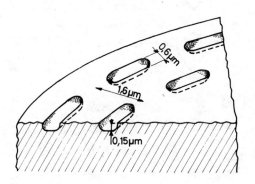

- Fig. 4 - Cut view of a Thomson Videodisc
with relief modulation.

Several demonstrations have been made with laboratory prototypes utilizing the type of modulation described above. Discs used were transparent flexible embossed discs. Since those demonstrations, studies have dealt with :

1 - Simplification of technology, improvement of performance, quality and reliability.

2 - Adaptation and optimization of technical options in relation with various fields of application : consumer market, institutional market and professional market.

This paper gives results concerning work being done in the field of optical readout.

The highest temporal frequency to be stored depends on the chosen coding scheme. This is typically 8 to 9 MHz for European TV standards for additive coding. In the case of a synchronous disc (1500 rpm) this leads to a maximum spatial frequency of the order of 1400 cycles/mm on the inner track at radius 7 cm. Such spatial frequencies can be written and read by optical means with a good signal to noise ratio when high numerical aperture optics is used (typically N.A. = 0.4 for readout). Only small depth of focus is attainable with high N.A. optics. Therefore a precise positionning of the information with respect to the focal point of the light beam is necessary. This light beam acts as a "reading stylus" without any mechanical contact (fig. 5), and its position has to be controlled very accurately in order to obtain a good readout (fig. 6) :

$$\pm \quad 2 \quad \mu m \text{ in the vertical direction}$$

$$\pm \quad 0.1 \quad \mu m \text{ in the radial direction}$$

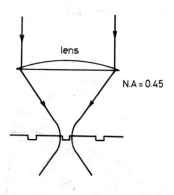

- Fig. 5 - Optical stylus.

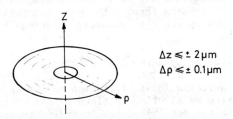

$$\Delta z \leqslant \pm 2\,\mu m$$
$$\Delta \rho \leqslant \pm 0.1\,\mu m$$

– Fig. 6 – Optical readout precision.

VERTICAL TRACKING

Two kinds of solutions appear to be suitable :
- Aerodynamic stabilization.
- Vertical servoing of the readout objective.

1. Aerodynamic stabilization

Aerodynamic stabilization is well adapted to flexible discs. Two types of systems, passive and active stabilizers have been considered. Two solutions have been particularly studied.

a – U-Shaped stabilizer (Fig. 7).

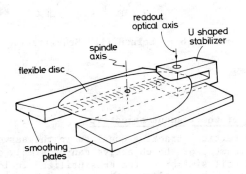

– Fig. 7 – U Shape stabilizer.

The first one to be used, consists of a U-Shaped guide in which the rotating disc is introduced. When the disc is rotating, air is forced between the upper and lower parts of the guide, creating a combination of aerodynamic forces. This rather simple and efficient solution has already been described and de-

monstrated several times. Its major drawback is inherent to the fact that it
is necessary to use a narrow gap between the two guides to obtain the desired
stabilization for all discs (typically 50 microns in addition to disc thickness). In dusty environment this may cause severe damages of disc surface when
large dust particles are passing through the gap. For this reason the U-Shaped
stabilizer has been given up and the following solutions have been proposed to
overcome this drawback.

b - Stabilization by Bernouilli effect (fig. 8)

In place of a passive aerodynamic stabilizer an active aerodynamic one is being
used. A properly distributed air flow is blown on the rotating disc through a
nozzle placed under the readout objective. According to Bernouilli effect, the
laminar flow created between the nozzle and the disc provides a local stabilization of the disc surface. This type of stabilization is more efficient than
the one obtained with the U-Shaped stabilizer. Furthermore dust particles,
which would otherwise pass under the readout lens and affect the quality of
the videosignal, are blown away due to the air flow. This solution however demands the realization of a compact pump.

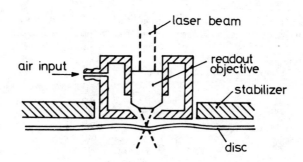

- Fig. 8 - "Bernouilli" Stabilizer.

2. Vertical servoing of the readout objective

An other approach to vertical tracking consists in the servoing of the readout
objective. The displacement of the objective can be simply obtained by the use
of an electrodynamic coil similar to the ones utilized in loudspeakers (Fig.9).
The driving signal of the servo loop can be obtained by several ways. Work has
been emphasized, on the use of simple optical sensors. The most interesting
ones appear to be those of the "single light spot" type (no need of extra light
beam in addition to the readout beam) using only few optical and electronic
components. Two sensors of this kind adapted to transparent or reflective discs
are particularly suitable. These ones use energy reflected by the information
carrying surface which is always sufficient to give a good signal to noise ratio for the error signal.

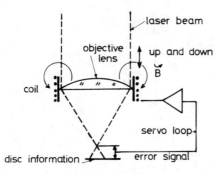

– Fig. 9 – Vertical focus servoing by a coil.

– Astigmatic sensor (fig.10a, b)

The error signal is derived by means of a system including a cylindrical lens
(placed on the beam reflected by the disc) and 4 photocells (connected by pairs,
oriented at 45° with respect to the axis of the lens and placed in the plane
of least diffusion, Po on the figure). When the disc surface is in focus, the
light spot centered on the 4 photocells is circular, and the electrical signal
obtained from the differential amplifier is equal to zero. When the disc surface
is out of focus, the light spot becomes elliptical. Depending on whether the
disc is positionned too low or too high, the long axis of the ellipse is in
one direction or at 90°. The signal obtained from the differential amplifier
is significant of the sign and of the amplitude of defocusing. The dynamic
range depends on the focal length of the cylindrical lens. It can be adjusted
at a typical value of $\pm$ 50 microns (fig. 11).

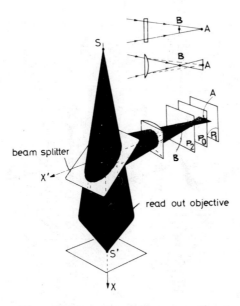

– Fig. 10a – Astigmatic sensor (system schematic).

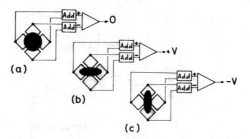

(a) disc in focus
(b) disc out of focus - too far -
(c) disc out of focus - too close -

– Fig. 10b – Astigmatic Sensor.
                Typical Photocells Illumination.

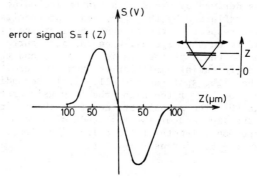

– Fig. 11 – Signal obtained from astigmatic sensor.

– Asymmetric sensor (fig. 12)

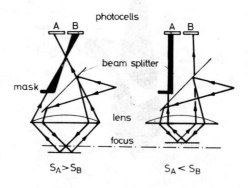

– Fig. 12 – Asymmetric Sensor.

This new solution is even more simple than the preceeding one. The only additional part consists of a mask (screen) placed on the reflected beam. Two photocells connected to a differential amplifier provide the error signal. When the disc is out of focus, depending on whether it is too low or too high, the mask projects a shadow on one or the other of the two photocells. Again the signal obtained from the differential amplifier provides us with desired informations concerning the sign and the amplitude of defocusing (fig. 13).

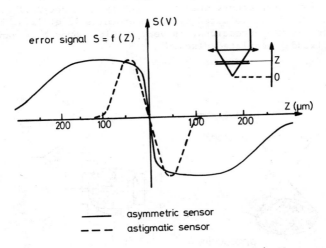

asymmetric sensor
astigmatic sensor

- Fig. 13 - Signal obtained from asymmetric sensor.

This simple system needs few elements and is easy to adjust. Moreover due to its very large dynamic range (typical value : $\pm$ 200 μm), it renders absolete the logical sequence, otherwhise necessary to find the right focus.

RADIAL TRACKING

The radial tracking can be accomplished by deflection of the reading light beam driven by a significant error signal (fig. 14).

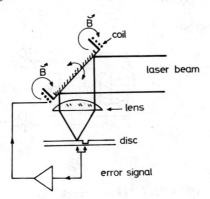

- Fig. 14 - Radial tracking.

Deflection of the light beam can be obtained by an electrodynamically actuated
tilting mirror. The error signal is derived in a way similar to the one used
to get the H.F. signal. For radial tracking, work has been again emphasized on
"single light spot" systems since they avoid the complexity inherent to the
use of additional light spots. The choice of the system depends on the type
of modulation used on the disc.

1) - In the case of "phase" discs (one particular type of discs with relief modulation) it is possible to make use of the asymmetry in the light beam intensity distribution which occurs when the light spot is not well centered on the
track. This asymmetry is detected by two photocells (fig. 15.a, b). This process, described on many occasions, is very simple. However it can not be used
to track "amplitude" modulated discs.

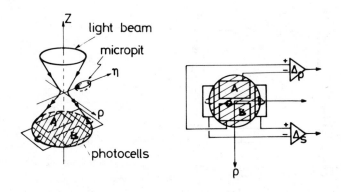

- Fig. 15a - "Far field" tracking.
Light spot on track axis.

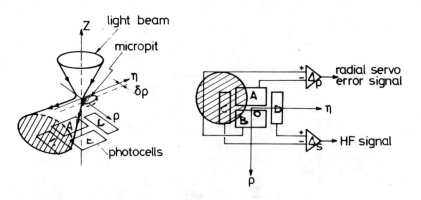

- Fig. 15b - "Far field" tracking.
Light spot out of track axis.

2) - An other simple system (fig. 16), well adapted to "amplitude" discs, detects the error of track position at the focal point of the reflected beam. When the track is not well centered there is an asymmetric light distribution detected by two photocells.

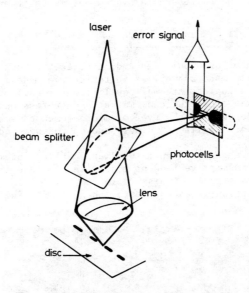

- Fig. 16 - Radial tracking sensor for amplitude disc.

3) A very interesting solution (fig. 17) well adapted for the two types of discs (amplitude and phase) and the two modes of readout (transmission and reflection) can also be used. The readout light spot is wobbled in the radial direction (at a typical frequency of 60 KHz) inducing an amplitude modulation

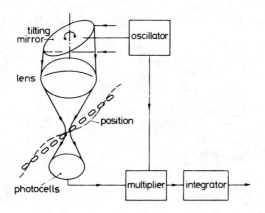

- Fig. 17 - Spotwobble tracking.

of the intensity of the transmitted or reflected light beam. This modulation
may be heterodyned by the wobbling signal, providing the radial error signal.
One can achieve simply light spot wobble of small amplitude, by actuating an
additional mirror, place on the light path, with piezoelectric element. Excellent results are obtained with such a system and makes it an interesting
choice.

MAIN FEATURES OF THOMSON PLAYERS

Several prototypes have been built according to the various solutions described
above (in transmissive mode as well as in reflective mode). It has been demonstrated that optical solutions are well qualified for various fields of applications (consumer market, institutional market and professionnal market). Choice
of right solutions depends on the type of microelements used to store information ("phase" discs or "amplitude" discs) and on foreseen applications.

One of the prototype is shown on figure 18. Image quality obtained, for example,
on a player using transmissive readout mode with flexible embossed discs is
characteristized by a signal to noise ratio in the luminance (3 MHz bandwidth),
and by an intermodulation, equal or better than 45 dB (fig. 19).

-Fig. 18 - Thomson prototype of videodisc player.

CARRIER	$\frac{signal}{noise}$	62 dB
(30 KHz bandwith)		
INTERMODULATION		50 dB
$(\frac{\omega_0 \pm \omega_1}{\omega_0})$		
LUMINANCE	$\frac{signal}{noise}$	45 dB
(3 MHz bandwith)		
$\frac{signal}{moire}$ (luminance)		42 dB

-Fig. 19 - Typical picture data.

CONCLUSION

The above example of a video player system gives only one aspect of optical video recording.

A whole series of new applications can be imagined with the development of new optical recording materials and more sophisticated recording techniques.

REFERENCES

(1) W. Bruch : A color Video Disc system,
     Journal of SMPTE, vol. 81, pp. 303-306 (1972)

(2) K. Compaan an P. Kramer : The Philips "VLP" system,
     Philips Tech. Rev., Vol. 33, pp. 178-180 (1973)

(3) K.P. Broadbent : A review of the MCA Disco Vision system,
     Journal of SMPTE, Vol. 83, pp. 554-559 (1974)

(4) G. Broussaud, E. Spitz, C. Tinet, F. Le Carvennec, A video disc optical
     design suitable for the consumer market,
     IEEE Trans. BTR, Vol. 20, pp. 332-337 (1974)

(5) R.L. Whitman : A transmission mode optical video disc system,
     SID 74, pp. 34-35 (1974)

(6) J. Kaufman Clemens : Information records and Recording/Play back system,
     U.S. patent 3842194 (1974)

SESSION 3.

IMAGE PROCESSING III

Chairman: N. ABRAMSON

ON PERFECT IMAGE CORRECTION BY UNSHARP MASKING

Leo Levi* and M. Mossel**

*Jerusalem College of Technology, P.O.B. 16031, Jerusalem
**National Physical Laboratory, Hebrew U. Campus, Jerusalem

ABSTRACT

Incoherent image enhancement, such as unsharp masking, is more economical
than the coherent techniques, but also more limited. Here the limitations
on perfect image de-blurring are investigated. Both linear and non-linear
processes are treated and extensive results, based on computer simulation,
are presented.

1. Introduction

Unsharp masking is, perhaps, the oldest of the sophisticated methods of
image enhancement. Its original implementation, 45 years ago [1], involved the combination of a blurred original negative with a positive copy
which was intentionally blurred further: the result was a sharper print. This
process was, in effect, a partial equalization of the spatial frequency spectrum. Since then, many processes have been devised, operating in widely
different ways, but all in a manner analagous to the above. [2] Some of
these involve scanning (with a spot which is either constant [3-7] or dyna-
mically controlled [8] [9]). Simultaneous methods have been developed based
on the Herschel effect of photography [10] and on the quenching of photophors.
[11, 12] With the exception of the scanning technique using a constant spot,
all of these are essentially non-linear processes. These we have analyzed
elsewhere [2]. They do not lend themselves to analysis in terms of the
transfer function (tf); only in the limiting case of vanishing modulation,
do they approach linearity. [2]

Because of this limit, however, and also because of the significance of
the constant-spot scanning techniques, the analysis of the linear unsharp
masking process is of interest.

2. Linear Unsharp Masking

In the linear version, the primary image spectrum S_1, is the product of
the object spectrum, S_o, with the tf, T_o, of the original imaging system.
Unsharp masking consists of adding to this spectrum the spectrum, S_m, of the
"mask," where the mask is an attenuated and blurred negative version of the
primary image. Specifically, $$S_m = -b \, S_1 T_m \quad ,$$

where T_m is the tf of the blurring process associated with making the mask,
and b is the attenuation factor.

It is convenient to express the image signal (s) in terms of

73

its mean value $\bar{s}$, and its normalized variation:

$$\tilde{s} = s/\bar{s} - 1 .$$

The spectrum of the primary image may then be written;

$$S_1 = \bar{s}_1 (\delta + \tilde{S}_1) = \bar{s}_1 (\delta + S_o T_o).$$

where δ stands for $\delta(\nu)$ and represents the Dirac delta function.

We may now write the spectrum resulting from linear masking:

$$S_2 = S_1 + S_m = \bar{s}_1 (\delta + S_o T_o) - b \bar{s}_1 (\delta + S_o T_o) T_m \qquad (1)$$

$$= \bar{s}_1 (1-b) (\delta + \frac{1-bT_m}{1-b} T_o S_o).$$

Here we have made use of the fact that

$$T_m(0) = 1,$$

so that

$$T_m \delta = \delta.$$

Note that the mean signal value has been attenuated by a factor $(1-b)$, showing that b must be maintained below unity. However, since the same attenuation is applied to the signal variation at the origin, the _modulation_ there has not been affected.

To compensate for the attenuation, the signal must be amplified, after masking, by a factor

$$g = 1/(1-b). \qquad (2)$$

3. Multiplicative Masking

In multiplicative masking, it is not the amplitude, but the modulation of the mask that is attenuated by a factor, b. Also, instead of adding it negatively, the sign of the modulation is reversed.

Consider a sinusoidal object function:

$$s_o = \bar{s}_o (1 + M \cos 2\pi\nu).$$

The primary image will be of the form:

$$s_1 = \bar{s}_1 (1 + MT_o \cos 2\pi\nu x),$$

and the blurred mask function:

$$s_m = \bar{s}_m (1 - bMT_m T_o \cos 2\pi\nu x).$$

The resulting image then has the form:

$$s_2 = \bar{s}_1 \bar{s}_m$$

$$= \bar{s}_1 \bar{s}_m [1 + T_o M(1-bT_m) \cos 2\pi\nu x - bT_m T_o^2 M^2 \cos^2 2\pi\nu x]. \qquad (3)$$

We immediately note two fundamental differences between this and the result (1):

(a) In (3) the modulation at the origin is reduced by a factor (1-b), whereas in (1) it remained unaffected.

(b) A term containing $\cos^2 2\pi\nu x = (1 + \cos 4\pi\nu x)/2$ appears. This represents a signal component of twice the object frequency and therefore implies a distortion of the original signal shape. It can be seen that this distortion approaches zero with M^2 and may be negligible for small values of M.

4. Complete Correction

Complete correction by unsharp masking is theoretically possible- but only with linear (i.e. additive) processing. Referring to (1) it is evident that the requirement for this is:

$$(1-bT_m) T_o / (1-b) = 1.$$

This is readily seen to imply

$$T_m = (T_o - 1 + b)/bT_o . \tag{4}$$

This tf becomes negative at the frequency at which T_o drops to the value of (1-b). In the usual optical system, having a low-pass tf, it then rapidly tends toward negative infinity and this limits the implementation of the ideal T_m.

To obtain a practical blurring function, the function (4) may simply be truncated at some convenient value, or it may be damped, for instance, by multiplying it with a bell-shaped curve.

The following considerations indicate the optimum truncation point. Sharp truncation is equivalent to multiplication with a step function and therefore tends to introduce extended oscillations into the corresponding spread function. These, in turn, make physical implementation difficult. This effect may be minimized by truncating the spectrum at its zero-crossing, that is at the spatial frequency, ν_o, where

$$T_o(\nu_o) = 1-b \tag{5}$$

[Cf. (4) with $T_m = 0$.]

The damping, to be effective, must be stronger than the divergence of the function (4). It has the advantage of suppressing the inconvenient oscillations, but must be expected to yield a less perfect enhancement.

The classical method for preparing the mask is by defocusing. In terms of our ideal correction spectrum (4), this means that we should pick a degree of defocusing, whose blur spread function approximates, as closely as possible, that of the ideal correction spectrum. Here the variables at our disposal are the amount of defocusing (d) and the relative aperture. Rather than the relative aperture we use another parameter derived from it: the (geometrical) blur diameter

$$a = d/F \tag{6}$$

where F=L/D is the effective F/number of the imaging system,

D is the diameter of the exit pupil of the imaging system and

L is the distance from this pupil to the image plane.

This should be the primary determinant of the mask performance.

5. Results of Computer Simulation

Image enhancement by linear unsharp masking was tested by simulation on a digital computer. Two objects were used:
   a. A unit step function
   b. A three-bar pattern with gaps between bars equal
      to the bar width.
These were blurred by a Gaussian spread function (to simulate atmospheric blurring which approaches a near-Gaussian shape on long exposure [13]):

$$T_o(\nu) = e^{-2\pi\sigma_o^2 \nu^2} . \qquad\qquad (7)$$

The blurring was normalized by setting the standard deviation, σ_o, equal to unity. The blurred images were then restituted by means of unsharp masks of three types:

   a. The ideal function (4) truncated at its zero:

$$T_m(\nu) = 0, \; \nu > \nu_o,$$

where ν_o is defined by (5).

   b. The function (4) again with T_o as given by (7), damped by multiplication with a Gaussian. The standard deviation of this Gaussian was chosen so that the total area under the resulting curve vanishes, i.e. so that the area element above the ν-axis equals that below the axis. The damped correction function then has the form

where

$$T_m = \{1-(1-b)\}e^{2\pi^2 \nu^2} e^{-2\pi^2 c^2 \nu^2 /b} \qquad (8)$$

$$c^2 = 1/[1-(1-b)^2]$$

is the standard deviation of the damping function.

   c. The function resulting when an aberrationless lens is defocused. This has the form: [14]

$$T(\nu_r,\Delta) = \frac{4}{\pi}\int_{\nu_r}^{-1} \sqrt{1-u^2} \; \cos [\; 2\pi\nu_r \, \Delta(u - \nu_r)] \; du, \nu_r > 1$$

$$= 0, \qquad\qquad\qquad\qquad \nu_r > 1$$

$$\nu_r = \lambda F\nu$$

and has been evaluated accurately over the range $(0 < \Delta < 100)$ where Δ is the defocusing, measured in units of Rayleigh's $\lambda/4$-tolerance on defocusing.[16]

Hence

$$d_R = 2\lambda F^2$$

$$\Delta = d/d_R = d/2\lambda F^2$$

where d is the amount of defocusing and
      F is the effective F/number of the system.

The geometrical blur (6) is then

$$a = d/F = 2\lambda F \Delta.$$

For each type of restitution mask, results were computed for four values of b:

$$b = 0, \ 0.5, \ 0.75, \ 0.9,$$

with the application of the corresponding gain values:

$$g = 1, \ 2, \ 4, \ 10.$$

The curve for b=0 represents the uncorrected primary image.

The results are shown in Figs. 1-6. Fig. 1 shows the results obtained with the "ideal" function (4) truncated at the zero-crossing. Figs. 2-5 depict the results obtained with a severely defocused aberrationless lens ($\Delta = 100$) with geometrical blur diameter varying from 1 - 10. In each of these figures are shown (a) the primary blurred step function image (b=0) and the results obtained with b = 0.5, 0.75, 0.9. Also shown are the results obtained with tri-bar patters of unit bar width (b) and of twice this width (c).

The method is not very sensitive to the amount of defocusing: in Fig. 6 we compare the results obtained with $\Delta = 10$ to these with $\Delta = 100$, both for the case of blur diameter a=4.

Results obtained with the damped "ideal" function (8) were quite similar to those shown in Fig.6. They are included in Table 1 described in the next section.

6. Quality Criteria

To compare and evaluate the results, we must first establish image quality criteria.

Classically, such criteria are based on the integrated squared deviation of the image from the object

$$B = 2 \int_o^\alpha [1 \ - \ s(x) \ ]^2 \ dx.$$

To obtain a criterion which correlates positively with image quality, we define the enhancement criterion Q_1, as the reduction in B, normalized with respect to B_o ,the value of B obtained in the primary blurred image:

$$Q_1 = (B_o - B \ )/B_o . \tag{12}$$

In terms related to the image parameters more directly accessible to measurement, we note that   in the case of the step function, the effect of blurring appears as a decrease in the angle of the transition. This angle is maximum at the origin and we may take its value there as a criterion of image quality. In terms of this criterion, all the proposed mask generating methods can yield substantial image enhancement. However, this is invariably accompanied by a distortion of the image contour – in the form of overshoot adjacent to the transition. In an actual system, such overshoots may mask weak (even strong) object detail near a pronounced contour. The amount of overshoot should therefore enter the quality criterion negatively. We have chosen as the criterion the slope angle Θ at the origin divided by the square of the peak height. [This factor is squared in order to permit it to be relatively insignificant for small values of overshoot (h) and to become duly important for large values]. Specifically, this second criterion is defined:

$$Q_2 = 2\Theta/\pi \ (1 + h \ )^2 \tag{13}$$

where the factor $2/\pi$ is introduced to normalize the quality criterion.

Discussion of Results

In Table 1 we list the integrated squared deviation, the angle at the origin and the overshoot for the cases illustrated in Figs. 1-6 and for the damped "ideal" enhancement method. Together with these are listed the quality criteria Q_1 and Q_2.

It can be seen that in all columns the truncated "ideal" filter yields the best results, the advantage becoming more pronounced with increasing enhancement - at b =0.9, the enhancement criterion is superior by about 50% to that of the best factor obtained with defocusing.

REFERENCES

1. G. Spiegler and K. Juris, "Ein neues Verfahren zur Herstellung ausgegliche-
ner Kopien nach besonders harten Originalaufnahmen, " Phot. Korr.67
4-9 (1931). See also ibid 69, 36-41 (1933)

2. L. Levi, "Unsharp Masking and Related Image Enhancement Techniques,"
Comp. Graph. Image Proc. 3, 163-177 (1974)

3. P. Dumontet, Machine elaborant des produits de composition en vue de la
correction de certain defauts dus a la diffraction," Opt. Acta 3
145-6 (1956)

4. R.V. Shack, "Image processing by an Optical Analog Device," Pattern Recog.
2, 123-6 (1970)

5. L. Swindell, " A Noncoherent Optical Analog Image Processor," Appl. Opt.9
2459-69 (1970)

6. S.W. Levine and H. Mate, "Selected Electronic Techniques for Image Enhancement, " Proc. SPIE Image Enhancement Seminar II (1963)

7. W.F. Schreiber, "Wirephoto Quality Improvement by Unsharp Masking, "Pattern
Recog. 2, 117-121 (1970)

8. P. Pargas, " The Principle of Velocity Modulation Dodging," Phot.Sci.Eng.9
219-227 (1965)

9. D.R. Craig, "The Log-Etron: A Fully Automatic Servo-Controlled Scanning
Light Source for Printing," Phot. Eng. 5, 219-226 (1954)

10.M. Johnson, "Use of the Herschel Effect in Improving Aerial Photographs,"
J.Opt.Soc. Am. 41, 248-751 (1951)

11. A.J. Watson, "The Fluoro-Dodge Method for Contrast Control," Photogram.
Eng.24, 638-643 (1958)

12. A.B. Clarke, " A Photographic Edge Isolation Technique," Photogram.Eng.28
393-9 (1962)

13. D.L. Fried," Optical Resolution through a Randomly Inhomogeneous Medium
for Very Long and Very Short Exposures," J. Opt.Soc. Am.56, 1372-9 (1966)

14. L.Levi and R.H. Austing, "Tables of the Modulation Transfer Function of
a Defocused Perfect Lens," Appl. Opt.7, 967-974 (1968)

15. L. Levi, Handbook of Tables of Functions for Applied Optics,CRC, Cleveland
(1974)

16. E.g. L.Levi, Applied Optics , Wiley (1968) Sect. 9,4,3.

TABLE 1

Blur Function	b = 0.5					b = 0.75					b = 0.9				
	D	2θ/π	h	Q₁	Q₂	D	2θ/π	h	Q₁	Q₂	D	2θ/π	h	Q₁	Q₂
Uncorrected (b=0)	.11685	.241	0	0	—	—	—	—	—	—	—	—	—	—	—
Ideal correct. trunc.	.0825	.327	.0625	.294	.290	.0694	.381	.075	.406	.329	.0593	.432	.0807	.493	.370
Ideal correct. damped		.353	.107		.288		.456	.253		.290		.639	.730		.213
Δ = 100, a = 1	.113	.248	0	.0329	.248	.107	.261	0	.084	.261	.089	.298	.0088	.238	.293
a = 2	.104	.265	.0013	.110	.265	.0854	.311	.021	.269	.299	.0773	.429	.1491	.338	.325
a = 4	.0861	.309	.0355	.263	.288	.0939	.425	.184	.196	.303	.671	.638	.7081	-4.74	.217
a = 10	.144	.377	.0212	-.232	.256	.983	.562	.699	-7.41	.195	9.77	.786	2.1878	-82.6	.077
Δ = 10, a = 4		.309	.0392		.286		.424	.191		.299		.637	.7198		.215

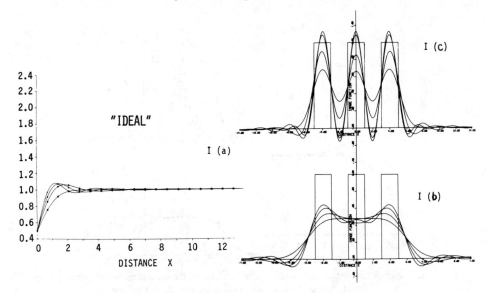

Fig. 1.   Results obtained with truncated ideal mask.
          a.   Unit step function (only upper half is shown)
          b.   Tri-bar pattern, unity bar width
          c.   Tri-bar pattern
          In each graph, curves shown are for subtraction weights
          b=0, 0.5, 0.75, 0.9

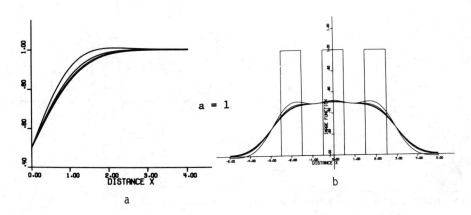

Fig. 2.   Results obtained with defocusing – blur diameter: unity
          a.   Unit step function (only upper half is shown)
          b.   Tri-bar pattern, unity bar width
          c.   Tri-bar pattern
          In each graph curves shown are for subtraction weights
          b=0, 0.5, 0.75, 0.9

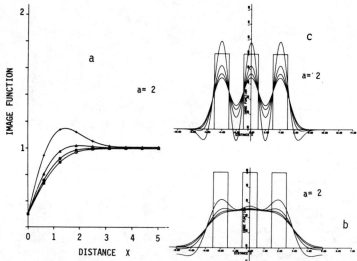

Fig. 3.   Results obtained with defocusing - blur diameter: two units
          a.   Unit step function (only upper half is shown)
          b.   Tri-bar pattern, unity bar width
          c.   Tri-bar pattern
          In each graph curves shown are for subtraction weights
          b=0, 0.5, 0.75, 0.9

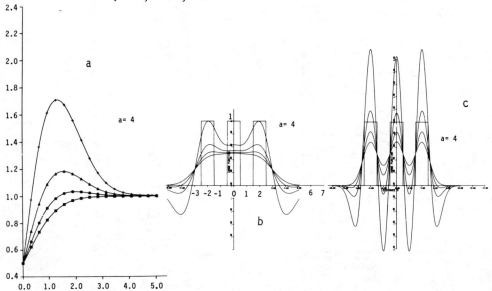

Fig. 4.   Results obtained with defocusing - blur diameter four units
          a.   Unit step function (only upper half is shown)
          b.   Tri-bar pattern, unity bar width
          c.   Tri-bar pattern
          In each graph curves are for subtraction weights
          b=0, 0.5, 0.75, 0.9

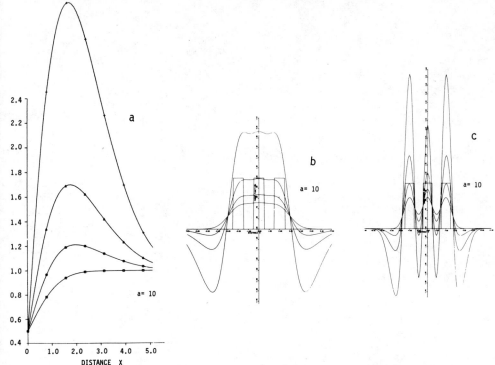

Fig. 5. Results obtained with defocusing - blur diameter: ten units
         a.   Unit step function (only upper half is shown)
         b.   Tri-bar pattern, unity bar width
         c.   Tri-bar pattern
         In each graph curves shown are for subtraction weights
         b=0, 0.5, 0.75, 0.9

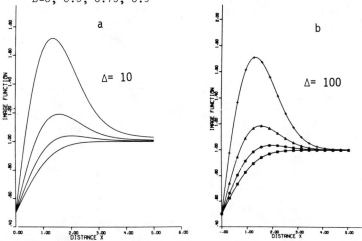

Fig. 6.   Comparison of results obtained with different degrees of defocusing
         a.   Defocusing 10 times Rayleigh's λ/4- criterion tolerance
         b.   Defocusing 100 times Rayleigh's λ/4- criterion tolerance
         In both cases results are for a unit step junction with blur diameters:
         4.   Curves are shown for subtraction weights
         b=0, 0.5, 0.75, 0.9

REAL TIME GRID CODING AND INTERLACING FOR IMAGE SUBTRACTION

N. Konforti and E. Marom

School of Engineering, Tel Aviv University, Israel

ABSTRACT

A technique for real time grid coding and interlacing applicable to image subtraction is described. The interlacing is accomplished optically without moving the coding grid by photographing incoherently the two images to be subtracted in perfect registration and by placing a Ronchi grating just in front of the image plane. If illumination directions of the two images are oblique and of opposite inclinations, an interlacing pattern is generated superposed on the image. Conventional spatial filtering of the composite image generates the desired subtraction Results obtained for the subtraction of the multi-tone images are presented.

INTRODUCTION

The technique of optical image subtraction consists in emphasizing the difference between two images under inspection, while suppressing their common (identical) parts. Since most practical cases for which image subtraction techniques are needed are two dimensional transparencies of variable density, only the amplitude space modulation of the illuminating beam is of significance and thus incoherent light sources are sufficient. Use of incoherent source illumination is even advantageous since it elliminates phase modulation interference and undesired phase effects.

On the other hand, the extraction of the desired difference information is done by spatial filtering and thus coherent light sources or electronic processing techniques are necessary.

Therefore, a two-step process has to be carried out for the complete subtraction operation. In the first step complimentary-coded images are combined on a common record forming an interlaced image, while in the second step the record is spatially filtered. Possibilities exist to perform these steps in real time by utilizing incoherent to coherent image convertors, whereby the modulated coherent beam could be utilized on line for purposes of filtering and display as well.

Several techniques for coding have been already described in the literature, utilizing sequential recordings of grid coded images (Ref. 1) or randomly diffused coders (Ref. 2-4). In this work the several (> 2) recording steps have been combined into one by utilizing a single coding grid, but providing inclined illumination beams for the two images, thus projecting a grid pattern over the recording plane, as further explained and illustrated.

CODING AND INTERLACING PRINCIPLE

Grid coding seems to be the most advantageous when building an interlaced

85

N. Konforti, E. Marom

image. Such a coding was implemented by Pennington et.al. (1) in a two-step process consisting of moving the coding grid to its complementary position between the exposure of the two objects. The set-up represented in Fig. 1 provides the possibility of grid coding interlacing without moving the coding grid. First, a transparency, T1, is illuminated from a light source S1

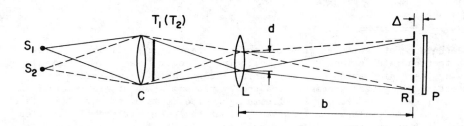

Fig. 1. Coding interlacing set up with a stationary grid.

through a field lens C. This lens forms an image of the light source S1 on the lens L at a distance d/2 from the center of the lens. The lens L images the transparency T1 on a Ronchi grating plate R. This image is recorded on a photographic plate P placed right behind the Ronchi ruling, at a distance Δ. So, a latent image T1 coded by the grid of the Ronchi ruling is recorded on P.

Substituing T1 with a second transparency T2 and moving the light source S1 to a new position, one generates the image of S2 at a second point on the imaging lens L, axially symmetric to the image of S1. The image of T2 formed

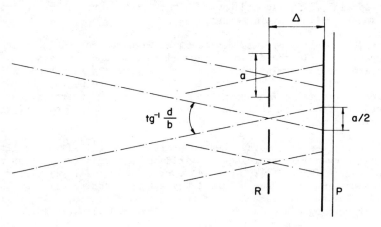

Fig. 2. Geometrical determination of the spacing between the grid and the photographic plate.

by the lens L through the Ronchi ruling on the photographic plate P will be
modulated by the same code as the image T1, but with a different aspect angle,
or equivalently, a lateral displacement.

Fig. 2 shows that the spatial displacement of half the period of the Ronchi
ruling will be achieved if the geometry of the set-up will be aligned
according to the formula:

$$\Delta = \frac{ab}{2d} \tag{1}$$

The distance Δ has to be constant throughout the whole area of the interlaced
image. The greater the distance Δ, the easier it is to achieve small
relative tolerances; however the extent of Δ is limited by diffraction
effects.

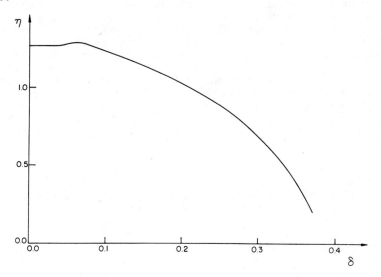

Fig. 3. Diffraction efficiency of first order beam as a function
of grid-photographic plate spacing.

Figure 3 describes the diffraction efficiency of the first harmonic component
of the grating pattern, as a function of the spacing Δ. This diffraction
efficiency has been normalized with respect to the effeciency of a sinusoidal
grating of maximum contrast. The spacing Δ is normalized with respect to
a^2/λ

$$\Delta = \delta \, a^2/\lambda \tag{2}$$

We can easily determine by inspection from Fig. 3, that $\delta \leq 0.2$ for adequate
diffraction efficiencies.   For an incident angle α normal to the grating
orientation, the effective Ronchi ruling period is the true grid period
multiplied by cos α, while the diffracting distance between the grating and
the photographic plate is the actual spacing Δ divided by cosαcosβ. β is the
maximum angle of incidence along the grating orientation.

Thus the effective value of δ at the extremity of the field of view is given
by

$$\frac{\Delta}{\cos\alpha\cos\beta} = \delta_{limit} \frac{a^2\cos^2\alpha}{\lambda} \tag{3}$$

or

$$\delta_{limit} = \frac{\delta(\alpha = 0)}{\cos^3\alpha\cos\beta} \tag{4}$$

Taking into account the curve shown in Fig. 3, the width of the field of view (α,β) and the ease of alignment, the recommended value for δ should be between 0.05 and 0.10.

Without any loss of energy or relevant image information an iris diaphragm inserted into the lens L and centered at the imaging point of the light source can limit its aperture, and thus the modulation transfer function to half the frequency of the Ronchi ruling. This iris has to be placed at the imaging point of the light source S1 for the first exposure and then moved to the imaging point of S2 for the second one. Furthermore, the iris may be substituted by a slit opening, the width of which is related to the Ronchi ruling space frequency, thus allowing higher frequencies in a vertical direction to the grid code modulation. If such an iris is used, limiting the light coming from one image to pass above the slit, while the light carrying information from the second image to pass below it, then the exact location of the light source is unimportant, and so there is no need to move this source. Moreover one can then use a diffuser in front of the input transparences with a consequence of lower energy efficiency.

It should be noted that transparences illuminated through ground glass plates can be substituted by two-dimensional opaque objects. Furthermore, 3-D scenes can be compared provided they are contained into the depth of-focus defined by distance d, since the described scheme is basically a two-dimensional one.

REAL TIME CODING AND INTERLACING SET-UP

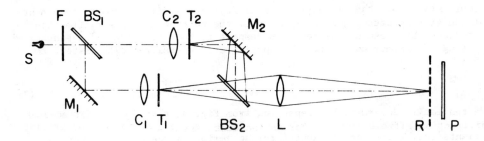

Fig. 4. Set up for simultaneous recording of the interlacing
        pattern onto the two images

A real-time set-up according to the principle described above is illustrated
in Fig. 4.   Here the two transparencies are inserted into the two arms of the
Mach-Zehnder interferometer.   The lens L images simultaneously both transparencies on the Ronchi ruling R.   In the initial alignment of the interferometer, the image of the light source S is formed by means of the field
lenses C_1 and C_2 in the center of the lens L.   Thus images T1 and T2 appear
in P with same code modulation and no lateral displacement of the grating.
The mirror M1 is then tilted so that the image of the light source S will
form in the upper part of the lens leaving the image through the second arm
of the set-up in its lower part.   A filter F is utilized to limit operation
at a narrower spectral region of the light source S.

According to the scheme described above the two images of transparencies T1
and T2 will now form at P, grid coded by the Ronchi ruling at different
lateral displacements.   This difference can be precisely aligned to half a
period by the tilting of the mirror M1 so that grid coding is achieved.

It should be noted that the beam splitter BS2 may be substituted by a front
surface mirror if placed sufficiently close to the lens L and if its size
will cover about one half of the lens size.   In that case the beam from T1
will pass unobstructed through one half of the lens, where most energy is
localized due to the focussing effect of C_1 and the tilt of M_1, while the
beam modulated by T_2 will be reflected by M_2 and the substituted mirror to
pass through the second half of the lens.

EXPERIMENTS

Two similar scenes consisting of the image of a metal closet containing
optical components and instruments were recorded on a regular black and white
35 mm film as shown in Fig. 5A and Fig. 5B.   The differences between
the two images are: a laboratory jack and a light-detector stand missing in
one of the scenes and a rotation in the position of a knob in the middle of

                 A   B

Fig. 5. Images to be subtracted.

the front panel of a high voltage supply. Its position is observed by a tiny
arrow extending beyond the knob's circumference.

Figure 6 shows an enlarged part of the center of the interlaced image
including the jack which appears in only one of the scenes. It can be easily

Fig. 6. Enlarged portion of the interlaced combined picture.

seen that the jack is defined by a grid code having half the spatial frequency
of the surrounding background. The spatial frequency of the code on the jack
is identical to the Ronchi grating frequency used during the recording, whilst
the background is coded at double frequency. Thus difference extraction can
be achieved by spatial frequency analysis. In this experiment, difference
extraction was done with a simple coherent processor illustrated in Fig. 7: a

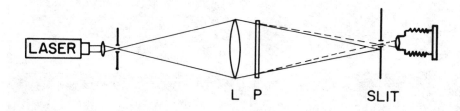

Fig. 7. Coherent processor for difference extraction display.

lens L directs a converging beam through the transparency of the interlaced
image P to a point on an opaque screen. This beam is deflected by the grid
coding in the transparency P. The deflection is affected by the grid
frequency: common parts of the images being coded by a double frequency of the
Ronchi ruling can be separated from the basic frequency of the Ronchi ruling
modulating only the difference information. A slit is positioned in the

spectral plane, so that only information related to the difference between
the images will pass through it , and form an image of the difference
structure.   A record of the differences between the objects in Fig. 5A and
Fig. 5B is shown in Fig. 8.   The jack and the light detector are clearly

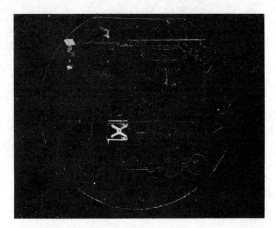

Fig. 8. Display of the subtracted image.

defined on the strongly attenuated background.   The knob's changed position
is hardly detected, being close to the noise level of the surrounding
structure. The difference created by turning the knob can be regarded as the
subtraction resolution limit.   The ratio of the knob's arrow area to
the whole scene is about 10^{-6}, making the information hardly detectable.

CONCLUSION

A technique for simultaneous recording two grid coded images suitable for
further processing for extracting the difference between them has been
described.   The coherently filtered record provides the desired subtracted
information.   It is possible to combine the two steps into one and obtain the
complete real time operation by using an incoherent to coherent light valve,
that are now emerging from scientific laboratories.

REFERENCES

(1) K.S. Pennington, P.M. Will and G.L. Shelton, "A Technique for Extraction
       of Differences from Scenes, Opt. Commun. 2, 113, (1970)

(2) S. Debrus, M. Francon and C.P. Grover, Detection of Differences Between
       Two Images, Opt. Commun. 4, 172, (1971).

(3) S. Debrus, M. Francon and P. Koulev, Extraction de la Difference Entre
       Deux Images, Nouv. Rev. Optique 5, 153,(1974)

(4) E. Marom and I.Kasher, Optical Distribution of Multiple Exposures in
       Speckled Image Subtraction Setups, Nouv. Rev. Optique, 7, (1976).

SESSION 4.

IMAGE PROCESSING IV

Chairman: A. MARECHAL

OPTICAL FILTERING OF TEMPORAL FREQUENCIES AND RELATED VISTAS

Jean-Charles Viénot and Jean-Pierre Goedgebuer

University of Besançon, France

ABSTRACT

The temporal frequency spectrum of a broad-band source is modulated
as the light beam passes through either any optical system (lens, chromatic filter, interferometer) or a pupil containing a message whose spatial frequency
spectrum contributes to the modulation. The result is displayed at the output
of a spectroscope as a function of ν (or λ). That can be applied to various
metrological problems as well as to the transposition to the chromatic domain
of some techniques of coherent and incoherent optics such as speckle, holography, information processing and coding.

INTRODUCTION

The major part of this paper deals with optical filtering of temporal frequency
spectra and the like. Most of the experimental work is based on two groups of
properties of optical systems including lenses, interferometers and dispersors:
1. An extension of the classical Abbe's theory applies to any linear, space and
time invariant system as the input message is a function of space and/or time;
an exchange between the spatial frequency u and temporal frequency ν leads
straightforward to impulse responses (spread functions) and transfer functions;
2. Any light disturbance incident onto a complex pupil being a function both of
space and time, the output derived from the diffraction integral expresses some
relationship between transverse and longitudinal coordinates, that is between
space and time, since along any θ-direction the path difference, Δ, between
wavelets is linked to a time delay $\tau=\Delta/c$, where c is the velocity of light.

TEMPORAL RESPONSES AND FILTERING

The transposition to the time domain of the Abbe's theory suggests some sort of
similarity between space and time variables in optics, at least as far as the
statistical aspect of light is concerned. However since time does not present
any symmetry no rigorous parallelism could be formulated in agreement with the
causality principle. Fortunately in the various situations we will encounter
here - as in those in electrical or electronic   signal processing - where both
space and time variables are combined at the technical level, such a principle
does not intervene and experimental results are independent of it. Nevertheless
attempting to use temporal together with spatial variables is no contradiction
to the assertion that the spatial frequency has been and remains one essential
quantity in Modern Optics. As a matter of fact it appears as a natural extension
as soon as an increase of the information capacity of the transmission channel
is wanted. One solution could be to work with several monochromatic channels
simultaneously, or with a polychromatic carrier that will be white light eventually.

Let us now enter upon dimensional considerations. In conventional optical image
formation as well as in coherent or incoherent optical processing, we have to
do with modifications in spatial distributions of the amplitude or intensity -
generally in the Fourier domain. These modified distributions are transverse.

Another approach considers how the information to be coded or processed acts on the longitudinal profile of the analytical signal which represents the carrier, i.e. the light disturbance itself. Here any transverse information is described through a function of time. These two possibilities of coding in u and v which are the spatial and temporal frequencies reciprocal of x and t respectively, express similar filtering properties. This is illustrated in Fig. 1. We

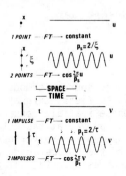

see that in both cases the second point, or the second impulse, causes a modulation or a sinusoidal filtering of the spectrum of the first element taken alone.

Subject to this remark we can recall the general mechanisms discussed earlier (Ref 1). They govern optical imagery and allow to define impulse and transfer functions, whether time or space is considered.

The following relations imply linearity and invariance, that is isoplanetism and stationarity of the system (S). They hold the comparison between space and time input functions in order to dwell on the temporal filtering aspect whose occurence comes out already in two trivial instances.

Fig. 1

$$\underset{object}{space\ domain} \rightarrow \underset{domain\ (entrance)}{spatial\ frequency} \quad (S) \quad \underset{domain\ (exit)}{spatial\ frequency} \rightarrow \underset{image}{space\ domain}$$

$$\delta(x) \rightarrow F_\infty(u)=const. \quad === \quad F'(u)=H(u) \rightarrow h(x')$$
$$f(x) \rightarrow F(u) \quad === \quad F'(u)=F(u)H(u) \rightarrow f'(x')=(f\circledast h)(x')$$

e.g. (S) limited by a rectangular aperture:

$$H(u)=rect\ \frac{u-u_0}{\Delta u} \rightarrow h(x')=\text{FT}\{H(u)\}=sinc(\pi\Delta ux').exp\{j2\pi u_0 x'\}$$

$$\underset{input}{time\ domain} \rightarrow \underset{domain\ (entrance)}{temporal\ frequency} \quad (S) \quad \underset{domain\ (exit)}{temporal\ frequency} \rightarrow \underset{output}{time\ domain}$$

$$\delta(t) \rightarrow F(\nu)=const. \quad === \quad F'(\nu)=H(\nu) \rightarrow h(t')$$
$$f(t) \rightarrow F(\nu) \quad === \quad F'(\nu)=F(\nu)H(\nu) \rightarrow f'(t')=(f\circledast h)(t')$$

e.g. (S) limited in the visible region (model for white light signal):

$$H(\nu)=rect\ \frac{\nu-\nu_0}{\Delta \nu} \rightarrow h(t')=\text{FT}\{H(\nu)\}=sinc(\pi\Delta\nu t').exp\{j2\pi\nu_0 t'\}$$

One comment can be added about the physical signification of the examples, as a distinction must be made between the two actual situations : (i) in the geometrical, or spatial, domain one has direct access to the square modulus of the impulse response solely, that is here $|h(x')|^2=sinc^2(\pi\Delta ux')$, the exponential term being lost - this not only due to the invariance of the diffraction pattern, but above all because the sensor is quadratic (that exponential can be taken into account and the phase recorded by holography); (ii) in the temporal domain the impulse response is not directly attainable but it can be shown that the only accessible quantity is either the autocorrelation function of $h(t')$, or the square modulus $|H(\nu)|^2$ of its TFT, the latter being performed by means of a spectroscope set in cascade with the system (S) as will soon be explained.

The above example, namely the gating of the visible spectrum inside a rectangular "aperture" is actually that of a very broad band chromatic filter whose TFT represents the average wave train of white light corresponding to the fre-

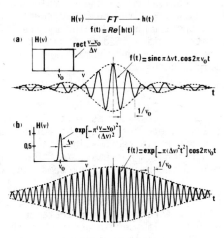

quencies ranging from 428 THz to 749 THz that is $\Delta\nu=321$ THz, or in more familiar notations, 0.4 μm < λ < 0.7 μm, centered at 0.55 μm ($\Delta\lambda=0.3$ μm). The shape of the wave train is given in fig. 2(a). For a gaussian filter centered in the green and isolating a band of about 12 nm, the shape is that shown in fig. 2(b).

In certain respects such chromatic filters could be termed static or passive temporal filters. Another category of temporal filters that we do use in information processing and coding is not without bringing matched filters to mind, though at the moment we still hesitate on the choice of the best suited expression to name them. Their behaviour will be mainly referred to in the discussion below.

Fig. 2

Let us then examine the Michelson interferometer of fig. 3(a) where the two mirrors are separated by the distance D. The imput impulse elicits a response $h(t)$ made of a Dirac pair (fig. 3(b)) whose separation is $\tau=\Delta/c$, c being the velocity of light. At time t_0 the amplitude of such responses is proportional to $\delta(t-t_0\pm\tau/2)$ and their exploitation is made secure by means of a spectroscope that achieves the Temporal Fourier Transform (TFT) in amplitude. Strictly speaking, one gets :

$$FT\left[\delta\left(t-t_0-\frac{\tau}{2}\right)+\delta\left(t-t_0+\frac{\tau}{2}\right)\right]=2cos\left(2\pi\nu\frac{\tau}{2}\right).exp\{-j2\pi\nu t_0\}$$

Behind the spectroscope, energies alone can be observed and the square modulus of the exponential term equals 1. The energy then obtained (fig. 3(c)) is proportional to the square cosine, namely $1+cos(2\pi\nu\tau)$. Such a filtering function generates black fringes periodic in ν, in the widespread uniform spectrum assumed for the ideal impulse at the input. Quite similar conditions are met with fig.4.

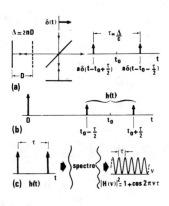

Fig. 3

Fig. 4 →

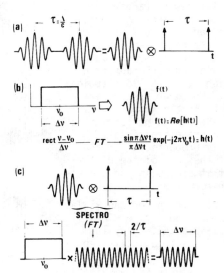

It is the actual physical case of a white light "pulse". Such two pulses emerging from the interferometer, shifted by τ, are equivalent to the convolution of
one of them - that is input (see fig. 4(a) and (b)) - with the Dirac couple. At
the output of the spectroscope our channelled spectrum is modulated by the TFT
of the impulse of white light, i.e. by the gate function itself (fig. 4(c)).
The phenomenon is nowhere observed but in the time frequency domain and the
fundamental relation *input* x *admittance* = *output* keeps valid :

$$F(\nu).H(\nu)=F'(\nu) \rightarrow |F'(\nu)|^2=B(\nu)=|F(\nu)|^2.|H(\nu)|^2$$

$B(\nu)$ is the power spectrum, observable as an intensity distribution. Its value
has been plotted *vs* ν for various other optical devices such as multiple beam
interferometer, two and several diffracting slits, etc. (ref. 1). We noted that
the Michelson and Fabry-Pérot interferometers offer limited possibilities inas-
much as modifying their intrinsic transfer functions is not easy. Yet they can
further permit simple measurements; no wonder if interferential systems should
fit metrological problems.

Devices dealing with diffraction allow to modify complex distributions (therefore the function H) inside any pupil along an arbitrary way, for instance by
using slits arranged regularly or according to some law set in advance. Such
systems will be relied upon in the coding of $f(t)$ functions, information transfer and the recording of temporal Fourier holograms.

TEMPORAL SPECKLE

Recent work has been carried out
about phenomena due to the scattering of white light by a rough
surface (Ref 2). Once more this
is a transposition $x \rightarrow t$, or an
extrapolation of the basic conc-
ept of speckle recording and analysis developed in the space
domain (fig. 5). The power spectra are given by the Spatial and
Temporal Fourier Transforms of
the amplitudes (square moduli
of them). In space the Fourier
transformer is a lens. In time
a spectroscope is required. As
the extension of the entrance is
infinite along the transverse
direction x, or the longitudinal
one, t, the power spectra are
constant. As the pupils are limited, the pseudo-periods, that
is the average graininess, is
inversely proportional to the
size of the part of the disturbance that occurs in the speckle
pattern formation.

Practically the experimental lay
out is shown in fig. 6. The filtering law is defined by the
surface (S) schematically represented according a simple step
model, yielding a random distri-

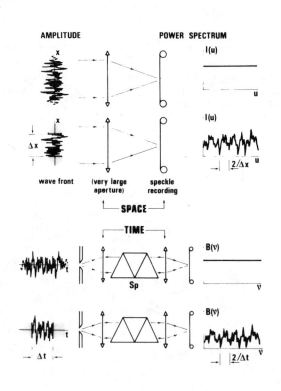

Fig. 5

bution of scattered wavelets which in turn produce a spectrogram that is actually called a temporal speckle (fig. 7). Moreover, by diffraction of a monochromatic wave through the speckled spectrogram, $B(\nu)$, taken as a transparency, one obtains the autocorrelation of the pupil, that is a representation of the average surface roughness (fig. 8).

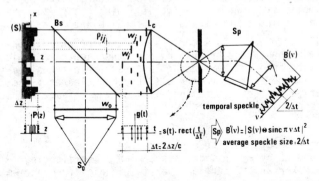

Fig. 6. Recording of temporal speckle from the surface (S); S_0 : white light source; w_0 : plane incident wave; $P(z)$: histogram of (S); $s(t)$: random distribution of wavelets $w_{,,}$, $w_{,j}$,... reflected back; $g(t)$: impulse response of (S) after the convolution theorem recalled in the next paragraph; Sp : spectrograph performing the TFT; $B(\nu)$: intensity distribution in the speckle pattern as a function of ν along x direction.

Fig. 7. Speckled spectrogram, the ν axis is horizontal.

Fig. 8. Diffraction pattern of a speckled spectrogram →

FILTERING BY GENERAL DIFFRACTING PUPILS

The description of diffraction phenomena at a given pupil from Kirchhoff integral brings forth the spectrum of the pupil expressed in terms of its Fourier Transform multiplied by a factor $-j\nu/c$. This factor is generally ignored in the

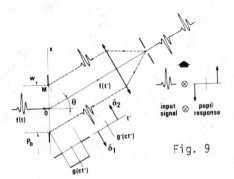

Fig. 9

calculation, whereas it has to be seen as introducing the time derivative (after the derivative theorem). Far from being difficult to interpret, this element serves as a pretence to a wider use of temporal filtering effect for an optical pupil conventionally described through its geometry as it was pointed out in Ref 1, 3, 4. If a disturbance $f(t)$ in the form of a plane wave is incident onto the most general complex pupil defined by $g(x)$ the time response, $f'(t')$, attached to the pupil is the convolution of the temporal input function with the first time derivative of the projection of the pupil along the considered direction of diffraction. This can be written as :

$$f'(t')=f(t)\circledast\frac{d}{dt}\{proj_\theta g(x)\}$$

t' is a retarded time; in the Fig. 9 the derivative of g is noted g'; the pupil is assumed rectangular for simplicity : the derivative is a Dirac pair ($\pm$); the resulting signal is represented behind a slit set at the focal plane of a lens.

All above means that any pupil, or object, is a temporal filter for any disturbance in the optical frequency range (and even beyond, over the electromagnetic spectrum). The result illustrated here with a rectangular aperture extends readily to any system of pupils. The concept of coding also becomes clearer : the response varies as the derivative of the information, i.e. of the message borne by the pupil and can be then conveniently matched.

Remarks

Before getting any further three points must be emphasized.
1. The derivative effect introduced by $-jv/c$ expresses a high pass filtering whose effect will be all the weaker as the carrier operates at high frequencies and will be then negligible in some cases.
2. The temporal transfer function $H(t)$ has been recorded on a quadratic receptor that makes it necessary to implement a reference impulse after or before the signal in order not to loose any information. As the whole is eventually analysed by means of a spectroscope, it can be said that an "interference" occurs between temporal spectra at the output of the spectroscopic device. This results in a temporal hologram wherefrom information can be easily reconstructed.
3. The field of optical metrology is often limited by the coherence of the sources and the cost of measurements : white light interferometry can be achieved if the previous considerations are taken into account.

METROLOGICAL APPLICATIONS

Let us go back to the output of the spectroscope placed in cascade with the two beam interferometer. The channelled spectrum of fig. 10 represents the coding of the delay between the two wave trains coming from the interferometer. The fringe function does indicate that the period is $1/\tau$ along the ν axis. This method applies to a whole category of measurements involving path differences, such as rugosity or profile testing and telemeasurements of displacements (Ref 5).

Fig.10. Channelled spectrum

Principle of profile determination. An automatic apparatus has been built, instantly providing the profiles of surface microaccidents by measuring the spectral interval between a set of dark bands recorded at the output of the spectroscope. Irregularities in the spacings evince defects along a direction perpendicular to the ν axis (fig. 11). The measures thus obtained range from one to several hundred microns, with an accuracy of 1% or about.

Fig.11. Defects display

Other applications dealing with profiles were presented elsewhere (Ref 1, 5).

FOURIER HOLOGRAPHY RELATED TO TEMPORAL FREQUENCY

The recording of a Fourier Hologram is that of the interference of spectra in the focal plane of a lens. It has been shown that a spectroscope is a temporal Fourier transformer, and that the channelled spectrum is the result of the spectrum analysis of two wave trains, that is of coherent superposition of two temporal spectra, one of them acting as a reference (fig. 12). The analogy in the temporal domain holds as well in image reconstruction by diffraction of a monochromatic beam through the Temporal Fourier Hologram (TFH). Figure 13 shows such an image reconstructed from the pattern of fig. 11; the departure of the actual surface from the ideal profile is 1 μm or about. Another example is that of a step surface hologram (fig. 14), leading to the very profile (fig. 15). The equivalent of double exposure holograms applied to displacement determination is given in fig. 16 (double exposure spectrograms) : in ±1 orders one gets successive positions of the surface (fig. 17).

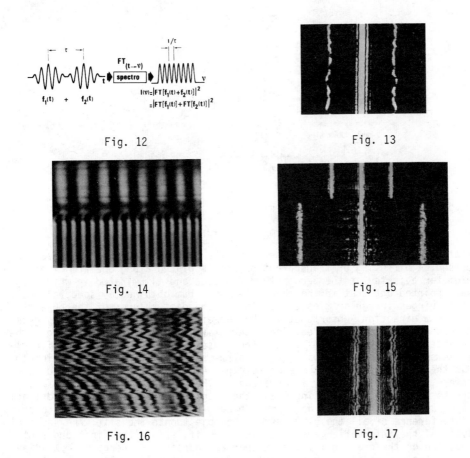

Fig. 12

Fig. 13

Fig. 14

Fig. 15

Fig. 16

Fig. 17

TFH recording can also be accounted for in a way illustrating the existing relation between a geometrical pupil distribution and the longitudinal information temporally exploited. On the diagram fig. 18 the geometrical object $g(y)$ and a

reference $K\delta(y-y_r)$ are illuminated in white light. In the diffraction plane a
dispersor isolates the monochromatic FH, spreading them along the y axis. The
spectroscope performs the analysis of the energy at $P(y_0)$. In that region the
contribution of each monochromatic component is related to a separate region
of the hologram. The spectroscope thus "perceive" each slice of the object in
a form coded by the colour parameter. The various distributions are drawn at
bottom of fig. 18 (amplitude at the input plane, energy for three chromatic
components at the F plane, i.e. at the entrance of the spectroscope, power spectrum at its output). Thus a biunivoque relation is already herewith confirmed
between the geometrical (spatial) and the chromatic (temporal) variables. At
the output of the spectroscope, the polychromatic hologram is displayed as a
function of λ. We presented monochromatic reconstructed images on various occasions (e.g. Ref 1, 3) as well as results obtained with extended self-luminous
objects (spatially and temporally incoherent); in this case the object is laid
out in front of an image doubling device allowing each couple of homologous
points to constitute an independent reference-information set.

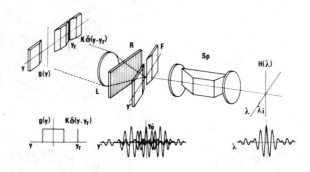

Fig. 18

3-D INFORMATION PROCESSING IN WHITE LIGHT

In what comes next we intend to generalize the concepts and experiments already
discussed. A diffusing or transparent object is placed along one arm of an interferometer adjusted at the zero path difference and illuminated in white light.
We have pointed out that in the channelled spectrum obtained by spectral analysis the fringes irregularities correspond to the disturbances of the object-wave
along the x direction. Unfortunately the geometrical information along y gets
lost. How could it be recovered? One alternative may consist in scanning either
the object or its image along y. The other possibility is summed up in fig. 19.
Some elements of homologous waves such as Δx_1 and $\Delta x_1'$ interfere at wavelength
λ_1, at the entrance slit of the spectroscope. At the same point, Δx_2 and $\Delta x_2'$
interfere, at λ_2. All the xy information is thus coded as a function of λ,
thanks to the dispersive grating R and the imaging lens L. It is yet advisable
to make sure that the source does provide spectral components allowing to bring
the interference of all the elements of the object onto the slit. The spectral
analysis of the information contained in the slit, set in the first order of
diffraction of the grating leads to a spectral density of energy $O(x',\lambda)$ displayed at the output plane of the spectroscope. λ being directly linked to y,
the description of the object is complete. In a spectrogram of the same kind as
in fig. 20, the contour of the phase object (triangle) can be observed together
with the fringes whose accidents , for each value of the wavelength, allow to
trace out the variations in the optical width. As another example, one could

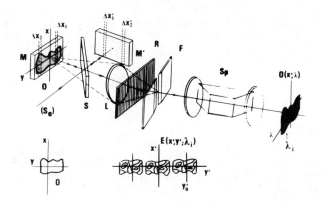

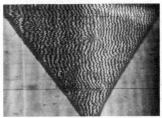

Fig. 19

Fig. 20   Fig. 21

determine thickness defects in a photographic film by examining the fringes
superimposed to the image (fig. 21), the coding relation being defined by the
light intensity E measured on the spectrogram, the spectral density of the source $B(\lambda)$, the function O' (a geometrical image of the object), the local path
difference $\Delta(x,y)$ and the lateral shift of the diffracted pattern for λ_i and
the grating of frequency N set at the distance D from the $x'y'$ plane :

$$E = O'(x',y').B(\lambda).[1+cos\frac{2\pi}{\lambda}\Delta(x',y')]\otimes\delta(y'-DN\lambda_i).$$

CONCLUSION

While most methods of image optical processing deal with quasi-monochromatic
wave trains, that involving a limitation in the possibilities of such carriers,
we thought that the information capacity could be increased by using the complete range of the visible light spectrum. Let us take an example. Assuming
that one works with a band $\Delta\lambda \simeq 0.3$ μm centered at 0.55 μm, any device capable
of resolving $\delta\lambda$ (i.e. whose resolving power is $R = \lambda/\delta\lambda$) will allow to carry
out a number of independent information elements $N = \Delta\lambda/\delta\lambda$, that is to say more

than half the figure expressing R. It means that the increase of the capacity of the processing system, or transmission channel, could be of several orders of magnitude - the multiplicative factor would be the number of spectral elements which behave as independent carriers working then simultaneously.

———————————

REFERENCES

1. J. Ch. Viénot, J. P. Goedgebuer, A. Lacourt, Space and Time Variables in Optics and Holography : Recent Experimental Aspects, *Proceedings ICO X, Int. Conf. Prague, 1975,* and *Applied Optics,* 15, n° 12, (1976).

2. J. P. Goedgebuer, J. Ch. Viénot, Temporal Speckle, *Optics Communications,* (in publication).

3. C. Froehly, A. Lacourt, J. Ch. Viénot, Notion de Réponse Impulsionnelle et de Fonction de Transfert Temporelles des Pupilles Optiques, Justifications Expérimentales et Applications, *Nouvelle Revue d'Optique,* 4, n°4, (1973).

4. A. Lacourt, J. P. Goedgebuer, J. Ch. Viénot, Re-assessing basic Landmarks on Space-Time Optics, *Optics Communications,* (in publication).

5. A. Lacourt, J. Ch. Viénot, J. P. Goedgebuer, Longitudinal Analysis of the Complex Transmittance of Optical Pupil by Spectral Modulation of White Light, *Japan Jl of Applied Physics,* 14, (1975).

VOLUME HOLOGRAM REPRESENTATION OF SPACE-VARIANT SYSTEMS

Robert J. Marks II, John F. Walkup and Marion O. Hagler

Department of Electrical Engineering, Texas Tech University, Lubbock, Texas 79409

ABSTRACT

Methods of characterizing linear space-variant systems by their responses to various sets of inputs are discussed. Recording these responses within a volume hologram results in a filter which is approximately equivalent, in an input-output sense, to the space-variant system.

The approaches considered are (1) storing the transfer functions of the system for point sampling of the input plane on playback: (2) a piecewise isoplanatic approximation approach based on division of the input plane into isoplanatic regions, with storage of the transfer function for each isoplanatic patch; and (3) storage of the system's responses to elements of an orthonormal basis set. The potential advantages and limitations of each of these approaches, as well as experimental results, are discussed.

INTRODUCTION

It is well known that a holographically recorded filter can be placed in the Fourier plane of a coherent optical processor and used to represent the transfer function of a linear, space-invariant system. This single filter displays the input-output characteristics of the corresponding system (1). Unfortunately, many optical systems which one might desire to represent holographically are space-variant. For example, even an ideal imaging system with non-unity magnification is rigorously space-variant. Space-variant systems may, however, be characterized by cataloging the system's responses to a number of separate inputs, as contrasted with the use of the single point spread function required to characterize the space-invariant system. If a thick medium is used to store the system responses, the resulting volume hologram can subsequently act as a space variant filter which exhibits the input-output characteristics of the original system. Such a representation can, in principle, significantly reduce the weight and size of a coherent processor, and should also improve its orientation stability.

For mathematical and notational simplicity attention will be restricted to one dimensional linear, space-variant systems. For additional details, the reader is referred to some recent papers by the authors and their colleagues (2-6). Space variant systems, characterized by the linear operator $S[\cdot]$, may be described by the superposition integral

$$g(x) = S[f(x)] = \int_{\infty}^{\infty} f(\xi)h(x-\xi;\xi)d\xi \qquad (1)$$

where $f(x)$ is the input, $g(x)$ is the output, and $h(x-\xi;\xi)$ represents the space-variant line spread function. That is

$$h(x-\xi;\xi) = S[\delta(x-\xi)] \qquad (2)$$

where $\delta(x)$ is the Dirac delta function. This form of the line spread function has some advantages in describing space-variant systems (6-7). In the event that S is space invariant, we find that

$$h(x-\xi;\xi) \to h(x-\xi) \qquad (3)$$

which says that $h(x-\xi;\xi)$ is independent of its second argument.

We now consider three approaches to representing the effects of the space-variant system S described by Eq. (1). Each of these approaches may, at least in principle, be implemented holographically. Some details on the holographic implementation of the sampling theorem approach are given in a later section.

THE SAMPLING THEOREM APPROACH

A modification of the Whittaker-Shannon sampling theorem (1) permits us to design a holographic representation for space-variant systems based on line source sampling of the input plane of S, and subsequent angle-multiplexed holographic storage of the transfer function of S for each input plane sample.

The sampling theorem for space-variant systems (5) is based on the concept of the system's variation spectrum, which is defined by

$$H_\xi(x;v) \overset{\Delta}{=} F_\xi[h(x;\xi)] \qquad (4)$$

where $F_\xi[\cdot]$ denotes Fourier transformation with respect to the input variable ξ, and where v is the frequency variable associated with ξ. The variation spectrum is a measure of how the line spread function changes form with respect to the input variable ξ. In brief, the theorem states that if the input, $f(\xi)$, is bandlimited to bandwidth $2W_f$, and if $h(x;\xi)$ has a variation spectrum of width $2W_v$ (i.e. $H_\xi(x; v) = 0$ for $|v|>W_v$ for all x), then their product $f(\xi)h(x;\xi)$ will have bandwidth $2W = 2W_f + 2W_v$. As a result, the sampling theorem states that by sampling the input line to the system S at a rate of 2W samples per unit length, the output $g(x)$ may be written as the infinite sum

$$g(x) = \sum_{n=-\infty}^{\infty} f(\xi_n)h(x-\xi_n;\xi_n)*\text{sinc }(2Wx) \qquad (5)$$

where $\xi_n = n/2W$. Equivalently, in the frequency domain,

$$G(f_x) = \frac{1}{2W} \sum_{n=-\infty}^{\infty} f(\xi_n) H_x(f_x; \xi_n) \exp(-j2\pi f_x \xi_n) \, \text{rect} \ (f_x/2W) \ (6)$$

where, $\text{sinc} \ x = \frac{\sin \pi x}{\pi x}$, and

$$\text{rect} \ x \overset{\Delta}{=} \begin{array}{l} 1, \ \ |x| < \frac{1}{2} \\[4pt] 0, \ \ |x| \geq \frac{1}{2} \end{array} \qquad\qquad (7)$$

Equation (6) states that $G(f_x)$ may be obtained as a weighted sum of individual transfer functions $H_x(f_x; \xi_n)$, where the nth of which is weighted by the nth input sample $f(\xi_n)$. The presence of rect $(f_x/2W)$ in Eq. (6) indicates the low pass interpolation filter (bandwidth 2W) function necessary to reconstruct the continuous output $g(x)$. One obvious problem with Eqs. (5) and (6) is that they require us to store a countably infinite number of holograms for an exact reconstruction of $g(x)$. In practice we would expect to approximate $g(x)$ by storing a finite number determined by space-bandwidth product considerations.

The advantage of the sampling theorem approach is that it specifies a technique for exact reconstruction of the continuous output $g(x)$ (i.e. low pass filtering). A disadvantage is that it requires that we sample the input to S at a minimum rate determined by the sum of the variation bandwidth of S and the input bandwidth. One technique for cutting down on the density of input plane samples required is to employ the piecewise isoplanatic approximation (PIA) approach described next.

THE PIECEWISE ISOPLANATIC APPROXIMATION (PIA) APPROACH

The piecewise isoplanatic approximation, or PIA approach (4) makes the assumption that the space-variant system S is piecewise space-invariant (see Ref. 1). It effectively divides the input line into segments (or the plane into "patches"), and each segment is characterized by its own line spread function. Mathematically we may rewrite the input $f(\xi)$ as

$$f(\xi) = \sum_m f_m(\xi - \xi_m)$$

$$= \sum_m f(\xi) \, \text{rect}(\xi; \ell_m, u_m) \qquad\qquad (8)$$

where rect $(\xi; \ell_m, u_m)$ represents a rectangle function of unit height, extending from $\xi = \ell_m$ to $\xi = u_m$, and $\ell_m \leq \xi_m \leq u_m$. Since $g(x) = S[f(\xi)]$ we obtain

$$g(x) = \sum_m S[f_m(\xi - \xi_m)] \qquad\qquad (9)$$

or equivalently

$$g(x) = \sum_m \int_{\ell_m}^{u_m} f(\xi) h(x-\xi;\xi_m) d\xi \qquad (10)$$

When compared with the sampling theorem approach, the PIA approach appears to offer the advantage of being independent of the bandwidth of the input function $f(x)$. Rather it is the manner in which the line spread function $h(x-\xi;\xi)$ changes which determines the number of holograms which must be stored to represent the system. Depending on the relative sizes of the system's variation bandwidth and the input bandwidth, this advantage may or may not be significant. It would appear to be difficult, in general, to compare the performance of a PIA-based implementation with one based on the sampling theorem in situations where systems are not truly piecewise isoplanatic.

In the next section we discuss the orthonormal response approach, where the nature of the input function must be considered, but where the nature of the space-variant system is in general, not a determining factor.

THE ORTHONORMAL RESPONSE APPROACH

A third approach to characterizing the space-variant system is to expand the input function $f(x)$ as a weighted sum of the elements of an orthonormal basis set. Thus we write

$$f(x) = \sum_n \alpha_n \phi_n(x) \qquad (11)$$

where the elements of the set $\{\phi_n(x): n = 1,2,\dots\}$ are assumed orthonormal, i.e.

$$\int_{-\infty}^{\infty} \phi_n(\xi) \phi_m^*(\xi) d\xi = \delta_{nm} \qquad (12)$$

with δ_{nm} being the familiar Kronecker delta function. Substituting Eq. (11) into the relationship $g(x) = S[f(x)]$ we obtain

$$g(x) = \sum_n \alpha_n S[\phi_n(x)] \qquad (13)$$

where the expansion coefficients are found by

$$\alpha_n = \int_{-\infty}^{\infty} f(\xi) \phi_n^*(\xi) d\xi \qquad (14)$$

Note now that if we let $\Phi_n(x)$ be the response of S to the input $\phi_n(x)$,

$$\Phi_n(x) = S[\phi_n(x)] \qquad (15)$$

then Eq. (13) may be rewritten

$$g(x) = \sum_n \alpha_n \Phi_n(x) \qquad (16)$$

It is worth noting here that the elements of the system response set $\{\Phi_n(x)\}$ are not necessarily orthonormal, as are the elements of the input set $\{\phi_n(x)\}$.

To illustrate this approach, if we assume that $f(x)$ is bandlimited to bandwidth $2W_f$, then by the sampling theorem

$$f(x) = \sum_{n=-\infty}^{\infty} f(x_n) \text{ sinc } 2W_f(x-x_n) \tag{17}$$

where $x_n = n/2W_f$. We can view Eq. (17) as an orthonormal expansion with

$$\phi_n(x) = \sqrt{2W_f} \text{ sinc } 2W_f (x-x_n) \tag{18}$$

and with

$$\alpha_n = \frac{1}{\sqrt{2W_f}} f(x_n) \tag{19}$$

The system's sinc response is then given by

$$S[\text{sinc } 2W_f(x-x_n)] = \frac{1}{\sqrt{2W_f}} \Phi_n(x) \tag{20}$$

and the system output, $g(x)$, may be written as

$$g(x) = \sum_n f(x_n) S[\text{sinc } 2W_f(x-x_n)] \tag{21}$$

Note here that the advantage of the orthonormal response characterization is that only the input bandwidth determines the minimum required input plane sampling rate, not the variation bandwidth of S. In addition, since the sinc function is just the Fourier transform of the rectangle function (Ref. 1), physical generation of a coherent system's sinc response is easily implemented. A possible approach to implementing the sinc response approach is discussed at the end of the next section.

We have presented three approaches to characterizing the performance of space-variant systems. In principle, each of these approaches can be implemented experimentally. In our experimental work to date we have assumed point sampling of the input plane of a system, with the idea of implementing the sampling theorem approach. In the next section we briefly discuss some of the practical limitations present when one attempts to angle multiplex the sampled transfer functions of a space-variant system within a thick recording medium.

IMPLEMENTATIONS

Our experimental work to date has concentrated on the holographic implementation of the sampling theorem approach (2-3). Figure 1 illustrates (in one dimension) the basic approach. Point sources are used to sample the input plane of the space-variant system S, resulting in the system's point spread function appearing at the output of S. The nth reference point source, shown offset by a, also lies in the output plane of S. After the lens L performs a Fourier transform, the interference of the reference plane wave and the transfer function of S for the nth point source input is holographically recorded in the thick medium. When we perform this operation sequentially, with a different reference point source for each object point source, we are angle multiplexing a number of transfer function holograms into the medium. By using the extinction angle concept we can guarantee essentially non-interfering holograms.

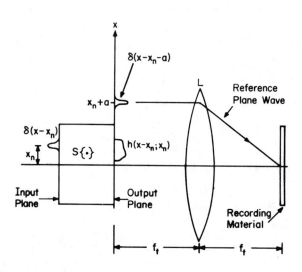

Fig. 1.   Recording volume hologram: sampling theorem approach

One places the volume hologram in the Fourier plane of a coherent optical processor for playback, as indicated schematically in Fig. 2. On playback one spatially samples the input plane using a duplicate of the reference array. Each input point accesses the hologram which represents the transfer function of S for that input point location. Neglecting crosstalk between the stored holograms, coherent addition of the outputs then gives the desired response. The experiments have mainly been performed using a DuPont holographic photopolymer (8) as the recording material, and an Argon laser operating at 5145 $\overset{o}{A}$ (2-3). Experimental implementations of simple one- and two-lens magnifiers for objects consisting of simple arrays of point sources have been produced and found to yield the correct magnifications though the images contained some aberrations. Additional experimental work is in progress.

The extinction angle, $\Delta\theta$, is the major factor affecting system
resolution (2-3,9). One can show that even in the worst case
where the system point spread functions associated with sampling
the input plane of S overlap each other completely, it is possible, by properly spacing the reference array point sources, to
obtain noninterfering holograms in the thick recording medium. If
we assume (see Fig. 1) that the transforming lens has focal length
$f_t >> a$, the reference array offset, we find that based on the

assumption that the extinction angle is essentially invariant as
one moves the reference and object beams over their respective
arrays (a reasonable approximation in many cases), the minimum
reference array element-to-element spacing is given by

$$\Delta x_{min} \cong f_t \, \Delta\theta \tag{22}$$

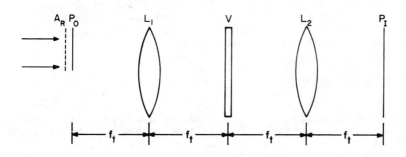

Fig. 2. Playback scheme: sampling theorem approach

This spacing will assure minimum hologram crosstalk on playback.
To illustrate, we found that for a 100 micron thick layer of the
DuPont holographic photopolymer, with f_t = 10 cm. and with an on

axis object array, plus a 30° reference array offset, the extinc-
tion angle was $\Delta\theta \cong 2°$. Based on Eq. (22) this predicts that Δx_{min}

= 3.5 mm. Since on playback one illuminates the input transparency through a duplicate of the reference array, the maximum
spatial frequency present in the object for which we could achieve
the Nyquist sampling rate would be $1/2\Delta x_{min}$ = 0.14 cycles per mil-

limeter = 3.6 lines per inch. While this is not a very high resolution, it should also be noted that thick recording media such
as photochromic glasses exist with extinction angles one and two
orders of magnitude smaller than the example just cited (10), so
that the state of the art predicts much higher system capabilities. It is clear, however, that storing large numbers of well
require, poses numerous problems. These include signal-to noise
ratio problems, crosstalk problems, and dynamic range problems,
to mention just a few. An additional problem involves the difficulty in obtaining resolutions, for two dimensional representations, which are equal to the one dimensional system resolutions
predicted on the basis of the extinction angle concept (11).

A comment should be made concerning implementation of the sinc response approach discussed in Eqs. (17)-(21). Based on Eq. (21) and Figs. 1-2, we see that by replacing the Dirac delta function by a sinc function at the input to S, one can effectively use the same recording geometry for implementing the sinc response approach as was used in implementing the sampling theorem approach.

SUMMARY

The three approaches presented for synthesizing and using volume holograms to represent space-variant optical systems appear promising, despite some obvious practical limitations. The potential savings offered by such holographic optical systems representations should, however, be sufficient to motivate further research into overcoming some of the problems identified to date. Additional work is underway at present to explore the potential and limitations of these and alternative techniques for optically representing space-variant optical processors.

ACKNOWLEDGEMENTS

The authors wish to express their appreciation to Dr. Thomas F. Krile of the Rose-Hulman Institute of Technology, Terre Haute, Indiana, for his suggestions during the course of the research. This work has been supported by the Air Force Office of Scientific Research, USAF, under Grant AFOSR-75-2855A.

REFERENCES

(1) J. W. Goodman, Introduction to Fourier Optics, McGraw-Hill, New York, 1968.

(2) L. M. Deen, J. F. Walkup and M. O. Hagler, Representations of space-variant optical systems using volume holograms, Appl. Opt., 14, 2438 (1975).

(3) J. F. Walkup and M. O. Hagler, Volume hologram representations of space-variant optical systems, Proc. of Tech. Prgrm., Electro-Optical Systems Design Conf.-1975, Anaheim, Calif., Nov. 11-13, 1975, pp. 31-37.

(4) R. J. Marks II and T. F. Krile, Systems theory for holographic representation of space-variant systems, Appl. Opt. (to appear).

(5) R. J. Marks II, J. F. Walkup and M. O. Hagler, A sampling theorem for space-variant systems, J. Opt. Soc. Am., 66 (1976-to appear).

(6) R. J. Marks II, J. F. Walkup and M. O. Hagler, On line spread function notation, Appl. Opt., 15 (1976-to appear).

(7) A. W. Lohmann and D. P. Paris, Space-variant image formation,

J. Opt. Soc. Am., 55, 1007 (1965).

(8)   B. L. Booth, Photopolymer material for holography, Appl. Opt., 14, 593 (1975).

(9)   H. M. Smith, Principles of Holography, 2nd edition, Wiley-Interscience, New York, 1975.

(10) A. A. Friesem and J. L. Walker, Thick absorption recording media in holography, Appl. Opt., 9, 201 (1970).

(11) R. J. Collier, C. B. Burckhardt and L. H. Lin, Optical Holography, Academic Press, New York, 1971.

INFORMATION CONTENT OF AN IMAGE

Lev Levitin

Tel-Aviv University, Ramat-Aviv, Israel

ABSTRACT

The physical limitations of the information content of an image caused by the wave and quantum nature of light are considered, provided the image is obtained by illumination of a plane reflecting or transparent object by incoherent light. The maximum information which can be obtained during a given time interval is found.

INTRODUCTION

One of the most general and widespread methods of information recording and reproduction is to assign optical properties of a certain object and then to read the information by viewing of the object during a certain time interval in natural reflected or transmitted light or in the light emitted by the elements of the object surface themselves. Such are all the methods related to photography, television, printing, etc. (Note, that it does not cover the case of holography where coherent and not natural (thermal) radiation is used). In the present paper we discuss the question of what are the fundamental limitations in those methods imposed by the wave and quantum nature of light and statistical properties of natural (thermal) radiation and what is the maximum amount of information in an optical image of the object obtained by observation during the given time interval.

FORMULATION OF THE PROBLEM

Now we specify the formulation of our problem. Let the object surface be a plane of area S. The optical properties of the surface in relation to observation in reflected light can be specified by the luminous reflectivity $a(x,y,\nu)$ as a function of the surface point (x,y) and the frequency of radiation, ν. (Here and below we imply the reflectivity in the direction to the observer. It is unessential for us, whether the reflection is diffuse or regular). So long as only the limitations related to the physical nature of light are the matter of our interest, we disengage ourselves from the material structure of the surface, owing to which the reflectivity can be defined correctly not for a point of the surface, but only for areas large enough in comparison with interatomic distances. The idealization is reasonable as the interatomic distances are much less than the length of waves contributing considerably to the spectrum of natural light sources.

115

The function $a(x,y,\nu)$ should be interpreted as the luminous transmittance in the case of observation in transmitted light and as the ratio of the spectral intensity of radiation in a given point of the surface to the maximum available spectral intensity in the case of self-luminous surface. We assume also that the functions $a(x,y,\nu)$ can be assigned independently for the two of polarization states of radiation

DEGREES OF FREEDOM OF RADIATION

The set of possible functions $a(x,y,\nu)$ is restricted only be inequalities $0 \le a(x,y,\nu) \le 1$. Obviously, the choice of a certain function $a(x,y,\nu)$ gives an infinite amount of information. But the use of electromagnetic radiation for the reading of information changes the situation in principle.

Indeed, suppose the area S is viewed by a detector of radiation (an optical system) with an entrance pupil, which is seen from any point of the area S under a solid angle Ω . Then, as is known, (e.g.,(Ref.1.) the number of "spatial"degrees of freedom, or, in other words, the number of field oscillators (quantum states of photons) differing in the directions of wave vectors (or, for another choice of the set of oscillators, by the spatial localization of photons) is equal to

$$G_s = \frac{\nu^2 \Omega S}{c^2} \tag{1}$$

where ν is the frequency of field oscillators, C is the light velocity.

If the object is observed during a time interval τ then the frequency uncertainty of the quantum states of photons is $\Delta\nu = 1/\tau$ and the number of the states differing in frequency in a frequency interval $\delta\nu$ is equal to

$$G_f = \frac{\delta\nu}{\Delta\nu} = \tau\delta\nu \tag{2}$$

Taking into account two possible polarization states, we obtain the total number of degrees of freedom of radiation:

$$G = \frac{2}{c^2} \nu^2 \Omega S \tau \delta\nu \tag{3}$$

(Assuming $\nu^2\Omega S\tau\delta\nu/c^2 \gg 1$, i.e., the approximation of geometrical optics is valid). Thus, the number of degrees of freedom of radiation is finite. There is a sense to assign the reflectivity, a, only in details, which do not exceed the spatial resolution (it means, over areas $\Delta S = c^2/\nu^2\Omega$) and the frequency resolution (it means, over frequency intervals $\Delta\nu = 1/\tau$). Hence, at such a method of recording, the information is specified by a set of values of G random variables, namely, of the reflectivities for each of the field oscillators. The amount of information is maximum when all the random variables are independent and is equal in this case to the sum of the amount of information in each of the variables.

STATISTICS OF RADIATION AND INFORMATION

The amount of information could be infinite even for one degree of freedom, if the reflectivity, a, could take on an infinite set of well distinguishable values. Here, however, the quantum nature of the light and the statistical properties of thermal radiation come into play. As is known (e.g., (Ref. 2), in the radiation of natural (non-coherent) sources the states of field oscillators are statistically independent and described by Gibbs's distribution of occupation numbers

$$p(n) = \frac{1}{\bar{n} + 1} \left(\frac{\bar{n}}{\bar{n} + 1} \right)^n \tag{4}$$

Here n is the occupation number, i.e., the number of photons in a given quantum state, $\bar{n}$ is the mean occupation number.

If the surface is illuminated by thermal (i.e., obeying Gibbs's distribution) radiation, the reflected light has also Gibbs's distribution, the states of the field oscillators being independent, provided the illuminating beam is rather wide, and has a sufficiently large bandwidth. It follows from the transformational properties of states of electromagnetic field with respect to a change of a complete set of field oscillators for another one. (See, e.g. (Ref. 3), (Ref. 4)).

Let $P(\nu)$ be the average energy of a field oscillator of frequency ν in the radiation of the illumination source, and let $r(\nu)$ be the part of the energy, which is registered by the detector, if the reflectivity, a, equals to unity. Then to assign the value of a means to assign the mean number of registered photons, $\bar{n}$, for the corresponding field oscillator:

$$\bar{n} = \frac{ar(\nu)P(\nu)}{h\nu} \tag{5}$$

The maximum mean number of registered photons is

$$\bar{n}_{max} = \frac{r(\nu)P(\nu)}{h\nu} \tag{6}$$

As is shown in (Ref. 5), when the mean number of photons, distributed according to (4), is limited by a maximum value $\bar{n}_{max} \lesssim 9$ (in the optical range it corresponds to temperatures $T < 3.10^5$ °K), it proves to be optimal to use as signals only two values of the mean number of photons: $\bar{n} = 0$ and $\bar{n} = \bar{n}_{max}$ (i.e., two values of reflectivity : a = 0 and a = 1). Then the maximum amount of information (per a field oscillator) is equal to

$$I_{max} = \ln\left[1 + \frac{r(\nu)P(\nu)}{r(\nu)P(\nu)+h\nu} \left(\frac{h\nu}{r(\nu)P(\nu)+h\nu} \right)^{h\nu/r(\nu)P(\nu)} \right] \tag{7}$$

When a spectral band from ν_1 to ν_2 is used, assuming $(\nu_2-\nu_1)\tau \gg 1$, with (3) and (7) taken into account, we obtain, that the maximim amount of information, which can be obtained by observation of an object, illuminated by incoherent light, is equal to

$$J_{max} = \frac{2\tau\Omega S}{c^2} \int_{\nu_1}^{\nu_2} \nu^2 \ln\left[1 + \frac{r(\nu)P(\nu)}{r(\nu)P(\nu)+h\nu} \left(\frac{h\nu}{r(\nu)P(\nu)+h\nu} \right)^{h\nu/r(\nu)P(\nu)} \right] d\nu \tag{8}$$

Thus, the amount of information increases proportionally to the area of the object, S, and to the time of observation, τ.

It seems to be interesting to get an explicit expression for the amount of information in the case, when the radiation registered by the detector obeys

Plank's spectral distribution (for instance, when the illumination source is a black body, and $r(\nu) = 1$), i.e. when

$$r(\nu)P(\nu) = \frac{h\nu}{\exp(\frac{h\nu}{kT}) - 1} \tag{9}$$

Let the solid angle take on the maximum possible value 2π, and the frequency band is infinite*. The maximum amount of information, R_{max}, per unit of time for a unit area is equal to

$$R_{max} = \frac{4\pi}{c^2} \int_0^\infty \nu^2 \ln[1 + e^{-\frac{h\nu}{kT}}(1 - e^{-\frac{h\nu}{kT}})^{(e^{\frac{h\nu}{kT}} - 1)}] \, d\nu =$$

$$= \frac{4\pi\,\kappa\,(kT)^3}{c^2 h^3} = \frac{\kappa\sqrt{2}}{\pi^2\sqrt{c}} (\frac{15P}{\pi h})^{3/4} \tag{10}$$

Where

$$P = \frac{4\pi}{c^2} \int_0^\infty \nu^2 \frac{h\nu}{\exp(\frac{h}{kT}) - 1} \, d\nu = \frac{4\pi^5 (kT)^4}{15 c^2 h^3} \tag{11}$$

is the energy flux of reflected radiation for a unit area, and

$$\kappa = \int_0^\infty x^2 \ln[1 - e^{-x}(1 - e^{-x})^{e^x - 1}] dx \approx 0,772$$

is a numerical constant.

ACKNOWLEDGEMENTS

The author appreciates the role of Dr. D.S. Lebedev (Institute of Information Transmission Problems, Moscow, USSR), who attracted the author's attention to this area, and thanks also R. Piotrkowski for a numerical calculation.

REFERENCES

(1) D. Gabor, Light and information, Progress in Optics, 1, 111 (1961).

(2) L.D. Landau and E.M. Lifshits, Statistical Physics, Nauka, Moscow, 1964, (in Russian).

(3) V.P. Morozov, Photon Statistics and Signal Transmission Theory, doctoral thesis, Gorky State University, Gorky, 1971, (in Russian).

(4) V.V. Mitiugov, Physical Principles of Information Transmission Theory, Sovietskoie Radio, Moscow, 1976, (in Russian).

(5) L.B. Levitin, Information transmission by thermal radiation, Proc. of the II National Conf. on the Coding Theory and its Applications, Section 5, 49, Nauka, Moscow, 1965. (In Russian)

* At low frequencies the geometrical optics is not valid, and one cannot use formula (3). In addition, for $(r(\nu)P(\nu)/h\nu) > 9$ formula (7) is not correct. But the contribution of low frequencies is small due to the factor ν^2 in the integrand, and it allows us to expand the range of integration to zero.

DECODING TECHNIQUES FOR USE WITH ANNULAR CODED APERTURES

R. G. Simpson, H. H. Barrett and H. D. Fisher

Optical Sciences Center and Department of Radiology,
University of Arizona, Tucson, Arizona

ABSTRACT

Several processing algorithms for annular coded-aperture images are discussed in this paper. The algorithms are logical extensions of simple correlation decoding (matched filtering). Each improved algorithm eliminates another undesirable deterministic feature of the final system point spread function (PSF). In addition, the results of a signal-to-noise ratio (SNR) calculation for one of the algorithms is presented.

INTRODUCTION

Coded aperture imaging techniques have been applied to imaging situations in nuclear medicine, x-ray astronomy and related areas in which quantum noise, resulting from photon counting, fundamentally limits image quality. The imaging process is basically a shadow-casting operation in which each point in the object forms a scaled version of the aperture pattern in the detector plane. The use of a coded aperture in lieu of a pinhole greatly increases the collection efficiency. However, the recorded image in a coded-aperture system must be processed in order to achieve the same resolution as the pinhole system. This processing affects the final image in two ways. Deterministically, the behavior of the system PSF depends on the processing that is performed. Attainable resolution, compactness of the PSF and artifacts are all determined by the aperture in conjunction with the processing algorithm. Statistically, since the recorded image contains fluctuations due to the quantum noise, the processing algorithm determines the statistical behavior of the processed image. As a result of the encoding and detection processes and subsequent processing, the noise in the image becomes dependent on object size and object distribution.

DECODING ALGORITHMS

For simplicity, only planar objects will be considered here so that all of the annular shadows are the same size. The decoding operation must then estimate the location and strength of the annular shadows produced by the various points in the object. When Walton[1] first used the annulus in a nuclear medicine imaging system, the decoding algorithm consisted, in effect, of correlating the coded image with a thin annulus. The PSF resulting from this decoding algorithm is shown in Fig. 1a. This function does not have the compactness desired in a PSF. A given point in the object will contribute to the reconstruction of another point whenever their associated shadows overlap. This is disastrous in many cases as can be seen from the reconstruction of a disc object using this algorithm shown in Fig. 3.

The PSF can be made more compact through the inclusion of a rho-filter in the decoding algorithm[2]. The need for the rho-filter can be seen through the following argument. Mathematically, the coded image g(x,y) is the result of convolving the object function o(x,y) with the aperture function ann(x,y) which can be written

$$g(x,y) = o(x,y)**ann(x,y) \qquad (1)$$

Geometric scale factors have been suppressed. If both sides of Eqn. 1 are Fourier transformed, the spectrum of the coded image $G(\xi,\eta)$ equals the spectrum of the object $O(\xi,\eta)$ times the spectrum of the annulus $ANN(\xi,\eta)$

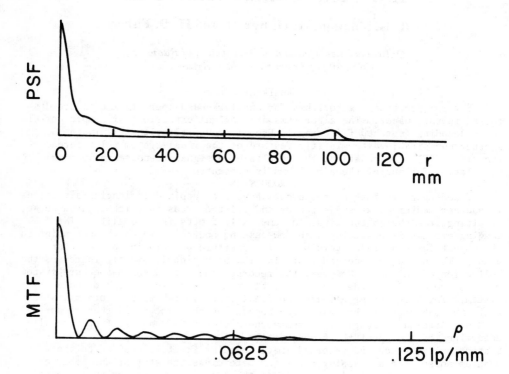

Fig. 1.   a)   PSF corresponding to autocorrelation of an annulus.   The
              annular shadow has a mean radius of 50 mm.   b)   Fourier
              transform of a).   This is the effective system transfer
              function when simple correlation decoding is used.

$$G(\xi,\eta) = O(\xi,\eta) \cdot ANN(\xi,\eta) \qquad (2)$$

Here ξ and η are Cartesian Fourier components conjugate to x and y respectively.   The spectrum of the aperture function, for a thin annulus, is close-
ly approximated by $ANN(\xi,\eta) = 2\pi\bar{r}J_0(2\pi\bar{r}\rho)$ where $\bar{r}$ is the mean annular shadow
radius, $\rho = \sqrt{\xi^2 + \eta^2}$ is the magnitude of the spatial frequency vector in
cycles (or line pairs) per unit length, and $J_0(\cdot\cdot)$ is the zero-order Bessel
function.   Beyond the first zero, the spectrum of the annulus can be further
approximated as

$$ANN(\rho) = 2\pi\bar{r}J_0(2\pi\bar{r}\rho)$$

$$\simeq 2(\frac{\bar{r}}{\rho})^{\frac{1}{2}}\cos(2\pi\bar{r}\rho - \pi/4) \qquad (3)$$

If correlation processing is considered such that a decoding annulus is
convolved with the coded image, the decoded image, f(x,y), is given by

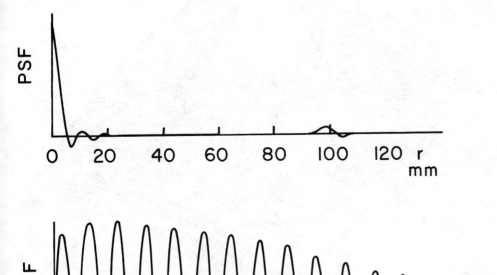

Fig. 2 a) PSF resulting from correlating with an annulus and rho-filtering.
        b) Fourier transform of a). This is the effective system transfer
           function when the rho-filter is included with the correlation
           decoding.

$$f(x,y) = o(x,y)**ann(x,y)**ann(x,y) \qquad (4)$$

The spectrum of the decoded image is then given by

$$F(\xi,\eta) = O(\xi,\eta)ANN^2(\xi,\eta) \qquad (5)$$

$ANN^2(\xi,\eta)$ can therefore be considered an effective transfer function (TF).
This function is shown in Fig. 1b. Using the above approximation for $ANN(\rho)$,
one has

$$ANN^2(\rho) \simeq (4\bar{r}/\rho)\cos^2(2\pi\bar{r}\rho - \pi/4)$$

$$= (2\bar{r}/\rho)\{1+ \cos(4\pi\bar{r}\rho - \pi/2)$$

$$= (2\bar{r}/\rho)(1+ \sin4\pi\bar{r}\rho). \qquad (6)$$

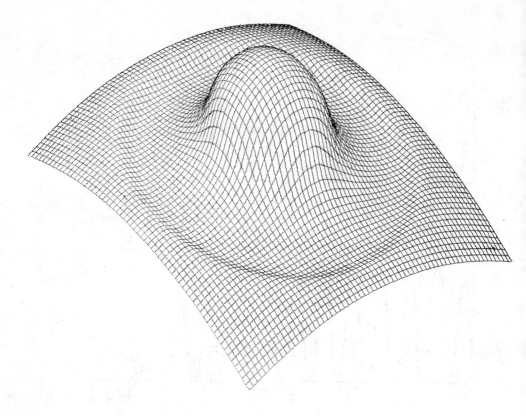

Fig. 3.   Reconstruction of a disc object using simple correlation decoding.
          The value at the origin depends linearly on the diameter of the disc
          as a result of the 1/r behavior of the PSF.

The decoded image spectrum is therefore related to the object spectrum by a
function that falls off as $1/\rho$. Suppose now that the decoded-image spectrum
is multiplied by ρ. One factor of $\sqrt{\rho}$ undoes the damage done by our choice
of correlation decoding and the other factor of $\sqrt{\rho}$ is an additional boost
that flattens out the effective transfer function. This factor of ρ is
referred to as a rho-filter. For a point object, the decoded-image spectrum
then becomes

$$F(\xi,\eta) = ANN^2(\xi,\eta)\cdot\rho$$

$$\approx 2\bar{r}(1+\sin 4\pi\bar{r}\rho) \tag{7}$$

The exact function is shown in Fig. 2b. The constant term results in a PSF
with a sharp peak at the origin and the sine term contributes to behavior
near $r = 2\bar{r}$ (see Fig. 2a). Except for the behavior at $2\bar{r}$, the PSF with the
rho-filtered correlation decoding is now much more compact. The benefits of
this are clearly shown in Fig. 4. The ringing is due to the abrupt cutoff
in the frequency plane and can be greatly suppressed through the use of an
apodizing function.

      The behavior of the PSF at values of r near $2\bar{r}$ is of no concern if the
object is confined to a circle of radius $\bar{r}$. If the object gets larger than

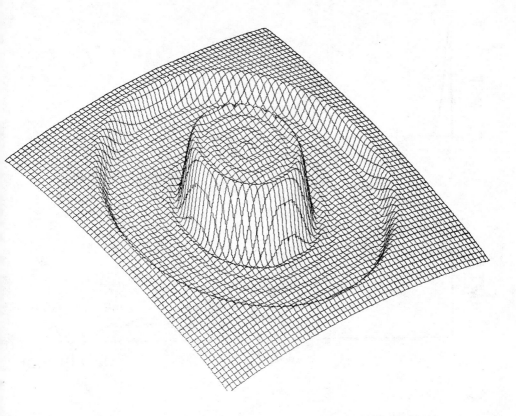

Fig. 4.   Reconstruction of the same disc as Fig. 3 with the rho-filter
          included in the processing. The value at the origin is now
          independent of disc diameter. The ringing is due to the sharp
          cut-off in frequency space (see Fig. 2b). The use of apodizing
          functions can eliminate this.

this, however, the glitch at $2\bar{r}$ can be troublesome. Is there a way to get
rid of it?
     The glitch results from the zeros in the TF for a single annulus. In
fact, the glitch may be regarded as an aliasing artifact resulting from the
incomplete sampling of the object spectrum. Without fundamentally changing
the encoding aperture, there is no way to eliminate the glitch. However, if
two coded images of the same object are recorded with two annuli of different
size, and each is processed by correlation followed by rho filtering, the two
TF's will have sine terms with different frequencies since $\bar{r}$ is changed.
Therefore, it should be possible to linearly combine information from the two
images, using one to fill in the gaps of the other. An implementation of
this scheme follows.
     The first parameter to fix is the relative size of the two annuli. If
an Anger camera is the detector, as is often the case in nuclear medicine,
the detector has relatively poor resolution which sets a limit on how far

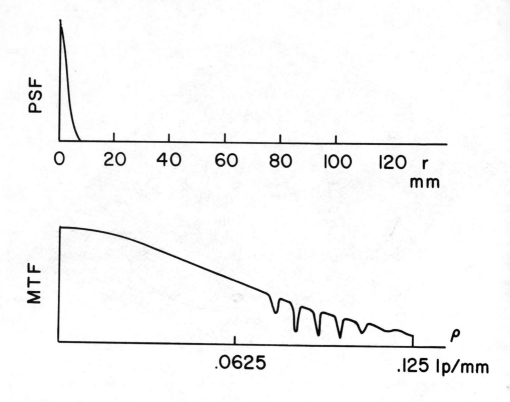

Fig. 5. a) PSF resulting from the combination of two coded images formed
         with two annuli of different size. The glitch at r = 100 mm in
         Fig. 2a has been eliminated. b) Fourier transform of a). This is
         the effective transfer function for the two annulus system.

out in the frequency domain the filter should be extended. It seems reasonable to choose the annuli so that the nth zero of the TF associated with one
annulus and the (n + 1)st zero of the TF associated with the other annulus
coincide near the maximum allowed frequency. In this way, no other zeros
can match up in the allowed frequency range and the two TF's are not radic-
ally different. If the twelfth and thirteenth zeros are matched, as in the
example to follow, the ratio of annular radii is 1.085.

Again, as a first step in the decoding, we will correlate each coded
image with an appropriately scaled annulus. Let TF, be the system transfer
function at this stage for the annulus with mean radius $\bar{r}$, and similarly for
TF_2, so that

$$TF_1 = ANN_1^2 (\rho)$$
$$TF_2 = ANN_2^2 (\rho)$$

When TF_1 and TF_2 are combined, the resulting transfer function should be
constant out to the maximum frequency. Letting $W_1(\rho)$ and $W_2(\rho)$ be the weight
given to the respective transfer functions at a given frequency, we have

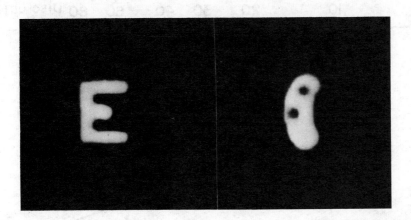

Fig. 6. a) Reconstruction of an E-shaped phantom. The E was four inches
         tall and three inches wide and imaged at unit magnification.
         The coded image was detected with an Anger camera. b) Recon-
         struction of a kidney shaped phantom with two regions of de-
         creased activity. The phantom was eight inches tall and the
         cold regions 1½ inches in diameter. The object was imaged at a
         magnification of ½. The coded image was detected with an Anger
         camera.

$$1 = W_1(\rho)\text{ANN}_1^2(\rho) + W_2(\rho)\text{ANN}_2^2(\rho) \qquad (8)$$

Whichever TF is larger at a particular frequency should be weighted more
heavily, so that

$$\left|\frac{TF_1(\rho)}{TF_2(\rho)}\right|^\alpha = \frac{W_1(\rho)}{W_2(\rho)} \qquad\qquad (9)$$

is an additional equation that can be used to constrain the solution. The
case $\alpha = \frac{1}{2}$ is selected here although other choices might be advantageous when
noise is considered. W_1 and W_2 can then be determined, using the
$2\pi\bar{r}J_0(2\pi\bar{r}\rho)$ expression for ANN. The Fourier transforms of the two coded
images are then added together, frequency by frequency.
    Two additional modifications were included in the algorithm to improve
its behavior. The first was a limit on how large the weights could be. This
was done by analogy to inverse filtering where a limit is set on how much
amplification is allowed so that a frequency component with a low SNR does
not get excessively amplified. The second modification was the multiplication of the combined TF by an apodizing function to alleviate the ringing
that occurs with a sharp cutoff. The combined TF produced by this algorithm
is shown in Fig. 5b. The dips in the curve are regions where the weights
were prevented from being as large as equations 8 and 9 demanded. The
resulting PSF is shown in Fig. 5a. Note that the glitch has been eliminated.
It must be added, however, that the glitch reappears when the scale of the
processing annulus does not match the scale of the annular shadow. This is
of interest if one is concerned with the out-of-focus behavior of the system

R. G. Simpson, H. H. Barrett, H. D. Fisher

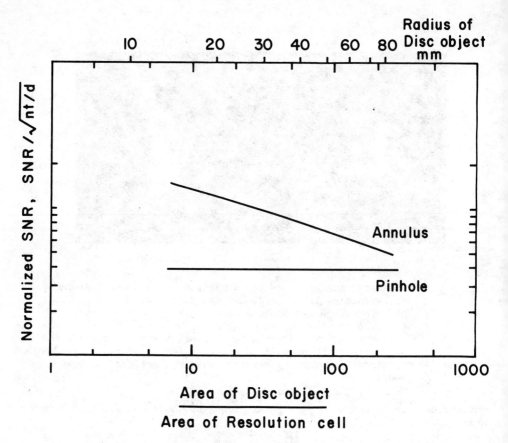

Fig. 7. SNR for a pinhole system and an annular aperture system as a function of object size. The two systems used the same recording geometry and the final images contained the same resolution. The object was a disc and the SNR was calculated at the center of the disc. n is proportional to the activity in the object, t is the exposure time and d is the object-to-detector distance.

in anticipation of its use as a tomographic imaging system. Reconstruction of the letter E and a kidney shaped phantom with two cold regions are shown in Fig. 6.

SIGNAL-TO-NOISE RATIO

The coded image detected in a coded-aperture system contains noise with the following properties. It is Poisson in nature and the fluctuations at one point in the coded image are independent of fluctuations at neighboring points, assuming that detection consists of the generation of an x-y coordinate for each event detected. Since the variance equals the mean in a Poisson process, and the mean value at a fixed point in the coded image, determined over an ensemble of images, varies with location, the noise variance will change with location. The noise is therefore non-stationary and power spectrum analysis is not appropriate. The calculation of the noise

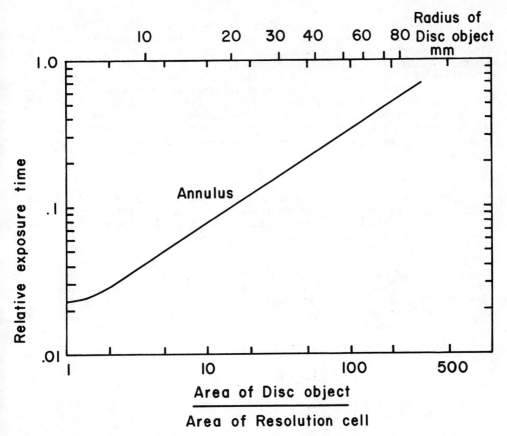

Fig. 8. Relative exposure time required for the annular aperture system to produce the same SNR as the pinhole system as a function of object size. Same physical constraints as listed in Fig. 7.

in the decoded image can be performed using techniques developed for non-stationary Poisson statistics.[3] Using such techniques, the SNR in the decoded image is given by

$$SNR = \frac{\langle o \rangle **ann**h}{\sqrt{\langle o \rangle **ann**h^2}}$$

where, h is the processing function that is convolved with the coded image to yield the decoded image, $\langle o \rangle$ is the mean object and ann is the annular shadow.

The calculation of the SNR was performed for a system where the annular shadow had a mean radius of 58mm and the maximum frequency, ρ_m, used in processing the data was 0.1 lp/mm. The object was a uniform disc of variable radius and the SNR was evaluated at the center of the reconstructed disc. The processing function h was assumed to be $\rho \cdot ANN(\rho)$ times an apodizing function. For comparison purposes, the SNR was calculated for a pinhole system with the same object-to-aperture and aperture-to-detector distances and with the size of the pinhole adjusted to give the same resolution as the decoded annular image. When normalized for the same activity in the object n, same exposure time t, and same object-to-detector distance d, the SNR as a function of object size for the pinhole and the annulus is shown in Fig. 7. This same information is reformulated as relative exposure time required for the annulus compared to the pinhole for images with the same SNR and plotted as a function of object size in Fig. 8. The exposure time advantage for the annular aperture system decreases as the disc becomes larger, until the advantage disappears just as the object becomes large enough for the glitch to be a problem.

SUMMARY

Simple correlation decoding of annular coded aperture images was shown to result in a net PSF that was unsatisfactory. The decoding was greatly improved through the inclusion of a rho-filter in the processing step. The PSF when rho filtering was included still suffered from an artifact at $r = 2\bar{r}$. A method for eliminating this artifact was described, using two coded images formed with apertures of two different sizes. Finally, the results of a SNR calculation for rho-filtered correlation decoding were presented. The results showed that the annulus still retained an exposure time advantage over the pinhole until the ring artifact at $2\bar{r}$ began to interfere with the re-construction.

REFERENCES

[1] P.W. Walton, J. Nucl. Med. __14__, 861 (1973).

[2] R.G. Simpson, H.H. Barrett, J.A. Subach, and H.D. Fisher, Opt. Eng., Sept.--Oct. 1975.

[3] C.E. Metz and R.N. Beck, J. Nucl. Med. __15__, 165 (1974).

SESSION 5.

INTERFEROMETRY

Chairman: W. STEEL

ACHROMATIC FRINGES FROM EXTENDED SOURCES

G. R. Wloch, S. M. Bose and J. R. Cozens

Electrical Engineering Department, Imperial College, London

ABSTRACT

A wavelength compensated wavefront folding interferometer is described which produces over 400 achromatic fringes from extended polychromatic objects.

INTRODUCTION

The inherently two-dimensional nature of optical wavefronts leads to the expectation that the potential of parallel optical signal processing would prove considerable. The technique most widely developed so far involves coherent light, its power deriving from the remarkable Fourier transform property of lenses. While there are many applications for which this system is admirable, there are many others for which the requirement that the information has to be impressed onto a coherent carrier is very restrictive. Furthermore, coherent processing relates the transforms of complex amplitudes, so that the Fourier transform of a signal cannot be recorded directly with a detector.

If we consider the form in which two-dimensional signals are likely to occur in practice, among the most obvious examples might be photographic transparencies, optical images (i.e. intensity distributions) and cathode ray tube displays. Only the first of these is readily imposed onto a coherent beam.

One useful development might therefore be towards parallel processing systems that can use direct signals as inputs. We have chosen to aim for a system that could use, for example, a C.R.T. screen display directly as an input signal. The system must therefore be able to operate with spatially incoherent, polychromatic signals in real time.

The system we are investigating is based on the wavefront folding interferometer (Refs. 1-4). In this instrument, the light from a two-dimensional incoherent source is amplitude divided, with one beam then being subjected to a relative two-fold rotation and shear. Thus the intensity g (x,y) at a point in the original source will become $g(x-X_0, y-Y_0)$ and $g(-x-X_0, -y-Y_0)$ in the two virtual sources for shears of $\pm$ (X_0, Y_0), respectively. Subsequent interference between the beams is effected in the plane of a screen or detector some distance from the source, Fig. 1. If we represent the source as an array of mutually independent point sources, then only corresponding points in (a b c d) and (a'b'c'd') can interfere in the output (ξ , η) plane. The complementary sources at (x,y) and $(-x,-y)$ will produce a sinusoidal intensity distribution in the output plane. These fringe patterns for each pair of points will add in intensity, leading to the intensity distribution in the

131

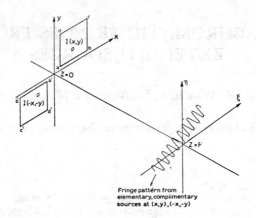

Fig. 1   Intensity distributions in the input and output
planes of a wavefront folding interferometer

output plane (for large F)

$$I^1(\xi,\eta) = \frac{2}{(\lambda F)^2} \int_0^\infty I(x,y) \left\{ 1 + \cos\frac{2k}{F}\left[\xi x + \eta y\right] \right\} dxdy \tag{1}$$

where $I(x,y) = g(x-X_o, y-Y_o)$. Thus the output contains a spatial representation of the cosine Fourier transform of the source intensity distribution which can be filtered, and also directly recorded.

We can obtain further insight into the nature of the output if eqn.(1) is written in the form:

$$I^1(\xi,\eta) = \frac{2}{(\lambda F)^2} \left[G_e(\frac{2\xi}{\lambda F}, \frac{2\eta}{\lambda F}) \cos\left\{ 2\pi\,(\frac{2\xi}{\lambda F}X_o + \frac{2\eta}{\lambda F}Y_o\right\} \right.$$

$$\left. - j\,G_o(\frac{2\xi}{\lambda F}, \frac{2\eta}{\lambda F}) \sin\left\{ 2\pi\,(\frac{2\xi}{\lambda F}X_o + \frac{2\eta}{\lambda F}Y_o)\right\} \right] + C_1$$

where G_e and G_o are the Fourier transforms of the even (g_e) and odd (g_o) parts respectively of the function $g(x,y)$ and C_1 is a constant. Thus it is evident that the output consists of three parts:

(a) a cosine fringe pattern modulated by the transform of the even part of the input signal,

(b) a sine fringe pattern modulated by the transform of the odd part of the input,

(c) a constant intensity which allows the sinusoidal terms to be directly observed.

Since the transform of g_e is real and the transform of g_o is imaginary, complete information about the complex transform is available in the form of a real signal in the output plane. This is illustrated, Fig. 2, by means of the calculated one dimensional outputs for input functions.

$$g(x) \;=\; \text{rect}\left(\frac{x}{b}\right). \;\; (1 + \cos \frac{x}{d}) \qquad \text{(purely even)}$$

$$g(x) \;=\; \text{rect}\left(\frac{x}{b}\right). \;\; (1 + \sin \frac{x}{d}) \qquad \text{(odd + even)}$$

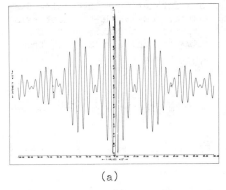

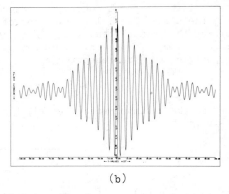

(a)   (b)

Fig. 2   Plots of output intensity for

(a) $g(x) = \text{rect}\left(\frac{x}{b}\right) (1 + \cos \frac{x}{d})$

(b) $g(x) = \text{rect}\left(\frac{x}{b}\right) (1 + \sin \frac{x}{d})$

both for $b = d$.

The ability of the interferometer to distinguish and display both transforms is clear.

A calculated two-dimensional fringe pattern for a rectangular slot is also shown, Fig. 3, where the phase information in the transform is apparent in the fringe displacements.

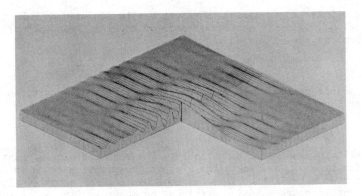

Fig. 3   2-dimensional fringe pattern for a rectangular slot input

It should be noted that the input to the interferometer is in the form of an intensity distribution and hence g(x,y) can only be real and positive. It is therefore not possible to have a truly odd input function and the output intensity distribution always has its maximum at the centre.

Bandwidth Restriction

One important restriction limits useful application of the instrument in the form so far described. As we move from the centre of the output plane, the visibility of the fringes from each pair of point sources will fall, due to their finite temporal coherence. Since regions further from the centre display the higher spatial frequencies of the source, some trade-off must exist between the optical bandwidth, $\lambda/\Delta\lambda$, and the maximum detectable spatial frequency, ω_s. From eqn.(1) we can simply show that the displacement from the origin $R(\omega_s)$ representing ω_s is

$$R(\omega_s) = \frac{F}{2k} \cdot \omega_s \tag{2}$$

whereas the maximum range, R(max), for fringe formation by a pair of sources at $\pm(x,y)$ is

$$R_{max} = \frac{\pi F}{k(x^2+y^2)^{1/2}} \quad \frac{\lambda}{\Delta\lambda} \tag{3}$$

Combining eqn.(2) and (3),

$$\omega_s = 2\pi \frac{1}{(x^2+y^2)^{1/2}} \frac{\lambda}{\Delta\lambda} \cdot$$

For fairly wide bandwidths, say $\lambda/\Delta\lambda \approx 10$, and spatial frequencies up to, say, 10 lines/m.m., source dimensions of only 1 m.m. can be accommodated. Alternatively we could use a large source and increase $\lambda/\Delta\lambda$ with a filter. Thus a T.V. screen, dimension say D_o, would have spatial frequencies up to about $2\pi^{300}/D$, which for $\lambda = 6000 \text{ A}$, would require $\Delta\lambda \approx 20 \text{ A}$. These restrictions would be clearly unacceptable in many cases.

The Achromatic Interferometer

A considerable increase in useful bandwidth can be achieved by making the fringe period, Λ, from any particular pair of sources independent for a one dimensional source. From eqn.(1)

$$\Lambda = \frac{F\lambda}{2x} ,$$

thus if all source points could be displaced so that their positions linearly shifted according to wavelength as

$$\delta x = x \cdot \frac{\delta\lambda}{\lambda} ,$$

then achromatic fringes would be obtained. The required displacement is thus not only a function of wavelength but also of the position of an object point from the interferometer axis. This can be achieved, for example, with a pair of mutually inclined transmission gratings, as shown in Fig. 4. A suitable choice of N_1, N_2 and θ leads to the formation of a large number of fringes.

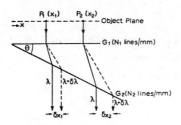

Fig. 4 The passage of two rays through a pair of inclined gratings

The ray representation indicated in Fig. 4 allows a simple illustrative
analysis to be performed, but contains an uncertain level of approximation.
A more rigorous treatment is outlined below.

A point source in the input plane is represented by a uniform spectrum of
plane waves. The passage of each component wave through the gratings can
then be described accurately by the grating equation. In Fig. 5, the component waves from a source $S(x_1, y_1, 0)$ propagate through gratings 1 and 2;

after diffraction these components are modified in amplitude and phase and are
expressed in the co-ordinate system (x_3, y_3, z_3). This system is then rotated
by θ_0 and translated by 'a' and 'b' to the final system (x_5, y_5, z_5), centre
O_5. The z_5-axis is the interferometer axis and the plane $z_5=0$ is the folding
plane.

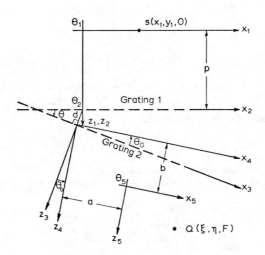

Fig. 5 Co-ordinate systems for plane wave analysis

The contribution to the field at a distant point $Q(\xi, \eta, F)$, in the output
plane, from any component wave can be written as

$$dE = dA \exp - jk[(\alpha_5 \xi + \beta_5 \eta + \gamma_5 F) - \phi_5]$$

where dA depends upon the source amplitude, grating dimensions and other standard factors; α_5, β_5, γ_5 are the direction cosines of the particular component and ϕ_5 is a phase term which depends on the system configuration and grating parameters. For small α_5, β_5, we have

$$\phi_5 \approx [\alpha_5 x_5 + \beta_5 y_5 + (W_0 - W_1 \alpha_5^2 - W_2 \beta_5^2) + \text{higher order terms in } \alpha_5, \beta_5],$$

where x_5, y_5, W_0, W_1, W_2 are dependent on the grating and system parameters N_1, N_2, d, θ_0 θ and λ. The dimensions x_5, y_5 can be interpreted as effective displacements from the interferometer axis.

By a suitable choice of parameters, we can make $W_1 \approx W_2 [= z_5/2,]$ so that

$$\phi_5 \approx [\alpha_5 x_5 + \beta_5 y_5 + z_5(1 - \frac{\alpha_5^2}{2} - \frac{\beta_5^2}{2}) + W + \text{higher terms}]$$

where W is a constant independent of α_5, β_5. This phase term implies that the elementary plane wave can be interpreted as originating from a virtual source at x_5, y_5, z_5 and propagating with an intrinsic phase determined by W and the higher order terms. For regions far away from the source, the stationary phase method (Ref. 3) can be used to determine the amplitude at Q. Denoting the stationary point by α_{o5}, β_{o5} and restricting our argument to one dimensional folding, we obtain

$$\alpha_{o5} = f(\xi, \eta, F) ; \quad \beta_{o5} = 0; \quad \gamma_{o5} = \sqrt{1 - \alpha_{o5}^2}$$

Thus the stationary phase solution leads to

$$E_Q = A \exp - jk[(\alpha_{o5} \xi + \gamma_{o5} F) - (\alpha_{o5} x_5 + \gamma_{o5} z_5) - W - \text{higher terms}]$$

where A is the complex amplitude. Interference between the wave from x_5, y_5, z_5 and its folded companion from $-x_5$, $-y_5$, z_5 occurs at Q. As Q varies, a fringe pattern is generated.

On substituting typical experimental values, the contribution of the higher order terms to the intensity at Q is found to be at least 3 orders of magnitude smaller than the dominant (sinusoidal) term. Thus it is evident that the aberrations introduced by the gratings are small, at least in the one-dimensional case.

The expression for x_5 is of the form

$$x_5 = [\{A + F_1(\lambda)\} + \{B + F_2(\lambda)\}] x_1$$

and describes the mapping of x_5 into the virtual object plane. The constants A, B and the functions F_1, F_2 depend on the system parameters and can be chosen to yield the maximum number of polychromatic fringes. The optimum condition is determined by a numerical analysis.

EXPERIMENTAL OBSERVATIONS

The achromatic system was tested in a one-dimensional version for simplicity.
The configuration of the interferometer is shown in Fig. 6. The source used
was a photographic transparency with a tungsten halide projection lamp
focussed onto it. The lamp, lens and transparency were mounted on the same
plate, along with the two gratings. This plate could be rotated and displaced
linearly with respect to the rest of the system.

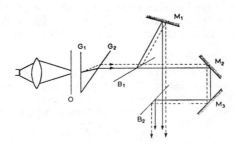

Fig. 6   The experimental interferometer arrangement

The arrangement of beam splitters and mirrors achieved the necessary 1-fold
rotation by introducing an even number of reflections on one path and an odd
number in the other. The careful positioning of the second beam splitter for
superposition of the two virtual objects is useful in aligning the interferometer. Mirrors 2 and 3 were mounted on a common plate which could be moved
with a micrometer drive so that the path lengths of the two beams were equalized to within $1\,\mu m$. The required shear, X_o, was introduced by driving the
complete object plate perpendicular to the interferometer axis. This ensured
that both virtual objects were displaced by the same distance from the axis
but in opposite directions, thereby maintaining overall alignment.

When an object transparency, in the form of a one-dimensional binary grating,
10 lines/m.m., mark-space ratio 4:1, is illuminated in the interferometer,
the output pattern is as shown in Fig. 7.

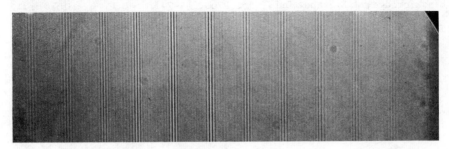

Fig. 7   Output fringe pattern for an input binary
grating of 10 lines/m.m., mark-space ratio 4:1

As expected, the transform of the input varies the visibility of a set of carrier fringes. The positions of the bands of fringes are determined by the grating periodicity, while their width is determined by the overall aperture. The maximum fringe visibility in each band is governed by the mark-space ratio. The extent of the transform is not in fact limited by loss of achromaticity, but by the exit aperture of the final beam splitter. The range over which fringes can be observed in the present system shows that approximately 250 black and white fringes can be formed from a white source. In other examples, up to 450 have been obtained.

For the purposes of demonstration, the object width was limited to 1.5 m.m., so that several fringes could be clearly seen in each patch. However, if the object size is increased to 8 m.m. a comparable number of fringes can be obtained, showing that achromaticity is maintained for extended objects. The limit of 8 m.m. was imposed purely by the size of the lamp filament.

It should be possible to use the output transparency subsequently as the input to the interferometer, hence reforming the original input directly. However, the normal output contains a large uniform background component, which, on a second transit through the interferometer, would swamp the useful signal. This difficulty could be overcome using electronic techniques in conjunction with a T.V. camera tube as the detecting element.

CONCLUSION

We have shown that a one-dimensional wavefront folding interferometer can be made essentially achromatic for extended objects. The extension of the technique to two-dimensions should not present any further difficulty. Since the interferometer output is a display of the cosine Fourier transform of the object as an intensity distribution, it can be recorded with a camera tube, the video signal then being available for electronic processing. This feature makes the system potentially attractive as a hybrid optical-electronic real-time processor.

REFERENCES

1. Mertz, L., Transformations in Optics, Wiley, New York, p.111, 1965.

2. Stroke, G.W., Restrick, R.C., App. Phys. Lett., 7, 229, (1965).

3. Dainty, J.C., Scaddon, R.J., Mon. Nat. R. Astr. Soc., 167, 69, (1974).

4. Breckinridge, J.B., App. Opt., 13, 2760, (1974).

5. Jones, D.S., Kline, M., J. Maths. Phys., 37, 1, (1958).

APPLICATION OF AN INTERFEROMETRIC METHOD TO MEASURE SURFACE DEFORMATIONS IN SEMICONDUCTOR MANUFACTURING

W. Jaerisch and G. Makosch

IBM Germany, Sindelfingen

ABSTRACT

In the photolithographic process, presently employed for the production of integrated circuits, sets of correlated masks are printed on silicon wafers. Since the electrical properties of the final integrated circuits depend to a great extend on the precision at which such patterns are superimposed, simple and fast overlay measurements and flatness measurements as well are very important(1).

A simple optical interference method for flatness measurements will be described which can be used under manufacturing conditions. This method permits testing of surface height variations by nearly grazing light incidence by absence of a physical reference plane. It can be applied to polished surfaces and rough surfaces as well.

INTRODUCTION

The measurement of in-plane distortions of wafer patterns is of great importance to eliminate overlay errors in the photolithographic process. Another very important factor that influences the printing quality is the wafer flatness. A poor image quality is obtained when using warped silicon wafers. Processing time and costs can be saved if wafers with surface deformations exceeding the tolerances are eliminated before entering the advanced process steps. Presently, in manufacturing lines mechanical, pneumatic and electrical methods or optical Fizeau interference methods with reference flats or prisms placed very closely in front of the surface to be tested are used to measure the topography of the wafers. The non-optical methods are partly destructive methods (contact, contamination) and provide surface profiles only along specific traces or at definite measuring points. The common interferometric methods produce interference

fringe patterns of the topography of highly polished wafers
whereby adjacent fringes represent contour height differences of
one half wavelength of the illuminating light. For testing wafer
warpages on the order of 10 um or more the $\lambda/2$-resolution
ranging from 0.25 - 0.35 um is too small. The evaluation of such
dense interference patterns is too difficult and time consuming.
A lower resolution can be achieved by oblique illumination(2-4).
This can be done for instance with prisms used instead of optical
flats. However, these methods are virtually contact methods and
that means damages of the test surface cannot be excluded.

In order to avoid these difficulties an optical surface tester
based on nearly grazing incidence object illumination was developed that works contactlessly. This method permits measuring
rough surfaces and highly polished surfaces as well. It provides
an adjustable and, therefore, problem adapted sensitivity.

GRAZING INCIDENCE CONTOUR MAPPING - BASIC CONCEPT

The principle of operation is illustrated in Fig. 1. An expanded
and collimated laser beam is reflected from a plane mirror down-
wards onto the measuring arrangement. This arrangement consists
of an adjustable wafer fixture. In difference to other methods
the reference surface is located beside the test surface and not
on top of the test surface. Therefore, this arrangement allows a

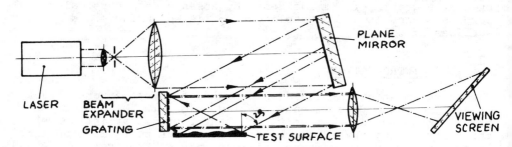

Fig. 1   Basic principle of the measuring method

free access to the measuring device without endangering the test
surface. This reference is a reflection grating with a line
spacing of several microns which is oriented perpendicularly to
the test surface. The direction of the grid bars is perpendicular
to the drawing plane. The collimated light beam illuminates the
wafer surface and the grating area simultaneously. The angle of
incidence ϑ is chosen such that light reflected at the wafer is
diffracted normally at the grating. This part of the illuminating
beam serves as object beam that is phase modulated by height
variations of the test object. Another part of the illumination
beam strikes the grating directly at an angle of incidence ϑ.
After being diffracted by the grating this light travels in the
same direction as the object beam. This second bundle with undeformed wavefronts serves as a reference beam.

The superposition of these two components, the object beam and reference beam, leads to an interference pattern with a fringe spacing corresponding to

$$\delta = \frac{\lambda}{2\cos\vartheta} \qquad\qquad (1)$$

An image of this pattern is formed on a screen properly tilted with respect to the optical axis to correct image distortions. At properly adapted intensity of the reference bundle a contrast of the visible interference patterns up to 100 % is achievable for all interesting surface qualities. A further advantage of this method is that reference and object beam are subjected to the same phase changes if the grating is not accurately flat.

This prooves that the flatness of the grating surface is less important as in common Fizeau interferometry with physical flat surfaces. In the method described the grating surface does not act as a reference surface. Here the undistorted wavefront of the reference beam itself serves as a reference. And therefore care has to be taken on high quality beam expanders and mirrors. The only purpose of the grating is the recombination of the unmodulated reference beam and the phase modulated object beam to obtain an interference pattern.

The generation of the interference pattern is explained in Fig.2. In an arbitrary point of the grating, two components S_1 and S_2 are superposed in the direction normal to the grating plane. S_1 is the reference ray being directly diffracted by the grating. S_2 is the object beam, reflected from the wafer surface and then diffracted by the grating. With complex amplitude notation these waves superimposed at the grating point P can be described as follows:

A phase variation $\Delta\varphi$ of the object beam S_2 caused by a height variation Δh of the object is given by:

$$\Delta\varphi = \frac{2\pi}{\lambda}\cdot(x_2 - x_1) = \frac{4\pi}{\lambda}\cdot\Delta h\cdot\cos\vartheta \qquad\qquad (2)$$

Considering S_1 as reference beam these components are:

$$S_1 = S_1$$

$$S_2 = S_1\, e^{i\cdot 2\cdot\Delta h\cdot K\cdot\cos\vartheta}, \qquad\qquad K = 2\cdot\pi/\lambda \qquad\qquad (3)$$

and their superposition is given by

$$S = S_1 + S_2 = S_1\cdot(1 + e^{i\cdot 2\cdot\Delta h\cdot K\cdot\cos\vartheta}) \qquad\qquad (4)$$

The resulting intensity is

$$J = 1/2\left|S\right|^2 = S_1\cdot(1 + \cos(2\cdot\Delta h\cdot K\cdot\cos\vartheta)) \qquad\qquad (5)$$

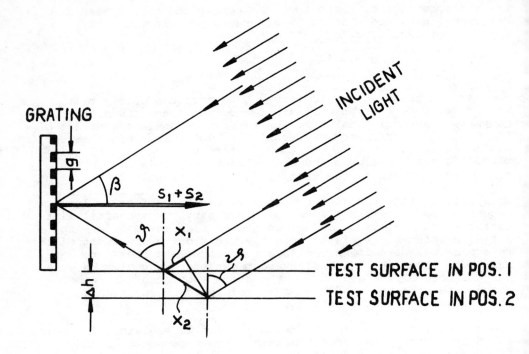

Fig. 2   Generation of the interference pattern

This is a periodic function with dark fringes for:

$$2 \cdot \Delta h \cdot K \cdot \cos \vartheta = (2n+1) \cdot \pi , \qquad (n = 0,1,2\ldots) \qquad (6)$$

and bright fringes for:

$$2 \cdot \Delta h \cdot K \cdot \cos \vartheta = 2 \cdot n \cdot \pi \qquad\qquad (n = 0,1,2\ldots) \qquad (7)$$

The separation δ of two adjacent dark fringes follows from (6) when

$$\frac{4\pi}{\lambda} \cdot \cos \vartheta \cdot \delta = 2\pi \qquad\qquad (8)$$

$$\delta = \frac{\lambda}{2 \cdot \cos \vartheta} \qquad\qquad (9)$$

By choice of proper grating constants all angles of incidence up to 90° can be realized. Hence, all measuring sensitivities from $\lambda/2$ up to several microns are practical.

In addition, the measuring sensitivity can be varied at a given grating constant if an angle of illumination is chosen corresponding to one of the higher diffraction orders given by

$$\sin \beta = \cos \vartheta_m = m \cdot \lambda/g, \qquad m = (1),2,3,4 \ldots \qquad (10)$$

Then the combination of   (9) and (10) leads to an expression for the fringe separation

$$\delta = g/2m \qquad\qquad m = (1), 2, 3 \ldots . \qquad\qquad (11)$$

Hence, the sensitivity increases with increasing angle of illumination.

These considerations show that the fringe separation of the method can be varied theoretically from $\lambda/2$ to infinite. In practice, however, a variation of the fringe separation between 0.5 um and 5 um is realistic.

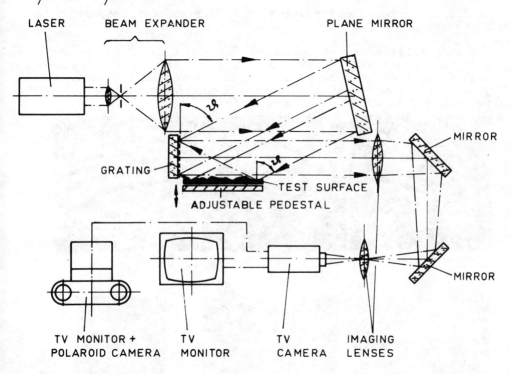

Fig. 3   Measuring arrangement

Based upon the principle described before manufacturing measuring tools were built with adaptable sensitivities from 0.5 um to several microns (Fig. 3). The diameter of an expanded 5 mW He-Ne laser beam can be chosen either 50, 100 or 150 mm. The wavefront distortions of the incident beams are smaller than a tenth of the wavelength. The flatness of the implemented mirrors is better than $\lambda/20$. The wafers are vacuum clamped onto $\lambda/10$ flat pedestals. The maximum-minimum distinction of extremes of an interference pattern is performed by a motor driven lifting mechanism of the pedestal by a small amount. Fringes which bulge

out indicate   maxima and fringes which disappear indicate
minima. The alignment of the wafer surface relative to the
grating surface that means the adjustment to obtain a minimum of
interference fringes in the field of view, is made by tilting the
pedestal with the aid of small electrical driving motors. The
observation and evaluation of the interference patterns can be
done on a screen or on high resolution TV monitors. The pattern
recorded by a TV camera with 875 scanning lines is also transferred to a second small TV monitor. 4"x5" Polaroid pictures of
the interference pattern can be taken from this screen. The
elliptical distortion of the optical image is corrected electronically at the TV camera.

As necessary for manufacturing tools all laser safety requirements are satisfied.

For simplicity in handling and evaluating this device allows a
very fast flatness test by counting a few fringes and therefore a
high thruput of test objects. Beside testing silicon wafers and
glass photo masks, metall plates and even very rough ceramic and
ferrite plates were tested successfully.

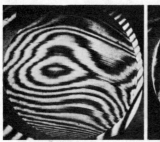  

a) wafer after slicing   b) after lapping   c) after polishing

 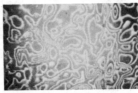

d) ferrite surface at   e) at lower sensitivity   f) ceramic plate
   high sensitivity

Fig. 4   Interference patterns of different surfaces
          a),b),c): semiconductor wafers (fringe spacing 1.5 /um)
          d),e):   ground ferrite material
          f):   dull ceramic plate (fringe spacing 2 /um)

Several measurement results are shown in the last figure (Fig.4).
It shows interference patterns of silicon wafers at different
process steps and the topography of a dull ferrite material at
different measuring sensitivities. Fig. 4f shows the fringe
pattern of a ground ceramic surface.

CONCLUSION

An interference method for measuring flatness variations of semi-conductor wafers is described. The method was developed with the goal to improve the printing quality of presently used photolithographic printing tools. Therefore, one major point by developing the method was its simplicity and flexibility which enables its application in IC-manufacturing.

Besides for testing wafers, the described method may be very useful for testing optical and non-optical surfaces as well. Its great advantage is seen in the variable measuring sensitivity which can be adapted to particular measuring problems encountered in the widely spread field of surface testing.

REFERENCES

(1).   W. Jaerisch, Flatness Measurements in Semiconductor Manufacturing, Feinwerktechnik-meßtechnik 83, #3 (1975).

(2).   P. Hariharan, Improved Oblique Incident Interferometer, Opt. Eng. 14, #3 (1975).

(3).   I. D. Briers, Interferometric Flatness Testing of Nonoptical Surfaces, Appl. Opt. 10, #3 (1971).

(4).   N. Abramson, The Interferoscope, Optik 30, #1 (1969).

EXPERIMENTAL PROCESS TO RESTORE SIGN AND VALUE OF THE PHASE IN CLASSICAL INTERFEROMETRY

Gérard Roblin

Institut d'Optique, Laboratoire associé au CNRS,

3 Boulevard Pasteur, 75015 Paris, France

ABSTRACT

An experimental process is described, capable to restore the sign and value of the phase of a pure phase object. It is based on a phase modulation interferometry combined with scanning of the object. Accuracy of this method is discussed in terms of experimental parameters.

INTRODUCTION

The modulus of the complex amplitude of a pure phase object is constant. So, classical square law detectors which are used in optics (eye, photographic plate, photoelectric cell) are not able to show any details of such an object. Interferometry alone is able to supply informations which are contained in the object because it replaces phase variations by illumination variations. Indeed, if two coherent beams of same amplitude are interfering, one of them crossing the object, for a point of the beam where the phase is p, there is a point in the interference pattern where the illumination is $\sin^2 p/2$. But this method has a great disadvantage which is well known by the users of the interferometers. The information on the sign of the phase and thus on the direction of the relief is lost because two same phases with opposite signs are giving equal illuminations. This information is very important as much in biology as in dynamics or mechanics and its knowledge is often sufficient, a measure of the phase not being necessary. Therefore it is interesting to propose a process which provides the whole and quick information everywhere in the field.

To restore the sign of the phase, it is necessary that the illumination of interference pattern, which is the only available optical signal, be expressed in a form which contains an odd part in terms of the phase and that this signal be treated in such a manner that only this odd part is extracted. One remarks, that if a constant phase k is added everywhere, the illumination of the interference pattern which is the optical signal becomes :

$$E_k = \tfrac{1}{2}\left[1-\cos(p+k)\right] = \tfrac{1}{2}\left[1-\cos(p)\cos(k)\right] + \tfrac{1}{2}\sin(p)\sin(k) \quad (1)$$

This signal contains an even and an odd part in terms of the phase. We can observe the reciprocity of rôles of p and k and then it is equivalent to extract the odd part of the signal in terms of p or of k (Ref. 1).So,if k is modulated in odd terms of a parameter u :

$$k = K(u) = -K(-u) \tag{2}$$

the illumination varies periodically as a function of this parameter :

$$E_k(u) = \tfrac{1}{2}A_0 + \cos(p)\sum A_{2n}\cos(2nu) + \sin(p)\sum B_{2n-1}\sin(2n-1)u \tag{3}$$

If it is possible to extract amplitude of one of the odd components in the form of a direct current of the same sign, the information on the sign of the phase is retained.

Practically sine wave modulation of the phase in terms of time

$$k = k_0 \sin(gt) \tag{4}$$

provides solution of this problem. Illumination of the interference pattern for each phase p becomes a periodic time signal :

$$E_p(t) = \tfrac{1}{2}\{1 - \cos(p)[J_0(k_0) + 2\sum J_{2n}(k_0)\cos(2ngt)] + 2\sin(p)\sum J_{2n-1}(k_0)\sin(2n-1)gt\} \tag{5}$$

Electrical signal, which is provided by the photoelectric cell receiving luminous energy, has the same form. By means of synchronous detection (Fig.1) it is possible to extract a direct current

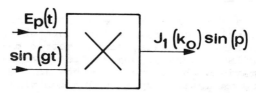

Fig. 1. Synchronous detection of the fundamental frequency

which is proportional to the imaginary part of the complex amplitude of the object. Because this part is odd, this current has the same sign as that of the phase p. For restoring the sign everywhere in the field, in other terms the direction of the relief which is given by the sign of the phase, it is necessary to extract such a signal everywhere in the field either simultaneously or successively.

POSSIBLE PROCESSES FOR RESTORING THE INFORMATIONS

Simultaneous Restitution
Practically this method alone, consisting in simultaneous extractions of the signals which have the same sign as the phase, is able to give whole the information at every moment. It can be applied for time variable objects (living preparations in biology, acceleration studies in wind-tunnels). Practically (Fig. 2) a photoelectric mosaic which is conjugated with the interference pattern given by the phase modulation interferometer (PMI) is used. By means of storage, the output signals of each cell can be treated afterwards at deferred time. Disadvantages of this method consist in the complexity of this receiver which must have a good resolving power and in the fact that each cell be followed by a synchronous detector, the output signal of which is being stored.

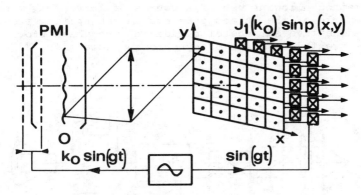

Fig. 2. Simultaneous restitution

Successive Restitutions by Scanning Method

If the object does not have time variations (inert preparations, surface structure) or if they are very slow, informations on the sign of the phase can be extracted successively at each point of the object by scanning. In this case there is only one receiver, in a pupil plane for instance. The object can be displaced mechanically and its image scans the plane of a small field stop. The output signal is recorded in terms of the coordinates of the object. The measuring area has a uniform transmission factor, but the range of scanning speeds is limited. It is also possible to scan the interference pattern by using flying-spot method (Fig. 3).

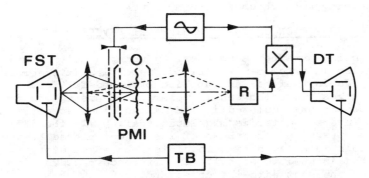

Fig. 3. Successive restitutions by flying-spot method

Output signal of the synchronous detector can be shown by a display tube synchronized with the scanning tube. This process should be very interesting because it provides a wide range of scanning speeds. We shall speak about this experiment afterwards. The disadvantages of this process are the quick wear of the phosphor (because the brightness of the spot is kept high so as to obtain an adequate level of signal) and the time variable brightness of the measuring area, which is the image of the spot, because it

G. Roblin

depends on the phosphor structure.

Another method allowing use of a source of large brightness to il-
luminate whole of the object consists in conjugating the object
with the sensitive surface of a vidicon tube which provides a signal to be treated by means of synchronous detection (Fig. 4). The

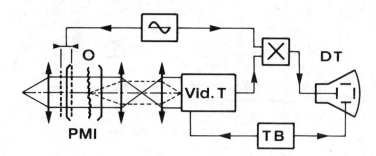

Fig. 4. Successive restitutions by using a vidicon tube

range of speeds can be large with an electrostatic vidicon tube,
but we have observed experimentally that its response to the luminous energy modulation which is produced by the phase modulation
is too small to have a good accuracy.

In order to make this method operational we are carrying out at
present an experiment with an image dissector tube which provides
all the necessary advantages : association of a source of large
brightness to have an adequate level of signal ; possibility of a
wide range of scanning speeds ; existence of a uniform transmission measuring area which is the aperture of the dissector tube.

DESCRIPTION OF THE EXPERIMENTAL DEVICE

We are testing an experiment (Ref. 2) using a lateral splitting
polarization interferometer. This differential interferometer described by Jamin (Ref. 3) and applied by Lebedeff (Ref. 4) in microscopy gives two orthogonally polarized beams and their phase
difference can be modulated if the tilt of the two crystals which
make the interferometer is modulated (Ref. 5).To the differential
variations $p(x,y)$ of the phase of the object placed between the
two crystals is added a time modulated phase $k=k_0\sin(gt)$. Flying-
spot method is used to scan the interference pattern. A microscope
conjugates the spot with the object (Fig. 5).

Two basic experiments are possible. First a static experiment
where phase difference is not modulated. The spot scans the interference pattern and the photocell signal is directly used by the
display tube which is synchronized with the scanning tube. The
signal may modulate, either, the Y-plates in which case we obtain
photometric analysis of the interference pattern ($\sin^2 p/2$) for
each scanned line, or it may modulate the intensity of its electron beam (Z modulation) in which case we obtain the interference
pattern image. Second possible experiment is a dynamic experiment

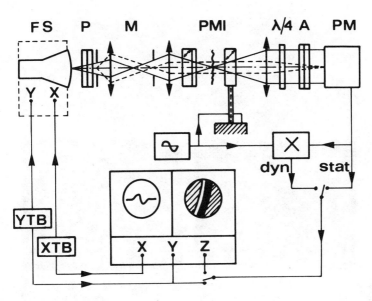

Fig. 5. Diagram of the experiment

where the phase difference is modulated. The output signal and a
reference signal, given by an oscillator driving modulation, enter
in a synchronous detector. Its output signal is used in the same
conditions by the display tube and we obtain curves giving imaginary part of the complex amplitude (sin p) for each scanned line
or its image in intensity if an electrical background is added in
order to compensate its negative values.

Figure 6 gives an example of the results. The object is a plane
parallel plate with circular grooves of triangular profile. So,
the transmitted beam presents also triangular profile. In subtractive interference Jamin's interferometer gives a derivative beam
which provides positive and negative values around a mean zero
value. When the Y-plates of the display tube are modulated, curves
giving photometric analysis of the interference pattern (a) or variations of the direction of the relief (b) are obtained according
as the experiment is static or dynamic. In the same conditions
"images" are obtained if we modulate the intensity of the electron
beam.

Thus the sign of the phase is obtained without necessarily measuring its value. Measurement is possible by displacement of a com-
pensator (in our case the rotation of an analyser) which can be
driven by the output signal. When the phase difference is compensated, the signal is zero and the movement is stopped. If the com-
pensator is rigidly locked with an optical encoder, variations of
phase can be recorded or displayed by counting the encoder impulses. Of course that requires sufficiently slow scanning speeds and
can not be applied for real time restitution.

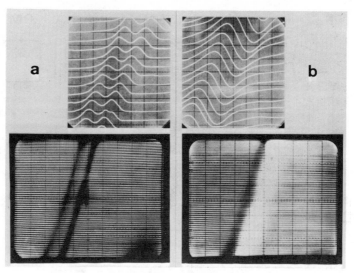

Fig. 6. Experimental results

INFLUENCE OF DIFFERENT PARAMETERS

Theoretically we have seen that the output signal of the synchronous detector is

$$s_0 = J_1(k_0)\sin[p(x,y)] \tag{6}$$

In fact this signal decreases on account of several factors. The
expression (6) for this signal is valid only at a point. But the
analysing surface has a finite area and a transmission or emission
factor $T(x,y)$ which can be non uniform inside this area. Thus the
luminous energy received by the cell at each point (x_0,y_0) of the
interference pattern is :

$$B(x_0,y_0) = \frac{\iint T(x-x_0,y-y_0)[1-\cos p(x,y)]\,dxdy}{\iint T(x-x_0,y-y_0)dxdy} \tag{7}$$

Consequently the restored phase is not the true phase p but is an
apparent phase p_a which is given by the correlation function :

$$\cos[p_a(x,y)] = \frac{\iint T(x-x_0,y-y_0)\cos[p(x,y)]\,dxdy}{\iint T(x-x_0,y-y_0)dxdy} \tag{8}$$

Moreover the previous assumptions do not take into account the
process of scanning. Taking into account scanning with a speed v
in the direction x and phase modulation, the resultant phase at a
time t is :

$$\begin{aligned} q(t) &= p(x+vt,y) + k_0\sin(gt) = \\ &= p(x,y) + p'_x(x,y)vt + k_0\sin(gt) \end{aligned} \tag{9}$$

where $p_x^!$ is the phase derivative in the scanning direction. Thus the instantaneous frequency of the time signal given by (5) is

$$(1/2\pi)(g+p_x^!v) = f + \Delta f \qquad (10)$$

which is different from the modulation frequency f on which the synchronous detector is tuned. Then the signal is reduced in the ratio of the by-pass value $F(\Delta f)$ of the detector. Let us suppose for instance that the reference signal is a pure sine function and that the time constant is T, this value is :

$$F(\Delta f) = \left|\frac{\sin(\pi T \Delta f)}{\pi T \Delta f}\right| \frac{1}{1 + \Delta f/2f} = \qquad (11)$$

$$= \left|\frac{\sin(\tfrac{1}{2}p_x^! v T)}{p_x^! v T}\right| \frac{1}{1 + p_x^! v/2g}$$

The output signal decreases with increasing speed and time constant and becomes :

$$s_o^! = F(\Delta f)J_1(k_o)\overline{\sin p_a(x,y)} \qquad (12)$$

A mean value of sin (p_a) must be introduced for two reasons : each elementary measurement is made during a modulation period and the mean of these elementary measurements is made during the time constant. So if l is the length covered during a modulation period $1/f$, the elementary value of sin(p) is :

$$\sin(p_E) = (1/l)\int_o^l \sin(p + p_x^!)dx =$$

$$= \frac{\sin(\tfrac{1}{2}p'l)}{\tfrac{1}{2}p'l}\sin(p + \tfrac{1}{2}p'l) =$$

$$= \frac{\sin(\tfrac{1}{2}p'v/f)}{\tfrac{1}{2}p'v/f}\sin(p + \tfrac{1}{2}p'v/f) \qquad (13)$$

If we have access to the sine value of the apparent phase at the centre of gravity of the length covered during a modulation period, the factor

$$C = [\sin(\tfrac{1}{2}p'v/f)]/(\tfrac{1}{2}p'v/f) \qquad (14)$$

reduces the output signal when the scanning speed increases. The mean over the duration of the time constant provides an analogous result. If the variations of p are linear on the length L covered during the time constant, a new factor appears which decreases with increasing scanning speed and time constant

$$M = [\sin(\tfrac{1}{2}p'vT)]/(\tfrac{1}{2}p'vT) \qquad (15)$$

Under these conditions, the mean value of sin(p_a) is :

$$\overline{\sin(p_a)} = \frac{\sin(\tfrac{1}{2}p'v/f)}{\tfrac{1}{2}p'v/f}\frac{\sin(\tfrac{1}{2}p'vT)}{\tfrac{1}{2}p'vT}\sin(p_a + p'L/2) \qquad (16)$$

Figure 7 shows for a single scanned line the variations of the output signal in terms of time constant T (speed being constant) and in terms of scanning speed (T being constant). We remark that

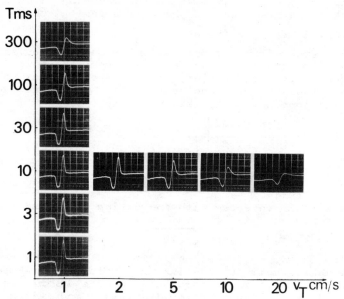

Fig. 7. Signal variations in terms of v and T

the signal decreases when speed and time constant increase (speed
influence is more important). The small signal fluctuations which
correspond to local large phase gradients vanish when these parameters increase. We remark again that the time constant must not be
less than modulation period (here 5ms) or else output signal has
residual modulations (which is evident). But if it is very large
the microdetails can not be extracted.

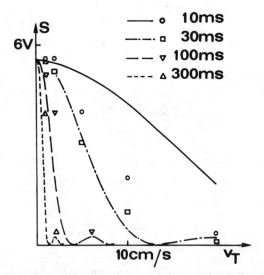

Fig. 8. Comparison between experimental and theoretical results

Figure 8 gives a comparison between the experimental results and the theoretical results. The theoretical curves are drawn according to the above assumptions.

CONCLUSION

The described method gives rapidly the direction of relief of a pure phase object without necessarily measuring the phase value itself, a measurement that requires a long time. This method can be applied with much interest in biology and medecine where it permits an accurate diagnosis when abnormal cells are studied. In the same way it is also very important to know this information in surface microinterferometry as well as in profile or stress pattern studies in dynamics or mechanics.

REFERENCES

(1) G.Roblin, Sur la restitution du signe de la phase en interférométrie,
     Opt. Commun. 15, 379 (1975).

(2) M.Allain, M.Prévost et G.Roblin, Procédé expérimental de restitution de la partie imaginaire de l'amplitude complexe d'un objet de phase,
     Opt. Commun. 15, 384 (1975).

(3) J.Jamin, Sur un réfracteur différentiel pour la lumière polarisée,
     C. R. Acad. Sci. (Paris) 67, 814 (1868).

(4) A.A.Lebedeff, L'interféromètre à polarisation et ses applications,
     Rev. Opt. 9, 385 (1930).

(5) G.Nomarski et G.Roblin, Sur une méthode de modulation du retard optique,
     C. R. Acad. Sci. (Paris) 276 B, 251 (1973).

INTERFERENCE SURFACE TESTING WITH QUARTER-WAVELENGTH SENSITIVITY

G. Makosch and W. Jaerisch

IBM Germany, Sindelfingen

ABSTRACT

A contactless interference surface testing method is described that is capable of creating fringe patterns with fringe separations corresponding to a surface deformation of a quarter-wavelength of the illumination light. Contrary to common Fizeau fringes where the interference pattern occurs by superposing two wavefronts, the described method is based upon a superposition of four wavefronts. These wavefronts are created by a diffraction transmission grating which is contactlessly placed over the surface to be tested. Basically, this technique provides a beat frequency pattern with adaptable fringe separation.

Practical applications of this method to semiconductor silicon wafers are presented.

INTRODUCTION

The flatness of optical surfaces is commonly specified in terms of a wavelength of light λ. For this purpose the surface to be measured is interferometrically compared with an accurately plane reference surface. This reference surface, the lower surface of an "optical flat", is placed over the test surface (Fig. 1a). If both surfaces are illuminated by a monochromatic parallel light beam, a Fizeau fringe pattern may be observed in reflected light[1]. Deformations of the test surface are determined by counting the number of Fizeau fringes and examining their shape. For each fringe the space between the test- and reference surface changes by one half wavelength at normal light incidence. Hence, $\lambda/2$ is a measure for the sensitivity of interferometric surface testing. The measuring sensitivity of the Fizeau method at a given λ can be decreased by increasing the angle of incidence ϑ_o. This leads to a fringe separation greater than $\lambda/2$. However, no simple method is known so far that allows to decrease the fringe separation below $\lambda/2$. This paper describes an interferometric surface testing method which provides an interference pattern with fringes corresponding to surface deviations of a quarter wavelength.

This new technique employs a transmission grating contactlessly placed over the surface to be tested (Fig. 1b). It is illuminated by a coherent parallel light beam at an angle of incidence ϑ_o.

The interference pattern is observed in the direction of the
first or a higher diffraction order. Generally, this pattern is
composed of macro- and micro-contour fringes. Both types of
fringes can be used for coarse and fine flatness measurements, re-
spectively(2). Under proper experimental conditions which will be
more closely examined later, this interference pattern changes
into a $\lambda/4$ fringe pattern (Fig. 1c). Compared with commonly
known Fizeau fringes this means a two fold increased measuring
sensitivity.

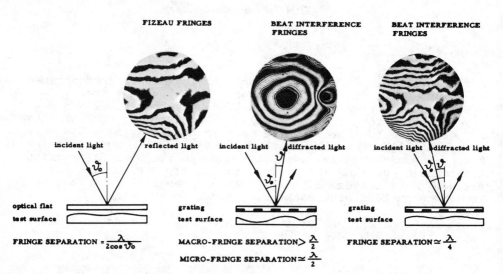

Fig. 1 Comparison of beat interference patterns
        with the Fizeau fringe pattern.

PRINCIPLE OF THE MEASURING METHOD

The fringe pattern which is observed in diffracted light by using
a grating as reference surface results from the interference of
light diffracted by the grating and reflected from the surface
being tested. According to the diffraction theory, at the grating
each incident wave divides into reflected and transmitted
diffraction components (Fig. 2). These diffracted waves propagate
in discrete angles which are defined by the equation (1):

$$\sin \vartheta_m = \sin \vartheta_o + m \cdot \lambda/g \, , \quad m = 0, \pm 1, \pm 2 \ldots \ldots \qquad (1)$$

where ϑ_o is the angle of incidence, λ is the wavelength of
illumination, g is the grating constant, and m denotes the order
of diffraction. If a transmission grating is placed over a
reflecting surface, the light passing through the grating is
reflected from the test surface and interacts with light
diffracted in reflection from the grating.

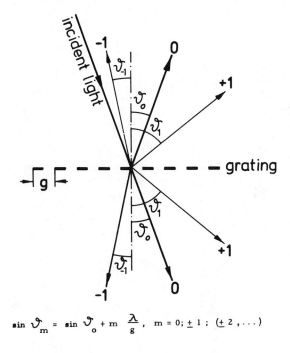

$$\sin \vartheta_m = \sin \vartheta_o + m \frac{\lambda}{g} , \quad m = 0; \pm 1 ; (\pm 2 , \dots)$$

Fig. 2 Principle of the diffraction grating

It is obvious, that the interference pattern formed in one particular diffraction order occurs by superposing all waves travelling in this particular direction. Figure 3 illustrates the formation of these different light components.

It is assumed that the whole grating is illuminated by a parallel coherent light beam at an angle of incidence ϑ_o. Four parallel rays S_1, S_2, S_3 and S_4 will then come together at an arbitrary point P at the grating surface. Their direction of propagation corresponds to one of the diffraction orders, preferably the plus or minus first order. The ray S_1 is a first reflection order produced by a ray incident at P. The ray S_2 passes as first transmission order the grating and is reflected from the test surface toward P. The ray S_3 results from a ray that passes the grating as zero transmission order, is reflected from the test surface, and strikes the lower surface of the grating. Here, S_3 is split   off as first reflection order and returns in reflection from the test surface to P. And finally, the ray S_4 is created as first transmission order by a ray that passes the grating as zero order and is reflected from the test surface toward P. In determining the intensity of light at P which results by superposing these four waves, it is convenient to use complex amplitude notation.

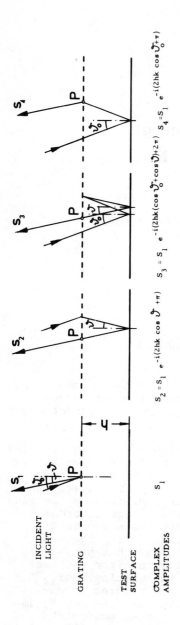

INCIDENT LIGHT

GRATING

TEST SURFACE

COMPLEX AMPLITUDES

S_1

$S_2 = S_1\, e^{-i(2hk\cos\vartheta\, +\pi)}$ $S_3 = S_1\, e^{-i(2hk(\cos\vartheta_o+\cos\vartheta)+2\pi)}$ $S_4 = S_1\, e^{-i(2hk\cos\vartheta_o\, +\pi)}$

Fig. 3   Formation of different light components passing through the grating at point P. ϑ_o – angle of light incidence, ϑ – angle of diffraction, $\kappa = 2\pi \cdot \dfrac{1}{\lambda}$ - propagation constant.

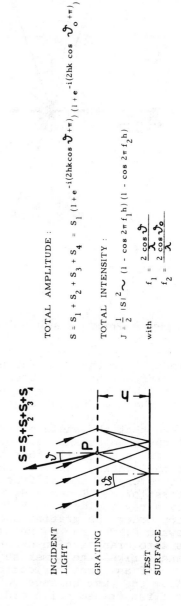

INCIDENT LIGHT

GRATING

TEST SURFACE

$S = S_1 + S_2 + S_3 + S_4$

TOTAL AMPLITUDE :

$S = S_1 + S_2 + S_3 + S_4 = S_1\,(1 + e^{-i(2hk\cos\vartheta\, +\pi)})(1 + e^{-i(2hk\cos\vartheta_o\, +\pi)})$

TOTAL INTENSITY :

$J = \dfrac{1}{2}\,|S|^2 \sim (1 - \cos 2\pi f_1 h)(1 - \cos 2\pi f_2 h)$

with $f_1 = \dfrac{2\cos\vartheta}{\lambda}$

        $f_2 = \dfrac{2\cos\vartheta_o}{\lambda}$

Fig. 4   Superposition of different light components at point P.

This leads to the equations for S_1, S_2, S_3 and S_4. The phase relationship of these waves at P was determined taking into account the phase change of π on reflection at the test surface and considering S , as reference wave. In addition, it is assumed that the intensities of the various components are equal. This can be closely achieved in practice by using a suitable diffraction grating. The superposition of these four waves at P leads to an expression for the total amplitude S (Fig. 4). By multiplication with the complex conjugate it follows the formula for the intensity J at P

with

$$J \sim (1 - \cos 2\pi f_1 h)(1 - \cos 2\pi f_2 h) \tag{2}$$

$$f_1 = 2\cos \vartheta / \lambda \quad ; \quad f_2 = 2\cos \vartheta_o / \lambda$$

This expression describes the intensity variation at any point P at the grating surface with varyings gap h between the grating and test surface. It corresponds to a multiplicative type of superposition of two oscillations with the frequencies f_1 and f_2. The calculated intensity variation versus air gap h between the grating and test surface is plotted in Fig. 5 for a grating constant $g = 0.37\,\mu m$, an illumination wavelength $\lambda = 6328$ Å and angle of illumination $\vartheta_o = 60°$. The angle of the minus first diffraction order is then $\vartheta = -57.9°$.

The resulting intensity distribution is a beat frequency pattern composed of narrow intensity peaks whose amplitudes are modulated by a slowly varying periodic function. This phenomenon may be compared with combining two tones having slightly different frequencies f_1 and f_2 which sound simultaneously. At certain equal intervals they are in phase and reinforce each other; at intermediate periods they are in opposite phase and tend to neutralize each other. The successive maxima of the envelope curve are at intervals

$$\Delta = \frac{\lambda}{2 \, |(\cos\vartheta - \cos\vartheta_o)|} \tag{3}$$

The minima of the fine peaks are splitted at a minimum of the envelope curve. Here, their frequency approaches twice the frequency found at a maximum of the macrostructure. Their separations are given by

$$\delta' = \lambda/4 \, (1/\cos\vartheta + 1/\cos\vartheta_o) \quad \text{(Beat Maximum)}$$
$$\delta'' = \lambda/8 \, (1/\cos\vartheta + 1/\cos\vartheta_o) \quad \text{(Beat Minimum)} \tag{4}$$

At small angles of incidence ϑ_o and grating constants g larger than $10\,\mu m$ the first order diffraction angle ϑ approaches zero. Under these conditions, these two fringe separations are about

$$\delta' = \lambda/2 \quad \text{and} \quad \delta'' = \lambda/4.$$

Similar to Fizeau fringe patterns, the present method provides fringes of equal air gap thickness.

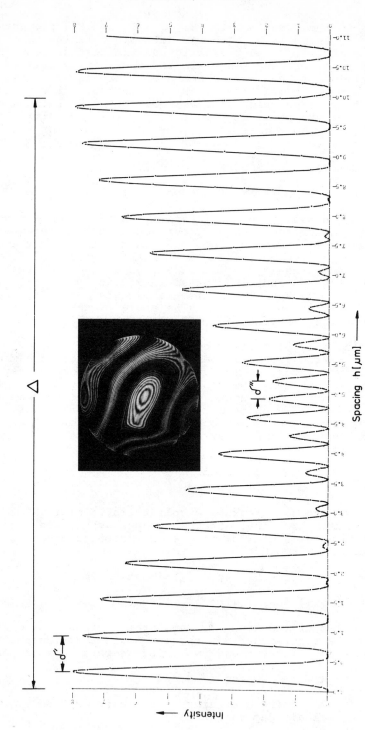

Fig. 5   Interference beat frequency pattern at various distances h between grating and test surface. $\vartheta_0 = 60°$, $\vartheta = 57.9°$, $g = 0.37$ µm, $\lambda = 6328$ Å.

However, different from the Fizeau pattern where the "contour
lines" have constant intensity and constant resolution, the
pattern obtained by this technique is a beat frequency interference pattern. Basically, it consists of macrofringes with a
separation $\triangle$ and microfringes. Both scales can be used for either
coarse or fine surface tolerance control. The separation of the
macrofringes can be varied between $\lambda/2$ and several microns by a
proper choice of the angle of incidence and diffraction. Therefore, this scale can be used when severely deformed surfaces are
to be tested. The picture shown in Fig. 5 is a beat interferogram
of a semiconductor wafer. In this case, the fringe separation of
the macrofringes corresponds to a 10 um surface deformation of
the wafer. The effect of line doubling at a minimum of the envelope curve can be utilized to obtain $\lambda/4$ fringe patterns of a
test surface. For this purpose, the separation of the macrofringes is increased. By suitable choice of the angle of
incidence, the separation $\triangle$ of the macrofringes can be stretched
arbitrarily. Accordingly, the minimum of the envelope curve is
extended and a large depth $\triangle h$ occurs with $\lambda/4$ fringe separation.
The test surface is brought into a mean distance h = $\triangle/2$ to the
grating or into a distance corresponding to one of the successive
minima of the envelope curve.

Under these conditions, the obtained interference fringe pattern
of the test surface is one with a fringe separation corresponding
to $\lambda/4$ surface deformations. Thus, the conditions for a $\lambda/4$
fringe pattern are given by

$$h = (m + 1/2)\triangle \ , \ m = 0,1,2,3....$$

$$\triangle \gg \text{surface deformations}$$

(5)

EXPERIMENTAL RESULTS

The formation of a $\lambda/4$ fringe pattern can easily be observed with
an optical arrangement shown in Fig. 6. A He-Ne laser beam
(λ = 6328 Å) is expanded and illuminates in reflection over a
plane mirror the transmission grating placed contactlessly over
the test surface. The first diffraction order is reflected back
through the collimator lens which now acts as imaging lens.
A small mirror separates the first order from other orders to
form an image of the interference pattern on the screen. By
placing the surface in a mean distance corresponding to one of
the minima of the envelope curve, it may be arranged that the
interference pattern observed on the screen is one with $\lambda/4$
fringe separation.

In practice, however, it is more convenient to create the $\lambda/4$
fringe pattern by adapting the period of the envelope curve at a
given air gap between grating and test surface (h$\simeq$0.5 mm). This
is achieved according to (1) and (3) by altering the angle of
illumination as shown in Fig. 7.

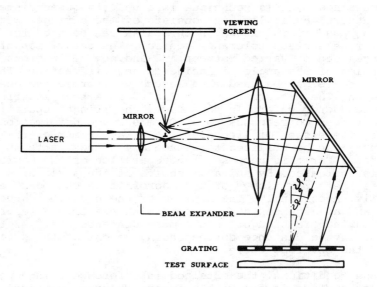

Fig. 6 Experimental arrangement

If both, the angle of illumination and the angle of diffraction
are equal, the period $\triangle$ of the envelope curve is infinitely
large. In this case, the fringe pattern is formed at a maximum of
the envelope curve (Fig. 7a). It corresponds to a common Fizeau
fringe pattern with $\lambda/2$ fringe separation. With increasing angle
of incidence $\triangle$ becomes smaller. Correspondingly, the interference
pattern changes its appearance (Fig. 7b). If the period of the
envelope curve is twice as large as the air gap h, a $\lambda/4$ fringe
pattern is observed on the screen (m = O in Equ. 5)(Fig. 7c). The
test surface in this case was a semiconductor silicon wafer.
A grating with a grating constant g = 16 µm was used. The wafer
was placed at a mean distance h of about 0.6 mm to the grating.
A He-Ne laser with a wavelength λ = 6328 Å was used as light
source. The angle of incidence was varied between 1.9° and
2.28°. For each angle of illumination, the theoretical sections
of the corresponding beat functions are plotted in Fig. 7.

These pictures demonstrate a good agreement of experimental results and theoretically calculated interference patterns.

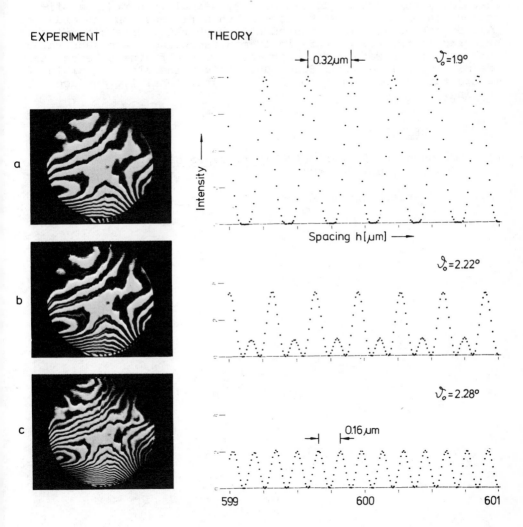

Fig. 7 Interference beat frequency patterns at various angles of illumination ν_0 g = 16.6 μm, λ = 6328 Å, h = 599 - 601 μm.

CONCLUSION

The described method is a practical and precise nondestructive
method for surface testing. It provides a measuring resolution a
factor two higher than commonly used Fizeau interference methods.
Therefore, this technique can be very useful in certain applications in which optical surfaces with a flatness on the order of
$\lambda/2$ or better are to be tested at a high rate. This situation is
expected in the near future in IC-manufacturing where substrates
such as silicon wafers and photomasks with a flatness better than
one micron will be used. Based upon this principle a flexible
manufacturing tool can be built for highly sensitive surface
testing.

REFERENCES

(1). M. Born and E. Wolf (1965) Principle of Optics, MacMillan,
     New York, pp 289, 403.

(2). W. Jaerisch and G. Makosch, Appl. Opt. 12, 1552 (1973).

SESSION 6.

HOLOGRAPHIC INTERFEROMETRY I

Chairman: S. LOWENTHAL

QUANTITATIVE STRAIN MEASUREMENT THROUGH HOLOGRAPHIC INTERFEROMETRY

R. Dändliker

Brown Boveri Research Center, CH-5401 Baden, Switzerland

INTRODUCTION

The quantitative determination of mechanical strain on the surface of an arbitrarily shaped object through holographic interferometry requires the solution of the following three basic problems (1):

- relation between surface strain and surface displacement,
- relation between derivation of the surface displacement and the interference fringes in the image plane,
- interpolation of the interference fringe pattern and quantitative determination of the interference phase.

Heterodyne holographic interferometry (2) allows to measure the interference phase φ with an accuracy and reproducibility of better than 1 degree, i.e. about 1/1000 of a fringe, at any point in the fringe pattern (3). Therefore one can determine locally first and second derivatives of the fringe pattern with sufficient accuracy to calculate the corresponding differential changes of the surface displacement.

The strategy for the evaluation of strain, rotation, bending and torsion through holographic interferometry is shown in Fig. 1. In one or more double exposure holograms the change of the object is recorded. The geometrical setup of object, hologram and illumination has to be known too. From the fringe pattern in the reconstructed image of the object the interference phase $\varphi_n(\xi,\eta)$ for different sensitivity vectors $\vec{E}_n$ is measured and the first and second differences are calculated. In a second step the displacement vector $\vec{u}(\xi,\eta)$ and its derivatives are determined in the image system (ξ,η,ζ), i.e. the components of $\vec{u}$ in the image system and their derivatives with respect to the image plane coordinates (ξ,η) are calculated. For this step the sensitivity vectors $\vec{E}_n(\xi,\eta)$ in the image system have to be known as a function of position in the image plane, either analytically or numerically. In a third step the displacement vector $\vec{u}'(\xi,\eta)$ and

R. Dändliker

its derivatives have to be transformed onto the object surface
to obtain the deformation of the object surface in a coordinate
system (x,y,z) with z parallel to the surface normal. For this
step the shape of the object and its position in the image system has to be known. In the last step the strain, rotation,
bending and torsion (change of curvature) are calculated from
the displacement $\vec{u}(x,y,z)$ and its derivatives on the surface.

SURFACE DISPLACEMENT, STRAIN, CHANGE OF CURVATURE

The object coordinate system (x,y,z) for the point P is selected
so that the z-axis is parallel to the surface normal and the x-
and y-axes are parallel to the tangential plane (Fig. 2). As
shown elsewhere (4) the strain ε and the rotation Ω can be calculated from the vector gradient $\nabla\vec{u}$ of the displacement
$\vec{u}(x,y,z)$. Assuming a free surface, i.e. no external forces, as
mechanical boundary conditions, the components of the surface
strain tensor ε are (1)

$$\varepsilon_{\alpha\beta} = \frac{1}{2} \, (\partial u_\alpha/\partial x_\beta + \partial u_\beta/\partial x_\alpha) , \tag{1}$$

where greek indicies (e.g. α,β) are used for in-plane components
(u_x, u_y) and in-plane coordinates (x,y). The components of the
rotation vector $\vec{\Omega}$ are (1)

$$\Omega_x = \partial u_z/\partial y, \qquad \Omega_y = -\partial u_z/\partial x$$
$$\Omega_z = \frac{1}{2} \, (\partial u_y/\partial x - \partial u_x/\partial y) . \tag{2}$$

Bending and torsion of a surface are given by the change of curvature due to the applied load. The original curvature of the
object surface is described by a tensor κ with the elements

$$\kappa_{\alpha\beta} = \partial^2 F/\partial x_\alpha \partial x_\beta , \tag{3}$$

where z = F(x,y) defines the surface and x,y are orthogonal coordinates in the tangential plane (5). The change of curvature
consists of two different contributions, namely the variation of
the surface tilt and the influence of the surface strain, which
reduces the curvature proportional to its original value (1).
The tensor components of the change of curvature are found to be

$$\Delta\kappa_{\alpha\beta} = \partial u_z/\partial x_\alpha \partial x_\beta - \kappa_{\alpha\gamma}\varepsilon_{\gamma\beta} - \varepsilon_{\alpha\gamma}\kappa_{\gamma\beta} , \tag{4}$$

where the convention has been used, that summation is performed
over repetitively used indicies (γ = 1,2 = x,y). In the case of
dominant bending and torsion the contributions of the second
term in Eq. (4) can be neglected if the products of the curva-
ture $\kappa_{\alpha\beta}$ and the in-plane strains $\varepsilon_{\alpha\beta}$ are small.

The bending (change of curvature) $\Delta\kappa_s$ and the torsion (change
of rotation angle) $d\Omega_s/ds$ for an arbitrary direction $\vec{s}$ on the

surface are obtained from the tensor of change of curvature $\Delta\kappa$ by the relations

$$\Delta\kappa_s = \Delta\kappa_{xx}\cos^2\theta + 2\Delta\kappa_{xy}\cos\theta\sin\theta + \Delta\kappa_{yy}\sin^2\theta$$

$$d\Omega_s/ds = \Delta\kappa_{xy}\cos2\theta + \frac{1}{2}(\Delta\kappa_{yy}-\Delta\kappa_{xx})\sin2\theta, \tag{5}$$

where θ is the angle between the x-axis and the direction $\vec{s}$ (Fig. 2).

CALCULATION OF DISPLACEMENT VECTOR FROM FRINGES IN THE IMAGE

The geometry of the setup for the holographic interferometry is shown in Fig. 3. It is assumed that the same imaging system is used for all the holographic interferograms with different sensitivity vectors $\vec{E}_n$. This can be realized by using either different illumination sources Q_n or different observation directions (6,7), e.g. different portions of the hologram and the lens aperture (8). The actual observation point $P'(\xi',\eta')$ in the image plane is simply related to the point $P(\xi,\eta)$ in the conjugate object plane by the magnification. Therefore the (ξ,η,ζ) are used as image coordinates. The object coordinate system (x,y,z) for the corresponding point on the surface is mainly determined by the surface normal $\vec{n}$.

Displacement Vector in the Image System

The interference phase measured in the image is given by

$$\varphi_n(\xi,\eta) = (\vec{k}_i-\vec{k}_o)_n\vec{u} = 2k(\vec{E}_n\vec{u}) = 2k\, U_n(\xi,\eta), \tag{6}$$

where $\vec{k}_i$ and $\vec{k}_o$ are the wave vectors of the illuminating and the observed light, and $U_n(\xi,\eta)$ is the observed component of the displacement $\vec{u}$ in the direction of the sensitivity vectors $\vec{E}_n(\xi,\eta)$. If the displacement vector is represented by its components $u'_k(\xi,\eta)$ in the image coordinate system (ξ,η,ζ), Eq. (6) can be written as

$$U_n(\xi,\eta) = e_{nk}(\xi,\eta)\, u'_k(\xi,\eta), \tag{7}$$

where e_{nk} are the components of the $\vec{E}_n(\xi,\eta)$ in the image system. With at least 3 different sensitivity vectors (6,8,9) one can in general calculate all three components of the displacement by the inverse relation of Eq. (7), which reads

$$u'_k(\xi,\eta) = f_{kn}(\xi,\eta)\, U_n(\xi,\eta), \text{ with } f_{mn}\, e_{nk} = \delta_{mk}, \tag{8}$$

where $\delta_{mk} = 1$ for $m = k$ and $\delta_{mk} = 0$ for $m \neq k$. The elements $f_{kn}(\xi,\eta)$ of the inverse sensitivity matrix can be calculated either analytically or numerically from the recording and imaging geometry, which determines the sensitivity vectors $\vec{E}_n(\xi,\eta)$.

The derivatives of the displacement $u'(\xi,\eta)$ in the image system are directly obtained from Eq. (8), taking into account that also the inverse sensitivity f_{kn} depends on the position ξ,η in the image plane. In general the contributions of the differential change of the sensitivity vectors $E_n(\xi,\eta)$ cannot be neglected (1,7,10). At least an estimation of these contributions is recommended in each special case. Therefore the derivatives in the image system read

$$\partial u_k'/\partial\xi_\alpha = f_{kn}(\partial U_n/\partial\xi_\alpha) + (\partial f_{kn}/\partial\xi_\alpha)\,U_n \tag{9a}$$

$$\partial^2 u_k'/\partial\xi_\alpha\partial\xi_\beta = f_{kn}(\partial^2 U_n/\partial\xi_\alpha\partial\xi_\beta) + (\partial f_{kn}/\partial\xi_\alpha)(\partial U_n/\partial\xi_\beta)$$
$$+ (\partial f_{kn}/\partial\xi_\beta)(\partial U_n/\partial\xi_\alpha) + (\partial^2 f_{kn}/\partial\xi_\alpha\partial\xi_\beta)\,U_n \tag{9b}$$

More than three sensitivity vectors may be used to eliminate the zero fringe ambiguity (6,11) or to improve the accuracy by least-squares solutions (8,11) of Eqs. (8) and (9).

Displacement and Deformation in the Object System

The relation between the image system (ξ,η,ζ) to the object system (x,y,z) can be approximately described by a linear transformation

$$x_m = R_{mk}\,\xi_k, \tag{10}$$

where R_{mk} is orthonormal, i.e. $R_{mk}R_{ek} = \delta_{me}$, corresponding to a general rotation in space. This approximation is valid as long as the distance between object surface and object plane (ξ,η) is small compared with the viewing distance. The rotation matrix R is mainly determined by the direction of the surface normal in the image system (z-axis parallel to $\vec{n}$). Using Eq. (10) the components of the displacement $\vec{u}$ in the object system are obtained by

$$u_m(x,y,z) = R_{mk}u_k'(\xi,\eta). \tag{11}$$

For the transformation of the derivatives, however, it has to be considered that the observed displacement $\vec{u}(x,y,z)$ is restricted to the surface, which means that x,y,z have to fulfill the mentioned equation $z = F(x,y)$, describing the surface. This leads to an extra contribution for the second derivative, containing the original surface curvature $\kappa_{\alpha\beta}$ as defined in Eq. (3):

$$\partial u_m/\partial x_\alpha = R_{mk}(\partial u_k'/\partial\xi_\gamma)\,R_{\gamma\alpha}^{-1} \tag{12a}$$

$$\partial^2 u_m/\partial x_\alpha\partial x_\beta = R_{mk}(\partial u_k'/\partial\xi_\gamma\partial\xi_\delta)\,R_{\gamma\alpha}^{-1}R_{\delta\beta}^{-1}$$
$$+ R_{mk}(\partial u_k'/\partial\xi_\gamma)\,R_{\gamma z}^{-1}\kappa_{\gamma\beta}, \tag{12b}$$

where R_{ke}^{-1} is the inverse rotation matrix and fulfills therefore the following relation: $R_{mk} R_{ke}^{-1} = \delta_{me}$, $R_{ke}^{-1} = R_{ek}$. Remember that in Eqs. (10) and (12) the greek indicies $(\alpha,\beta,\gamma,\delta)$ refer only to the in-plane coordinates (x,y) or (ξ,η), respectively.

Surface strain ε and rotation Ω are obtained from the measured interference phase by consecutively applying Eqs. (6), (8), (9a), (12a) and (1) or (2). For the calculation of bending and torsion (change of curvature $\Delta\kappa$) Eqs. (9b), (12b) and (4) have to be used additionally.

MEASUREMENT OF INTERFERENCE PHASE

It has been shown elsewhere that with heterodyne holographic in- terferometry (2) the interference phase (ξ,η) can be measured with an accuracy and reproducibility of better than 0.3° at any position within the interference pattern (1,3). This corresponds to an interpolation accuracy of better than 10^{-3} of one fringe or to an accuracy of $\delta U_n = 2 \times 10^{-4}$ μm for the observed displace- ment component. The basic idea of this method (Fig. 5) is to re- cord the two object fields with two different reference waves R_1 and R_2 on a hologram plate (12,13), such that the object fields O_1 and O_2 are reconstructed independently. This allows the in- troduction of a small frequency shift, e.g. $\Omega = 100$ kHz, between the two reconstructed and interfering light fields, resulting in an intensity modulation at the beat frequency $\omega_2 - \omega_1 = \Omega$ for any given point of the interference pattern. The optical phase dif- ference is converted into the phase of the beat frequency (14).

Compared with classical double exposure holographic interfero- metry the use of two different reference beams introduces some additional difficulties with unwanted reconstructions and repo- sitioning of the hologram with respect to the reference beams (13). The repositioning requirements can be drastically reduced by using reference sources or reference directions which are very close together. As a consequence, however, the undesired cross-reconstructions, i.e. the hologram $O_1 R_2^*$ reconstructed by R_2 and $O_2 R_2^*$ reconstructed by R_1, do overlap with the desired re- constructions $R_1 R_1^* R_1$ and $O_1 R_1^* R_1$, which give rise to the inter- ference pattern. But as long as the lateral offset between these images is larger than the speckle size, the statistical phases Φ of the speckle pattern of the desired and undesired image are not correlated and therefore their superposition does not pro- duce any macroscopic interference fringes (15). This means that the corresponding beat signals will have arbitrary phase shifts and their contributions to the signal of a photodetector with an area larger than the speckle size only add up incoherently.

The number N of independent speckles within the detector area F_D can be calculated as

$$N = F_D/F_S = F_P F_D/(\lambda d_o)^2 , \tag{13}$$

where F_S is the speckle area, F_P is the area of the pupil of the imaging lens, λ is the wavelength and d_o is the distance between lens and image plane. The detector output signal is then given by

$$i_s(P,t) = N i_o \left[\cos(\Omega t + \psi) + \sqrt{3/2N} \cos(\Omega t + \Phi) \right] , \tag{14}$$

where i_o is the average signal from one speckle. The first part is the desired signal from $(O_1 R_1^* R_1 + O_2 R_2^* R_2)$ with the interference phase ψ. The second part is the result of the uncorrelated contributions of the unwanted reconstructions with an arbitrary phase Φ. That secondpart can be considered as phase noise and causes an error in the measurement of the interference phase of

$$\delta\psi = \sqrt{3/4N} , \qquad (N > 10) , \tag{15}$$

which decreases with increasing number of speckles. This means that even in the case of overlapping cross-reconstructions the interference phase can be measured accurately, if the detector averages over a sufficient number of speckles. Note that for non-overlapping cross-reconstructions the number of averaged speckles has no direct influence on the phase error.

These findings have been verified experimentally in a setup with a detector diameter of 1.3 mm, i.e. $F_D = 1.33$ mm^2, an imaging lens of focal length f = 300 mm and apertures between f/9 and f/64, a distance to the image plane of d_o = 427 mm, and λ = 514 mm. The corresponding speckle numbers N and phase errors $\delta\psi$, calculated from Eqs. (13) and (15), are shown in Fig. 5. The minimum expected phase error is $\delta\psi = 0.3°$, which is of the same order as the reproducibility of the heterodyne phase measurement in the case of nonoverlapping images (1,3). The measured values for $\delta\psi$ are obtained as the standard deviation of 40 independent measurements taken at different positions in the reconstructed image of an object, which has not changed between the two exposures. The measured phase variations are even smaller than expected. This would indicate that the contribution of the uncorrelated images is smaller due to a better averaging. The residual modulation signal for completely uncorrelated images has been checked to fall also somewhat below the value expected from the calculated number of speckles.

The angular difference between the directions of the two reference waves was about 5×10^{-4}, which is 4 times the resolution limiting diffraction of the lens aperture at f/90. For this separation of the reference beams the most critical repositioning around the axis normal to the hologram plate is calculated, using Eq. (15) of Ref. (13), as $\Delta\gamma = 2°$ for the maximum aperture f/9. With this type of setup two-reference-beam holographic interferometry can be made nearly as compact and uncritical as classical double exposure holography.

SENSITIVITY FOR SURFACE STRAIN MEASUREMENT

The sensitivity of the described method for the measurement of the surface strain ε is estimated for the special case of three sensitivity vectors symmetrically distributed around the ζ-axis at the angle β and a simple rotation of the object system around the ξ-axis by the angle ϑ (Fig. 6).

The sensitivity vectors and the rotation of the object system are given by

$$
e_{nk} = \begin{vmatrix} \sin\beta & 0 & \cos\beta \\ -\sin\beta/2 & \sqrt{3}\,\sin\beta/2 & \cos\beta \\ -\sin\beta/2 & -\sqrt{3}\,\sin\beta/2 & \cos\beta \end{vmatrix} \; ; \quad R_{mk} = \begin{vmatrix} 1 & 0 & 0 \\ 0 & \cos\vartheta & \sin\vartheta \\ 0 & -\cos\vartheta & \cos\vartheta \end{vmatrix}.
$$

Assuming that the errors of the phase measurement $\delta\varphi$ are independent and equal for U_1, U_2 and U_3 the accuracy $\delta\varepsilon$ for the strain components can be calculated versus the angles β and ϑ through Eqs. (1), (9a), (12a).

Figure 5 shows the numerical results for $\delta\varphi = 0.4^\circ$ and intervals $\Delta\xi = \Delta\eta = 3$ mm between adjacent points in the image plane. The sensitivities $\delta\varepsilon$ for the different strain components and for different rotation angles ϑ of the surface normal are nearly the same. The main dependence is on the angle β between the $\vec{E}$ and the ζ-axis. A sensitivity of $\delta\varepsilon = 10^{-6} = 1$ μstrain is already obtained for $\beta = 5^\circ$, which can even be realized by observing the object through different portions of the same hologram (6,8).

PURE BENDING AND TORSION

In the case of pure bending and torsion the change of curvature $\Delta\kappa$ can be determined approximately using a single sensitivity vector $\vec{E}$ (1). For pure bending and torsion it is assumed that $\varepsilon_{\alpha\beta} << \Omega_{x,y}$ and $\partial\varepsilon_{\alpha\beta}/\partial x_\gamma = 0$, which is strictly true for the neutral line where $\varepsilon_{\alpha\beta} = 0$, holds also good enough on the observed object surface. These assumptions allow to express the second derivatives of the in-plane displacement by the tilt $\Omega_x = \partial u_z/\partial y$, $\partial_y = -\partial u_z/\partial x$ and to determine the tilt directly from only one measured displacement component U in the direction of the sensitivity vector $\vec{E} = (e_x, e_y, e_z)$:

$$
\partial^2 u_\gamma/\partial x_\alpha \partial x_\beta \overset{\sim}{=} -\kappa_{\alpha\beta}\, (\partial u_z/\partial x_\gamma) \tag{16a}
$$

$$
\partial u_z/\partial x_\alpha \overset{\sim}{=} (\partial U/\partial x_\alpha)/e_z, \tag{16b}
$$

where $e_z = (\vec{E}\vec{n})$ is the sensitivity component normal to the surface.

Following Eq. (4) in the case of negligible in-plane strain
($\varepsilon_{\alpha\gamma}\kappa_{\gamma\beta} = 0$) the change of curvature can be approximately determined as

$$\Delta\kappa_{\alpha\beta} \cong (\partial^2 u_z/\partial x_\alpha \partial x_\beta) \cong$$ (17)

$$\frac{1}{e_z^2}\frac{\partial^2 U}{\partial x_\alpha \partial x_\beta} + \frac{1}{e_z^2}\kappa_{\alpha\beta}\frac{\partial U}{\partial x_\gamma}e_\gamma - \frac{1}{e_z^2}\left[\frac{\partial U}{\partial x_\beta}\frac{\partial e_z}{\partial x_\alpha} + \frac{\partial U}{\partial x_\alpha}\frac{\partial e_z}{\partial x_\beta}\right].$$

This expression is the more general and 3-dimensional form of
Eq. (29) in Ref. (1). The derivatives of U in the object system
(x,y) are obtained from the corresponding derivatives in the
image system (ξ,η) similarly as shown in Eq. (12):

$$\frac{\partial U}{\partial x_\alpha} = \frac{\partial U}{\partial \xi_\gamma}R_{\gamma\alpha}^{-1}, \qquad \frac{\partial^2 U}{\partial x_\alpha \partial x_\beta} = \frac{\partial^2 U}{\partial \xi_\gamma \partial \xi_\delta}R_{\gamma\alpha}^{-1}R_{\delta\beta}^{-1} + \frac{\partial U}{\partial \xi_\gamma}R_{\gamma z}^{-1}\kappa_{\alpha\beta}.$$ (18)

It is convenient to use an experimental setup with the illumination
tion source Q (Fig. 3) close to the imaging lens, i.e. illumination
tion and observation direction nearly the same. To reduce the
contributions of the differential change of the sensitivity vector, the distance d between lens and object should be large compared with the transverse extension of the object (d/ξ_{max},
$d/\eta_{max} > 5$). Under this experimental conditions and for a rotation of the object system around the ξ-axis by the angle ϑ (Fig.
5) one obtains from Eqs. (17) and (18) the following simple explicit expressions for the change of curvature:

$$\Delta\kappa_{xx} = (\partial^2 U/\partial \xi^2)/\sin\vartheta$$

$$\Delta\kappa_{xy} = (\partial^2 U/\partial \xi \partial \eta) - (\partial U/\partial \xi)/d \, \tan\vartheta$$ (19)

$$\Delta\kappa_{yy} = (\partial^2 U/\partial \eta^2)\sin\vartheta - 2(\partial U/\partial \eta)\cos\vartheta/d.$$

The two terms containing the original curvature $\kappa_{\alpha\beta}$ in Eqs. (17)
and (18) compensate each other as long as the sensitivity and
the imaging directions are close enough.

The Eqs. (17) and (19) have been applied experimentally to a
number of mechanical structures subjected to pure bending and
torsion. Equations (19) turned out to be very useful for fast
evaluation of experiments and to give quite accurate results for
simple object geometries. The sensitivity $\delta(\Delta\kappa)$ of this method
can be estimated from Eqs. (19) for the same parameters as for
$\delta\varepsilon$ (Fig. 6), viz. $\delta\varphi = 0.4°$, $\Delta\xi = \Delta\eta = 3$ mm. For $\vartheta = 0$ one gets
$\delta(\Delta\kappa) = 5\times10^{-5}$ m^{-1}, which corresponds to a bending induced
strain of $\varepsilon_B = 5\times10^{-7} = 0.5$ µstrain on the surface at 10 mm
distance from the neutral line.

In Fig. 7 experimental results are shown for a so-called

Bourdon-tube, commonly used as pressure gauges (16). This is a circular tube of elliptical cross-section, closed at one end. By applying variable pressure inside the tube its surface will be subject to a known displacement (16). The tube always remains circular. This corresponds to a constant bending moment for each radial section, i.e. to a constant change of curvature $\Delta\kappa$. The results obtained from Eq. (17) are in good agreement with the expected constant bending. Neglecting the last term in Eq. (17), containing the change of the sensitivity vector $(\partial e_z/\partial x_\alpha)$, leads to a systematic error for increasing s, as shown in Fig. 7 by the second, thin line. Vertical bars indicate the estimated experimental error. The local deviations of the bending can be easily explained by variable sections of the tube, since the change of curvature depends on the third power of the wall thickness and on thefourth power of the linear dimensions of the cross-section (16).

In Fig. 8 experimental results are shown for a turbine blade under static bending load. The bending along the blade axis (Fig. 8a) shows good agreement with mechanical theory. Note that the spatial resolution of the measurement is about 3 mm and that the local variations of bending stiffness can be detected. The change of curvature $\Delta\kappa_s$ of the blade-sections normal to the axis have been measured in two independent experiments. The results in Fig. 8b show the excellent reproducibility of the measurements within the expected limits. These results are not compared with theory, since the mechanical model used does not include the change of cross-section shape as a reaction on bending load.

CONCLUSIONS

It has been shown that heterodyne holographic interferometry is a powerful tool for fringe interpolation, which allows to determine local derivatives of the surface displacement with sufficient accuracy for strain measurements. Because this technique is very well suited for automatisation and electronic data processing, it might have some advantage over other elegant methods (17), which ask for more experimental skill. Fringe interpolation, determination of the displacement vector and calculation of the derivatives, including the variation of the sensitivity vector, can be made independent of the object geometry, if the same imaging is used for all fringe measurements. Shape and position of the object with respect to the imaging system has to be known to transform afterwards the displacement vector and its derivatives onto the surface. It turns out that with heterodyne holographic interferometry one can determine even with small angles between the different sensitivity vectors, i.e. from a single hologram recording, strain values as small as 1 μstrain. In addition one can determine the local change of curvature, which is useful in case of pure bending and torsion, and might allow to extrapolate strain and stress distributions into the material below the surface (18).

Acknowledgments are mainly due to B. Ineichen, who has performed
most of the experiments and developed the heterodyne holographic
interferometry to a reliable technique, but also to F. M.
Mottier, E. Marom, W. Schumann and M. Dubas for many helpful
and clarifying discussions.

REFERENCES

(1)   R. Dändliker, B. Eliasson, B. Ineichen, F. M. Mottier, The
      Engineering Uses of Coherent Optics, Cambridge U.P.,
      Cambridge, 1976, p. 99-117.
(2)   R. Dändliker, B. Ineichen, F. M. Mottier, High resolution
      hologram interferometry by electronic phase measurement,
      Opt. Commun. 9, 412-416 (1973).
(3)   R. Dändliker, B. Ineichen, F. M. Mottier, International
      Optical Computing Conference, IEEE Inc., New York, 1974,
      p. 69-72.
(4)   W. Schumann, Some aspects of optical techniques for strain
      determination, Exp. Mech. 13, 225-231 (1973).
(5)   A. Duschek, Vorlesungen über höhere Mathematik; Zweiter
      Band, Springer, Wien, 1963, p. 345-361.
(6)   J. E. Sollid, Holographic interferometry applied to mea-
      surements of small static displacement of diffusely
      reflecting surfaces, Appl. Optics, 8, 1587-1595 (1969).
(7)   D. Bijl, R. Jones, A new theory for the practical interpre-
      tation of holographic interference patterns resulting
      from static surface displacements, Optica Acta, 21,
      105-118 (1974).
(8)   S. K. Dhir, J. P. Sikora, An improved method for obtaining
      the general-displacement field from a holographic inter-
      ferogram, Exp. Mech.,12, 323-327 (1972).
(9)   T. Matsumoto, K. Iwata, R. Nagata, Measuring accuracy of
      three-dimensional displacements in holographic inter-
      ferometry, Appl. Optics, 12, 961-967 (1973).
(10) R. Pryputniewicz, K. A. Stetson, Holographic strain analy-
      sis: extension of fringe-vector method to include per-
      spective, Appl. Optics, 15, 725-728 (1976).
(11) V. Fossati Bellani, A. Sona, Measurement of three-dimen-
      sional displacement by scanning a double-exposure holo-
      gram, Appl. Optics, 13, 1337-1341 (1974).
(12) G. S. Ballard, Double exposure holographic interferometry
      with separate reference beams, J. Appl. Phys., 39,
      4846-4848 (1968).
(13) R. Dändliker, E. Marom, F. M. Mottier, Two-reference-beam
      holographic interferometry, J. Opt. Soc. Am., 66, 23-30
      (1976).
(14) R. Crane, New developments in interferometry: V. Inter-
      ference phase measurement, Appl. Optics, 8, 538-542
      (1969).
(15) R. Dändliker, E. Marom, F. M. Mottier, Wavefront sampling
      in holographic interferometry, Opt. Commun., 6, 368-371
      (1972).

(16) W. Wuest, <u>VDI-Forschungsheft 489</u>, VDI-Verlag, Düsseldorf, 1962.

(17) M. Dubas, W. Schumann, On direct measurement of strain and rotation in holographic interferometry using the line of complete localization, <u>Optica Acta</u>, 22, 807-819 (1975).

(18) M. Dubas, W. Schumann, Sur la détermination holographique de l'état de déformation à la surface d'un corps non-transparent, <u>Optica Acta</u>, 21, 547-562 (1974).

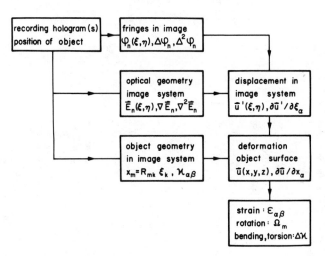

<u>Fig. 1</u> Strategy of evaluation.

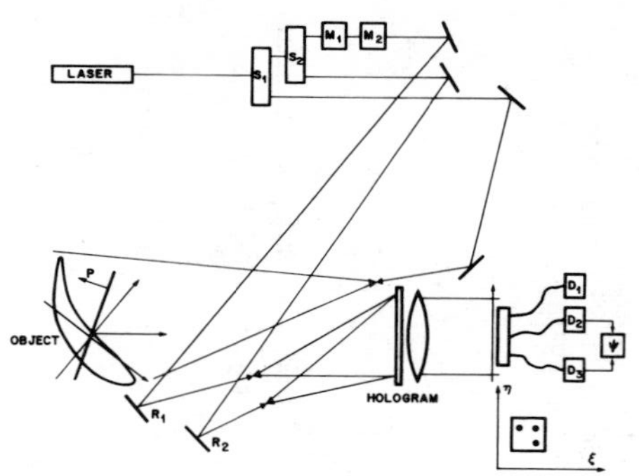

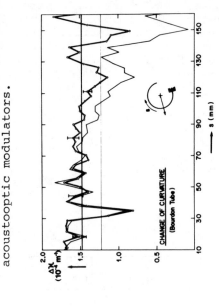

Fig. 4 Setup for heterodyne holographic interferometry. S: beam-splitters, M:
acoustooptic modulators.

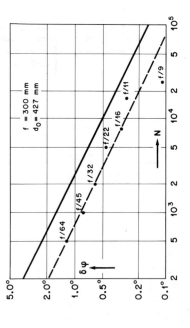

Fig. 7 Change of curvature $\Delta\kappa$ of the Bourdon
tube vs. surface coordinate s.

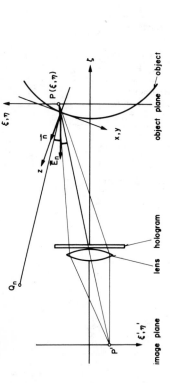

Fig. 3 Geometry of the holographic interferometry with image system (ξ,η,ζ), object
system (x,y,z) and sensitivity vector
$\vec{E}(\xi,\eta)$.

Fig. 5 Phase error $\delta\varphi$ vs. speckle number
N for overlapping cross-reconstruc-
tions.

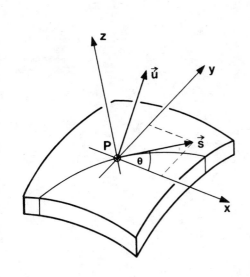

Fig. 2 Object coordinates
(x,y,z) at point P: z
has the direction of the
surface normal.

Fig. 6 Sensitivity $\delta\varepsilon$ for in-
plane strain measurement
vs. angle β between ζ-
axis and sensitivity vec-
tors.

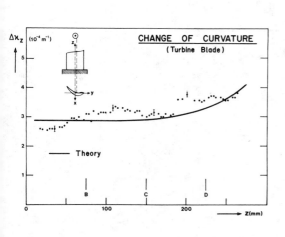

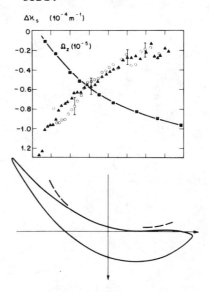

Fig. 8 Change of curvature $\Delta\kappa$ of a turbine blade under static
bending load.

a) Bending along blade axis.

b) Change of curvature Δk_s and
rotation Ω_z of the cross-
section.

TWO REFERENCE BEAM HOLOGRAPHIC INTERFEROMETRY FOR AERODYNAMIC FLOW STUDIES

Jean Surget

Office National d'Etudes et de Recherches Aérospatiales (ONERA), 92320 Châtillon (France)

SUMMARY

Double exposure holographic interferometry with two reference beams makes it possible to obtain, after processing, many different interferometric presentations from a single hologram of an aerodynamic flow field.

A double-reference-beam holographic set-up has first been tested in the laboratory, then applied to interferometric gas flow studies in a wind tunnel. This technique is particularly adapted to transient phenomena studies such as aerodynamic flows in shock tubes.

Following the good results obtained from the first wind tunnel experiments, a more sophisticated dual-reference-beam holographic bench has been built. This apparatus permits both studies of transparent media (such as aerodynamic flows) and scattering object deformations (such as gas turbine blades). An optical scheme is presented as well as interferograms obtained in the laboratory.

INTRODUCTION

The holographic interferometry benches presented in this paper are different from conventional holographic set-ups in that a second reference light source is used (Ref 1 through 4). The position of this additional reference source can be adjusted and its radiation phase can be continously shifted. This improvement introduces more flexibility into double exposure technique . The possibilities of application of interferometry are increased and experiments otherwise impossible with conventional interferometers or single reference beam holographic instruments, are thus feasible. The double exposure double reference source holographic method is particularly suited for obtaining multiple interferograms from a single aerodynamic flow field hologram. It is well adapted to visualization of transient phenomena such as aerodynamic streams in shock tubes.

DESCRIPTION OF THE METHOD

The double exposure technique using a single reference beam holographic apparatus is well known. In the case of aerodynamic flow studies, two light waves Σ_E and Σ_E' are successively recorded on the same photographic plate H, using the same reference light wave Σ_R to create the necessary interference to make the holograms. Σ_E and Σ_E' are the test beam light waves transmitted through the wind tunnel test section without and with the stream.

183

After processing, when H is illuminated by the Σ_R reference light wave emitted by the common reference source, Σ_E and Σ_E' are simultaneously reconstructed and their resulting interference pattern is observed. However, the two waves Σ_E and Σ_E' are inseparable, being both reconstructed by the same reference wave : the particular interferogram aerodynamic flow field presentation obtained is inflexible.

On the contrary, a double reference holographic set-up allows separate wave recording and reconstruction by using the proper reference source such as is schematically shown in Fig. 1.

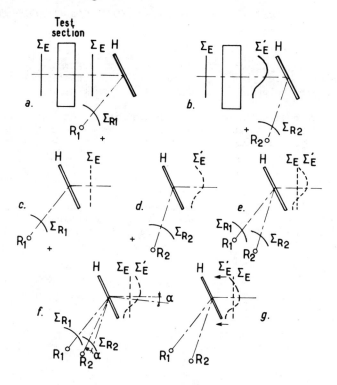

Fig. 1. Double exposure technique with dual reference
        source holographic system.

        O Reference source ON

        +   "   "   OFF

        Figure 1a : First exposure of photographic plate H. The wind tunnel is not running and the test section is illuminated by the test light wave Σ_E. The corresponding information is stored on H by means of the reference light wave Σ_{R1} given by the first reference source R_1 (a shutter is placed before the second reference source R_2.

Figure 1b : Second exposure with aerodynamic flow. The test light wave Σ_E become Σ_E' across the test section. The corresponding information is stored on the same photographic plate H, using reference light wave Σ_{R2} from the second reference source R_2 (the shutter is now placed before R_1).

After processing, H is replaced in the holographic bench slide holder. When H is illuminated by Σ_{R1} only, the undisturbed test light wave Σ_E is reconstructed Fig. 1c. Using Σ_{R2} alone, the test light wave containing flow field information Σ_E' is reconstructed Fig. 1d. Simultaneous illumination of H by Σ_{R1} and Σ_{R2} allows both Σ_E and Σ_E' reconstructions Fig. 1e. When the photographic plate H is replaced in its previous position, Σ_E and Σ_E' are combined, and the aerodynamic flow field zero order interferogram appears as with a single reference holographic apparatus. However, many different interferometric presentations can be obtained by means of a displacement or a phase modification of one of the reference light sources, such as R_2 for example :

a) A small displacement of R_2 (in its plane) tilts the reconstructed test light wave Σ_E' Fig. 1f. A finite fringe interferogram is observed (horizontal fringes for vertical displacement and inversely). Fringe spacing changes in inverse ratio to the displacement.

b) A phase shift of R_2 translates the whole fringe pattern providing other data points in the reconstructed flow field interferogram (Ref 5).

Experiments

Holographic bench for transparent media interferometric studies

A double-reference-beam holographic set-up has been tested first in the laboratory (Ref 5), then applied to gas flow visualization in the S8 wind tunnel of the fluid mechanics laboratory at ONERA Chalais-Meudon. For this experiment, several components are added to the single reference beam holographic bench used for transonic flow studies (Ref 6). The modified bench optical arrangement is schematically shown in Fig. 2. Adjustments of the additional reference source R_2 can be obtained by means of :
1) a displacement of L_2, to obtain equal displacement of R_2 and consequently tilting of Σ_{R2} such as shown in Fig. 1f;

2) a translation of p in its plane to shift the phase of R_2, as shown in Fig. 1g.

The double reference source holographic bench is shown Fig. 3. Optical pathlengths of the test beam and the reference beams are approximately balanced. The angular spacing between the two reference beams and the test beam is sufficient to avoid overlapping of wanted and unwanted reconstructions; each reference source reconstructs 4 images, giving a total of 8 images (Ref 4,5). The light source, not shown in Fig. 3, is a helium-neon laser of 15 mW power. With Agfa-Gevaert 10E75 photographic plates, the exposure time is 1/250 second for a rectangular optical field size of 100 x 140 mm. When a ground glass is placed between the test section and the photographic plate, the exposure duration becomes 1/30 second. For the sake of clarity, only experiments without ground glass are described.

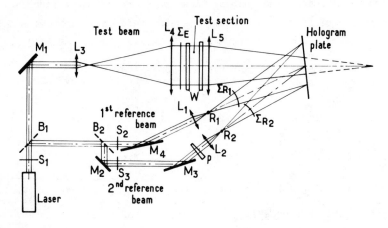

Fig. 2 Two reference source holographic arrangement

W = Tunnel windows
B_1, B_2 = Beam splitters
M_1, M_2, M_3, M_4 = Plan mirrors
L_1, L_2, L_3 = Beam expander lenses
L_4, L_5 = Collimated lenses
R_1 = First reference light source
R_2 = Additional reference light source

p = Prismatic glass plate for optical
        pathlength adjustment
S_1, S_2, S_3 = Shutters
Σ_E = Test light wavefront
Σ_{R_1}, Σ_{R_2} = Reference light wavefronts

Interactions of a shock with turbulent boundary layers were studied in a wind tunnel at low supersonic Mach number with a bump on a wall Fig. 4. Interferograms obtained from a single aerodynamic flow field hologram by means of the method described above are shown in Fig. 5 and 6. Reconstructions are made on the same bench as for hologram recording.

Figure 5a is the reconstructed image of the operating wind tunnel section obtained with reference source R_2 only. This image is related to the air stream shadowgraph.

The zero-order interferogram obtained by using simultaneously reference sources R_1 and R_2 is shown on Fig. 5b. Double exposure technique applied by means of a single reference source holographic set-up produces exactly the same result, but nothing else, while the double reference beam bench makes it possible to obtain other interferometric presentations such as vertical finite-fringe interferograms Fig. 5c to 5e or horizontal finite-fringe interferograms Fig. 5f through 5i.

Phase changes of R_2 by means of displacing the prismatic glass plate p(Fig. 2) gives the presentations shown in Fig. 6.

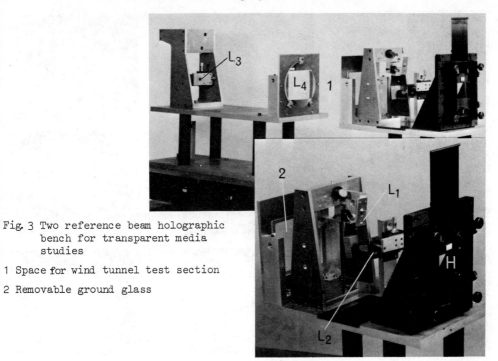

Fig. 3 Two reference beam holographic
bench for transparent media
studies

1 Space for wind tunnel test section

2 Removable ground glass

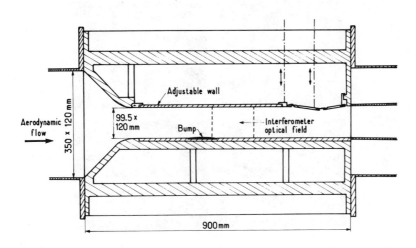

Fig. 4 Wind tunnel test section

J. Surget

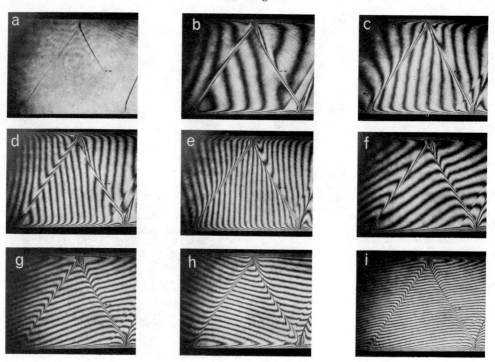

Fig. 5 Additional reference source tilting effect

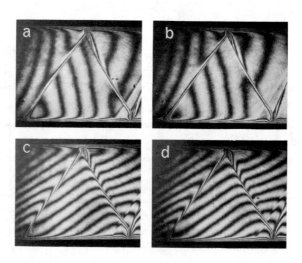

Fig. 6 Additional reference source shifting effect

Figure 6a shows the zero-order interferogram similar to that of Fig. 5b.

In Fig. 6b, a half light wave phase shift of R_2 translates the whole fringe pattern.

In Fig. 6c and 6d, the same effect occurs from finite-fringe interferogram.

Holographic Bench for Interferometric Studies of both Transparent Medias and Scattering Objet

The optical scheme of this apparatus is shown Fig. 7 for double-glass transparent media interferometric examination , and Fig. 7b for scattering object deformation studies.

The improvements of this apparatus give the following advantages :

- The processed hologram plate alignement with respect to the reference sources is facilitated by the use of parallel light to form the reference beams.

- Both reconstructed image storage and pattern fringes analysis are simplified, because a real image is formed in the plane P' by objective O_2.

Indeed it is easy to put in this image plane P' :

- a photographic plate,
- or a movable photoelectric cell for punctual light analysis,
- or two cells for heterodyne detection subject to introduce a small frequency shift between the optical frequencies of the two reference beams (Ref 3).

On the ground-plan of Fig.7, the two reference beams appear symmetric with respect to the test beam (angular value 14°). In fact, the three beams are not coplanar in order to avoid undesired overlapping of reconstructions. The reference beams are in a plane inclined at 8° to the horizontal plane containing the test beam.

The optical field diameter is 100 mm. The light source is the same 15 mW output helium-neon laser used for the previously presented bench. For the same photographic plates, the exposure time is 1/60 second in the case of this apparatus.

At the present time, experiments are performed in two ways : transparent media examination and scattering object deformation studies.

Figure 8 shows several interferograms obtained, by means of the described above method, from a single hologram of air disturbance created by a hot metallic cylinder. The original zero-order interferogram shown Fig. 8a can be modified to obtain vertical finite-fringe interferograms Fig. 8b through 8d, or horizontal finite-fringe interferograms Fig. 8e through 8g or inclined finite-fringe interferograms Fig. 8h and 8i by means of reference beams tilting. This is obtained by small displacement of the collimating objective O_6 in its plane (Fig. 7). The fringe pattern of the interferogram Fig. 8j can be translated Fig. 8k by shifting the second reference beam, by means of the prismatic glass plate p adjustment (Fig. 7).

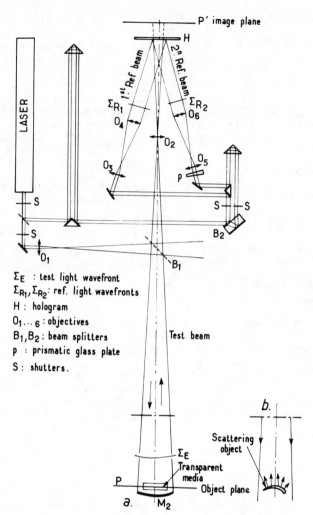

Fig. 7 Holographic bench for interferometric studies of both transparent medias
and scattering objects

A sample recording of scattering object recording is presented Fig. 9. It shows
a double exposure holographic interferogram of a deformed gas turbine blade
with a beaded reflective surface. For this experiment , the blade is substituted for the spherical mirror M_2 as shown in Fig. 7b.

The first exposure is made with the blade at rest and the second exposure with
the blade strained, respectively with use of the first and the second reference
beam.

When an undesired whole blade displacement is induced between the two exposures
by the strain, the reconstructed fringe pattern appears outside the reconstructed blade image.

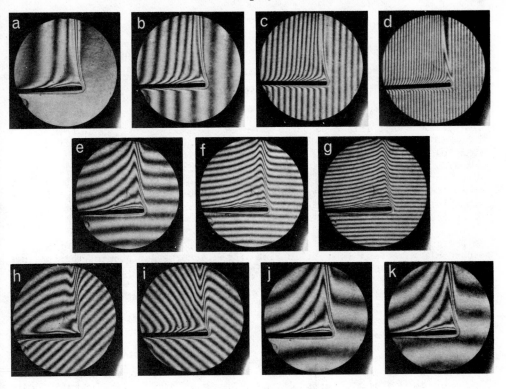

Fig. 8 Air disturbance near a hot metallic cylinder

Fig. 9 Strain pattern in gas turbine blade

In that case, an additional reference beam angular adjustment moves the corresponding reconstructed image into coincidence with the other, and proper location of the fringe pattern is restored. Moreover, additional phase shifting of the reference beam translates the fringe pattern on the blade surface.

CONCLUSION

Two double reference beam holographic benches have been constructed and tested:
- the first (for transparent media studies) in the aerodynamic laboratory,
- the second (both for transparent media and scattering object studies) in the optics laboratory only.

Clear interference fringe patterns were obtained, mainly for gas flow visualization. The versatility provided by the dual beam holographic system has been exploited to obtain many different interferometric presentations from the processed hologram of a single aerodynamic flow field. Infinite and finite-fringe interferograms were readily realized, with suitable fringe spacing and inclination to observe avantageously the aerodynamic phenomenon. This result is obtained by displacement and phase adjustment of the additional reference source.

The present work was done to assess the feasibility of two-reference-beam holographic interferometry application in the adverse environment surrounding a wind tunnel. However, experiments were improved in a small wind tunnel where vibration is no problem, allowing continuous low power laser and a cheap optical set-up.

REFERENCES

1   Tadao Tsuruta, Novio Shiotake and Yoshinobu Itoh, Hologram interferometry using two reference beams, Japanese J. Appl. Phys. 7, 1092 (Sept. 1968).

2   G.S. Ballard, Double-exposure holographic interferometry with separate reference beams, J. Appl. Phys. 39, 4846 (1968).

3   R. Dändliker, B. Ineichen and F.M. Mottier, High resolution hologram interferometry by electronic phase measurement, Opt. Commun. 9, 412 (1973).

4   R. Dändliker, E. Marom and F.M. Mottier, Two-reference-beam holographic interferometry, J. Opt. Soc. Am. 66, 23 (1976).

5   J. Surget, Schéma d'holographie à deux sources de référence, Nouv. Rev. Opt. 5, 201 (1974).

6   J. Delery and J. Surget, Interférométrie holographique quantitative en écoulement transsonique, Rech. Aérosp., to be published.

NON-LINEAR EFFECTS IN
HOLOGRAPHIC INTERFEROMETRY

J. Katz,* E. Marom* and R. Dändliker**

*School of Engineering, Tel-Aviv University, Israel
**Brown Boveri Research Center, Baden, Switzerland

ABSTRACT

The analysis of non-linear effects in the holographic process and their influence on the accuracy of double exposure holographic interferometric data is
presented. It is shown that these effects are significant in high resolution
holographic interferometry by electronic phase measurement, but almost negligible otherwise.

INTRODUCTION

Studies published on the subject of non-linear effects in holographic imaging
predict distortions and loss of contrast of the image, as well as the formation of false images (Ref 1-6). Holographic interferometry, which is employed
extensively in the area of non-destructive testing, is mainly concerned with
fringe patterns superimposed upon the image and their interpretation (Ref 7-8).
It is the intent of this article to evaluate non-linear effects in holographic
interferometry, to determine to what extent they affect the accuracy of the
measurements and also to point out the precautions needed for circumventing
them.

NON-LINEAR CHARACTERISTIC OF PHOTOGRAPHIC RECORDINGS

A common part of all holographic set-ups is the photographic plate, which is
characterized by the t_a-E curve (or, alternatively, by the t_a-I curve),
where t_a is the amplitude transmittance of the plate and E is the exposure to which the hologram was exposed.

Practically, the reference waves illuminations establish the bias point E_0
along the t_a-E curve, to which t_0 , the average amplitude transmittance,
corresponds. If the range of exposure fluctuations around E_0 is kept
within the linear portion of the t_a-E curve, the following linear relation
can be obtained.

$$t_a = t_o - \beta_1(I - I_o) \tag{1}$$

where I is the illumination intensity, T the exposure time, and
$I_o = E_0/T$. To be in the linear region of the t_a-E curve, one should limit
operation to small modulation indices, thus yielding poor efficiencies. Hence
in practical applications non-linearities cannot be avoided.

Some possible representations of the t_a-E curve exist. In this article, the first few terms of the polynomial approximation are used, i.e.

$$t_a = t_o + \sum_{n=1}^{3} \beta_n (I - I_o)^n \tag{2}$$

SPECKLE STATISTICS FOR DIFFUSE OBJECTS

In almost all situations encountered in holographic interferometry, the object wave illuminates the hologram diffusely. A diffuse object scatters light over a wide angular range. The random scattering surface combines with the coherent illumination in forming the speckle phenomenon (Ref 9). Because scatterers in a diffuse object are numerous and randomly distributed, statistical tools are employed.

In this paper the approach described in Ref 10 is adopted. The wave leaving the diffuse surface is given by

$$W(\xi,\eta) = V(\xi,\eta)\rho(\xi,\eta) \tag{3}$$

where $V(\xi,\eta)$ describes the macroscopic shape of the object and $\rho(\xi,\eta)$ is the complex statistical correlation function affected by the surface roughness. In most instances ρ is described by its autocorrelation function which is very narrow compared to the system impulse response and thus can be represented by the Dirac delta function, i.e., the statistical properties of ρ are defined for the ensemble averages as

$$< \rho(\underline{\xi}_1) \; \rho^*(\underline{\xi}_2) > = \delta(\underline{\xi}_1 - \underline{\xi}_2) \tag{4a}$$

$$< \rho(\underline{\xi}_1)\rho^*(\underline{\xi}_2)\rho(\underline{\xi}_3)\rho^*(\underline{\xi}_4)> = \delta(\underline{\xi}_1 - \underline{\xi}_2)\delta(\underline{\xi}_3 - \underline{\xi}_4)$$
$$+ \delta(\underline{\xi}_1 - \underline{\xi}_4)\delta(\underline{\xi}_2 - \underline{\xi}_3) \tag{4b}$$

with similar relations for the higher order averages.

We can further assume that the speckle distribution ρ is identical for both the object's initial state (V_1) and its final state (V_2), since the latter is just a slightly distorted version of the former and the amount of the displacement is small enough not to change the microstructure (if the displacements are larger, holographic interferometry is much too sensitive to be used).

NON-LINEAR EFFECTS IN HOMODYNE INTERFEROMETRY

A two beam holographic interferometric set-up in which no frequency shift between the two reconstructing beams is introduced will be further referred to as "Homodyne Interferometry."

The set-up configuration and terminology are shown in Fig. 1. Waves in the object plane are denoted by W, in the hologram plane by O and in the image plane by U. The coordinates are (ξ,η), (μ,ν) and (x,y), respectively. The imaging lens is placed behind the hologram plane in the reconstruction stage. Same results apply also if the lens is not in the hologram plane, as long as it is the smallest aperture in the system.

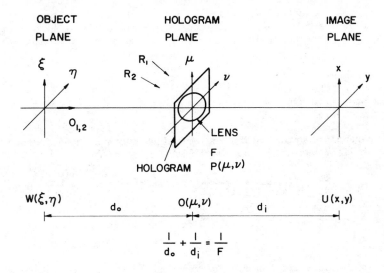

Fig. 1. Set-up configuration used for calculations of wave
propagation in an optical system. (Fresnel-type hologram).

Upon recording a double reference beam double exposure hologram (Ref. 11),
$I - I_o$ is given by

$$I-I_o = |0_1|^2 + |0_2|^2 + R_1 0_1^* + R_1^* 0_1 + R_2 0_2^* + R_2^* 0_2 \qquad (5)$$

where R_1 and R_2 are the reference beams used in the first and the second
exposures, respectively, and 0_1 and 0_2 are the waves corresponding to the
object's initial and final states, respectively. (The single reference beam
case can be derived by substituting $R_1 = R_2 = R$ in the relevant expressions).

Effects due to Reconstruction with One of the Reference Beams

From Eqs. (1) and (5) it is realized that using the linear approximation, only
0_1 is reconstructed when illuminating the hologram with only one of the
reference beams, i.e. R_1. In order to obtain an interference pattern, both
R_1 and R_2 should thus be used. However, when higher powers of Eq. (5) are
taken in consideration, one obtains that the total wave distribution propagating in the direction of the primary image when reconstructing with only one
of the original reference beams (R_1) is given at the hologram plane by

$$0_{TOTAL}(1) = \delta_0 0_1 + \delta_1 0_1 |0_1|^2 + \delta_2 0_1 |0_2|^2 + \delta_3 0_1 |0_1|^2 |0_2|^2 + \delta_4 0_1 |0_1|^4 +$$

$$+ \delta_5 0_1 |0_2|^4 + \ldots \qquad (6)$$

where δ_j are constants depending upon β_n and R_n [see Eqs. (2) and (5)].

Generally $\delta_o \gg \delta_j$ $(j \neq o)$. It is realized that this wave distribution [Eq.(6)] contains information about O_2, which is not predicted by the linear assumption. Although the phase of O_2 is not directly displayed, it will manifest its existence in the image plane, when objects under investigation are of a diffused nature.

In order to demonstrate this non-linear effect, it is assumed for a moment that only two of the terms of Eq. (6) exist, namely $\delta_o O_1$ and $\delta_2 O_1 |O_2|^2$. Using this simplification one gains more insight into the processes without reducing the generality of the results.

Using the parabolic approximation and omitting constant phase factors, the primary wave distribution in the image plane is given by

$$U_1 = \frac{1}{\lambda d_i} \iint \left\{ \frac{1}{\lambda d_o} \iint v_1(\xi,\eta)\rho(\xi,\eta)e^{i \frac{K}{2d_o}[(\mu-\xi)^2+(\nu-\eta)^2]} d\xi d\eta \right\}$$
$$\times P(\mu,\nu)e^{-i \frac{K}{2F}(\mu^2+\nu^2)} e^{i \frac{K}{2d_i}[(x-\mu)^2+(y-\nu)^2]} d\mu d\nu \quad (7)$$

where P is the aperture function of the lens (d_o, d_i and F are shown in Fig. 1).

Assuming that the system point spread function is given by

$$\tilde{h}(x,y;\xi,\eta) = \frac{1}{\lambda^2 d_i^2} \iint P(\mu,\nu)e^{-iK\{\mu[\frac{\xi}{d_o}+\frac{x}{d_i}]+\nu[\frac{\eta}{d_o}+\frac{y}{d_i}]\}} d\mu d\nu \quad (8)$$

and applying the imaging condition $1/d_o + 1/d_i = 1/F$ one obtains

$$U_1 = Me^{i \frac{K}{2d_i}(x^2+y^2)} \iint v_1(\xi,\eta)\rho(\xi,\eta)e^{iK \frac{\xi^2+\eta^2}{2d_o}} \tilde{h}(x,y;\xi,\eta)d\xi d\eta \quad (9)$$

where M is the magnification ratio ($M = d_i/d_o$).

Following the same steps, the wave $O_1|O_2|^2$ yields the following wave in the image plane

$$U_1|U_2|^2 = M\frac{e^{i \frac{K}{2d_i}(x^2+y^2)}}{\lambda^2 d_o^2} \iiiiint v_1(\xi_1,\eta_1)v_2(\xi_2,\eta_2)v_2^*(\xi_3,\eta_3)$$
$$\times \rho(\xi_1,\eta_1)\rho(\xi_2,\eta_2)\rho^*(\xi_3,\eta_3)e^{i \frac{K}{2d_o}[(\xi_1^2+\xi_2^2-\xi_3^2)+(\eta_1^2+\eta_2^2-\eta_3^2)]}$$
$$\times \tilde{h}(x,y;\xi_1+\xi_2-\xi_3,\eta_1+\eta_2-\eta_3)d\xi_1 d\eta_1 d\xi_2 d\eta_2 d\xi_3 d\eta_3 \quad (10)$$

Since the terms described by Eqs. (9) and (10) are displayed simultaneously and are mutually coherent, their intensity is given by:

$$I(x,y) = <|\delta_o U_1 + \delta_2 U_1 |U_2|^2|^2> \approx \delta_o^2 \{<|U_1|^2> + \frac{2\delta_2}{\delta_o} Re<U_1^*;U_1|U_2|^2>\} \quad (11)$$

Using Eq. (4) after substituting Eqs. (9) and (10) into Eq. (11), one readily obtains

$$I(x,y) = \delta_o^2 \{ \overline{|V_1|^2} + \frac{2\delta_2}{\delta_o} \, Re[I_2 \overline{|V_1|^2} + I_{12} \overline{V_1^* V_2}] \} \quad (12)$$

where

$$\overline{V_m^* V_n} \equiv \overline{V_m^*(x,y) V_n(x,y)} = M^2 \iint_{Object} V_m^*(\xi,\eta) V_n(\xi,\eta) |\tilde{h}(x,y;\xi,\eta)|^2 d\xi d\eta$$

$$n,m = 1,2 \quad (13)$$

denotes the "incoherent imaging operation" $[|\tilde{h}(x,y;\xi,\eta)|^2$ being the optical transfer function – OTF – of the optical system] and

$$I_{12} = I_{21}^* = \frac{1}{\lambda^2 d_o^2} \iint V_1(\xi,\eta) V_2^*(\xi,\eta) d\xi d\eta \triangleq |I_{12}| e^{i\theta} \quad (14a)$$

$$I_m \triangleq I_{mm} = \frac{1}{\lambda^2 d_o^2} \iint_{Object} |V_m(\xi,\eta)|^2 d\xi d\eta \qquad m = 1,2 \quad (14b)$$

are the second order mutual intensity functions.

If it is further assumed that $\tilde{h}$ is much narrower than the fringe pattern (otherwise the fringe pattern could not have been satisfactorily exhibited), one can approximate $\tilde{h}$ by

$$\tilde{h}(x,y;\xi,\eta) = \frac{1}{M^2} \delta(\xi + \frac{x}{M}, \eta + \frac{y}{M}) \quad (15)$$

As mentioned above, holographic interferometry can be employed only when small displacements are encountered, so the following relation exists between the final and the initial states of the recorded object:

$$V_2(\xi,\eta) = V_1(\xi,\eta) e^{i\psi(\xi,\eta)} \quad , \quad |V_2|^2 = |V_1|^2 \triangleq |V|^2 \quad (16)$$

Substitution of Eqs. (15) and (16) into Eq. (12) yields

$$I(x,y) = \delta_o^2 \frac{|V(-\frac{x}{M},-\frac{y}{M})|^2}{M^2} \{1+2 \frac{\delta_2}{\delta_o} [I_2 + |I_{12}| \cos(\psi[-\frac{x}{M},-\frac{y}{M}]+\theta)]\}$$

$$(17)$$

i.e., if $|I_{12}|$ is not null, a spurious fringe pattern with a contrast of about $2\delta_2|I_{12}|/\delta_o$ will be formed, while under the linear approximation no fringe pattern would have been observed!

If now the contributions of all the dominant terms (i.e. mainly terms of the form $<0_1^*$; (a non-linear term)$>$) in Eq. (6) to the primary image intensity distribution are taken into account, one obtains the general expression

$$I(x,y) = \frac{\left|V(-\frac{x}{M}, -\frac{y}{M})\right|^2}{M^2} \ A_1\{1 + m_1 \ \cos[\psi(-\frac{x}{M}, -\frac{y}{M}) + \theta]\} \qquad (18)$$

where A_1 and m_1 are nearly constant over the image plane. For small non-linearities $m_1 = 2\delta_2|I_{12}|/\delta_o$ as obtained above (Eq. 17). Under the linear approximation one obtains, as expected, that $I(x,y)$ equals $\delta_o^2|V(-\frac{x}{M}, -\frac{y}{M})|^2/M^2$, i.e. no fringe pattern is displayed.

The crosstalk fringe pattern m_1 is proportional to $|I_{12}|$. If the portion of the recorded scene that does not change between the two exposures is vanishingly small, $|I_{12}|$ tends to zero since the integral over the phase $\psi(\xi,\eta)$ [performed in Eq. (14a)] averages to zero for many cycles, i.e. many fringes. Thus most of the contributions to $|I_{12}|$ come from the unchanged parts of the recorded scene. This effect is demonstrated in Fig. 2. Both in Fig. 2a and in Fig. 2b the primary image obtained upon reconstruction with only one of the reference beams is shown. In Fig. 2a a large portion of the object does not change between exposures, thus a crosstalk fringe pattern is exhibited. In Fig. 2b all the recorded scene was deformed between exposures and thus no fringe pattern is observed.

When $|I_{12}|$ is significant (due to the contributions of the unchanged parts of the recorded scene) so that the crosstalk contrast is observable, θ decreases and there is no evident shift in the fringe pattern as described in Eq. (18). This is shown in Fig. 3. In Fig. 3a and Fig. 3b, the images obtained upon reconstruction with two and one reference beams,respectively, are displayed.

As an example, the relation between m_1, the crosstalk contrast and the modulation index I_o/I_R is shown for several values of parameter K (K is the portion of the illumination coming from the unchanged parts of the recording scene. $K \approx |I_{12}|/I_{1,2}$). In Figs. 4 and 5 calculated values and measured results are exhibited. The photographic plates used were Agfa Scientia 10E75 and the bias point chosen was 20 erg/cm^2 .

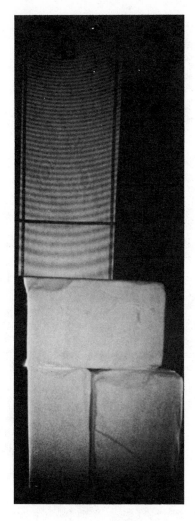

a   b

Fig. 2.   Image obtained upon reconstruction with one reference beam.
a) With unchanged portions in the object. b) Without
unchanged portions in the object. (The contrast in Fig.
2a was enhanced by the photographic process).

(Results of experiments performed by B. Ineichen at
Brown Boveni Research Center)

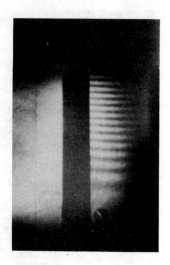

a   b

Fig. 3.   Reconstruction of double reference beam double
          exposure hologram. a) Reconstruction with both
          reference beams. b) Reconstruction with one
          reference beam.

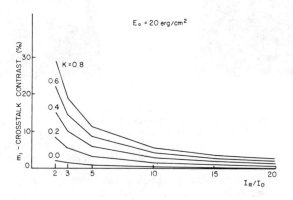

Fig. 4. Calculated values of crosstalk contrast

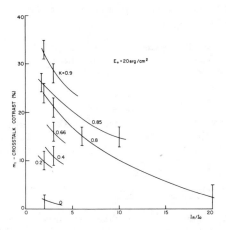

Fig. 5. Measured values of crosstalk contrast

Effects due to Reconstruction with Two Reference Beams

Reconstructing a double reference beam double exposure hologram with the two original reference beams leads [in a similar way to the derivation of Eq.(6)] to the following wave at the image plane:

$$U_{TOTAL} = \delta_o U_1 + \delta_1 U_1 |U_1|^2 + \delta_2 U_1 |U_2|^2 + \cdots$$

$$+ \varepsilon_o U_2 + \varepsilon_1 U_2 |U_2|^2 + \varepsilon_2 U_2 |U_1|^2 + \cdots \qquad (19)$$

δ_j and ε_j are constants depending upon β_n and R_n [See Eqs. (1) and (5)]. If $|R_1|^2 = |R_2|^2$, $\delta_j = \varepsilon_j$.

After performing similar calculations to the ones shown in the previous section, one obtains that the intensity distribution in the image plane is given by

$$I(x,y) = 2 \cdot \frac{\left| V(-\tfrac{x}{M}, -\tfrac{y}{M}) \right|^2}{M^2} A_2 \{1 + m_2 \cos[\psi(-\tfrac{x}{M}, -\tfrac{y}{M}) + \psi_e]\} \qquad (20)$$

(In the linear approximation, if $|R_1|^2 = |R_2|^2$, then $A_2 = \delta_o^2$ and $m_2 = 1$). Calculations show that the non-linear effects do not decrease the contrast considerably, thus $m_2 \lesssim 1$. Unlike the single reference beam reconstruction (Eq. 18), the phase error ψ_e does not equal θ . If this had been the case, considerable shifts of the fringe patterns would have been observed. ψ_e is found to be proportional to $|I_{12}| \sin\theta$ and is also a function of I_1, I_2, I_{12}, δ_j and ε_j. Under all practical situations, ψ_e tends to zero; if no part of the recorded scene remains unchanged between exposures, $|I_{12}|$ tends to zero. On the other hand, if there exists a constant unchanged illumination, θ hence also $\sin\theta$ - tends to zero. Calculations made for Scienta 10E75 plates of AGFA, indicate phase errors up to a few degrees at most, hence not detectable by ordinary means. These calculations were checked by experiments in which no shift in the fringe pattern was detected even for recordings taken at highly non-linear conditions.

It is interesting to note that same results apply (though with slightly different coefficients) also for the case of the conventional double exposure holographic interferometry (i.e., $R_1 = R_2 \triangleq R$).

Considering the calculations carried out in the previous section, one realizes that non-linear effects produce no crosstalk fringe patterns and no phase errors in the plane where they are generated. Thus a method that can virtually eliminate all phase errors (thus being significant for the heterodyne detection scheme discussed in the next section) is accomplished by using focused image holographic arrangements.

NON-LINEAR EFFECTS IN HETERODYNE DETECTION

The method of double reference beam double exposure holography with heterodyne detection (reconstruction) as in Ref. 12, is a method of high resolution capable of detecting phase differences of a few tenths of a degree. To obtain this, a frequency shift is introduced between the two reference beams in the reconstructing stage, the interferometric information thus appearing as the phase of this frequency shift at the detector output (Fig.6).

When utilizing this technique, no visible fringe pattern can be observed due to the high temporal frequency difference between the two images, thus smearing the fringe pattern.

Heterodyne Detection – Linear Model

The wave leaving the hologram upon reconstruction with frequency shifted reference beams is given by

$$O_{HET.}^{(LINEAR)} = \{R_1 e^{i(\omega_o + \Omega)t} + R_2 e^{i(\omega_o - \Omega)t}\} t_a \qquad (21)$$

where t_a is described in the linear model by Eq. (1), R_1 and R_2 are the reference beams, ω_o is the angular frequency of the recording reference beams and $\pm\Omega$ is the frequency shift of the reconstructing reference beams.

The intensity distribution of the primary image as measured in the image plane is given by:

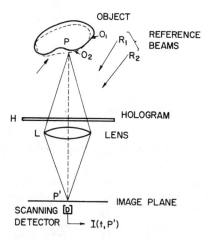

Fig. 6. Double reference beam double exposure set-up (after Ref. 12).

$$I_{HET}(x,y,t) = < U_{HET} \cdot U_{HET}^* > =$$
$$\text{(LINEAR)} \quad \text{(LINEAR)} \quad \text{(LINEAR)}$$

$$= \beta_1 \frac{|V(-\frac{x}{M}, -\frac{y}{M})|^2}{M^2}$$

$$X \quad \{ |R_1|^4 + |R_2|^4 + 2|R_1|^2|R_2|^2 \cos[2\Omega t - \psi(-\frac{x}{M}, -\frac{y}{M})] \}$$

$$+ \text{ (terms not in the direction of the primary image).} \quad (22)$$

Only the AC component which contains the needed phase information is of interest.

Heterodyne Detection – Including Non-linear Effects

In the heterodyne detection scheme, the phase errors introduced by non-linear effects are significant. The fact that aggravates the situation is that, unlike the homodyne detection method, in the heterodyne detection scheme the phase error is inherently a function of the coordinates.

Generalization of Eq. (21), to account also for non-linear terms, leads to the following expression for the wave distribution in the primary image plane:

$$U_{HET} = (\sum_j \pi_j f_j) e^{i(\omega_o + \Omega)t} + (\sum_\ell \pi_\ell f_\ell) e^{i(\omega_o - \Omega)t} \quad (23)$$

where π_j and π_ℓ are constants depending upon the theoretical model of the generation of the non-linear terms [i.e., δ_m and ε_m in Eqs. (6) and (19)] and f_j and f_ℓ are linear and non-linear terms like U_1, U_2, $U_1|U_2|^2$, $U_1|U_1|^2|U_2|^2 \ldots$ etc.

The intensity pattern measured by the detector is given by

$$I_{HET}(x,y,t) = <U_{HET} \cdot U_{HET}^*> = <|(\sum_j \pi_j f_j) e^{i(\omega_o + \Omega)t}|^2>$$

$$+ < | \sum_\ell \pi_\ell f_\ell) e^{i(\omega_o - \Omega)t}|^2 > +$$

$$+ 2Re\{<[(\sum_j \pi_j f_j) e^{i(\omega_o + \Omega)t}] \cdot [(\sum_\ell \pi_\ell^* f_\ell^*) e^{-i(\omega_o - \Omega)t}]> \} \quad (24)$$

The first and second terms in Eq. (24) produce a DC component not affecting the measurement. The last term is the component of interest which contains the wanted phase information and equals

$$I_{AC}(x,y,t) = 2Re\{e^{i2\Omega t} <(\sum_j \pi_j f_j)(\sum_\ell \pi_\ell^* f_\ell^*)> \} \quad (25)$$

Substituting the appropriate expressions for the various terms we obtain that this component of the detector current is proportional to

$$I_{AC}(x,y,t) \; \alpha \; \cos\{2 \; \Omega t - \tan^{-1}[\; \frac{\sin[\psi(-\frac{x}{M},-\frac{y}{M})] + a_1}{\cos[\psi(-\frac{x}{M},-\frac{y}{M})] + a_2} \;] \; \} \qquad (26)$$

where a_1 and a_2 are functions much smaller than unity, depending on $I_1, I_2, I_{12}, \delta_j$ and ε_j, as well as of ψ itself. When no non-linearities are assumed, $a_1 = a_2 = 0$ and we obtain the linear expression [Eq. (22)].

In Fig. 7 the dependence of the maximum phase error versus the modulation index for several values of parameter K $(K \approx |I_{12}|/I)$ is displayed

Two main conclusions may be drawn.
a) In order to eliminate the phase error, it is desirable to work at a small index of modulation and to decrease the unchanged part of the recorded scene to a minimum.
b) Total displacement measurements taken at average conditions (i.e. $I_R/I_O \approx 5$, $K \approx 0.2$) are not severely affected by the existence of the phase error, since it is not cumulative. However, when making local difference measurements, i.e., first and second order differentiations (which can be accomplished, for example, with a simultaneous scan of several detectors), the phase error can introduce irregular and more emphasized disturbances.

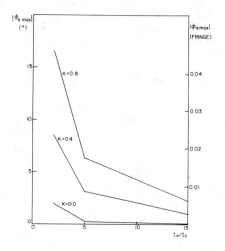

Fig. 7. Maximum phase error in heterodyne detection method.

CONCLUSIONS

Non-linear effects introduce two phenomena in holographic interferometry. The first one is the formation of crosstalk fringe patterns. The second one, more relevant to conventional holography, is the introduction of phase errors in the obtained fringe patterns. In conventional holography, however, these phase errors are insignificant and need not be taken into consideration.

On the other hand, these phase errors are highly significant in double reference beam double exposure holographic interferometry with electronic heterodyne detection. In this case one has to take several precautions, i.e.

to work at large ratios of reference beam intensity to object beam intensity and to decrease to minimum the illumination on the parts of the recorded scene that do not change between exposures. Another possible method to avoid these phase errors is by using the focused image holography set-up. This latter method has, however, some inherent practical difficulties.

REFERENCES

1. A. Kozma, Photographic recording of spatially modulated coherent light, J. Opt. Soc. Am. 56, 428 (1966).

2. A.A. Friesem and J.S. Zelenka, Effects of film nonlinearities in holography, Appl. Opt. 6, 1755 (1967).

3. J.W. Goodman and G.R.Knight, Effects of film nonlinearities on wavefront-reconstruction images of diffuse object, J. Opt. Soc. Am. 58, 1276 (1968).

4. O. Bryngdahl and A. Lohmann, Nonlinear effects in holography, J.Opt. Soc. Am. 58, 1325 (1968).

5. A. Kozma, G.W. Jull and K.O. Hill, An analytical and experimental study of nonlinearities in hologram recording, Appl. Opt. 9, 721 (1970).

6. C.H.F Velzel, Image contrast and efficiency of non-linearly recorded holograms of diffusely reflecting objects, Opt. Acta, 20, 585 (1973).

7. R. J. Collier, C.B. Burkhardt and L.H. Lin, "Optical Holography " Ch. 15, Academic Press, 1971.

8. R.K. Erf, "Holographic Nondestructive Testing," Ch. 4, 7, 8, Academic Press, 1974.

9. Ref. 7, pp. 345-351.

10. R. Dändliker and F.M. Mottier, Determination of coherence length from speckle contrast on a rough surface, J. Appl. Math. Phys. 22, 369 (1971).

11. R. Dändliker, E. Marom and F.M. Mottier, Two-reference-beam holographic interferometry, J. Opt. Soc. Am., 6, 23 (1976).

12. R. Dändliker, B. Ineichen and F.M. Mottier, High resolution hologram interferometry by electronic phase measurement, Opt. Commun. 9, 412, (1973).

ACCURACY AND REPRODUCIBILITY OF HETERODYNE HOLOGRAPHIC INTERFEROMETRY

B. Ineichen, R. Dändliker and J. Mastner

Brown Boveri Research Center, CH-5401 Baden, Switzerland

ABSTRACT

Heterodyne holographic interferometry can be used to determine very accurately the deformation of mechanical structures under load. The overall accuracy and reproducibility are theoretically and experimentally investigated. The reproducibility for the detection of mechanical torsion is verified to be better than 3%.

With heterodyne holographic interferometry (1) the interference phase can be measured with high resolution at any point in the fringe pattern. This technique can be used to determine very accurately the deformation of mechanical structures under load. A corresponding setup is shown in Fig. 1. The basic idea of this method is to record the two object fields before and after deformation with two different reference waves R_1 and R_2 on a hologram plate (2), such that the object fields O_1 and O_2 are reconstructed independently. This allows to introduce a small frequency shift Ω between the two reconstructed and interfering light fields, resulting in an intensity modulation at the beat frequency $\omega_2 - \omega_1 = \Omega$ for any given point of the interference pattern. The frequency shift of $\Omega = 100$ kHz is realized by two acoustooptic modulators M_1 and M_2 (Fig. 1) in cascade to give opposite frequency shift. During recording both modulators are driven with 40 MHz, so that the net shift is zero. During reconstruction one modulator is driven with 40 MHz and the other one with 40.1 MHz, so that the net shift is the desired 100 kHz. The optical phase shift between O_1 and O_2 is converted into the phase of the beat frequency (3). Therefore the difference of the interferometrically measured displacement between two points in the image is directly obtained from the phase difference ψ of the signals at the frequency Ω from the two detectors at this points (Fig. 1). The overall accuracy and reproducibility of the deformation measurements using this method depend on the specification of the phasemeter, the signal-to-noise ratio (SNR) of the detector signals, the mechanical stability of the optical setup, and the reproducibility of the mechanical deformation. The performance of the system with respect to these parameters has been theoretically and experimentally investigated.

ACCURACY AND STABILITY OF PHASE MEASUREMENT

The relative phase of the two detector signals is measured with

a zero crossing phase meter. The stability and accuracy of the
phase meter is electronically tested to be better than $\pm$ 0.1 for
clean input signals at 100 kHz. The phase meter requires for
proper operation, i.e. to avoid multiple zero crossings, a signal-to-noise ratio of at least 20 dB and a noise bandwidth of
less than the signal frequency. The signal noise introduces a
phase error $\delta\psi$ due to the fluctuations of the zero crossings.
This phase error is found to be

$$\delta\psi = (SNR)^{-1/2} N^{-1/2} = (SNR)^{-1/2} (\tau/T)^{-1/2}, \qquad (1)$$

where $N = \tau/T$ is the number of zero crossings observed during
the integration time τ of the phase meter. This means that a
measurement $(N = 1)$ with SNR = 20 dB yields $\delta\psi = 6^{\circ}$. This is
reduced to $\delta\psi = 0.06^{\circ}$ for $\tau = 100$ ms and a frequency of 100 kHz
$(T = 10$ μs).

The SNR of the detector signals can be estimated from the holographic setup, the hologram efficiency, and the laser power. It
is assumed, that a photomultiplier is used as detector and that
the light intensity is large enough to get shot-noise limited
detection. In the case the signal-to-noise ratio is found to be

$$SNR = Pm^2 \eta_H \eta_D F_D / 2F_o h\nu B, \qquad (2)$$

where P = power of the reconstructing reference beams, m =
fringe contrast, η_H = hologram efficiency, η_D = quantum efficiency of the detector, F_D = detector area, F_o = illuminated object
surface, $h\nu$ = photon energy, B = detection bandwidth. The power
P of the reference beam is defined as the power falling on that
area of the hologram, which contributes to the reconstructed
image.

For the described experiment this parameters were $P = 20$ mW,
$m = 0.4$, $\eta_H = 0.25$, $\eta_D = 0.1$, $F_D = 10^{-7}$ m^2, $F_o = 0.1$ m^2, $h\nu =$
3.8×10^{-10} Ws $(\lambda = 514$ nm), $B = 10$ kHz, yielding a SNR of 40 dB.

Besides the phase error due to amplitude noise, as shown in
Eq. (1), additional phase fluctuation may occur in the signal.
These are mainly caused by optical instabilities, e.g. in the
path length of the two reference waves, and by mechanical instabilities of the position of detector and reconstructed image.
Both effects are reduced by averaging over the integration time
τ of the phase meter. The overall accuracy of phase measurement,
including these phase fluctuations, are determined experimentally. A double exposure, two-reference-beam hologram of an object,
which has not been deformed between the two exposures, is recorded. The measured phase difference between the signals of two
detectors at a constant separation of 3 mm is measured at different positions in the object image. The result is shown in Fig.
2. The vertical bars indicate the digital resolution of $\pm$ 0.05°
of the phase meter reading. The measured phases show statistical

variations with $\delta\psi = 0.22^{O}$ around an average value of $\bar{\psi} = 1.05^{O}$.
This means that the interference phase can be measured with an
accuracy of $\delta\psi = 0.22^{O}$, corresponding to an interpolation of
$6x10^{-4}$ of a fringe.

ACCURACY AND REPRODUCIBILITY OF DEFORMATION MEASUREMENTS

The accuracy and reproducibility of the measurement of mechanical
deformation has been experimentally investigated for the case of
a turbine blade under static torsional load. The tested blade is
about 1.2 m long. The torsional load is realized by the couple of
two symmetric weights at the tip of the blade. The holograms were
recorded at 1 m distance from the blade. The reconstruction is
imaged with a magnification of 0.39 by a f/9 aperture objective,
closely located behind the hologram (2).

The detector array in the image plane is shown in Fig. 3. The
ends of two circular fibre bundles of 1 mm diameter are covered
with 0.1 mm wide slits, parallel to the blade axis. The 5 mm separation of the slits corresponds to a distance of $D_{obj} =$
12.8 mm on the object. The fibre bundles feed the light to two
photomultipliers.

The fringe pattern caused by the torsion of the blade is also
shown in Fig. 3. The fringe density varies from zero to a maximum value of 8 fringes/mm on the object. The measured phase difference ψ, including multiples of 2π, between the slits is proportional to the tilt $\theta(z)$ of the blade around its axis. The
corresponding relation reads (4)

$$\theta(z) = \partial u_n / \partial x = \psi\lambda/\pi\Delta x, \tag{3}$$

where u_n is the normal component of the surface displacement $\vec{u}$,
λ is the wavelength, and $\Delta x = D_{obj}$ is the separation of the detector slits on the object. The differential change $d\theta/dz$ of the
tilt along the z-axis of the blade is the torsion. In a first
approximation $d\theta/dz$ is related to the torsional moment M by (5)

$$d\theta/dz = M/GJ, \tag{4}$$

where GJ is the torsional stiffnes, given by the modulus of
elasticity in shear G and a gometry factor of the cross-section
J. On the other hand $d\theta/dz$ is experimentally obtained from the
difference of the phase measurements $\Delta\psi(z_n) = \psi(z_{n+1}) - \psi(z_{n-1})$
along the blade axis:

$$d\theta/dz = \Delta\psi\lambda/\pi\Delta x\Delta z, \tag{5}$$

where $\Delta z = z_{n+1} - z_{n-1}$. The relative error of the torsion due to
the measurement error $\delta\psi$ of the interference phase is found to
be

$$\delta(d\theta/dz)/(d\theta/dz) = \sqrt{2} \, \delta\psi/\Delta\psi \tag{6}$$

and depends therefore also on the spatial resolution Δx, Δz. Experimental results for $d\theta/dz$ are shown in Fig. 4, as triangles. There is good agreement with mechanical theory, shown as circles. The estimated absolute accuracy of about $\pm 5\%$ for the torsion is rather limited by the uncertainty of the mechanical and geometrical parameters, such as load, magnification factor, detector slit separation, than by the phase measurement itself. This becomes also clear from the following results for the reproducibility of the measurement of $d\theta/dz$.

Three independent double exposure experiments have been performed. The mechanical and optical setup for these experiments was the same. The three double exposure holograms, corresponding to three independent mechanical deformations of the blade, are evaluated and compared. The relative deviations of the three different values obtained for the torsion are calculated for each position z along the blade axis. The results are shown in Fig. 5 as circles and triangles. In addition a second evaluation of one of the holograms is made with 1 mm offset of the detector array along the z-axis. All the individual measurements are within a maximum relative error of less than 3%. The 3 percent error at z = 75 mm corresponds to a phase error of $\delta\psi = 0.36°$, since $\Delta\psi = 17°$ for $d\theta/dz = 1.5 \times 10^{-7}$ mm^{-1} in the described setup. This value compares very well with experimentally determined phase error for an undeformed object. This means that the mechanical conditions, e.g. the load, are reproducible enough in this experiment to use the full accuracy of heterodyne holographic interferometry.

CONCLUSIONS

It has been shown experimentally, that heterodyne holographic interferometry allows to interpolate interference fringe patterns with an accuracy of about $0.2°$, corresponding to 6×10^{-4} of one fringe. Moreother it has been experimentally verified, that this high sensitivity can be used reasonably to determine local deformations of mechanical objects, e.g. torsion of a turbine blade with high accuracy and reproducibility of better than 3%.

REFERENCES

(1) R. Dändliker, B. Ineichen, F. M. Mottier, High resolution hologram interferometry by electronic phase measurement, Opt. Commun. 9, 412-416 (1973).
(2) R. Dändliker, E. Marom, F. M. Mottier, Two-reference-beam holographic interferometry, J.Opt.Soc.Am. 66, 23-30 (1976).
(3) R. Crane, New developments in interferometry V. Interference phase measurement, Appl. Optics 8, 538-542 (1969).
(4) R. Dändliker, B. Eliasson, B. Ineichen, F. M. Mottier, The Engineering Uses of Coherent Optics, Cambridge U.P., Cambridge, 1976, p. 99-117.
(5) F. R. Shanley, Strength of Materials, McGraw-Hill, New York, 1957.

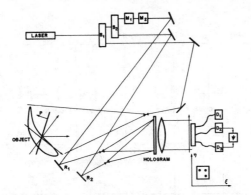

<u>Fig. 1</u> Setup for heterodyne holographic interferometry.
S: beam-splitters, M: acoustooptic modulators.

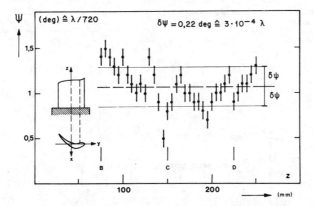

<u>Fig. 2</u> Variation $\delta\psi$ of phase measurement for an un-
deformed object.

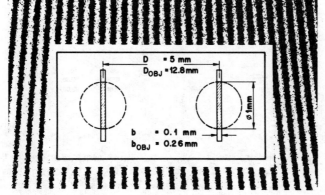

<u>Fig. 3</u> Detector array and fringe pattern for the
measurement of turbine blade torsion.

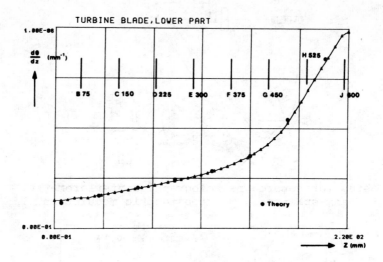

Fig. 4   Measured torsion dθ/dz of a turbine
         blade compared with theory

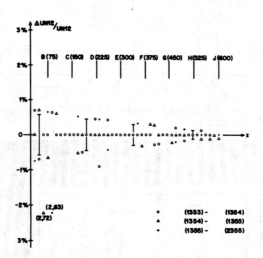

Fig. 5   Reproducibility of the measurement of
         torsion for three different holograms
         1353, 1354, 1355.

USE OF HOLOGRAPHIC INTERFEROMETRY FOR STUDIES OF PHASE CHANGES IN HELIUM NEAR THE ABSOLUTE ZERO

J. Pipman, S. G. Lipson and J. Landau

Physics Department, Technion — Israel Institute of Technology, Haifa, Israel

ABSTRACT

We are using real-time and double-exposure holographic interferometry to study the properties of ^{3}He-^{4}He mixtures at very low temperatures, as a function of temperature, pressure and ^{3}He concentration. The effects so far investigated are diffusion, surface tension and crystallization. The experimental method is described together with several examples of results obtained.

INTRODUCTION

This paper describes the use of holographic interferometry for the study of helium isotope (^{3}He and ^{4}He) mixtures at very low temperatures (0.4°K to 2°K) which are of great fundamental interest to solid-state physicists. The two isotopes, because of the difference in their zero-point energies, occupy atomic cells of different dimensions, and so their refractive indices are different despite their atomic similarity. The refractive index n is related to the molar volume by the relation

$$n - 1 = 0.1234 \times 2\pi/v$$

where v is the molar volume for that isotope. Since helium is a compressible fluid, the molar volume depends on temperature (T) pressure (p) isotropic concentration (x) and condensation state. In the experiments to be described below, all the above parameters have been varied. In the pure isotopes, n-1 has the values 4×10^{-2} and 3×10^{-3} for ^{3}He and ^{4}He respectively.

Most physics at low temperatures is carried out by electronic means. Although optical methods have been applied in several instances, in general the expected difficulty in achieving mechanical stability sufficient for interferometric work in a cryostat surrounded by large mechanical pumps has deterred research workers from using such techniques. In addition, a low-temperature cryostat has stringent requirements for the absence of heat leaks by conduction or radiation. It has to be designed carefully to allow optical measurements to be carried out without violating the insulation.

The advantage of interferometric observation in this field is that of spatial resolution. The effects which we have investigated are concerned with spatial variation of properties, which is almost inaccessible to electronic instrumentation which essentially has to be duplicated for every resolution element. The particular examples which we shall describe here are:

    1) diffusion
    2) surface tension

3) crystallization and annealing
4) phase separation

It is important to mention, if briefly, some of the outstanding features of
the ^{3}He-^{4}He phase diagram[1] in this region of temperature (Fig. 1). At temperatures above 0.85°K, the mixture is a homogeneous liquid at low pressures
(Fig. 1a). If the (x,T) point is to the left of the λ-line, the liquid is a
superfluid (very high theraml conductivity, virtually no viscosity); if to the
right of the λ-line, normal fluid with less thermal conductivity and more viscosity. On the λ-line, the specific heat peaks. Below 0.85°K, the mixture
separates into two phases which exist in equilibrium, in a similar way to many
metallic alloys. The heavier liquid is ^{4}He-rich; the lighter is ^{3}He-rich, and
there is a sharp boundary between them (see, for example, Fig. 9). As the
pressure is increased there are only slight quantitative changes in the phase
diagram until the solidification pressure is reached. An x = .50 mixture
solidifies at the lowest pressure (25 atmospheres), and a surprisingly complicated (for the two simplest isotopes in existence) phase diagram follows. An
example is given in Fig. 1(b).

EXPERIMENTAL METHOD

The helium is cooled by a ^{3}He refrigerator capable of reaching 0.4°K. The
experimental mixture is introduced into a copper cell with two parallel glass
windows, 11 mm in diameter, separated by a 3.6 mm space. At constant pressure,
a cell of this thickness give us one fringe of λ6328 Å for a concentration
difference of δx = 0.016. The cell is surrounded by heat shields at various
temperatures which prevent inflow of heat, and in those at 4.2°K, 77°K and
room temperature there are glass windows which transmit only visible light.
(The windows are not quite parallel, so that reflexions from them can be subsequently eliminated by spatial filtering). The cell temperature is measured
by a calibrated germanium resistance thermometer. The optical apparatus is
mounted on a heavy marble slab supported on soft inner tubes, which give an
excellent degree of mechanical isolation from the cryostat. We chose to use
the holographic method for two reasons. Firstly, the number of windows (eight)
in the light path made it unlikely that very high optical quality would be
achieved, so that the subtractive aspect of holographic interferometry would
be welcome. Secondly, we have shown independently[2] that double exposure
holography can be a very convenient tool in diffusion studies because of its
analogy to differentiation. We observe the cell in transmitted light, so that
unavoidable vibrations of the cryostat would have as little effect as possible.

The optical system is shown in Fig. 2. Up to the present we have used mainly
real-time holographic interferometry since it allows direct observation of the
interferograms during the experiment. This is essential when one has little
idea of what to expect. We intend to use double exposure methods to repeat
critical experiments now that we have identified them.

A few details of the optical method may be of interest. For real-time observation we prepare a hologram of the empty cell and replace it in its original
position before each cooling to liquid He temperature. The cell is imaged
into a plane just behind the hologram so that the illumination in the hologram
plane is fairly uniform; this allows us to use a rather high ratio of object-
beam to reference beam (about 1=2) without introducing serious photograhic non-
linearity. In addition the hologram uses almost all the transmitted light.
These points are important since we must photograph quickly-changing phenomena
without introducing too much energy into the cell.

The image of the cell is formed in a reflex camera which is situated as closely behind the hologram plane as possible, consistent with avoiding the reference beam. This allows recording of sharply focussed interferograms without any additional optics. The camera is, of course, not in mechanical contact with the optical table. Using a 3.5 mW He-Ne laser we can photograph the interferograms in 1/250 - 1/125 sec on Kodak Plus-X film. Photographing double-exposure holograms involves replacing the hologram plate-holder by the camera, and the plate by 35-mm holographic film. We hope soon to devise a setup which allows observation of the real-time interferograms at the same-time as photographing double-exposure holograms.

SOME RESULTS OF THE EXPERIMENTS

It is not our intention in this article to discuss the physics of the obser-vations which we have made, but rather to describe the optical techniques. A first report of the results has recently been presented[3] and also a short description of the cryogenic apparatus[4].

The most surprising results we have obtained concern diffusion processes. We create a concentration inhomogeneity in the liquid or solid helium by one of two methods - by using the phase separation or by partial solidification. The physical conditions (p or T) in the cell are then changed abruptly so that the inhomogeneity is no longer in equilibrium. Using the interferograms we can follow the equilibrating process which is dominated by diffusion. In simple cases (Fig. 3) the diffusion behaves sensibly according to Fick's law, and the mass-diffusion constant can be determined. In other cases (Fig. 4) the behaviour is more like a wave propagation. As yet we have no complete expla-nation of the phenomenon but we think it is due to coupled mass and thermal diffusion in the region of the λ-line, where the specific heat is very high. This gives an effective diffusion constant which reaches a minimum on the λ-line, and essentially prevents the flattening-out of the concentration gradient at that point. Quantitative analysis of the pictures is carried out by measurement of the shape of the interference fringes using picture-analysis equipment belonging to our colleagues working in elementary-particles. To deduce the diffusion constant from a set of such concentration profiles re-quires differentiation in both time and position. Now that we have discovered the phenomenon, we intend to carry out the differentiation processes holo-graphically using the double-exposure method for time-differentiation and shearing interferometry for positional differentiation. So far we have used numerical methods in both cases and, while there is no doubt about the exis-tence of the phenomena, the accuracy leaves a lot to be desired.

A second set of experiments has been carried out on surface tension, particu-larly at the interface between the two coexisting liquid phases below $0.85^{\circ}K$. Here we use interferometry to investigate the exact shape of the meniscus where the two phases meet the glass cell window (Fig. 5). The measured shape is then compared with a calculation, which involves ray-tracing the incident light through two menisci on opposite sides of the cell (Fig. 6). The method gives very accurate results for the interphase surface tension, compared with capillary rise and surface-wave methods[5,6]. The order of magnitude of sur-face tension is 10^{-2} dyne cm^{-1}; for surface tension of the order of 1 dyne cm^{-1} and larger this method is no more accurate than conventional ones. A sensitive qualitative indication of the surface tension is given by the dark line appearing along the interface . This is caused by a cylindrical lens effect (rays 3 and 4 in Fig. 6) and its width can be adjusted by the dia-phragm S in the back focal plane of the imaging lens. As a mixture is heated through the critical point at 0.85°K, it is instructive to watch the black line get thinner and disappear in a continuous manner.

We have also carried out several qualitative experiments on crystallization, annealing and phase equilibrium. Single crystals can be grown very easily from the pure isotopes (Fig. 7). The mixture solidifies as snow, and anneals in a time dependent on p and x, although independent of the temperature. This is indicative of atomic interchange by tunnelling, not thermally activated. The process is followed by observing the optical homogeneity of the helium, which improves as the annealing continues (Fig. 8). We have succeeded in producing two annealed solids in contact, with different concentrations, and tried to observe solid diffusion, but the time scale is apparently too slow for any changes to be visible within an hour or so. Phase equilibrium experiments have also provided some interesting observations. Fig. 9 shows what is apparently a system of four phases coexisting, which is impossible according to the Gibbs phase rule in this system which has three degrees of freedom (p, T, x). Apparently each of the two coexisting liquid phases has partially solidified giving us two liquid and two solid phase all having different concentrations. It is most likely that the system is not in true equilibrium, since the time necessary to equilibrate a solid with a liquid would depend on diffusion in the solid, which we have already mentioned as being very slow. Just the same, the observation is intriguing.

CONCLUSION

This article is intended to illustrate some of the observations which we have made using holography on solid and liquid ^{3}He-^{4}He mixtures. We intend to develop the quantitative aspects of the work considerably by the use of double-exposure methods, but have already made several qualitative observations of great interest.

REFERENCES

1. P.M. Tedrow and D.M. Lee, Phys. Rev. 181, 399 (1969).

2. N. Bechner and J. Pipman, "A Simple Method of Determining Diffusion Constants", J. Phys. D (in press).

3. J. Landau, S.G. Lipson, J. Pipman, Bull. Am. Phys. Soc. p. 643 (April 1976.

4. J. Pipman, S.G. Lipson and J. Landau, Laser and Electro-optik, p.24 (May 1976).

5. D.P.E. Dickson, D. Caroline and E. Mendoza, Phys. Lett. 32A, 419 (1970).

6. S.T. Boldarev, V.P. Peshkov, Physica 69, 171 (1973).

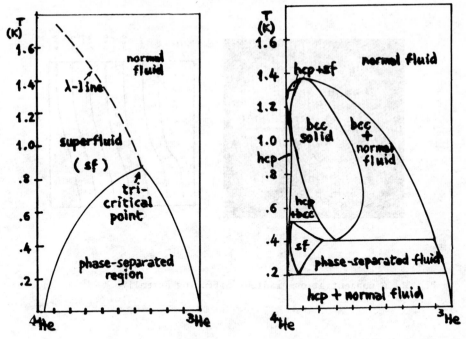

Fig. 1: ^{3}He-^{4}He phase diagrams at (a) 0 atm. and (b) 25.6 atm.

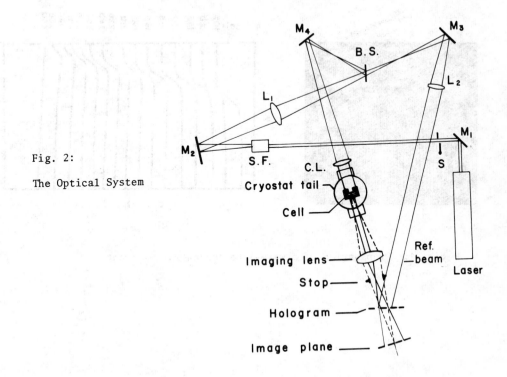

Fig. 2:

The Optical System

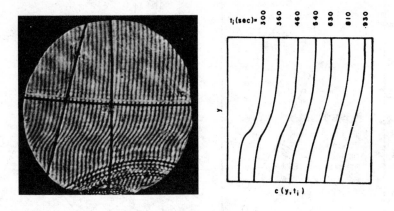

Fig 3: A concentration gradient diffusing normally

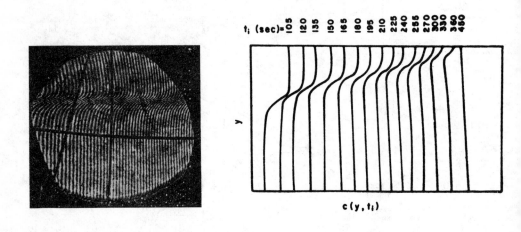

Fig. 4: Propagation of a wave-like concentration front

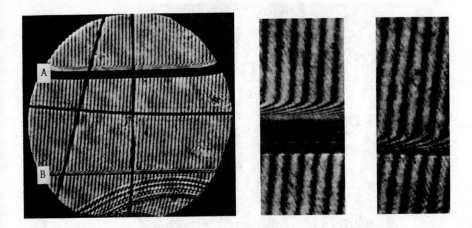

Fig. 5:   Interferogram showing the vapour-liquid meniscus (A) and the
          phase-separation meniscus (B). Also expanded views of the two
          menisci

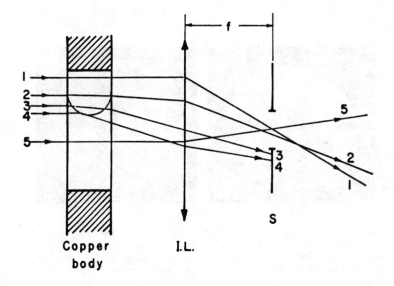

Fig. 6: Ray-Tracing through the meniscus

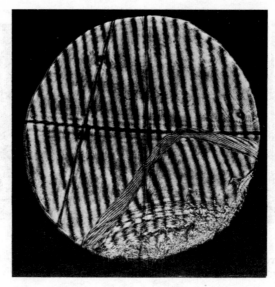

Fig. 7: A Single Crystal of ^{3}He

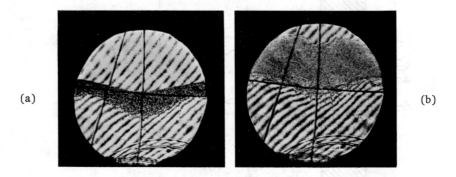

(a)   (b)

Fig. 8: Annealing of mixture crystals: (a) The upper phase is liquid,
         and the lower is solid and in process of annealing from below.
         (b) The lower phase has now annealed completely, while the
         upper has solidified and is starting to anneal from above

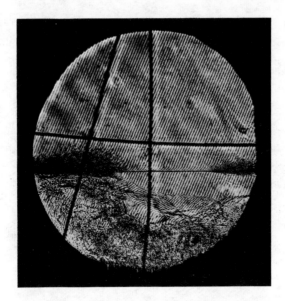

Fig. 9: Four phases in "equilibrium". The boundary between superfluid and normal liquids is the horizontal line in the centre. Each fluid has precipitated a solid phase

SESSION 7.

HOLOGRAPHIC INTERFEROMETRY II

Chairman: E. N. LEITH

AUTOMATIC MEASUREMENT OF 3-D DISPLACEMENTS BY USING THE SCANNING TECHNIQUE IN DOUBLE EXPOSURE HOLOGRAMS[†]

V. Fossati Bellani

C.I.S.E., P. O. Box 3986, 20100 Milano, Italy

ABSTRACT

An automatic read-out system for extracting information from double-exposure holograms exploiting the scanning technique is presented. Data processing is achieved by means of an ADC and a minicomputer.

INTRODUCTION

The development of a practical technique for obtaining quantitative information from double exposure holograms is still one of the most interesting problems in holographic interferometry.
A first step towards the achievement of this result has been the introduction of the scanning technique (1): this provides a practical means for counting fringes and can be used by moderately skilled operators. However the related data collection and processing is not fast enough for the requirements of,say, a testing and measuring laboratory, especially for routine work. Therefore our research has been aimed to find out an entirely automatic procedure for extracting information from double exposure holograms by the scanning technique. It has now been demonstrated the possibility of achieving quite accurate results in short time.
After recalling the formulas which allow the use of the scanning technique, the procedure presently used for collecting experimental data is described and a discussion on preliminary results is given. Finally the automatic procedure is illustrated. While software has already been brought up to an advanced stage, hardware is still under development and we will give here only an outline of the design of the automatic read-out system. Both software and hardware will be described in detail elsewhere.

THEORY

In principle by using the scanning technique two different approaches can be followed for obtaining the three components of the displacement $\vec{d}$ of a point under test. The first one requires the use of really "perfect" holograms with elimination of speckle and other noise sources. We briefly deal with this method because it might be usable in a next future. The second one is the presently used technique and exploits the available recording and detecting materials. Both methods exploit the recostructed real image which is analyzed by a photodetector while the laser beam sweeps across the hologram (1).
In the first case the photodetector, placed at the point being examined, records

---

†Work carried out by CISE on behalf of ENEL.

.225

a light intensity given by

$$I = \cos 2 \pi N \tag{1}$$

plus a d.c. term which we disregard, where N is the fringe order (in general not integer) related to the displacement $\vec{d}$ of the point by

$$\vec{d} \cdot (\hat{\rho} + \hat{\rho}_{ill}) = N \lambda \tag{2}$$

where $\hat{\rho}$ and $\hat{\rho}_{ill}$ are the unit vectors in the directions of observation and illumination, respectively, and λ is the wavelength of the light used to record the hologram. Inserting eq.(2) into eq.(1) the intensity results

$$I = \cos \left\{ \frac{2 \pi}{\lambda} \vec{d} \cdot (\hat{\rho} + \hat{\rho}_{ill}) \right\} \tag{3}$$

Referring to Fig. 1, we can write the components of the unit vectors $\hat{\rho}$ and $\hat{\rho}_{ill}$ as follows

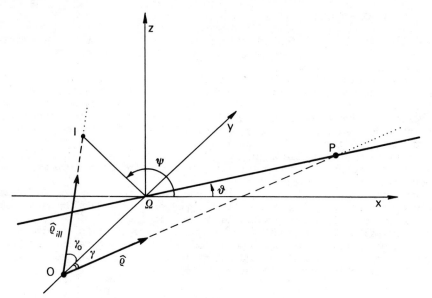

Fig. 1. Coordinate system and geometrical parameters: γ and γ_0 are respectively the angles of viewing and illuminating direction with respect to y a-xis, θ is the angle of scanning direction ΩP with x axis, I is the intersection point of the illuminating direction with the hologram plane (x,z) and ψ is the angle formed by IΩ direction with x axis.

$$\rho_x = \sin \gamma \cos \theta \qquad\qquad (\rho_{ill})_x = \sin \gamma_0 \cos \psi$$
$$\rho_y = \cos \gamma \qquad\qquad\qquad (\rho_{ill})_y = \cos \gamma_0$$
$$\rho_z = \sin \gamma \sin \theta \qquad\qquad (\rho_{ill})_z = \sin \gamma_0 \sin \psi$$

where γ, γ_0, θ, ψ are defined in the figure.
Eq. (2), then, becomes

$$I = \cos\left\{\frac{2\pi}{\lambda}\left[(\sin\gamma\ \cos\theta + \sin\ \gamma_0\ \cos\psi)\ d_x + \right.\right.$$
$$\left.\left. + (\cos\gamma + \cos\ \gamma_0)\ d_y + (\sin\gamma\ \sin\theta\ + \sin\gamma\ _0\ \sin\psi)d_z\right]\right\} \quad (3')$$

where d_x, d_y, d_z are the components of $\vec{d}$.
In eq (3) γ is the angle which determines the observation direction and is a
function of time, $\gamma = \gamma(t)$, through the position of the scanning mirror.

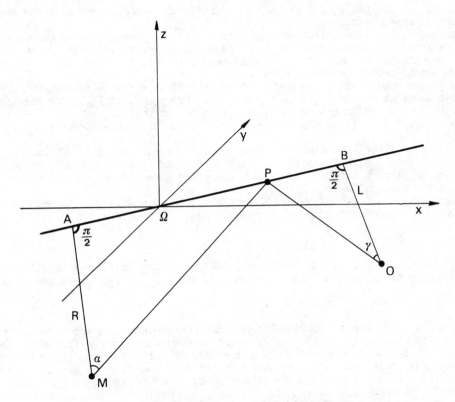

Fig. 2. Relation between viewing direction and scanning
beam position: M and O are, respectively, the
scanning mirror and object point positions, R
and L are respectively the distance of M and O
from scanning line, α is the scanning beam angular position. Note that M,O and the scanning line do not lie necessarily in the same plane and
angles α and γ are assumed positive when P is
at the right of A and B respectively.

Referring to Fig. 2, we see that

$$R \operatorname{tg} \alpha - L \operatorname{tg} \gamma = Q \tag{4}$$

where R, L are defined in the figure, Q is the distance AB and $\alpha = \alpha_0 + \omega t$ is the angle formed by the beam illuminating the hologram with an axis normal to the scanning direction axis and the signs depend upon the choice of the positive angles. From eq.(4), recalling the expression for α

$$\operatorname{tg} \gamma = -\frac{Q}{L} + \frac{R}{L} \operatorname{tg} (\alpha_0 + \omega t) \tag{4'}$$

where α_0 is the value of α at the beginning of the scanning and ω is the angular scanning velocity.

A number of experimental values of I for different times t_i can be determined as we shall see in the next section. These values can be inserted, along with the geometrical parameters of the experiment, in a best fitting computer program which yields the values of d_x, d_y, d_z.

The theory given here takes into account not only the fringe number, but the whole waveform as given by eqs.(3) and (3'). Provided that $\hat{\rho}_{ill}$ does not lie in the same plane of the scanning direction and of point O, eqs.(3) or (3') completely define I(t), which depends only on the unknown components d_x, d_y, d_z of $\vec{d}$.

Besides the linear scanning any other kind of scanning can be used, substituting the proper time dependence law $\gamma = \gamma(t)$. Unfortunately this is a non linear problem, which requires a long computing time and very good waveforms I(t), not easily available.

Therefore we have introduced an alternative technique which can be used with holograms recorded on standard materials. Eqs.(2) and (3) yield

$$N = \frac{\vec{d}}{\lambda} \cdot (\hat{\rho} + \hat{\rho}_{ill}) \tag{5}$$

where N is unknown. But, according to Aleksandrov and Bonch-Bruevich (3), if we choose another viewing direction defined by the unit vector $\hat{\rho}_i$, we can write

$$N - N_i = \frac{1}{\lambda} \vec{d} \cdot (\hat{\rho} - \hat{\rho}_i) \tag{6}$$

where N_i, also unknown, is the fringe order corresponding to the new viewing direction. The difference $N - N_i$ can be obtained by counting the fringes as in Ref. (1). But we choose in this case a number of points of the I(t) waveform, whose phase difference is known and constant (for instance the minima, corresponding to black fringes) for a number of different scannings and measure the times t_i at which they occur, and the other geometrical parameters of the scanning line. Between the ith and j^{th} minima the fringe number N changes from N_i to $N_j = N_i + j - i$ so that $N_j - N_i = j - i$.

All these data are given to the computer along with the number of minima n. The computer now implements $n(n-1)/2$ equations for each scanning, and solves the resulting overdetermined linear system in the three unknowns d_x, d_y, d_z. The equations can be written

$$j - i = \frac{\vec{d}}{\lambda} \cdot (\hat{\rho}_j - \hat{\rho}_i) \qquad (i, j = 1, n) \tag{7}$$

for each scanning and in scalar form

$$\frac{1}{\lambda}\left\{\left[(\sin \gamma_j - \sin \gamma_i) \cos \theta\right]d_x + (\cos \gamma_j - \cos \gamma_i) d_y + \right.$$
$$\left. + \left[(\sin \gamma_j - \sin \gamma_i) \sin \theta\right] d_z = j - i \right. \tag{7'}$$

where the meaning of γ_i and γ_j is obvious.
The minima have been chosen for practical reasons due to the computer program
which has been developed, but other points, like the zero-crossing points of
$I(t)$ might be more accurate if a good automatic selection technique could be
devised.

EXPERIMENTAL SET-UP AND PRELIMINARY RESULTS

The experimental set-up is shown in Fig. 3. The laser output is focused on an
oscillating mirror which reflects the beam onto a spherical mirror whose cen-

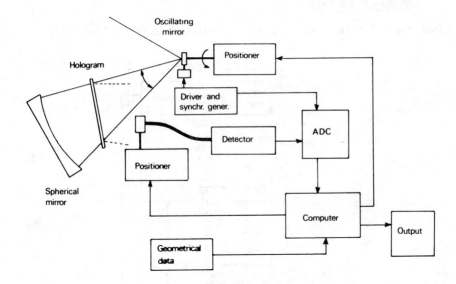

Fig. 3. Experimental set-up for the automatic read-out
        system.

ter of curvature lies on the oscillating one. The spherical mirror generates
the conjugate reference beam which, by crossing the hologram, yields the real
image. An alternative set-up exploits a pellicle beam splitter to avoid the dou
ble crossing of the plate. A photodetector records the signal at the examined
point: its output is monitored by an oscilloscope and sent to an analog to di-

gital converter. This way each numerical value of I(t) is related to the sampling time t. The values of t closest to the minima are assumed for deriving γ_i, γ_j of eq.(7').

Since the hologram has a non-constant diffraction efficiency, the sinusoidal output displays a slowly varying amplitude modulation which may cause some problems in data processing. We get rid of part of it by collecting light on a relatively large surface around the point under test. This large aperture detector washes out the fringes and records an average signal which can be subtracted from the previous one removing the amplitude modulation.

The first experiments have been performed for a given rigid body displacement to test the method. The results show accuracies of a fraction of wavelength in any direction, which means that displacements of a few tens of microns can be measured with errors of less than 0.5 microns by exploiting directly the digital data. These accuracies can be greatly improved by a better definition of the waveform phase. This can be achieved in an analog way as discussed by Biedermann and Ek (2) or by using a sampler with a better time resolution as intended in the final system.

AUTOMATIC SET-UP

The final goal of this work, as previously stated is to build a fully automa-

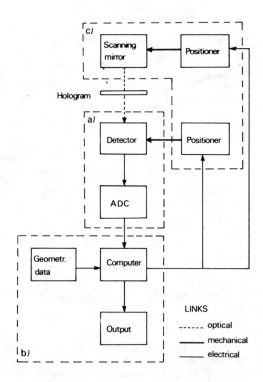

Fig. 4. Block diagram of the automatic read-out system.

tic read-out system which needs only a limited number of manual operations.
In order to achieve this result the following steps shown in the block diagram
of Fig. 4 must be scheduled:
a) detection of signal and analog to digital conversion
b) collection of digital data and processing by a computer
c) positioning of detector and scanning mirror.
For step a) we use an analog to digital converter driven by an external clock.
This allows tne operator to choose the proper part of the signal to be analyzed. The digitized data are then processed by a minicomputer working on line
(step b). The only operator's task here is to supply the computer with all the
geometrical information about the experiment. This is achieved by means of a
teletype, which acts also as the output unit. After processing the data of one
scanning, the computer drives a stepping motor for repositioning the oscillating mirror till the total number of required scannings have been performed
for one point. Then the detector is moved by another computer driven stepping
motor to a new point and the whole procedure starts again (step c).

CONCLUSION

Based on the scanning technique we have devised a set-up which will provide a
means for automatically processing holographic information. The preliminary
experiments using only part of the final instrumentation and software tested
on an IBM 370/125 computer, have been quite successful. They prove that such
a system can be actually used in research and industrial laboratories for very
fine quality control measurements, stress and vibration analysis, etc.
Since the method is based on the relative shift of the fringes while changing
the viewing direction, two or more holograms can be recorded for one object,
taken from different points of view. This improves the accuracy of the technique so that it can be scaled up to relatively large objects. Details will be
discussed in a next paper.

ACKNOWLEDGMENTS

The author wishes to acknowledge A. Sona for discussions and helpful advices,
L. Rizzi for discussions and continuous assistance in setting-up the computing
program and P. Delvò for experimental work.

REFERENCES

(1) V. Fossati Bellani and A. Sona, Measurement of Three-Dimensional Displace-
    ments by Scanning a Double-Exposure Hologram, Appl. Opt 13, 1337 (1974).

(2) K. Biedermann and L. Ek, paper presented at this conference.

(3) E.B. Aleksandrov and A. M. Bonch-Bruevich, Investigation of Surface Strains
    by the Hologram Technique, Sov. Phys.-Tech. Phys. 12, 258 (1967).

HOLOGRAM INTERFEROMETRY WITH A CONTINUOUSLY SCANNING RECONSTRUCTION BEAM

Leif Ek and Klaus Biedermann

Department of Physics II and Institute of Optical Research,
The Royal Institute of Technology, S-100 44 Stockholm 70, Sweden

ABSTRACT

We studied a system for fringe evaluation in hologram interferometry, where a thin conjugate reference beam scans the hologram, and the resulting motion of the fringes in the real image of the deformed object is recorded. The concept of the K-sphere is introduced for an investigation of the sensitivity of the system and its dependence on geometric parameters and the question whether the absolute fringe order is known or not. Experimental results show that a rigid body displacement vector, orthogonal to the observation vector, could be

determined with a typical error of $\frac{1}{8} \cdot \frac{\lambda}{2}$.

INTRODUCTION

The interference fringes in a holographic interferogram resulting from an object deformation are generally localized in space, and thus the fringes exhibit parallax when the direction of observation is changed. This phenomenon can be used to evaluate the displacement vector, as first pointed out by Alexandrov and Bonch-Bruevich [1], and later analysed in more detail by Sollid [2]. Another possibility for fringe evaluation is to determine the absolute fringe order by counting the number of fringes from an a-priori known zero-order fringe. This latter method is the most accurate one, but it gives the displacement component in the direction of the sensitivity vector only. Hence, if we are to evaluate a three-dimensional displacement from a single hologram, we have to make use of the fringe parallax in one way or another.

Among the problems fringe evaluation poses is how to measure the fringe phase in a simple and relatively accurate way. One possibility, first published by Fossati-Bellani and Sona [3], is to reconstruct a real object image by a thin conjugate reference beam, to the effect that interference fringes will then be projected down onto the image plane. When the beam position on the hologram is changed, due to the fringe parallax, the fringes will move across the stationary image. These fringe variations can be recorded by a photodetector in the real image. From the fringe phase measurements, together with data on the geometry, we can calculate the displacement vector at the detector posi-

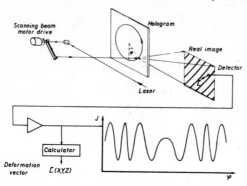

Fig. 1 A scanning reconstruc-tion beam gives moving fringes recorded by a photodetector in the real image. From the fringe phase, the displacement can be computed.

tion, preferably with a least square fit applied to the equations in order to minimize the effect of random errors in the observations [4].

2. CHARACTERISTICS OF THE EVALUATION PRINCIPLE

For the proper use of such a scanning system for fringe evaluation, we have to study in some detail the possibilities and limitations of the evaluation principle.

2.1 Scanning Geometry

From the curve measured during a scan of the reference beam, a set of fringe phase observation data Ω_i and corresponding data on the K_{2i} observation vectors are inserted into the equation system

$$(\overline{K}_{2i} - \overline{K}_1) \cdot \overline{L} = \Omega_i \qquad i = 1,2,\ldots,N \qquad (1)$$

where $\overline{K}_1$ is the illumination vector and $\overline{L}$ is the unknown displacement vector. The difference between observation vector and illumination vector $(\overline{K}_2 - \overline{K}_1)$ is often called sensitivity vector, for obvious reasons. Eq. (1) is valid when the absolute fringe order is known, and, if it is unknown, Eq. (1) should be replaced by

$$(\overline{K}_{2i} - \overline{K}_1) \cdot \overline{L} = \Omega_i + \Omega_o \qquad i = 1,2,\ldots,N \qquad (2)$$

where Ω_o is the absolute fringe order, an unknown integer times 2π.

As a means to study how the geometry chosen influences the measurement sensitivity, we can introduce a small displacement change in direction $\hat{u}$, namely $\delta L \cdot \hat{u}$, into Eq. (1). The corresponding change of the fringe argument will be

$$\delta\Omega_i = [(\overline{K}_{2i} - \overline{K}_1)\cdot\hat{u}]\delta L \qquad (3)$$

When the absolute fringe argument is unknown we can subtract one of the equations from all the others to eliminate the unknown fringe order, but at the same time the $\overline{K}_1$-vector disappears and Eq. (3) is now replaced by

$$\delta(\Omega_i - \Omega_N) = [(\overline{K}_{2i} - \overline{K}_{2N})\cdot\hat{u}]\delta L \qquad (4)$$

The relation between changes in the displacement and the fringe argument is thus governed by the scalar product in the square bracket of Eq.(3) or Eq.(4).

As a means for comparing the effect of different geometries we found a graphic presentation by aid of the K-sphere useful. As the K-vectors have the constant length $\frac{2\pi}{\lambda}$, they will define a sphere around the object point where the displacement is to be determined. The scanning K_2-vector will draw a track on the surface of the sphere, similar to the scanning curve on the hologram, while the K_1-vector will give a single point as indicated in Fig. 2.

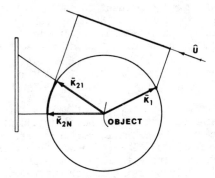

Fig. 2 *The sensitivity of a displacement measurement in the $\hat{u}$ direction is given by the projection of the K-sphere figure upon the $\hat{u}$-vector. Total fringe argument known.*

This track and point define a cone in space, and the length of the projection of this figure onto a line parallel to $\hat{u}$ is a measure of the sensitivity in this direction, when the absolute fringe order is known. To optimize the sensitivity for general displacements, the base of the cone should be as wide as possible, and the observation points on the hologram should be chosen close to the four corners of the hologram. In this way, a sensitivity as large as possible is obtained for directions normal to the sensitivity vector.

With unknown fringe order, on the other hand, the K_1-point is missing on the K-sphere, and what is left is a very flat figure as indicated in Fig. 3.

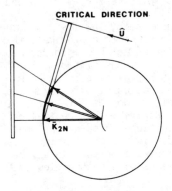

Fig. 3 *With the total fringe argument unknown, the K_1-point is missing, and the sensitivity in the critical direction is poor.*

The sensitivity will be very poor for directions in the vicinity of the average K_2-vector, which we may call the critical direction. For the sensitivity to displacement components perpendicular to the critical direction, the four corners of the hologram are still the important observations areas of the hologram. If we are to determine also the displacement component in the critical direction, roughly half the number of observations should be chosen

close to the center of the hologram in order to maximize the average sensitivity in this critical direction. Nevertheless, the sensitivity in the critical direction will be rather poor and it is instead recommended to introduce
a second view of the object by means of a mirror, or a second hologram [5].
An evaluation with unknown fringe order from each view will thus give two
sets of observations where the displacement in the critical directions can be
determined from the other observation set. By this means, a general three-
dimensional displacement can be computed.

If we, in this way, can determine the displacement component in the direction
of the sensitivity vector with an error of less than about 80 nm,we can deduce the absolute fringe order number. Then we can repeat the evaluation
according to Eq. (1) and obtain the component in the direction of the sensitivity vector with increased accuracy (Fig. 2).

2.2 Speckle Noise

With the scanning technique for fringe evaluation, we have relatively small
apertures in the system, both at the illuminated spot on the hologram and at
the light receiving detector in the real image. These small apertures will
introduce speckle noise into the detector signal [6]. The influence from this
noise can, in principle, be decreased by introducing electrical low-pass
filtering. The noise can also be reduced directly by using larger apertures,
with, however, a decrease of the cosine fringe modulation as a result. An analysis [7] shows that the best signal-to-noise ratio is obtained when the
resulting fringe modulation in the detector is around 0.4. This means that we,
in general, obtain satisfactory signal-to-noise ratios with aperture diameters
in the region of 1 to 4 mm, data which are feasible without introducing too
severe restrictions onto the equipment.

3. EXPERIMENTS

To examine our scanning system for fringe evaluation, some experiments have
been performed, where we determined the static displacement of a rigid body.

In the analysis it was shown that, for the scanning path, the most important
regions of the hologram were at the periphery. In order to approximate this
condition and at the same time have a simple scanning we choose to use a beam
scanning along a circular path. This was realized by means of two mirrors
mounted to a slowly rotating axis, as sketched in Fig. 1, where the beam
enters the scanner on the axis and leaves it at a radius of 34 mm from the
axis. By the choice of this circular scanning path we do not have access to
the information from the center of the hologram. However, this central part
is most essential when we are to determine the complete three-dimensional
displacement vector when the absolute fringe order is unknown. As indicated
earlier, the sensitivity in the critical direction will be low, in any case,
so we preferred to calculate this component by means of a second view or by
an evaluation with known fringe order.

When the reconstruction beam scans over the hologram the cosine fringes are
recorded by a photodetector in the real image. Unfortunately, we cannot expect
to have a constant envelope to the curve, since it is most unlikely that the
diffraction efficiency can be made sufficiently uniform over the hologram
surface. To reduce the effect of this variation, a second detector is placed

at a point of the image where the object is stationary, i.e. a point without fringes, and the fringe signal is divided by this reference signal.

The speckle noise was kept on a low level by suitable choice of aperture sizes. Typical values for the beam diameter and the detector diameter were 4 and 2 mm, respectively, giving a signal modulation of around 0.4. These data resulted in a ratio between cosine signal and speckle noise of about 20:1, giving sufficient accuracy in the fringe phase measurements.

The analogue cosine signal was divided by the reference signal by an analogue divider and fed to a x-t plotter.

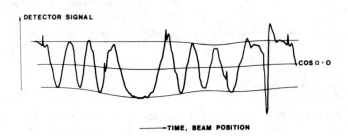

Fig. 4 Resulting fringe curve after division by refer-
ence signal. The points used in the least square fit
are those where $\cos\Omega_i = 0$.

From the plotted curve, Fig. 4, the points were chosen where the fringe argument $\Omega_i = n \cdot \pi + \frac{1}{2}\pi$, i.e. halfway between the light and dark fringes, as the derivative is maximum there, and least influence from the noise is to be expected. In these points, the corresponding observation vector, $\overline{K}_2$, was calculated and introduced into a least-square fit to Eq. (1) or Eq. (2) for the determination of the in-plane displacement at the signal detector position. With an error of the fringe phase measurements of about $\frac{\pi}{20}$ (three times as large as the error from the speckle noise) we can expect an uncertainty of the in-plane displacement components of about $\frac{1}{8} \cdot \frac{\lambda}{2}$ from the statistics of the least-square evaluation [6] for the geometry used.

The test object was a flat white painted surface given a rotational displacement by aid of a rotation table [7], in order that the displacement components be linear functions of space. Results obtained by this method for the displacement orthogonal to the critical direction for a test series are given in Fig. 5. The displacement values calculated deviate from the straight line by about $\frac{1}{8} \cdot \frac{\lambda}{2}$, or 40 nm if He-Ne laser light is used.

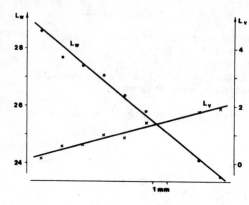

Fig. 5 *The calculated rigid body displacements of the test object in directions orthogonal to the critical direction. Distance hologram - object 220 mm.*

Displacements in units of $\frac{\lambda}{2}$.

4. CONCLUSIONS

The scanning system for fringe evaluation can give rather accurate results when the object displacement field contains at least a component of in-plane motion to ensure that the fringes exhibit sufficient parallax. The evaluation of the fringe pattern can be performed by hand using a desktop calculator for the least square fit. For large-scale application of the evaluation method in practice, the calculator should have access to the detector output directly.

REFERENCES

[1]   E.B. Alexandrov and A.M. Bonch-Bruevich, Investigation of Surface Strains by the Hologram Technique, Soviet Phys. Tech. Phys. 12, 258 (1967)

[2]   J.E. Sollid, Holographic Interferometry Applied to Measurements of Small Static Displacements of Diffusely Reflecting Surfaces. Appl. Opt. 8, 1587 (1969)

[3]   V. Fossati Bellani and A. Sona, Measurement of Three-Dimensional Displacements by Scanning a Double-Exposure Hologram, Appl. Opt. 13, 1337 (1974)

[4]   S.K. Dhir and J.P. Sikora, An Improved Method for Obtaining the General-displacement Field from a Holographic Interferogram, Exp. Mech. 12, 323 (1972)

[5]   P.W. King III, Holographic Interferometry Technique Utilizing Two Plates and Relative Fringe Orders for Measuring Micro-Displacements, Appl. Opt. 13, 231 (1974)

[6]   J.C. Dainty, Detection of Images Immersed in Speckle Noise, Opt. Acta, 18, 327 (1971)

[7]   L. Ek and K. Biedermann, Analysis of a System for Hologram Interfero-
      metry with a Continuously Scanning Reconstruction Beam, submitted to
      Appl. Opt.

[8]   N-E. Molin and K.A. Stetson, Measurement of Fringe Loci and Localiza-
      tion in Hologram Interferometry for Pivot Motion, In-Plane Rotation
      and In-Plane Translation, Optik, 31, 157 281 (1970)

MEASUREMENT OF THREE-DIMENSIONAL TEMPERATURE FIELDS BY HOLOGRAPHIC INTERFEROMETRY

C. M. Vest and P. T. Radulovic

*Department of Mechanical Engineering, The University of Michigan,
Ann Arbor,Michigan 48104, U.S.A.*

ABSTRACT

Multi-directional holographic interferometry can be used to measure three-dimensional, asymmetric temperature or density fields in transparent fluids. Experimental and computational techniques are described and used to measure temperature distributions in interacting natural convection plumes.

INTRODUCTION

In heat and mass transfer experiments the objective often is to determine the distribution of temperature throughout a fluid. This is commonly done by inserting thermocouple probes into the test section. In many experiments, however, it is undesirable or impossible to insert a probe into a test fluid without disrupting the field under study. This is particularly true in low-scale motions, such as natural convection, which are susceptible to instability, and in high-speed compressible flows where shocks may be induced by the probe. Optical techniques are an attractive alternative to the use of probes, because they require only the passage of a low-intensity optical wave through the fluid. Classical techniques, such as schlieren photography and Mach-Zehnder interferometry have proved useful for studying temperature and density distributions; however, they can be used for quantitative measurement of only those fields which are radially symmetric, or which have no variation in the direction of the optical axis. Optical holography, however, makes it possible to record multi-directional interferometric data, which provides a basis for measuring three-dimensional, asymmetric fields. In this paper we briefly discuss techniques for measuring three-dimensional temperature fields by holographic interferometry, and report on an experiment in which the accuracy of the technique was assessed by comparison with measurements made with several calibrated thermocouples. The experimental and computational methods discussed in this paper are also applicable to other important research areas such as combustion-, plasma-, and aerodynamic diagnostics.

When interferometry is used to study transparent objects, fringe
data provides measurements of the optical pathlength of a collection of rays which traverse the object. From this data, the distribution of refractive index throughout the object can be determined. Refractive index must in turn be related to physical
properties of interest: temperature in the case of heat transfer
experiments, density in the case of aerodynamics, and electron
number density in the case of plasma diagnostics. If the object
is radially symmetric, or has no variation along the direction of
the optical axis, the refractive index distribution can be determined by well-known methods using the data from a single interferogram (Ref. 1). When the object being studied is asymmetric,
a single interferogram does not provide sufficient data; in this
case multi-directional interferometric data must be utilized.
Data of this type can be obtained from a holographic interferogram in which the transparent object is back-illuminated by a
diffuser. The fringe pattern changes as an observer varies the
direction from which such a hologram is viewed or photographed.
The fringe pattern observed from a given orientation is equivalent to that which would be recorded by a classical interferometer such as the Mach-Zehnder with its optical axis aligned in
that direction. In this sense a single holographic interferogram contains the same information as a multitude of Mach-Zehnder
interferograms.

The problem that must be considered in analyzing multi-directional interferometric data is: given a set of interferograms, each
recorded from a different viewing direction, determine the refractive index distribution of the transparent object. The data
are a set of measurements of optical pathlength difference along
a collection of rays through the test section. If refraction is
negligible each ray is a straight line. The pathlength difference along a single straight ray S_k is

$$\Phi_k = \int_{S_k} f(x,y,z) \, dS_k \qquad (1)$$

where

$$f(x,y,z) = n(x,y,z) - n_o \qquad (2)$$

is the distribution of refractive index relative to a reference
value n_o. Usually, n_o is the uniform refractive index in a
quiescent test section at the time of the initial holographic
exposure. The problem of determining, or "reconstructing",
$f(x,y,z)$ from a set of measured values of Φ_k requires inversion
of a set of integral equations of the form (1). The underlying
problem in continuous mathematics is that of determining a scalar
field from a knowledge of its projections onto lines or planes,
and leads to the Radon transform (Ref. 2).

The three-dimensional field $f(x,y,z)$ can be determined by reconstructing the refractive index on several horizontal planes, z=
constant. If one had continuous, noise-free data for all possible rays in this plane the exact solution could be found (Ref.
3). In practice, of course, data are noisy and discrete, and
the range of viewing directions may be limited by the aperture
of the hologram or test section windows. The interpretation of

data of this type has been considered by Matulka and Collins (4),
Junginger and van Haeringer (5), Sweeney and Vest (6,7), and
others. The simplest approach is to divide the field by a grid
into a number of square elements, each of which is considered to
have a constant, unknown, value of refractive index. The optical pathlength of each ray is then approximated as a summation of
products of these unknown values and the length of the segment of
the ray within each element. Alternate approaches involve expansion of the field in orthogonal series with unknown coefficients,
or the use of interpolating series in which the unknown values
of refractive index at a set of sample points in the field appear
as coefficients. In all of these methods, equation (1) is replaced by a set of linear algebraic equations relating the measured optical pathlength values ϕ_k to the unknown coefficients.
The details of these equations are described in Ref. 6   and will
not be repeated here.

The problem of reconstructing refractive index fields from data
collected over a limited range of viewing directions is ill-posed.
This is reflected in the poor condition of the set of equations
which must be solved when experimental data is analyzed. Computer studies carried out in preparation for the experiment described in the following section indicated that large amounts of
redundant data must be measured in order to reduce amplification
of errors to an acceptable level. This effect is also documented
in Ref. 6.

EXPERIMENT

Multi-directional holographic interferometry was used to measure
the temperature field in the natural convection plume above two
heated disks. This temperature field is asymmetric and cannot be
measured by classical techniques. The primary objective of this
paper is to describe the holographic interferometer and to present typical experimental results for evaluation of the accuracy
of the measurement. This experiment was carried out in preparation for a parametric study of interacting thermal plumes.

Optical System

The holographic interferometer is shown schematically in Fig. 1.
The beam from a 50 mW He-Ne laser was divided by a beamsplitter
into an object beam and a reference beam. The reference beam was
expanded and filtered to form a spherical reference wave. The
object wave was collimated and passed through two holographic
phase gratings in order to analyze it into a set of plane waves
propagating in various directions through the test section. These
phase gratings were formed by recording the interference pattern
between two plane waves on a holographic plate. This plate was
then developed and bleached to form a sinusoidal phase grating
of transmittance (Ref. 8)

$$\underline{t}(x,y)=\text{rect}(\tfrac{x}{\ell})\text{rect}(\tfrac{y}{\ell})\sum_{q=-\infty}^{\infty} J_q \ (\tfrac{1}{2}\Delta\phi)e^{i2\pi qfx} \tag{3}$$

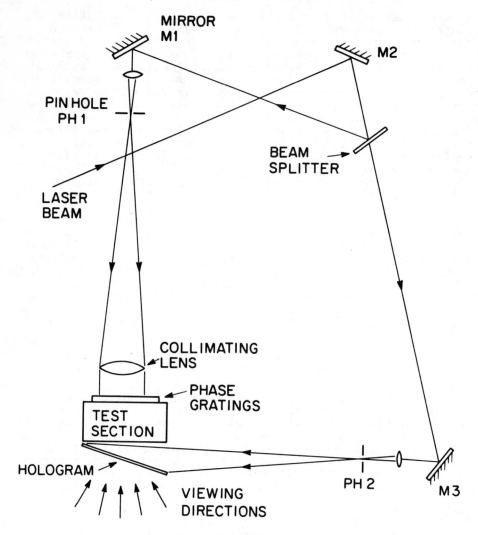

Fig. 1. Optical apparatus for recording multi-
directional holographic interferograms

where J_q is the Bessel function of the first kind of order q, ℓ
is the dimension of the plate, f is the spatial frequency of the
fringes, and $\Delta\phi$ is the peak-to-peak phase delay. A grating of
this type analyzes a plane wave into a set of plane waves travel-
ing at angles

$$\theta_q = \sin^{-1}(qf\lambda), \quad q = 0, \pm 1, \pm 2, \ldots .$$

Computer simulations indicated that data should be collected over
at least a 90° range of viewing angles (or 45° if symmetry about
the center line is assumed). The angular separation between

consecutive views should be between 2º and 8º to provide data with sufficient redundancy. By an empirical procedure a pair of gratings was produced that enables us to generate 17 plane wave components of satisfactory quality and intensity distributed over an angular range of 62º. These gratings were used in preference to a diffuser because with diffuse illumination the fringe resolution was insufficient for this experiment (Refs. 9,10).

The test section was a rectangular glass tank of dimensions 30x 20x17cm. The set of plane waves leaving this test section was recorded on 20x25 cm Kodak 120-02 holographic plates. The holograms were replaced in the interferometer for reconstruction by illumination with the same spherical wave originally used as the reference wave. The holograms were imaged with a telecentric system consisting of two identical lenses separated by twice their 60 cm focal length. A small aperture at the center of the system was used to select the particular component plane wave of interest. This system was found to be convenient, and has the useful characteristic of assuring 1:1 magnification. Fringe data was recorded by scanning the image of each interferogram with a precision vernier microscope. The entire system, except for the microscope, is shown in the photograph in Fig. 2.

Fig. 2. Photograph of the experimental apparatus.

Test Section

The glass test section was filled with deionized distilled water. Two horizontal circular disks, 10.5 mm in diameter were submerged in the water with their centers separated by 16 mm. Each disk

was heated by 11 turns of 28 gauge Chromel-A wire. The heaters
were connected in series to a filtered dc power supply adjusted to
give a heating rate of 9.22 W.

The objective of the experiment was to measure the temperature
distribution in a horizontal plane 3 cm above the plane of the
heated disks. Ten small thermocouple probes were inserted into
the water to provide a test of the accuracy of the interferometric
measurement. Each thermocouple had a chromel-constantan junction
formed from 40 gauge wires. The thermocouple beads were approximately 0.25 mm in diameter. The outputs of the thermocouples
were measured with a Honeywell Visicorder with flat-response galvanometers. This system was calibrated at six points against a
standard platinum resistance thermometer. Its accuracy during
the experiments was estimated to be better than $\pm$0.8^{o}C, including
allowance for drift.

Procedure and Results

A holographic interferogram was recorded by exposing the photographic plate to the object and reference waves simultaneously
before the heaters were energized. After the heaters were energized and a steady state was attained a second exposure was made
on the same plate. When the resulting hologram was developed, the
object waves were reconstructed by illuminating it with the reference wave. Along each viewing direction a fringe pattern could
be observed with the telecentric viewing system. The position of
each fringe was measured by scanning each view with the vernier
microscope. 17 views were used, and approximately 55 fringe locations were recorded in each view. Due to refraction, the angular range of the views was reduced to 46° in the test section.

Two views of the holographic interferogram are shown in Fig. 3.
These views are separated by 23°. Three pins and a sphere which
were used as fiducial marks for determining positions and viewing
directions are visible at the bottoms of these photographs. The
thermocouple probes can be seen extending downward into the plume.

The computer code used to analyze the interferometric data was
based on the grid method described in Ref. 6. A total of 933
fringe positions were recorded for reconstruction of the temperature field in one plane. From this data the refractive index was
determined at 27 points in the plane of interest by solving 933
algebraic equations in 27 unknowns by a least mean square algorithm. This high redundancy was required for good accuracy because the effective viewing angle was only 92°. Once the refractive index change f(x,y) was determined the temperature was calculated using a formula based on the data of Tilton and Taylor
(11):

$$T=5.6975[(1-0.18132 \times 10^5 xf)^{1/2}-1], \tag{4}$$

where f=n-n(0°C), and T[^{o}C].

A computer plot of the isothermal contours in a horizontal plane
3.0 cm above the surface of the heated disks is shown in Fig. 4.
The location and size of several thermocouple beads are also

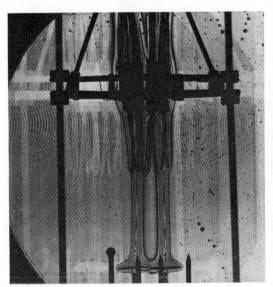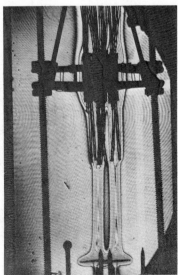

Fig. 3.   Two views of the holographic interferogram.

noted in this figure.   Table 1 indicates the temperature at these
locations computed from the interferometric data and the values
measured with the thermocouples.   The average difference between
the optical and thermocouple measurements is about 0.8°C, which
is the same as the uncertainty in the thermocouple measurements.
The maximum difference is about 1.3°C which is approximately 10
percent of the maximum temperature change in the plume.   The four
small isotherms at the edges of Fig. 3 are artifacts due to
"ringing".   These represent errors of about 0.5°C.

CONCLUSIONS

A holographic interferometer was constructed to provide multi-
directional interferometric data for measuring the temperature
distribution in an asymmetric natural convection plume in water.
Two phase gratings were used to produce 17 plane waves which
travelled through the test section in different directions.   The
data was analyzed by a method described in Ref. 6.   Comparison of
these measurements with simultaneous measurements by small, calibrated thermocouples indicated an accuracy of at least $\pm 1^\circ$C with
good spatial resolution.   Holographic interferometry is a practical and accurate method of measuring three-dimensional, asymmetric
temperature fields.   Its accuracy was found to be equivalent to
that of a good system of thermocouples, even when the angle of
view was rather restricted.

C. M. Vest, P. T. Radulovic

LABEL	TEMP.
1	23°C
2	24
3	25
4	26
5	27
6	28
7	29
8	30
9	31
10	32
11	33
12	34
13	35

1 мм

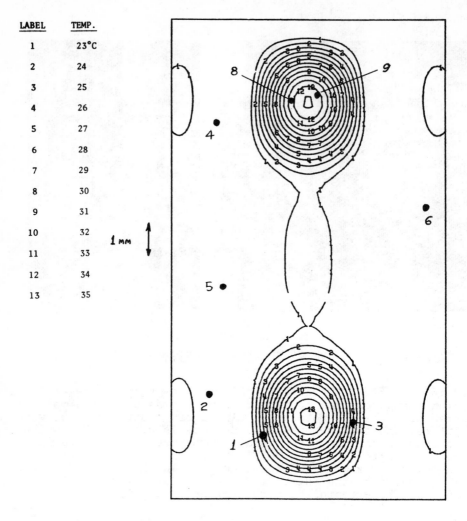

Fig. 4. Computer plot of isothermal contours.

ACKNOWLEDGMENT

This research was supported by the National Science Foundation under grant ENG-7400571.

REFERENCES

(1) W. Hauf and U. Grigull, Optical Methods in Heat Transfer, in Advances in Heat Transfer, J.P. Hartnett and T.F. Irvine, Jr., Eds., (Academic Press, New York, 1970), Vol. 6, pp. 134-36.

(2) Gel'Fand, I.M., Graev, M.I., and Vilenkin, N. Ya., General-
     ized Functions, Vol. 5, Academic Press, New York, 1966.

(3) M.V. Berry and D.F. Gibbs, The interpretation of optical
     projections, Proc. Royal Soc. A314, 143 (1970).

(4) R.D. Matulka and D.J. Collins, Determination of three-dim-
     ensional density fields from holographic interferograms,
     J. Appl. Phys. 42, 1109 (1971).

(5) H.G. Junginger and W. van Haeringer, Calculation of three-
     dimensional refractive index field using phase integrals,
     Opt. Commun. 5, 1 (1972).

(6) D.W. Sweeney and C.M. Vest, Reconstruction of three-dimen-
     sional refractive index fields from multi-directional
     interferometric data, Appl. Opt. 12, 2649 (1973).

(7) D.W. Sweeney and C.M. Vest, Measurement of three-dimension-
     al temperature fields above heated surfaces by holograph-
     ic interferometry, Int. J. Heat Mass Transf., 17, 1443
     (1974)

(8) Goodman, J.W., Introduction to Fourier Optics, McGraw-Hill,
     New York, 1968, pp. 69-70.

(9) C.M. Vest and D.W. Sweeney, Holographic interferometry of
     transparent objects with illumination derived from a
     phase grating, Appl. Opt., 9, 2321 (1970).

(10) M.M. Butsov, Use of a moire scatterer in holographic inter-
     ferometry, Sov. Phys.-Tech. Phys., 17, 325 (1972).

(11) L.W. Tilton and J.K. Taylor, Refractive index and disper-
     sion of distilled water for visible radiation, at tem-
     perature 0 to 60°C, J. Res. Natl. Bur. Std., 20, 490
     (1938).

TABLE 1   Comparison of Interferometric and Thermocouple Measurements

Thermocouple Number	Thermocouple Reading [°C]	Interferometric Measurement [°C]	Difference
1	24.4	25.6	1.2
2	22.6	21.8	0.8
3	27.0	25.7	1.3
4	21.6	20.4	1.2
5	22.0	21.4	0.6
6	22.2	22.4	0.2
7	34.2	33.8	0.4
8	34.5	34.2	0.3

INVESTIGATION OF CYLINDRICAL SYMMETRIC PHASE OBJECTS WITH HOLOGRAPHIC INTERFEROMETRY

Ervin Tanos

Research Institute of the Electrical Industry, 1158 Budapest, Hungary

ABSTRACT

A mathematical inversion which can be used to calculate
the data of every symmetrical phase objects.

INTRODUCTION

In our Institute we have been dealing with holographic interferometry of high temperature flowing gases, for example
technical plasmas. Our general outline is to determine the three
dimensional distribution of refractive index of plasmas and from
this the other physical parameters. The light rays pass through
these mediums without losses only the phases of the light change.
Consequently these plasmas may be named in a means phase objects.
To say generally the objective of this paper is to present a
mathematical technique by which the interferometric data can be
used to determine the distribution of refractive index of phase
objects.

The plasmas we investigated were flowing in a closed ceramic
tube. The possibility of the transmission of the light beam was
ensured by optical quality quartz windows which were smaller
than the total cross section of the tube. Beacuse of this it was
not possible to view the total cross section of the phase object
to be measured. Technically this problem is unsolvable. Since we
could not to use the well-known method of inversion we worked
out a mathematical method for inversion named by us "approximate
Fourier inversion". Using our method it is not necessary to know
the total cross section of the phase object. But the condition
of the application of the method is that the phase object must
have some symmetry.

THE METHOD

Let us consider Fig. 1. In this way we have a double exposure holographic interferogramm of the phase object bounded with planes. If the centers of the light and dark lines on the interferogramm are used as data points in the refractionless limit on the optical pathlengths S we obtain:

$$S_1 = \int_{L_1} u\ (\underline{r}_1)\ ds = (K+1)\frac{\lambda}{2}$$

$$S_2 = \int_{L_2} u\ (\underline{r}_2)\ ds = K\ \frac{\lambda}{2}$$

$$u(\underline{r}) = n(\underline{r}) - n_o \qquad\qquad /1/$$

where K is an integer, λ is the wavelength of the light and n/$\underline{r}$/ is the refractive index of the object relative to some reference field n_o. In the course of the derivation we will use the assumption that $S_1 > S_2$, to say the optical pathlengths change monotonically. Expanding in series S_1 we get in linear approximation:

$$S_1 = \int_{L_2} u(\underline{r}_2)ds\ +\ \int_{L_2} \nabla u(\underline{r}_2)\cdot\triangle\underline{r}\cdot ds \qquad\qquad /2/$$

where $\triangle\underline{r} = $ /$\triangle_x$; 0; $\triangle_z$/ can be seen on Fig. 1. After all we get the basic principle of the method:

$$S_1 - S_2 = \frac{\lambda}{2}\ =\int\nabla u\triangle\underline{r}\ ds \qquad\qquad /3/$$

Let us now use the symmetry feature of the object. Let be

$$y = g\ (r,x) \qquad\qquad /4/$$

the equation of the curves along which the function u is a constant u /r,z/. Here r is a parameter which is the distance from the origin along the y axis. In another manner for a given point /x;y/ the formule

$$r = f (x,y) \qquad\qquad /5/$$

gives the parameter r and the value of the refractive index
u/r,z/. It follows from this that the refractive index u/r/
depends only on two coordinates: r and z. Introducing the
coordinate transformation /4/ into the basic integral equation
yields:

$$\frac{\lambda}{2} = \int_{-\frac{L}{2}}^{\frac{L}{2}} \nabla u (x,y,z) \triangle \underline{r} \; dy = \int_{f(x,-\frac{L}{2})}^{f(x,\frac{L}{2})} \nabla u \left[x, g(r,x), z \right] \triangle \underline{r} \cdot$$

$$\cdot \frac{\partial g(r,x)}{\partial r} \; dr \qquad /6/$$

If there is a symmetry then necessary:

$$u \left[x, g(r,x), z \right] \underset{=}{=} u(r,z) \qquad\qquad /7/$$

Let us expand the function u /r,z/ in terms of a complete set
of orthogonal steady convergent function:

$$u(r,z) = \sum_{mn} a_{nm} \; \varphi_n (r) \psi_m (z) \qquad /8/$$

The usual formules of transformations are the following:

$$\frac{\partial u}{\partial x} = \frac{\partial u}{\partial r} \frac{\partial r}{\partial x} = \frac{\partial u}{\partial r} \propto (r,x) \qquad /9/$$

$$\frac{\partial u}{\partial r} = \sum_{mn} a_{nm} \frac{d\varphi_n (r)}{dr} \psi_m (z) \qquad /10/$$

$$\frac{\partial u}{\partial z} = \sum_{mn} a_{nm} \; \varphi_n (r) \frac{d\psi_m (z)}{dz} \qquad /11/$$

We recall the equation /6/ which may be rewritten by use of
equations /8/ - /11/ to the next form:

$$\frac{\lambda}{2} = \int\limits_{f(x,-\frac{L}{2})}^{f(x,\frac{L}{2})} \Big[\Delta_x \alpha(r,x) \sum_{nm} a_{nm} \frac{d\varphi_n(r)}{dr} \psi_m(z) +$$

$$+ \Delta_z \sum_{nm} a_{nm} \varphi_n(r) \frac{d\psi_m(z)}{dz} \Big] \frac{\partial g(r,x)}{\partial r} \, dr$$

/12/

Interchanging the order of integration and summation yields:

$$\frac{\lambda}{2} = \sum_{nm} a_{nm} \Big[\Delta_x \psi_m(z) A_n(x) + \Delta_z \frac{d\psi_m(z)}{dz} B_n(x) \Big] = \sum_{nm} a_{nm} C_{nm}(x,z)$$

$$A_n(x) = \int\limits_{f(x,-\frac{L}{2})}^{f(x,\frac{L}{2})} \alpha(r,x) \frac{d\varphi_n(r)}{dr} \cdot \frac{\partial g(r,x)}{\partial r} \, dr$$

$$B_n(x) = \int\limits_{f(x,-\frac{L}{2})}^{f(x,\frac{L}{2})} \varphi_n(r) \frac{\partial g(r,x)}{\partial r} \, dr$$

/13/

The functions A_n, B_n, ψ_m, $\frac{d\psi_m}{dz}$ can be computed and the
values of Δ_x and Δ_z can be measured from interferometric
fringe pattern. Of course in practice the upper limits of the
summation both for n and m are finite numbers: N an M.
If different Δr values are measured equation /13/ yields a
set of /M x N/ linear algebraic equations which can be solved
for the /M x N/ unknown a_{nm}.

We have tested the method for a cylindrical symmetric phase
object. Because of the requirement of the plane boundary we

must assume that:

$$n(\underline{r}) = n(r,z) \qquad r \leqq R$$

$$n(\underline{r}) = n_o \qquad r > R$$

/14/

where R is the radius of the phase object. It follows from this that:

$$u(\underline{r}) = u(r,z) \qquad r \leqq R$$

$$u(\underline{r}) = 0 \qquad r > R$$

/15/

To the equations /4/ - /6/ correspond the next formules:

$$y = \sqrt{r^2 - x^2}$$

$$dy = \frac{r}{\sqrt{r^2 - x^2}}\, dr_i \quad \int_{-\frac{L}{2}}^{\frac{L}{2}} \longrightarrow 2\int_{0}^{\sqrt{R^2 - x^2}} \longrightarrow 2\int_{x}^{R}$$

/16/

and

$$\frac{\lambda}{2} = 2 \int_{x}^{R} \left[\frac{\partial u(r,z)}{\partial r} \frac{x}{r} \triangle_x + \frac{\partial u(r,z)}{\partial z} \triangle_z \right] \frac{r}{\sqrt{r^2 - x^2}}\, dr$$

/17/

The function u /r,z/ may be written in the form:

$$u(r,z) = u(r) \cdot u(z)$$

/18/

Introducing this formule to the equation /17/, make partial integration in the first term of the equation and transforming the second term yields:

$$\int_{x}^{R} \frac{du(z)}{dz} \triangle_z u(r) \frac{r}{\sqrt{r^2 - x^2}}\, dr =$$

$$= \frac{du(z)}{dz} \triangle_z \left[u(R) \sqrt{R^2 - x^2} - \int_{x}^{R} \frac{du(r)}{dr} \sqrt{r^2 - x^2}\, dr \right]$$

/19/

and

$$\int_x^R u\,(z)\,\triangle_z \frac{du(r)}{dr}\ \frac{r}{\sqrt{r^2-x^2}}\ dr = u(z)\triangle_x \int_x^R \frac{\partial}{\partial x}\left[-\frac{du(r)}{dr}\sqrt{r^2-x^2}\right]\,dr=$$

$$= -\,u(z)\triangle_x \frac{d}{dx}\int_x^R \frac{du(r)}{dr}\,\sqrt{r^2-x^2}\,dr \hspace{3cm} /20/$$

Expanding the function u /r/ into a Fourier series we obtain a formule correspond to the equation /8/:

$$u(r) = \frac{a_o}{2} + \sum_{k=1}^{\infty}\ a_k\ \cos\left(k\,\pi\frac{r}{R}\right) \hspace{3cm} /21/$$

By this

$$u(R) = \frac{a_o}{2} + \sum_{k=1}^{\infty}\ a_k\ \cos\,(k\,\pi) \hspace{3cm} /22/$$

and

$$\frac{du(r)}{dr} = \sum_{k=1}^{\infty}\ a_k\left[-\frac{k\pi}{R}\ \sin\left(k\,\pi\,\frac{r}{R}\right)\right] \hspace{2cm} /23/$$

Introducing the formules /21/ - /23/ to the equations /19/ - /20/ and the next functions:

$$S_k(x) = -\frac{k\pi}{R}\int_x^R \sqrt{r^2-x^2}\ \sin\left(k\pi\frac{r}{R}\right)\,dr \hspace{2cm} /24/$$

and

$$C_k(x) = -\frac{dS_k(x)}{dx} \hspace{4cm} /25/$$

we obtain the final formule correspond to the equation /13/:

$$\frac{\lambda}{2} = a_o \frac{du(z)}{dz} \triangle_z \sqrt{R^2 - x^2} + 4 \sum_{k=1}^{\infty} \left[u(z) \triangle_x c_k(x) + \right.$$

$$\left. + \frac{du(z)}{dz} \triangle_z \sqrt{R^2 - x^2} \cos(k\pi) - \frac{du(z)}{dz} \triangle_z S_k(x) \right] a_k \qquad /26/$$

To use the equation /26/ is necessary to know the function
u/z/. If the function u/z is unknown then it must be written
in the form of a series for example: $u/z/ = 1 + \alpha z + \beta z^2 + \dots$
Then we have a set of equations for the unknown:

$a_k; \quad \alpha \cdot a_k; \quad \beta \cdot a_k; \quad \dots$

In the course of our calculations we had a model object which
was possible to describe by the next refractive index:

$$u(r,z) = (1 - \alpha \cdot z) \cdot (ar^4 + br^2 + c) \qquad /27/$$

The capacity of the computer at our disposal was relatively
small therfore we could calculate only a limited number of the
terms in the equation /26/. This way we can get the coefficients
by accuracy 5 - 10 %. In our opinion the high accuracy depends
only on the capacity of the computer and the method can be used
to calculate the data of every symmetrical phase objects.

The author would like to thank to his colleaques for their
assistance in this work.

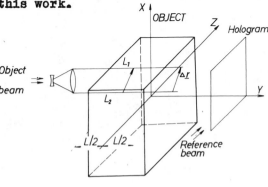

Figure 1.

HOLOGRAPHIC CONTROL OF DIFFUSION COEFFICIENTS IN WATER SOLUTIONS: CRYSTAL GROWTH FROM SOLUTIONS

Federico Bedarida,* Livio Zefiro* and Carlo Pontiggia**

**Istituto di Mineralogia dell'Università, Corso Europa,
Palazzo Scienze, 16100 Genova, Italy*
**Istituto di Scienze Fisiche dell'Università, Viale Benedetto XV, 16100 Genova, Italy

ABSTRACT

The present work is an application of transmission holographic interferometry to crystal growth from solution. In other words it is intended to control the mechanisms of crystal growth by checking also the concentration gradients and the convective movements of the solution.

INTRODUCTION

The idea for this work has been given by the interest the crystallographers have in crystal growth from solution. Many theoretical **hypothesis** on the crystal growth from solution have been put forward up to now, but the experimental works are rather few (1,2,3 4,5).

To resolve the differential equation that rules crystal growth from solution two terms have to be known

$$dM/dt \qquad \text{and} \qquad dC/dt$$

where dM/dt is the amount of matter deposited on the surface in a unit time interval, and dC/dx is the component of the concentration gradient normal to the face. dM/dt is rather easily measured at the microscope, since it depends upon the velocity of growth of the crystal faces. To measure dC/dx one needs optical interferometry to check the optical path variation related to the concentration gradient.

In the first works a crystal was observed with a normal microscope. The crystal was kept between two nearly parallel half silvered glass plates in order to obtain interferometric fringes in mono-

chromatic light. In those experiments, even with rather strict
conditions, the experimental results were not fitting with the
theoretical interpretation of the phenomenon, since convective
movements effective also in reduced volumes were not taken into
account, nor were they checked experimentally.

For these reasons in our laboratory such experiments have been
remade using microscopical holographic interferometry that have
allowed to study greater amounts of solution. Working with the
vessel in vertical position it has been possible to show very
clearly the convective movements in the solution.

EXPERIMENTAL RESULTS

A schema of the apparatus is drawn in Fig. 1 and photographed in
Fig. 2.

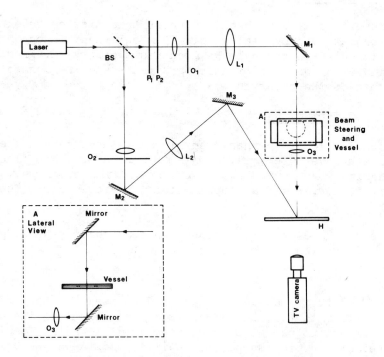

Fig. 1. Schema of the apparatus used with horizontal settlement of the vessel: BS=beam splitter,
P=polarizers, O=objectives, L=lenses, M=mirrors,
H=holographic plate, A=beam steering and vessel:
a vertical view of A is given on the left bottom
in dotted lines.

Fig. 2. General view of the experimental
apparatus.

The blue line ($\lambda = 4880$ Å) of a Spectra-Physics cW argon ion
laser model 162 has been used. The laser beam illuminating the
object and the reference beam had plane wave fronts (plane has
the meaning that the wave fronts are obtained through an objective and a lens, placed at a distance that is the focal length of
the lens). By this way the holographic interferometric technique
in real time was rather effective.

The hologram is done of a vessel, drawn in Fig. 3 filled with
distilled water. A 6x microscopic objective put between the vessel
and the holographic plate gives an enlarged image of a portion of
the vessel that may be registered by a macro TV camera system.
The hologram accurately put in its previous position gives an
image of the vessel which is compared with the direct image of
the vessel containing a solution of variable concentration or a
crystal which grows in the solution, since the concentration has
a strong gradient around the crystal.It was not sure at the very
beginning of the experiment that convective movements in the
liquid were not disturbing it, but in the experimental conditions
used no disturbance has been observed.

As it was done in the first experiments of the previous Authors,
the growth of $NaClO_3$ crystals has been studied. These crystals

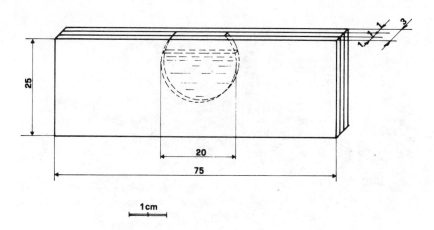

25

20

75

1cm

Fig. 3. Schema of the vessel.

are cubic and give without difficulty strongly supersaturated solutions in which it is possible to follow the growth from a small
crystal seed put in the solution.

Firstly the vessel was settled horizontal between the mirrors of
a beam steering (see detail in dotted lines in Fig. 1) obtaining
the results shown in Fig.4 and Fig. 5.

0.5 mm

Fig. 4. Fringes showing skeletal growth of a
$NaClO_3$ crystal (horizontal arrangement).

Fig. 5. Fringes showing regular growth of NaClO$_3$
crystal (horizontal arrangement).

During the growth the crystal takes solute out of the supersaturated solution in its neighbourhood more than what is replaced
from the outer solution. All around the crystal a zone is formed
where the concentration is lowered. This fact is checked by the
formation of interference fringes, corresponding to the variation
of the refractive index.

In a second experiment the vessel was settled vertical, and the
results are shown in Fig. 6 and in Fig. 7. From the surface of
the crystal the less concentrated solution goes up along a tubu-
lar channel, while a descending current outside this channel
brings fresh material on the faces of the crystal from more
concentrated regions. For this reason convective movements enhance
the velocity of growth of the crystal faces. The diameter of the
zone of the convective movements is some mm.

It must be remembered that convective movements are always present
even in very small volumes: for this reason theoretical calculations of crystal growth based only on diffusion are deceivable.

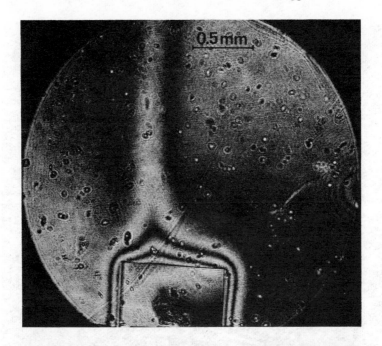

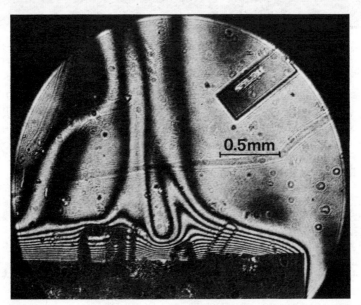

Fig. 6 and Fig. 7. Fringes showing convective movements in solution during growth of $NaClO_3$ crystal (vertical arrangement).

MEASUREMENT OF THE DIFFUSION COEFFICIENT

In the case of diffusion from a horizontal boundary which is initially sharp at x = a and which separates the solution and the solvent in a vertical liquid column of total length b, the resolution of the Fick's second law:

$$\frac{\partial c}{\partial t} = D \frac{\partial^2 c}{\partial x^2}$$

is a Fourier series of the form:

$$c\left(x,t\right) = c_0 \left\{ \frac{a}{b} + \frac{2}{\pi} \sum_{n=1}^{\infty} \frac{\sin\left(\frac{n\pi a}{b}\right)\cos\left(\frac{n\pi x}{b}\right)}{n} \, e^{-\left(\frac{n\pi}{b}\right)^2 Dt} \right\}$$

where c is the concentration at the level x and at the time t, and D is the diffusion coefficient.
The variations of the index of refraction correspond to variations in the concentration and are shown through holography with a system of horizontal fringes, Fig. 8.

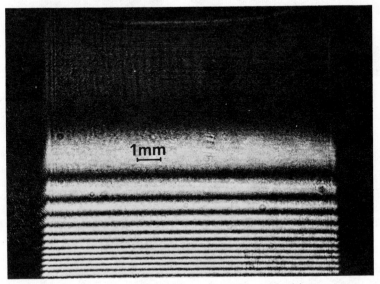

Fig. 8 Diffusion fringes in KCl solution.

By a computer minimization program it has been possible to obtain a reasonable value of $D = 2 \cdot 10^{-5}$ $cm^2 sec^{-1}$ in the case of KCl diffusion. A further refinement of these results may be achieved in strict conditions of very sharp meniscus, that implies a technical refinement of the experimental equipment of our laboratory.

REFERENCES

1) Berg, W. F., Crystal growth from solution, Proc. Roy. Soc. A 164, 79 (1938).

2) Bunn, C. W., Concentration gradients and the rate of growth crystals, Disc. Faraday Soc. 5, 40 (1949).

3) Goldsztaub, S., Role de la diffusion dans la croissance de cristaux à partir de solutions. Croissance de composes mineraux monocristallins, Masson, Paris, 1969.

4) Humphreys Owen, W. S., Crystal growth from solution, Proc. Roy. Soc. A 197, 218 (1949).

5) Kern, R., Mesures de la concentration de la solution pendent la croissance des cristaux, Bull. Soc. Franc. Minér. Crist. 76, 338 (1953).

SESSION 8.

HOLOGRAPHIC INTERFEROMETRY III

Chairman: K. P. MIYAKE

INTERFEROMETRIC INFORMATION

Nils Abramson

Div. Production Engineering, The Royal Institute of Technology,
100 44 Stockholm 70, Sweden

ABSTRACT

Interference fringes are the result of phase differences, often caused by changes in optical pathlengths. The evaluation of interferograms therefore depends on the study of distances and changes of distances. The holo-diagram is based on a bi-polar coordinate system and aimed at simplifying and unifying methods to study the distances from the object to the points of illuminations and the points of observations. The interference fringes are defined as intersections of the object by a set of interference surfaces in object space. When no real fringes exist in object space we introduce the concept "wavefronts of observation" which cause imaginary interference fringe surfaces intersecting the object. Introduced is also the new method of using holographic interference fringes as carriers of confidential information.

INTRODUCTION

Ordinary interferograms are two-dimensional representations of three-dimensional space. The third dimension is visualized by the interference fringes seen on the flat image of the object which like the level lines of a map represent loci of equal height.

Holographic interferograms (e.g. a double exposed hologram of a deformed object) are three-dimensional representations of a four-dimentional space. The fourth dimension is visualized by interference surfaces in space that intersect the three-dimensional image of the object and represents e.g. loci of equal displacement.

If the interference fringes do not exist in the object space but only in the image of the object space (as e.g. in hologram interferometry) they can be visualized as formed by interference between the illumination waves and the lightwaves used for the observation. The latter consist of only those lightwaves that from the object move towards the observer and thus their wavefronts appear to be spherical and moving inwards to the point of observation. Their phase is determined by the phase of the reference wave. For these waves we have introduced the concept "spherical wavefronts of observation".

If both the point of illumination and the point of observation are within finite distances the interference surfaces consist of a set of ellipsoids with those points at the foci.[2,3] (The stationary three-dimensional moiré fringe surfaces formed by one set of spherical expanding wavefronts and one set of spherical wavefronts that move inwards.)

269

The fringes in double exposure holography are caused by the difference in fringe patterns formed when the object in its two different positions intersects the ellipsoidal (primary) fringe surfaces.[1] These secondary interference fringe surfaces are analogous to the three-dimensional moiré fringe surfaces formed in empty space by two sets of ellipsoids one of which is displaced and rotated the same way as the object was changed between the exposures. The fringes exist everywhere in the reconstructed image space but are only seen where they intersect the image of the object.

Let us fix a point detector in the space among the hyperboloids formed by the reference beam and the beam from one single object point. If this point is moved along one of the ellipsoids no phase change will be found between object and reference beam at the detector. If, however, one ellipsoid is crossed by the object point also one hyperboloid will cross the detector.

If instead the object point is fixed and the point detector is moved along one of the hyperboloids there will be no phase shift between object and reference beam at the detector. If, however, one hyperboloid is crossed by the detector the result will be equivalent to the object crossing one ellipsoid.

Consequently there exists a one-to-one relation between the intersections of the ellipsoids by an object point and the intersection of the hyperboloids by points on the hologram plate. The result is that interference fringes caused by displacement vectors can be calculated by studying the moiré patterns formed by addition of their corresponding Fourier transforms. This statement is the Fourier transform of the more accepted statement that the moiré patterns formed by two sets of fringes can be calculated by addition of the corresponding vectors in the Fourier space.[4]

If we accept the statements given here we also have to accept the unifying method for evaluating fringe patterns and calculating the fundamental resolution of optical systems which was introduced in Ref. 1. The method also explains how the manipulation of the hyperbolas on the hologram plate transforms the interference fringes in the reconstructed image space e.g. as described in the experiments of Ref. 5 and 6.

THREE-DIMENSIONAL MOIRÉ OF TWO SETS OF SPHERES

Let A and B be the centers of two sets of concentric equidistant spheres (separation ΔR) and try to visualize the moiré effect as the spheres intersect in space. [1] The moiré patterns have to be rotational symmetric around A and B so let us study a cross section through this axis. (Fig. 1)

If every second space between the spherical shells is filled with darkness, there will be a set of dark and bright rhombs. (In the three-dimensional case: rings with a rhombic section). The rhombs form one set of ellipses (rotational symmetric ellipsoids) and one set of hyperbolas (rotational symmetric hyperboloids). A and B are in both cases focal points. The rhombs (rhombic rings) are identical along circles (sort of toroids) passing through A and B, because the peripheral angle 2α is constant. Thus the separation of the set of ellipsoids ($k \cdot \Delta R$) and the separation of the set of hyperboloids ($k^{*} \cdot \Delta R$) are constant along these toroids. ($k = 1/\cos \alpha$ $k^{*} = 1/\sin \alpha$)

The ellipsoids and the hyperboloids intersect everywhere at right angles. If one of the focal points is moved away to infinity the set of ellipsoids and the set of hyperboloids are transformed into two identical sets of

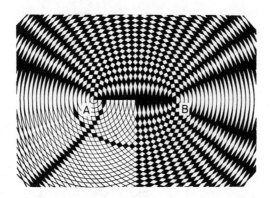

Fig. 1 Moiré fringes caused by two sets of concentric
spheres.

paraboloids intersecting everywhere at right angles. If both focal points are
moved away to infinity the ellipsoids and the hyperboloids are transformed
into flat surfaces intersecting at right angles.

DYNAMIC MOIRÉ

If the two sets of spheres around A and B both start expanding outwards with
the speed of light (c) the hyperboloids will remain stationary while the
ellipsoids start moving outward, away from AB, with a speed of k · c (faster
than light).

If the sets of spheres around A and B both start contracting inwards with the
speed of light the hyperboloids will also this time remain stationary while
the ellipsoids start moving with the speed of k · c, but this time inwards.

If the spheres around A are expanding while those around B are contracting,
both with the speed of light, the ellipsoids will remain stationary while
the hyperboloids move to the right (Fig. 1) with the speed of $k^* · c$.

If finally the spheres around A are contracting while those around B are
expanding, both with the speed of light, the ellipsoids will still remain
stationary while the hyperboloids move with the speed of $k^* · c$ but this time
to the left (Fig. 1)

INTERFEROMETRY

If the separation of the spherical shells in half the wavelength of light
($\Delta R = \lambda/2$), the sets of hyperboloids and ellipsoids represent interference
surfaces in space which are seen when they are intersected by an object.
Because our detection systems are slow compared to the frequency of light
only the stationary fringes have as yet been of any practical interest. The
ellipsoids have such angles that they could reflect light from A to B and
their separation is such that the Bragg conditions are fullfilled everywhere.

The same is true for the hyperboloids but they reflect light from A as if it
came from B.

Young's Fringes

If A and B are two mutually coherent point sources producing expanding spherical waves of illumination the intersection of the hyperboloids by a screen
represent the wellknown Young's fringes which appear projected on to the
screen. [1]

If a point object (C) is moved around in space it produces one pulse of light
each time it intersects one hyperboloid. When it is in a dark fringe the
signal will be zero whether C exists or not. These fringes of darkness determine the fundamental resolution of any system based on this configuration.

Young's Fringes of Observation

If A and B are two mutually coherent points of observation e.g. two closely
spaced pinholes in front of the eye, the retina will only receive light from
two sets of spherical wavefronts of observation which contract towards those
two points and have the same phase. (The illumination should be monochromatic).
The observer will therefore see fringes as if there existed hyperboloids in
space which were intersected by the object. These imaginary fringes could be
named Young's fringes of observation.[1]

If a point object (C) is moved around in space it produces one pulse at the
retina each time it intersects one of the hyperboloids. When it is positioned
in a "fringe of blindness" the signal at the retina will be zero whether C
exists or not. These fringes of blindness determine the fundamental resolution
of any system based on this configuration.

Holographic Fringes

If finally A is a point source of illumination while B is a point of observation which is made phase sensitive (coherent) by receiving also a direct
reference beam from A we have the case of one set of expanding spheres and
one set of contracting spheres. Thus the stationary fringes consist of a set
of ellipsoids with A and B as focal points. Also these fringes are imaginary
(or subjective [6]) as they depend on the position and phase of the observation
point B.

If a point object (C) is moved around among the ellipsoids there will be one
pulse at the detector point B each time an ellipsoid is intersected. There
exist ellipsoidal surfaces along which C will transmit light from A to B with
such a phase that the intensity at point B (which also receives the direct
reference beam from A) will be the same whether C exist or not. Thus there
exist also in this case a set of imaginary interference fringes in empty space
which are observed when they are intersected by the object. These "fringes
of blindness" determine the fundamental resolution of any measuring system
based on this configuration.

MEASURING SYSTEMS

If the Young's fringes are projected on to an object and doubly exposed using
an ordinary camera, the moiré of the two intersections of the hyperboloids can
be used to evaluate the relative motion between the object and the fringes.

If an object is doubly exposed using a camera the lens of which is blocked but for two diametral holes, the moiré of the two intersections of the hyperboloids can be used to evaluate the motion between the object and the "Young's fringes of observation".

If finally a hologram is double exposed the moiré of the two intersections of the ellipsoids can be used to evaluate the relative motion between object and the imaginary fringes.

As described in Ref. 1 and Ref. 9 the moiré of the hyperboloids or ellipsoids can be used as a unifying method for evaluating the results and resolution of a great number of optical systems. The "fringes of blindness" are probably the key to any imaging and measuring system. They determine the resolution whether they are produced by the moiré effect from conventional gratings, from diffraction fringes, from shadows or from real or imaginary interference fringes (including speckles).

MOIRÉ SIMPLIFICATION RULES

Let us study holographic contouring, which happens to be a good example for the demonstration of the generality of the described methods.

Contouring by Illumination Shearing

A holographic plate is double exposed. Between the two exposures the illumination point is displaced laterally. The first exposure is in the object space represented by one set of ellipsoids the focal points being the point of illumination A_1 and the point of observation B_1 (Fig. 2). Second exposure is similar to the first but A_1 is displaced to A_2 while B_1 is unchanged.

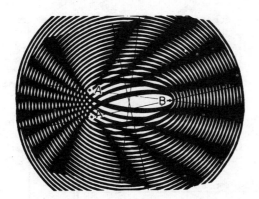

Fig. 2   Two identical sets of ellipsoids one of which has
been rotated around the common focal point B.

The resulting moiré pattern appears to be a set of hyperboloids with A_1 respectively A_2 as focal points. This result is quite natural as the ellipsoids were in the first place caused by the moiré of two sets of spheres. As one set of spheres (wavefronts of observations around B) was never changed we only have to deal with the moiré of the two sets of spheres around A_1 respectively A_2 (wavefronts

of illumination). Thus the resulting fringes are simply the ordinary Young's
fringes projected on to the object.

The simplification rules can be described by the block-diagram of Fig. 3.

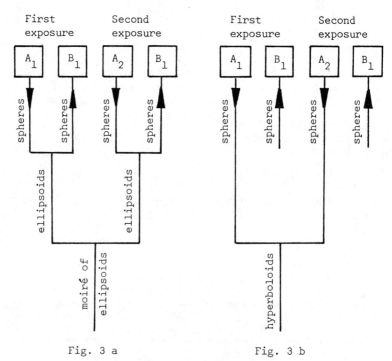

Fig. 3 a   Fig. 3 b

Fig. 3 a   Block-diagram of the formation of fringe sur-
           faces in the object space when the point of
           illumination A is displaced laterally between
           the two exposures of a hologram.

Fig. 3 b   Simplified block diagram of the same situation
           as in Fig. 3 a.

All the components which have not changed between the two exposures are of no
interest for the fringe formation. Thus we can exclude B_1 from the evaluation
and only study the moiré of A_1 and A_2. The block-diagram is simplified to that
of Fig. 3 b in which the holographic process is not even involved.

Contouring by Illumination and Observation Shearing

A hologram plate is doubly exposed. Between the two exposures the illumination
point and the holographic plate are both displaced laterally. The first exposure is as before represented by one set of ellipsoids, the focal points being
the point of illumination A_1 and the point of observation B_1. Prior to the
second exposure A_1 is displaced to A_2 and B_1 to B_2.

From the block-diagram of Fig. 4 a and 4 b it can be seen that the resulting
contouring fringes can either be represented by the moiré of two sets of ellipsoids or two sets of hyperboloids which if the distances are large can be

approximated into flat surfaces (fig. 5).

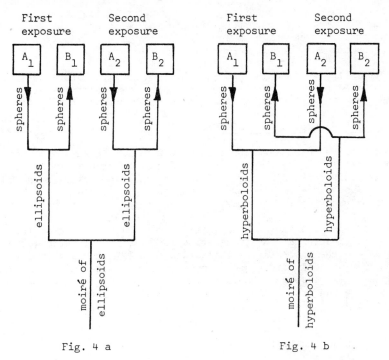

Fig. 4 a

Fig. 4 b

Fig. 4 a Block-diagram of the formation of fringe surfaces
in the object space when the point of illumination
A and the point of observation B are both displaced
between the two exposures of a hologram.

Fig. 4 b Alternative block-diagram of the same situation as
in Fig. 4 b.

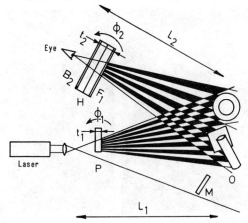

Fig. 5   Contouring by independent illumination and
observation shearing.

Practical results of this method are published in Ref. 6 in which the shearing
of the observation was made during reconstruction by tilting a sandwich hologram (H of Fig. 5). The contouring fringes are represented by the chains of
rhombs produced by the moiré of Young's fringes of illumination and Young's
fringes of observation.

Observe that Fig. 5 could just as well be used to explain shadow-moiré topography[7] of the object O. In that case the hologram plates (H) and the shearing
are taken away and instead a grating is inserted between the letters L_1 and L_2.

FRINGES CAUSED BY OBJECT MOTIONS

Contouring fringes can also be produced by translating an object between the
two exposures of a hologram [8]. A motion of the object C in relation to fixed
A and B is equivalent to moving the rigidly connected points A and B in relation to C. Thus either of the two methods of Fig. 4 a and Fig. 4 b can also in
this case be used for the evaluation of the contouring fringes in object space.

Finally there are all the motions that can be measured using ordinary hologram
interferometry: deformation, translation, rotation and vibration; for which the
ellipsoids are the most general evaluation tool as described in e.g. Ref. 9.
In that paper was only discussed hologram interferometry simply because this
field appears to represent the most general interference problems. The methods
presented here can however be used for conventional interferometry, stellar
interferometry, doppler measurement, speckle and moiré methods, radar, Doppler,
microscopy etc. as described in e.g. Fundamental Resolution of Optical Systems
(Ref. 1). Since its publication some new papers have been published in which
results are presented that support my statements. (Ref. 10 and 11.)

APPENDIX I

Example of Moiré Equivalence

A sandwich hologram was made of a one meter high centrifugal pump. Between
the two exposures the internal pressure was changed. To study the deformation

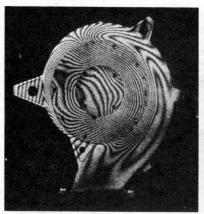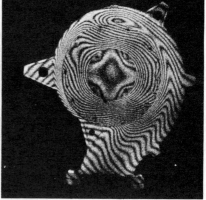

Fig. 6   Fringe manipulation by shearing interferometry
          (moiré at the hologram plate).

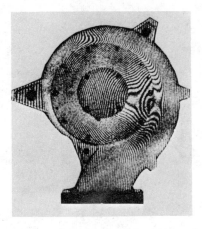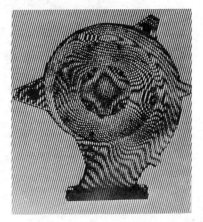

Fig. 7 Fringe manipulation by moiré at object image plane.

of the center of the plate the fringes were manipulated by interferometry at
the hologram plate (Fig. 6) and by moiré at the object image (Fig. 7). Observe the identical results. With both methods it is possible to minimize the
number of fringes intersecting the center plate and thus eliminate the influence of rigid motion.(From Nils Abramson, Proceedings of International Electro-
Optics International 74 conference in Brighton, England, page 35.)

APPENDIX II

Interference Fringes as Carriers of Confidential Information

We would like to introduce a new method to use holograms as carriers of confidential information. There exists since a long time a number of optical
methods (e.g. holography) to scramble an image of the confidential information.
Using the same methods in reverse it is possible to "unscramble" the information. Our new method, however, is based on the use of hologram interference
fringes from two combined hologram plates as carriers of the hidden information.

When this new method is used it is not even theoretically possible to find
any hidden information using only one of the plates because no information,
coded or uncoded, exists on any of the two plates. The information is for the
first time produced by interference phenomena when the difference between two
images is measured with interferometric precision during reconstruction. The
phase differences carrying the hidden message are much smaller than the random
phase variations existing on the diffuse surface, the deformation of which
was used to record the information. Because of its simplicity, large informations capacity and high security level we believe this method to be of high
practical value e.g. for identification cards. Figure 8 is from one of our
earlier experiments in which the letters N A are recorded in the form of interference fringes from two combined hologram plates.

Fig. 8   Hologram interference fringe letters.

References

1. Nils Abramson, Fundamental Resolution of Optical Systems, Optik, 39,
    141, December (1973).

2. C.L. Andrews, C.R. Charpentier, Confocal Conic Sections as Interference
    Patterns, American Journal of Physics, 40, 994, July (1972).

3. C.L. Andrews, D.P. Margolis, Elementary Use of Spheroidal Coordinates,
    American Journal of Physics, 4, 697, May (1973).

4. Olof Bryngdahl, Moiré: Formation and Interpretation, Journal of Optical
    Society of America, 64, 1287, October (1974).

5. Nils Abramson, Sandwich Hologram Interferometry 2: Some Practical
    Calculations, Applied Optics, 14, 98, April (1975).

6. Nils Abramson, Sandwich Hologram Interferometry 3: Contouring,
    Applied Optics, 15, 200, January (1975).

7. Hiroshi Takasaki, Simultaneous All-Around Measures of a Living Body
    by Moiré Topography, Journal of the American Society of Photogrammetry,
    XLI, No 12, 1527, December (1975).

8. Nils Abramson, Holographic contouring by translation, Applied Optics,
    15, 1018, April (1976).

9. Nils Abramson, The Holo-diagram IV: A Practical Device for Simulating
    Fringe Patterns in Hologram Interferometry, Applied Optics, 10, 2155,
    September (1971).

10. F. Durst and W.H. Stevenson, Visual modelling of laser Doppler anemometer
    signals by moiré fringes, Applied Optics, 15, 137, January (1976).

11. D. Joyeux, Real time measurement of very small transverse displacements
    of diffuse objects by random moiré, Applied Optics, 15, 1241, May (1976).

REAL AND NON-REAL TIME HOLOGRAPHIC NON-DESTRUCTIVE TESTING

Y. Katzir, A. A. Friesem and Z. Rav-Noy

Department of Electronics, The Weizmann Institute of Science,
Rehovot, Israel

ABSTRACT

A holographic non-destructive test procedure for identifying small defects and anomalies in both brazed and epoxi-bonded honeycomb panels is described. The tested panels include aluminum, steel, and composite materials, and appropriate stressing techniques were selected for each material; the stresses were introduced by means of pressure, vacuum and vibration. In addition to conventional photographic recording materials, photoconductor-thermoplastic devises, capable of rapid and in-situ development were incorporated into the holographic set-up to allow real-time operation. Experimental results demonstrating that both large and single cell disbonds can be detected in non-real and real time are presented.

INTRODUCTION

The application of holographic techniques to non-destructive testing (HNDT) has received a great deal of attention in recent years (1,2,3). Because of its extreme sensitivity, holographic interferometry allows for the detection of small defects and anomalies in diffuse three dimensional objects. Many holographic and stressing techniques have been considered for general HNDT, but most often these techniques must be especially adapted for each item to be tested, and in some cases for the nature of flaws that are sought.

In this paper, we describe the procedure and results of HNDT investigations for identifying disbonds in multilayered honeycomb panels. These investigations were conducted within the framework of a feasibility study, aimed at implementing HNDT in industry.

We will present specific examples where the holographic interferometry technique is used in a nondestructive test of honeycomb panels to detect flaws of areas ranging from several square millimeters to several square centimeters. The emphasis will be on recent investigations at our laboratories to detect disbonds in multilayered metallic and composite honeycomb and laminated structures which have progressed to the point where actual manufactured parts with dimensions of 0.5 meter by 0.5 meter brazed honeycomb structures can be readily tested. The holographic and various stress techniques will be described, and experimental results demonstrating that large as well as single cell disbonds can be easily detected will be presented.

Another important aspect of our work is the development of a complete system for performing the holographic tests in real time. The salient feature of the system is the photoconductor-thermoplastic devices on which holograms may be recorded in-situ and in virtually real time. The use of such recording material avoids the awkward wet chemical processing of conventional photographic films and ensures that the hologram occupies precisely the same position during readout as it did during recording.

TEST SAMPLES AND EXPERIMENTAL PROCEDURE

The basic honeycomb structure is made of two thin skin layers separated by a layer of honeycomb core; epoxy-bond or brazing at the interfaces joins the layers together. The tested samples consisted of three distinct series of honeycomb panels:

(a) Epoxy-bonded aluminum to aluminum square test panels of 25 cm sides, with skin thickness of 0.52 mm and core thickness of 1.3 cm. Controlled disbonds of various sizes were introduced by means of thin teflon sheets between the adhesive layer and either the skin or core.

(b) Bonded composite skin to aluminum core test panels having areas of 20 x 10 cm and 14 mm overall thickness. The skin was 1 mm thick glass fibre reinforced polyester, where 90% of the fibres lie along the smaller dimension of the panel. Disbonds were introduced into each plate by inserting thin teflon discs of varying diameters between the adhesive layer and either the skin or core.

(c) Brazed stainless steel actual aircraft panels shaped in the form of parallelogram with 50 cm sides. Skin thickness was 0.2 mm, and overall thickness of the panel varied between 7.5 and 2.7 cm. The panels were tested for either disbonds due to absence of braze material, or accumulations of excessive braze material.

Three basic stressing techniques were employed: (1) vacuum stressing. Applied by attaching a transparent plexiglass chamber to the flat surface of the tested panel so the panel was illuminated and viewed through the vacuum chamber. In some cases it was necessary to apply vacuum to both surfaces of the test panels.

(2) Internal pressure stressing. This was possible with the brazed samples of the "breathing" type of honeycomb structure where small holes were drilled in the cell walls of the honeycomb core. Internal gas pressure was applied through these holes to "inflate" the panel.

(3) Vibration stressing. The test samples were excited by a piezoelectric transducer, driven by an oscillator through a power amplifier. The appropriate resonant frequencies, at which a maximum number of disbonds were excited, were first recorded using either speckle or real time inspection, and then time average recordings were made.

The usual holographic recording and readout arrangements necessary for double exposure, time average, and real time interferometry were employed; the illumination source was an argon ion CW laser emitting green light of 514.5 nm wavelength. The holograms were recorded on both conventional high resolution photographic emulsions and photoconductor-thermoplastic recording devices.

PHOTOCONDUCTOR-THERMOPLASTIC DEVICES

Photoconductor-thermoplastic (PC-TP) devices are a family of erasable storage
materials on which holograms are recorded as surface deformations. The materials are sensitized by charging to a positive or negative potential, exposed
to actinic radiation, and developed by heating to the flow temperature. Electrostatic forces produce deformations which are proportional to the light
intensity used for exposure; a stable hologram is formed before surface
tension and elastic forces of the plastic layer restore the original smooth
surface. The hologram can then be erased by heating the thermoplastic layer
beyond the softening point. The application of greater heat releases residual charges and restores the original smooth surface of the thermoplastic
layer so the record sequence can be repeated.

The basic PC-TP device is composed of a glass substrate coated with a transparent conductor, a photoconductor layer, and a thermoplastic layer. The
photoconductive layer of our device was the   organic polymer poly-n-vinyl
carbazole (PVK) sensitized with trinitrofluorenone (TNF), and the thickness
of the layer was about 2μm. The thermoplastic was vinyl-toluene with a
layer thickness of about 0.5μm.

The salient features of photoconductor-thermoplastic recording materials which
makes them most attractive for HNDT are: (1) write-read-erase capabilities;
(2) in-situ recording, development, and fast readout; and (3) virtually
real time recording cycle (typically less than 3 seconds). As such, the use
of these materials, which avoids the awkward wet chemical processing, ensures
that the processed hologram occupies precisely the same position during
readout as it did during recording. Thus, in addition to allowing rapid hologram recording for double exposure and time average holography, these materials
are excellent for real-time holographic interferometry.

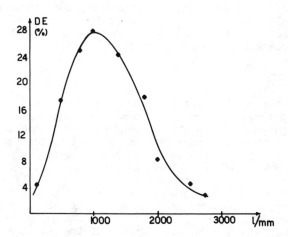

Fig. 1:   Spatial frequency response of photoconductor-thermoplastic
          recording sample, using modified heating development process.

Although PC-TP devices have been considered for hologram interferometry in the past, they suffered from limited recording bandwidth and thus were confined to recording holograms of relatively small objects. (4,5,6) We found that this limited bandwidth response may be due, in part, to improper development heating techniques. For example in conventional development a single voltage or current pulse of about 100 milliseconds is applied to the transparent conductor to heat the thermoplastic layer beyond its softening temperature, Tg (typically 75°C). The rise in temperature decreases the viscosity of thermoplastic so it deforms in a manner related to the desired holographic interference pattern. Rapid cooling caused by the termination of the pulse stabilizes the deformation until intentional erasure.

Unfortunately as the viscosity of the thermoplastic layer decreases, its ionic conductivity increases by the same order of magnitude-generally 3 to 4 - obeying the Nernst-Einstein-Townsend relation,

$$\eta\mu = \text{constant} , \qquad\qquad\qquad (1)$$

where η is the viscosity and μ the ionic conductivity. The sharp reduction of the conductivity causes partial collapse of the electric field across the thermoplastic layer, reducing the tension and consequently the deformation; thus the holographic response is decreased.

We found that by developing the PC-TP device with a series of narrow heat pulses, rather than a single long pulse, it is possible to improve the holographic response. The duration of each pulse in the series is about 100 microseconds and the repetition rate is selected so that the thermoplastic is cooled well below the Tg between successive pulses. Each successive pulse increases the deformation, and thereby the diffraction efficiency, obeying the Boltzmann superposition principle of quasi-elastic deformation. We have experimentally measured the modulation transfer function of the device with this mode of development and the results are shown in Figure 1. As shown, the bandwidth is about 1800 lines/mm, which is significantly greater than with conventional development. Although the exact mechanism distinguishing this mode of development from the conventional mode is not yet certain, we hypothesize that it is due to the inter-dependence between the mechanical and the dielectric relaxations of the thermoplastic layer.

EXPERIMENTAL RESULTS

Some examples of the results that we obtained with the various holographic and stressing techniques and the two recording media are given in Figures 2 through 6. Figure 2(a) shows the fringes produced by double exposure holographic interferometry on an epoxy-bonded aluminum to aluminum core test panel when subjected to vacuum stressing; the hologram was recorded on conventional photographic emulsions. All the disbond areas, having approximately square shape of sides ranging from 2 cm to 5 cm, are clearly evident; the coarse fringes are due to bulging of the entire test panel. For comparison, Figure 2 (b) shows the reconstructed image of a similar test panel, under the same stress conditions, except that the hologram was recorded on the photoconductor-thermoplastic device. Because of the relatively poor noise characteristics of PC-TP devices, the holographic imagery were, in general, inferior to those taken with photographic emulsions, but as shown in Figure 2(b), the results are adequate for interpretation of the fringe pattern.

To illustrate the wide bandwidth capabilities of the PC-TP device, we recorded
holograms of a relatively large, epoxy-bonded aluminum-to-aluminum core, test
panel. The recording geometry and the size of the panel, 20 cm by 30 cm,
were such that the effective recorded bandwidth was about 1000 lines/mm;
the center frequency was approximately 700 lines/mm. Some examples of real
time interference, made by changing the pressure difference, under vacuum
stressing conditions, are shown in Figure 3. A photograph of the test panel
under zero stress conditions is shown in Figure 3(a); because the thermoplastic layer does not shrink, there are no spurious fringes formed when the

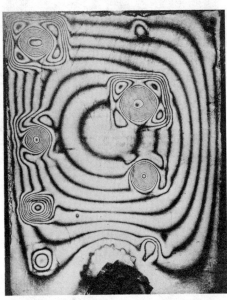

(a)   (b)

Fig. 2:   Controlled disbonds detected in aluminum-to-aluminum panels.
          Pressure differential, Δp = 20 torrs. a) Conventional
          photographic emulsions; b) Photoconductor-thermoplastic
          device.

reconstructed wave interferes with the undeformed object wave. The remaining photographs show the panel with superimposed fringes demonstrating the
effect of varying the pressure difference on the fringe density at the location of the disbonds.

Similar test panels were also investigated by time average holographic
technique. Figure 4 shows the reconstructed image from a hologram recorded
on a PC-TP device. The drive frequency of the vibration was 8500 Hz, just
sufficient to excite the smaller disbonds into the fundamental 1-1 mode,
while the larger disbonds were excited to higher vibration modes. As
shown, the coarse fringe pattern is absent with this holographic technique,
indicating that only the disbond areas are excited while the rest of the
panel remains stationary.

An example of double exposure technique for composite-to-aluminum core honeycomb structure is shown in Figure 5. Vacuum stressing with large pressure

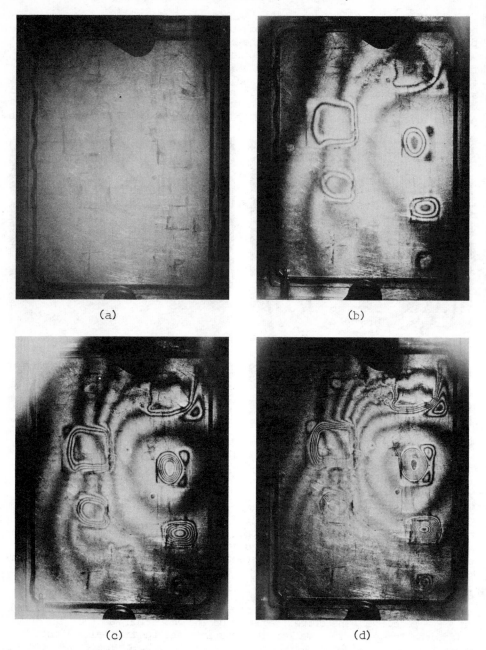

Fig. 3: Results from real-time interferometry on Aluminum-to-Aluminum panel. Pressure differentials: a) noll fringes; b) 6 torrs, c) 12 torrs, d) 20 torrs.

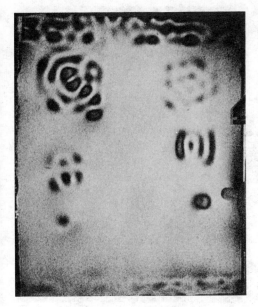

Fig. 4: Vibration stressing of
         Aluminum-to-Aluminum panels
         recorded in "time average"
         techniques on PC-TP.

Fig. 5: Composite-to-Aluminum panel
         tested by vacuum stressing.
         Pressure differential,
         Δp = 200 torr.

differences were necessary because of the thickness of the composite layer.
The larger pressure differences cause greater uniform bulging of the entire
panel resulting in a denser fringe pattern. As shown, the larger disbonds
of 1 mm and 8 mm diameters are easily detected, whereas the smaller disbond
of 5 mm diameter might be revealed as a slight curvature of the coarse fringe
structure in the upper part of the test panel. Note, that in the central
portion of the panel, where the coarse fringe patterns are more widely
spaced, it is possible to identify single cells of the honeycomb core, indicating the resolution possible with this NDT method.

Finally, Figure 6 shows the results of tests on actual manufactured brazed
honeycomb panels, each shaped as a parallelogram with dimensions of 0.5
meter by 0.5 meter. Double exposure holographic techniques and internal
pressure stressing were employed, while the recording medium was conventional photographic emulsions. Figure 6(a) shows a large disbond area
when the pressure difference between exposure was 5 torrs. Note the
absence of coarse fringes with the internal pressure stress techniques.
If the pressure difference is increased to 75 torrs it is even possible
to identify single cell disbonds, as shown in Figure 6(b); the fringes
encircle two adjacent cells, forming an 8-like pattern. The small regions,
characterized by the absence of fringes, correspond to accumulation of excessive brazing material. In these locations, the effective thickness is
large so that the deflection is minimal.

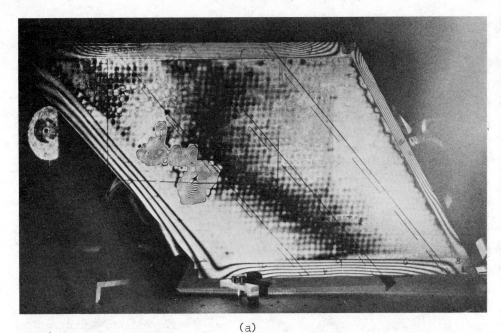

(a)

(b)

Fig. 6: Brazed stainless steel panel - stressed by internal pressure.
         Pressure differentials a) 5 torrs. b) 75 torrs. The fringes
         indicating the large disbond become so dense in Fig. (b) that
         they cannot be resolved.

DISCUSSION

In our investigation, the disbonds or mis-braze occurred between the skin of panels and the honey comb core. The disbond area may be treated as a clamped circular diaphragm. The maximum deflection, d_m, at the center of the disbond of radius r, for a skin thickness t, and for a pressure difference Δp, is given by

$$d_m = \frac{3 \, \Delta p \, (m^2 - 1) \, r^4}{16 \, E \, m^2 \, t^3} \qquad\qquad /2/$$

where E is Young's modulus and m is the inverse of Poisson's ratio for the material of the skin. In principle, the pressure difference, Δp, could be increased to detect a smaller disbond in a given panel. In practice, however, the maximum pressure difference must be limited by the mechanical constraint of the panels or by the mode of applying the pressure; if vacuum is used, Δp can not exceed 760 torrs. Thus, the smallest theoretical disbond size usually cannot be detected.

If, when recording the hologram, the panel is illuminated with a beam making an angle θ_1 with the normal to the surface, and during reconstruction it is is viewed at an angle θ_2, the reconstructed image has superimposed fringe pattern corresponding to a displacement of the surface, d, in the normal direction, given by

$$d = \frac{n \, \lambda}{\cos\theta_1 + \cos\theta_2} \qquad\qquad /3/$$

the term n is the number of fringes and θ is the wavelength of the illumination light. In general, the angles θ_1 and θ_2 are sufficiently small so that $d = \frac{n \, \lambda}{2}$.

From this relation and a count of the relevant fringes, we may quantitatively determine the deflection of the disbond area and compare these values to those predicted by Eq. (1); a deflection of $\lambda/2$ would result in one fringe. Indeed, such measurements of the disbonds in the test panels were in good agreement with the predicted values, confirming the validity of the diaphragm model.

When employing the vacuum stressing techniques we found that for pressure differences exceeding 50 torr the entire test panel bulges causing a "bias" fringe pattern which obscures the fringes at the disbond areas. The bias fringes were particularly excessive with thinner or larger panels, and would become severe when testing honey comb structures having relatively thick skins. To alleviate this bias fringes problem we also applied vacuum to the other side of the panel, thus compensating in part, for the one sides bulge. This modified arrangement permitted pressure difference up to 150 torr. More recently, an even greater improvement was achieved when the entire panel was inserted into a large vacuum chamger; the pressure is thus equalized on all sides, eliminating the mechanical stresses caused by the weight of attached vacuum chamber panels.

For the breathing type panels, where internal pressure stressing techniques are possible, the bias fringes in general, do not present a problem in the interpretation of the data.

CONCLUDING REMARKS

Our investigations established a feasible test procedure which may be incorporated into HNDT of honeycomb structures with skin thicknesses not exceeding
1mm. Moreover, the use of photoconductor-thermoplastic devices permits
virtually real time recording and measurement - thus making HNDT not only
a valuable laboratory technique, but perhaps a practical industrial one
as well.

The authors express their appreciation to M. Meron and E. Segal for supply-
ing the test samples and their encouragement. This research has been spon-
sored in part by the U. S. Air Force, AFOSR grant 76-3004.

REFERENCES

1. Robertson, E. R. & Harvey, J. M. (editors), The Engineering Uses of
    Holography. Cambridge University Press, 1970.

2. Robertson, E. R. (editor), The Engineering Uses of Coherent Optics,
    Cambridge University Press, 1976.

3. Erf, R. K., (editor), Holographic Nondestructive Testing, Academic Press,
    New York, 1974.

4. Bellamy, J. C., Ostrowsky, D. B., Poindron, M., & Spitz, E., In-situ
    double exposure interferometry using photoconductive thermoplastic
    film, Appl. Opt., 10, 1458 (1971).

5. Colburn, W. S. and Tompkins, E. N., Improved thermoplastic-photoconductor
    devices for holographic recording, Appl. Opt., 13, 2934 (1974).

6. Lee, T. C., Holographic recording on thermoplastic films, Appl. Opt.,
    13, 888 (1974).

MEASUREMENT OF THE AREA OF REAL CONTACT BETWEEN, AND WEAR OF, ARTICULATING SURFACES USING HOLOGRAPHIC INTERFEROMETRY

J. T. Atkinson and M. J. Lalor

Department of Mechanical Engineering, Liverpool Polytechnic, Byrom Street, Liverpool L3 3AF, England

ABSTRACT

Methods for measuring the area of real contact between, wear of, and change in surface finish of, articulating surfaces using holographic interferometry are proposed.

INTRODUCTION

When a human joint is affected by rheumatoid arthritis or osteo-arthritis it causes great pain at the site of the joint and the movement of the joint is restricted until eventually it ceases to function normally. This condition can be relieved by the insertion of an artificial joint or prosthesis at the joint site with the object of relieving pain and restoring movement. The first joint to be studied from the biomechanical standpoint was the hip joint and eventually a ball and socket joint of stainless steel and high density polyethylene (H.D.P.) evolved giving good function and low wear characteristics ensuring long life (about 10 to 15 years). It has been estimated that 20,000 of these and very similar devices are inserted successfully every year.

The design of knee joint prostheses has received a lot of attention recently and it is estimated that about 270 different prostheses are being developed worldwide. The prime objectives in their design have been to achieve maximum flexion and good joint stability together with pain relief. Some work has been done on estimating the life of these prostheses [1,2]. Assuming correct insertion of the prosthesis the most likely cause of failure will be the wear of one or other of the parts of the prosthesis. Thus there is a need for accurate quantitative measurement of the wear under the loading conditions experienced during normal activities. This information would enable a prediction of the relative lives of the different prostheses to be made.

Work to date has shown that all the wear in stainless steel-H.D.P. types of prostheses takes place in the H.D.P. component. Most wear measurements have been made using gravimetric techniques, these techniques suffer from major drawbacks in that although the H.D.P. component may change shape during normal use, there is very little change in weight, and the gravimetric results do not give a clear picture of the wear process.

289

It is proposed that wear tests be carried out on several of the designs of knee prostheses using holographic interferometry. The generation of contour fringes on the surfaces of the H.D.P. component using dual index holographic contouring (D.I.H.C.) provides a simple and accurate method for determining wear rates. Measurement of fringe visibility may also give additional information about the change in surface finish of the component.

It is the aim of this study to assess the applicability of D.I.H.C. to the measurement of wear rates of H.D.P. and discuss the possibility of measuring the area of real contact between the abrading surfaces using double exposure holographic techniques.

DUAL INDEX HOLOGRAPHIC CONTOURING

Calculation of the Contouring Depth

Dual Index Holographic contouring was first demonstrated in 1967 by Tsuruta et al [3] and has been widely researched since then, [4,5].

The arrangement commonly used is shown in Fig.1. The holographic plate is double exposed while the object is sequentially immersed in liquids of refractive index n_1 and n_2. Upon reconstruction of the hologram, contour fringes can be seen localised on the surface of the object. The contour depth Δh is given by [6]:

$$\Delta h = \lambda / [n_1 (1 + 1/\cos r_1) - n_2 (1 + 1/\cos r_2)] \tag{1}$$

where $r_1 = \arcsin [(\sin \theta_1)/n_1]$

and $r_2 = \arcsin [(\sin \theta_2)/n_2]$

Equation (1) differs from that normally quoted (3, 5) which is

$$\Delta h = \lambda / [n_1(1 + \cos r_1) - n_2 (1 + \cos r_2)] \tag{2}$$

Equation (1) is found using a ray tracing method as follows. Consider the optical path difference for P_1 (as shown in Fig.1), where the intensity is a maximum, between two similar rays travelling from the source to the observer, through liquids of indices n_1 and n_2 respectively,

$$m_1\lambda = n_1 h(1 + 1/\cos r_1) - n_2 h(1 + 1/\cos r_2) \tag{3}$$

where m_1 is an integer.

The optical path difference at P_2 (an adjacent bright fringe) between another set of similar rays is given by a similar expression

$$m_2\lambda = n_1(h\pm\Delta h)(1 + 1/\cos r_1) - n_2(h\pm\Delta h)(1 + 1/\cos r_2)$$

where $m_2 = m_1 \pm 1$. $\tag{4}$

Subtraction of equation (3) from equation (4) leads to equation (1).

Figure 2 shows equations (1) and (2) plotted together with experimental data from contouring an accurately known wedge. The wedge was made up of a flat diffusely reflecting surface (the object) clamped together with two slip gauges and an optical flat, made up as shown in Fig.1. the choice of slip gauges gave the very accurate value for the wedge angle. It should be noted that equations

(1) and (2) give the same value of Δh for $\theta = 0$, that is

$$\Delta h = \lambda/2.\Delta n \qquad (5)$$

where $\Delta n = n_1 - n_2$

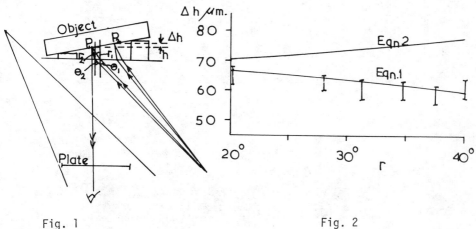

Fig. 1   Fig. 2

Measurement of Surface Roughness

It was proposed by Ribbens in 1974 [5] to use two wavelength holographic contouring to measure surface roughness. Similar theoretical results were presented by Tsuruta et al [3] when the dual index technique was first demonstrated. An approximate value for the degradation in fringe visibility for a perfect reflector was quoted as [3].

$$\nu = 1 - 8\pi^2 \,\Delta n^2 \,\sigma^2/\lambda^2 \qquad (6)$$

where σ = the root mean square value of surface roughness.

Ribbens states his results in terms of fringe contrast (R) :

$$R = [(1+\rho)^2+(2\pi\sigma\Delta n/\lambda)^2] / [(1-\rho) +(2\pi\sigma\Delta n/\lambda)^2]$$

$$= I_M/I_m \qquad (7)$$

where ρ = the reflection coefficient.

Rewriting equation (7) in terms of, ν, fringe visibility.

$$\nu = [(1+\rho)^2-(1-\rho)^2]/[(1+\rho)^2+(1-\rho)^2+1.(2\pi\sigma\Delta n/\lambda)^2]. \qquad (8)$$

For a perfect reflector equation (8) reduces to

$$\nu = 1/[1+ 2(\pi\sigma\Delta n/\lambda)^2] \qquad (9)$$

Comparison of equations (6) and (9) shows that, when Ribbens's result is expressed as the first two terms of the binomial expansion, apart from a difference of a factor of four in the term containing the roughness, the results are identical.

Figure 3 shows plots of the two different theories, for an Aluminium sample ($\rho = \cdot 96$) plotted as ν vs σ/λ_f where $\lambda_f = \lambda/\Delta n$. It can be seen that for the range $\cdot 05 \geqslant \sigma/2\Delta h \geqslant \cdot 3$ the curves are approximately linear.

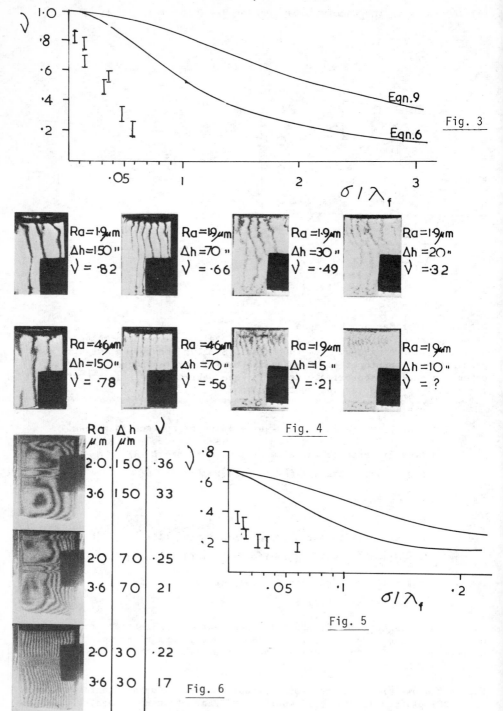

Fig. 3

Fig. 4

Fig. 5

Fig. 6

The experimental results shown in Fig.3. were found by contouring the wedge described previously. Two aluminium objects with Ra values of 1·9μm and 4·6μm respectively were contoured using liquids made up of methanol and a mixture of methanol and ethanol to give contouring depths of 15μm, 20μm, 30μm, 70μm and 150μm. The fringe visibilities were found by scanning the real images of the reconstructions, shown in Fig.4, with a pinhole and photomultiplier arrangement, the output from the photomultiplier was fed into an oscilloscope, the output trace was photographed and analysed.

The experimental results shown are plotted as ν vs Ra/2.Δh. It is assumed throughout that Ra $\simeq$ σ. The surfaces used were generated by grinding on a lapping block with rough and smooth paste to give Ra = 4·6μm and 1·9μm respectively.

It can be seen that the experimental results do not fit the theoretical curves, nevertheless they show that some simple relationship exists and that surface finish could be measured using this technique. A more extensive calibration could result in better accuracy.

The above experiment was repeated using an H.D.P. sample as the object. The results are shown in Fig.5. The H.D.P. used in most prostheses is transparent and holograms taken of H.D.P. in this state show a very noisy background.

To overcome this several coloured samples of the H.D.P. were obtained and tested by illuminating them with He Ne laser light. The blue sample appeared to give the best holograms. The results shown in Fig.5. are for blue H.D.P. samples, prepared by rubbing with different grades of emery paper. The lay of the surface finish of the H.D.P. was parallel to the contour fringes, and the Ra value used to plot the experimental points was found by taking twenty readings with a Talysurf across the sample perpendicular to the lay. The sample was contoured with contour depths of 30μm, 70μm, and 150μm. Photographs of the contoured objects are shown in Fig.6.

Wear Tests
Some preliminary wear tests were made on the blue H.D.P. using the Manchester knee designs, see Fig.7, the H.D.P. (tibial) component was mounted in the chuck of a shaping machine, the femoral component was mounted in place of the machine tool, loaded with the average body weight (70 Kg) and made to reciprocate over a 2 cm portion of the tibial component at a rate of 45 cycles/minute. The tests were unlubricated.

Contour maps of the tibial component before and after 1 min and 2 min. wear are shown in Fig.8. The contour depth is 100μm. It can be seen that the H.D.P. has plastically deformed (in the central portion) by $\sim$800μ in the first instance and $\sim$1200μ in the second. There was no wear debris created in these experiments and so it is assumed, no loss in weight.

Fig. 8

Δh=100mμ No wear   1 min   2 min

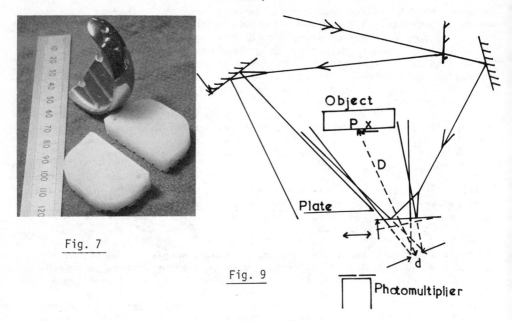

Fig. 7

Fig. 9

MEASUREMENT OF THE AREA OF REAL CONTACT.

From the tribological standpoint the area of real contact between two abrading surfaces would be a very useful measurement to make.

It is well known that the area of real contact between two surfaces (Ar) is much less than the apparent area and is governed by the inequality,

$$Ar \geqslant \frac{L}{p}$$

(10)

Where L is the applied load and p is the microhardness of the softer material [8] This is due to the fact that no two surfaces are flat. When two surfaces come into contact the surface asperities deform both elastically and plastically until the applied load is supported by a large number of these asperities. The ratio of the area of contact due to plastic and elastic deformation is a point of much discussion [9]. However it is generally agreed that inequality (10) gives the correct order of magnitude for the real contact area. Measurement of the area of real contact due to plastic deformation of the surface asperities maybe possible using the proposed method outlined below.

The technique could be carried out in real time, or using double exposure method Consider the real time set up shown in Fig. 9. A hologram of the specimen to be loaded is made in the usual way (exposed for ∿ 2s and developed for 20s) and returned to its original position using a dry real time plate holder. Residual fringes are removed by fine adjustment of the position of the reference beam source.

Now, if the object beam mirror is rotated through a small angle, a set of Youngs fringes will be observed localised on the object surface. The visibility of these fringes will depend upon the coherence length of the source, the surface finish of the object [10] and other (depolarisation) factors [11], and can be measured by scanning the image directly using a photomultiplier or indirectly by taking and analysing photographs of the interference pattern. The object is then removed and loaded or worn as required, and returned to its precise original position on the kinematic mounting (Fig.10). The visibility of the fringes now observed on the surface of the object (ν_B) will be altered compared to the initial value (ν_A) since the spatial coherence of the reflected and reconstructed waves contributing to the fringe pattern will suffer due to the plastic deformation of the surface asperities undergone during loading.

In general the light intensity at a point on the surface of the object will be given by (see Fig. 9) [12]

$$I = \rho^2 [a_1^2 + 2a_1a_2 \, |\gamma| \cos \delta] \tag{11}$$

where a_1 is the amplitude of the reflected wave.

a_2 is the amplitude of the reconstructed wave.

$|\gamma|$ is the degree of coherence.

and $\delta = 2\pi d./\lambda D$.

now, let $a_1 = \beta a_2$, and the fringe visibility will be given by:

$$\nu = 2\beta|\gamma| / [1 + \beta^2] \tag{12}$$

Now consider the loading of the object. Plastic deformation of the surface asperities before the summation of the amplitudes can be considered as an alteration of $|\gamma|$ (and to a lesser degree, β).

If the virtual image of the interference pattern is scanned with a slit attached to a photomultiplier, as shown in Figs. 9 and 11, the output from the photomultiplier will be given by

$$V = k \alpha S I = k^1 I \tag{13}$$

where k and k^1 are constants of proportionality, $\alpha.S$ is the slit area, and I is the average intensity over the slit area.

The initial (no loading) photomultiplier output will be given by

$$V_A = k^1\rho^2a_2^2 [1 + \beta^2 + 2\beta|\gamma| \cos \delta] \tag{14}$$

Now if we assume that:

(i)   At points of plastic deformation $|\gamma| = 0$

(ii)   At points of no plastic deformation $|\gamma| = 1$

(iii) The points of plastic deformation are small compared to the slit area.

(iv)   The points of plastic deformation are distributed evenly across the loaded area.

(v)   The aperture of the viewing lens in infinite

then the photomultiplier output after loading will be:

$$V_B = k^1\rho^2a_2^2 \left[(1-A)[1+\beta^2+2\beta|\gamma| \cos\delta] +A[1 + \beta^2] \right] \tag{15}$$

and the corresponding fringe visibility will be:

$$\nu_B = 2\beta|\gamma| (1-A)/[1+\beta^2] \tag{16}$$

where A = the ratio of the real contact area to the apparent area.

Examination of equations (12) and (16) shows that, if β remains constant, the
visibility of the fringes decreases after loading.

EXPERIMENTAL

The experiments to be described herein were carried out on the surface of an
aluminium sample (ρ = 89 Kg/mm^2) (shown in Fig.10) prepared by grinding on a
lapping block to give a rough but flat surface of Ra = 2μm.

Several methods were investigated, the real time technique described previously
was discarded in favour of the double exposure method now outlined, the
holographic arrangement used is shown in Fig.12. One experiment consists
of: doubly exposing a hologram (hologram A) with object beams 2 and 1
illuminating the object sequentially. Another hologram (hologram B) is then
made, one exposure with beam 2 falling on the object. The object is then
removed and loaded as required. This loading was carried out by placing the
flat surface of a slip gauge on each of the raised sample surfaces (shown in
Fig.10.). The required pressure was applied via a self aligning device (to
ensure that the loads were axially applied) on a hydraulic press. The object
is then replaced on the kinematic mounting and the second exposure, using
beam 1, made. The two holograms are then processed together.

The source positions of the two object beams were adjusted to give a fringe
width (ω) of ∿ 1mm. The visibilities ν_A and ν_B were found by scanning the
virtual images of the fringe patterns, using the system described earlier.
The ratio of the slit width to the fringe width (α/ω) was chosen such that
the correction term for the maximum and minimum intensities was less than 1%
and was ignored in these experiments.

Fig. 10

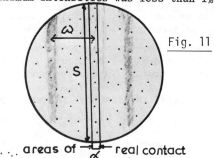

Fig. 11

areas of real contact

Fig.13. shows two reconstructions from holograms A and B, together with samples
of photomultiplier outputs obtained from scanning across the sample areas shown
Sample areas in column L, N and P were not loaded between exposures, it was
therefore possible to establish whether or not the visibility of unloaded areas
remained constant throughout the experiment. Table 1 shows how the fringe
visibility of the loaded and unloaded areas altered during the course of the
experiment. ν_A/ν_B is the ratio of the visibility of the loaded area to that of
the adjacent unloaded area, and $[\nu_A/\nu_B]$ is the ratio of the visibility of the
loaded area to that of the same surface unloaded. In general the fringe
visibility on the unloaded areas is different for the two holograms A and B.
This could be due to a change in; the position of localisation, and or shape of
the fringes between holograms A and B. It can be seen that in some cases the
fringe visibility of the loaded areas is increased with respect to that of the
adjacent unloaded area, this could be caused by a change in β; β was measured
by taking and scanning holograms with the object illuminated with only one of
the object beams and in all cases is close to unity.

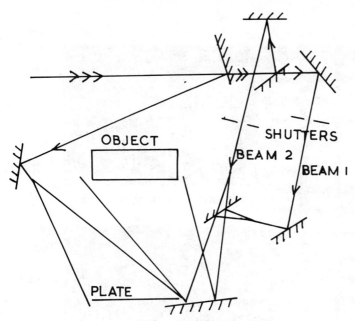

SHUTTERS

BEAM 2

BEAM I

OBJECT

PLATE

Fig. 12

TABLE 1 Sample Area	ν_A	ν_B	Load (kN)	Ar = L/P (mm)	A = Ar/50mm	(1-A)	$\dfrac{\overline{\nu_B}}{\nu_A}$	$\left[\dfrac{\nu_B}{\nu_A}\right]$
1	·48±·015	·36±·015	0					
2	·50±·015	·29±·015	1·0	1·15	·023	·977	·81	·59
3	·51±·015	·32±·015	1·5	1·7				
4	·51±·015	·42±·015	0		·034	·966	·76	·61
5	·43±·015	·22±·015	2·1	2·3	·046	·954	·63	·52
6	·50±·015	·35±·015	0					
7	·62±·015	·61±·015	0					
8	·57±·015	·63±·015	·52	·58	·012	·988	1·03	1·1
9	·54±·015	·46±·015	0					
10	·51±·015	·45±·015	1·1	1·3	·025	·975	·98	·88
11	·51±·015	·41±·015	0					
12	·49±·015	·46±·015	1·5	1·7	·034	·966	1·12	·94
13	·58±·015	·59±·015	0					
14	·51±·015	·54±·015	·8	·92	·018	·982	·915	1·06
15	·63±·015	·60±·015	0					
16	·60±·015	·59±·015	·3	·34	·007	·983	·983	·983

CONCLUSIONS

It is hoped that the work outlined here will lead to the development of long term and short term wear tests for knee prostheses. Although the area of contact experiments have not yet led to any quantitative results, it is felt that this technique may prove useful as a short term wear test. Any correlation between the short term tests and the long term tests (using D.I.H.C.) could be used to save time in any future tests. The authors hope to improve the area of contact measurements by, using a flat surface as the object and loading the specimens in situ.

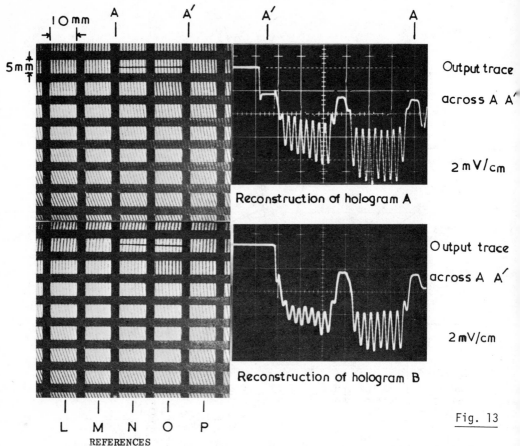

Fig. 13

REFERENCES

1. Seedhom, B.B. et al 'Wear of solid phase formed HDP in relation to the life of artificial hips and knees'. <u>Wear</u> 10 35 (1973)
2. Dumbleton, J.H. and Shen, C. 'The wear behaviour of ultrahip molecular weight polyethylene'. <u>Wear</u> 37 279 (1976)
3. Tsuruta et al 'Holographic generation of contour map of diffusely reflecting surface by using immersion method' <u>Japan J. Appl. Phys.</u> 6 661 (1967)
4. Zelenka, J.S. and Varner, J.R. 'Multiple index holographic contouring' <u>Appl. Opt.</u> 8 1431 (1969)
5. Marrrone and Ribbens 'Dual index contour mapping over a large range of contour spacings'. <u>Appl. Opt.</u> 14 23 (1975)
6. Roberts, I. private communication.
7. Ribbens, W.B. 'Surface roughness measurement by two wavelength holographic interferometry' <u>Appl. Opt.</u> 13 1085 (1974)
8. Bowden, F.P. and Tabor, D. 'Friction and Lubrication' Methuen 1967.
9. Archard, J.F. 'Surface topography and tribology' <u>Tribology</u> 7 213 (1974)
10. Ribbens, W.B. 'Interferometric surface roughness measurement' <u>Appl. Opt.</u> 8 2173 (1969).
11. Kellie, T.F. and Stevenson, W.H. 'Expt. techniques in real time H.I., II minimisation of depolarisation effects' <u>Optical Engineering</u> 12 131 (197?
12. Walles, S. 'Visibility and localisation of fringes in holographic interferome of diffusely reflecting surfaces' Arkiv for Fysik Bd 40 nr 26.

A HOLOGRAPHIC SHEARING
INTERFEROMETER

Kiyofumi Matsuda

*Mechanical Engineering Laboratory, 4-12-1 Igusa, Suginami-ku,
Tokyo 167*

ABSTRACT

This report describes a new type of shearing interferometer
using two three-beam holograms as its elements. Here it is shown
that the amount of shear can be changed by varying the seperation
between the two holograms. Also the effect of inhomogeneous film
thickness on interferograms is discussed. This interferometer is
not only so optically stable and convenient as a conventional
common-path interferometer but also can be easily constructed.

INTRODUCTION

So far several methods have been reported in applying holography
to shearing interferometry (Ref 1, 2, 3, 4, 5). In some of these
reports holograms are used as an element of a shearing
interferometer (Ref 3, 4, 5). This report describes a new type of
shearing interferometer using two three-beam holograms as its
elements.

A holographic shearing interferometer has some advantages over
a Ronchi-type interferometer in that the former can be easily
constructed (Ref 4). Also the interferograms produced by the
holographic shearing interferometer result in a large region of
overlap and have good contrast (Ref 3). The interferometer
described here has these advantages and moreover an amount of
shear can be easily changed by varying only the seperation
between the two three-beam holograms.

DESCRIPTION OF INTERFEROMETER

Figure 1 shows the principle of the interferometer. In Fig. 1
(a), two plane waves f_{H2} and $f_{H(2+\Delta)}$, slightly different in
direction from each other, are recorded in a double exposure on
a photographic plate placed at H_1, using a plane wave f_{H1} as a
reference light. After processing the photographic plate, the
hologram H_1 is formed and placed in its original position. In
the second stage, the two plane waves f_{H2} and $f_{H(2+\Delta)}$ are
reconstructed by the reference light f_{H1} and the hologram H_1.

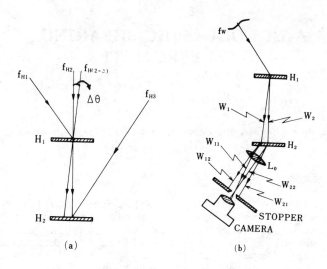

Fig. 1 Principle of the holographic shearing
        interferometer.

This time f_{H2} and $f_{H(2+\Delta)}$ are recorded on a photographic plate
placed at H_2, using a plane waves f_{H3} as a reference light. If
the constructed hologram H_2 is placed in its original position,
the pair of holograms H_1 and H_2 gives a holographic shearing
interferometer.

In the case of optical testing, an arrangement shown in Fig. 1
(b) is used. It is assumed that the wavefront f_w under test has
some deformations, for example, lens aberrations or atomospheric
disturbance. When the hologram H_1 is illuminated by f_w instead of
f_{H1}, the produced wavefronts W_1 and W_2 form the virtual images of
f_w which are sheared and inclined to each other. However in order
to eliminate the effect of the inclination, the hologram H_2
operates on the reconstructed wavefronts W_1 and W_2, and the
wavefronts W_{11}, W_{12}, W_{21}, and W_{22}. The wavefronts W_{11} and W_{22}
produce a lateral shear interference pattern in the region of
overlap. When f_w is a plane wave, W_{11} and W_{22} becomes f_{H3}. The
other unwanted wavefronts W_{12} and W_{21} are eliminated by a stopper
placed in the focal plane of a lens L_0. In this interferometer,
the amount of shear is determined by the angle $\Delta\theta$ between the
wavefronts f_{H2} and $f_{H(2+\Delta)}$, and the seperation D between the two
holograms, that is, the amount of shear ΔS can be represented as

$$\Delta S = D \Delta \theta \tag{1}$$

From the equation (1), it follows that any desired amount of shear can be obtained by varying $\Delta \theta$ and/or D. However in the case of varying $\Delta \theta$, the hologram H_1 and H_2 must be reproduced each time when one wants to change the amount of shear. On the contrary, the change of D can be achieved easily by only varying the seperation between the two holograms H_1 and H_2, and thus any desired amount of shear can be easily obtained.

It is significant at this stage to consider the effect of in homogeneous film thickness of the two holograms used in the interferometer. First the case where there is no change in the seperation D in recording and reconstruction of the plain waves must be considered. In this case the inhomogeneous film thickness has no effect on the interferograms due to: 1) The deformations of the wavefronts W_1 and W_2 due to the variation of the film thickness of the hologram H_1 are cancelled just behind the hologram H_2, because on passing through it W_1 and W_2 are multiplied approximately by their complex conjugates which is

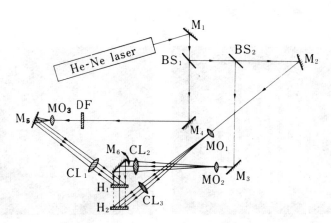

Fig. 2 Experimental arrangement for producing holograms: $M_1 \sim M_6$ reflecting mirror, BS_1 and BS_2 beam splitter, $MO_1 \sim MO_3$ microscope objectives lens, $CL_1 \sim CL_3$ collimator lens, H_1 and H_2 hologram, DF neutral density filter.

already recorded on the hologram H_2 and 2) the effect of the film thickness of the hologram H_2 is cancelled, because the produced interferograms are obtained by the interference between two beams originating from the identical part of the hologram H_2.

In the second place, the case where the amount of shear is changed by varying the seperation D should be considered. It is evident from the previous discussion that the inhomogeneous film thickness of only hologram H_1 has effect on the interferograms.

However an amount of shear is usually small and the interferograms are produced by the light which passes through any two points with the same seperation ΔS as the amount of shear on the hologram H_1. Therefore, if the film thickness is regarded as being homogeneous in the region of this seperation, the interferometer can be used.

The two holograms in this interferometer are desired to give a high diffraction efficiency. Therefore in practice the holograms have been produced by the reversal breach process with little flare light and high diffraction efficiency (Ref 6).

EXPERIMENTAL RESULTS

Figure 2 shows the arrangement used in this experiment and a

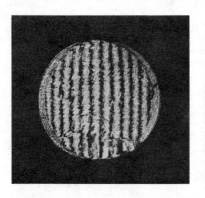 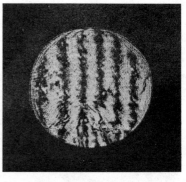

(a)   (b)

Fig. 3 Holographic shearing interferograms of
the wavefront with radius of the curvature.
The interferograms of (a) and (b) have
different seperation: (a) D=180mm ΔS=3mm, and
(b) D=90mm ΔS=1.5mm.

Fig. 4   A shearing interferogram of the
atmospheric disturbance around the flame of a
candle (ΔS=1.5mm).

He-Ne laser (λ=0.6328μm) was used as the coherent light source.
In this arrangement, the two plane waves f_{H2} and $f_{H(2+\Delta)}$,
slightly different in direction from each other, are obtained by
slightly rotating the mirror M_6. Figure 3 shows shearing
interferograms of the wavefront with 20 m radius of curvature.
The interferograms in Fig. 3 (a) and (b) had different separation,
D = 180 mm and 90 mm respectively, between two holograms. By
comparing two photographs, it is evident that the amounts of
shear (ΔS = 3 mm and 1.5 mm respectively) can be changed.

It is often found that the interferograms are irregularly
deviated from straight lines of the interference fringes (i.e.
lower part of Fig. 3 (b)), the interferograms are irregularly
deviated from straight lines of the interference fringes. This is
because of the inhomogeneous film thickness in this part of the
hologram H_1 which frequently occurs when it is dried after the
photographic process. The same section of the interferogram
pictured in Fig. 3 (a), is much better, because the separation D
between two holograms in reconstruction is the same as that
(D = 180 mm) in making the hologram H_2. Thus , as mentioned in
the previous discussion, the film's inhomogeneous thickness is
cancelled in that case.

Figure 4 shows a shearing interferogram of the atmospheric
disturbance around a candle flame. In this case, such a hologram
(H_1) has been used so that its film thickness can be regarded as
being almost homogeneous in the region previously mentioned (i.e.
the lower part of Fig. 3 (a)).

CONCLUSION

The lateral shearing interferometer using two three-beam holograms, described in this report is a simple, stable and easily constructed device. Moreover this interferometer provides a convenient way of obtaining any desired amount of shear by varying the seperation between the two holograms. In general, the common-path interferometer presented here is both practical and convenient, and can be applied to the measurement of lens aberrations and index gradients of any medium like gases or liquids, to the detection of abnormal index in glass plates and so on. In such applications, if it is required to use holograms with higher diffraction efficiency than those used here, then one will be able to use photoresists.

The author would like to thank Prof. J. tsujiuchi, Dr. K. Shimizu and Mr. S. Tsunekane for discussion and usefull suggestions, and Mr. H. Asano for his assistance during the experiments.

REFERENCES

(1) A. F. Belozerov et al., Production of a shearing interferogram and schlieren pictures of an optical inhomogeneity from a singly exposed hologram, Optics and Spectrosc. 27, 187 (1969).
(2) Y. Doi et al., Shearing interferometry by holography, Japan J. Appl. Opt. 12, 1036 (1973).
(3) J. C. Wyant, Double frequency grating lateral shear interferometer, Appl. Opt. 12, 2057 (1973).
(4) J. C. Fouere et al., Holographic radial shear interferometer, Appl. Opt. 13, 2035 (1974).
(5) O. Bryngdahl et al., Shearing interferometry in polar coordinates, J. Opt. Soc. Am. 64, 1606 (1974).
(6) R. L. Lamberts et al., Rversal bleaching for low flare light in holograms, Appl. Opt. 10, 1342 (1971).

INTERFEROMETRIC METHODS FOR MEASURING THE DISTRIBUTION OF CURVATURE AND TWIST OF PLATES

A. Assa,* J. Politch** and A. Betser***

*Formerly a graduate (D.Sc.) student at the Department of Aeronautical Eng.,
Technion I.I.T., Haifa, Israel.
**Department of Physics, Technion I.I.T., Haifa, Israel
***On leave from the Department of Aeronautical Eng., Technion I.I.T., Israel. Now at
the Department of Theoretical and Applied Mechanics, Univ. of Illinois at Urbana,
Urbana Champaign, Urbana, Illinois 61801, U.S.A.

ABSTRACT

Four methods of shearing interferometry in 'frozen' and 'real' time were
developed for recording and measuring the distribution of curvature and twist
of statically   loaded plane reflecting thin plates.

The diffraction gratings were applied as the shearing interferometric beam
splitter in this study.

INTRODUCTION

Herewith are described few methods concerned with the application of shearing
interferometry to the measurement of curvatures and twist of statically loaded
plane reflecting models. In all the systems described here, the diffraction
gratings are applied as the shearing interferometric beam-splitter. In the
general recording system, as described in Fig. 1, a plane light wave is
reflected from the model's surface. The curvatures of the reflected
(distorted) wave are then determined and related to the curvatures of the
model. The importance of this work is in the small deflection theory of
thin plates under lateral loading [1], where there were found the relations
between the "moment per unit length" (m_x, m_y, m_z) and the deflection $\bar{w}$:

$$m_{\bar{x}} = - D(\frac{\partial^2 \bar{w}}{\partial \bar{x}^2} + \mu \frac{\partial^2 \bar{w}}{\partial \bar{y}^2}) \tag{1}$$

$$m_{\bar{y}} = - D(\frac{\partial^2 \bar{w}}{\partial \bar{y}^2} + \mu \frac{\partial^2 \bar{w}}{\partial \bar{x}^2}) \tag{2}$$

$$m_{\overline{xy}} = - m_{\overline{yx}} = D(1-\mu) \frac{\partial^2 \bar{w}}{\partial \bar{x} \partial \bar{y}} \tag{3}$$

where the modulus of rigidity 'D' is defined by:

$$D = \frac{Eh^3}{12(1-\mu^2)} \tag{4}$$

305

here 'E' is the Young's modulus, 'h' is plate thickness and 'μ' is the Poisson ratio.

Equations (1) and (2) describe 'curvature components' in the x and y directions, and eq. (3) describes the 'twist' component. The fiber stresses existing in the plate are $\sigma_x = 6m_x/h^2$, $\sigma_y = 6m_y/h^2$, $\tau_{xy} = 6m_{xy}/h^2$, from where it is seen that the stresses are related to the values of the second derivative of the deflection.

Single and double frequency gratings, displaced from the focal plane are applied here [2,3]. Double exposure technique is required to obtain the "Record Step Interferogram" (RSI), followed by spatial filtering operation. Both exposures are obtained with the specimen loaded. Between the two exposures, the photographic plate or the model is translated in the 'x' or 'y' direction — depending on the derivative of interest [4].

It was found that if a distorted grating is recorded with either of the present systems, and the resulting RSI is then spatially filtered, the distortion of the first order diffracted wave is described by a function, which is proportional to the derivative of the function describing the distortion of the original wavefront, as reflected from the model. During the spatial filtering step, as described in Fig. 2, two identical waves arrive at the output plane, each having a distortion proportional to the derivative of the original wave. These two waves form a shearing inter-ferogram, due to the translation between the exposures.

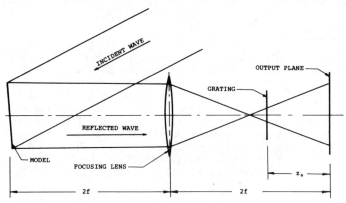

Fig. 1. General optical recording system.

Single Frequency Grating

The intensity $I(x,y)$ of the first exposure, incident on the photographic plate is described by:

$$I_1(x,y) = a^2 - 2ab\cdot\cos(2\pi\nu[\ x \ \frac{f-z_0}{f} + z_0 \frac{\partial w}{\partial x}]) +$$

$$+ b^2\cos^2(2\pi\nu(\ x \ \frac{f-z_0}{f} + z_0 \frac{\partial w}{\partial x})) \tag{5}$$

where 'a' and 'b' are amplitude components of the cosinusoidal grating, having a transmittance function: $t_g(x,y) = a - b \cdot \cos(2\pi\nu x)$; here ν is the spatial frequency and x is the coordinate in the grating plane, and z_o is the grating distance from the output plane, as described in Fig. 1.

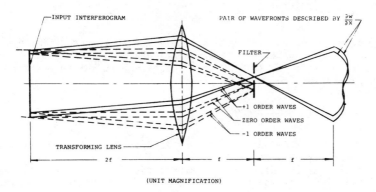

(UNIT MAGNIFICATION)

Fig. 2. The filtering procedure of a double exposure RSI.

Between the exposures, the photographic plate or the object is translated in the 'x' direction by an amount Δx. The intensity incident on the plate, with the translation taken into account, is given by:

$$I_2(x,y) = a^2 - 2ab \cdot \cos\left(2\pi\nu\left[(x+\Delta x)\,\frac{f-z_o}{f} + z_o\,\frac{\partial w}{\partial x}\,(x+\Delta x,y)\right]\right) +$$

$$+ b^2\cos^2\left(2\pi\nu\left[(x+\Delta x)\,\frac{f-z_o}{f} + z_o\,\frac{\partial w}{\partial x}\,(x+\Delta x,y)\right]\right) \qquad (6)$$

The function $(\partial w/\partial x)(x+\Delta x,y)$ can be expanded in a Taylor series. After carrying out this operation and retaining terms through the first order, eq. (6) can be written in the form:

$$I_2(x,y) = a^2 - 2ab \cdot \cos\left(2\pi\nu\left[(x+\Delta x)\,\frac{f-z_o}{f} + z_o\,\frac{\partial w}{\partial x} + z_o\Delta x\,\frac{\partial^2 w}{\partial x^2}\right]\right) +$$

$$+ b^2\cos^2\left(2\pi\nu\left[(x+\Delta x)\,\frac{f-z_o}{f} + z_o\,\frac{\partial w}{\partial x} + z_o\Delta x\,\frac{\partial^2 w}{\partial x^2}\right]\right) \qquad (7)$$

After the plate was developed, it is assumed that its amplitude transmission is proportional to the sum of the exposure intensities, as described in eq. (5) and eq. (7). The developed plate is spatially filtered, and the intensity distribution in the output plane is given by:

$$I_o(x,y) = 2a^2b^2\left(1+\cos\left(2\pi\Delta x\nu\,\frac{f-z_o}{f} + 2\pi\Delta x\nu z_o\,\frac{\partial^2 w}{\partial x^2}\right)\right) \qquad (8)$$

In order to obtain bright fringe, when the curvature is zero (i.e. bright background), the translation Δx has to satisfy the following relation:

$$\Delta x = m'\,\frac{1}{\nu}\,\frac{f}{f-z_o} \qquad (9)$$

A. Assa, J. Politch, A. Betser

where: $m' = 0, \pm 1, \pm 2, \ldots$

and in order to obtain a dark field:

$$\Delta x = (m' \pm \frac{1}{2}) \; \frac{1}{\nu} \; \frac{f}{f-z_o} \qquad (10)$$

The fringe order m' of eq. (8) arises solely due to the translation Δx, and is termed the "Fringe order Due to Translation" (FDT). For the above two cases, it follows from eq. (8) that the relation between the observed order and the curvature is:

$$\frac{\partial^2 \bar{w}}{\partial \bar{x}^2} = m \; \frac{1}{\Delta x \cdot \nu \; z_o} \qquad (11)$$

where the fringe order 'm' defines bright fringes, for the condition of eq. (9), and dark fringes for the condition of eq. (10). By changing the grating direction or the direction of translation, the complete set of second order derivatives can be measured. The following set of equations was developed, which relates the values of the second order derivatives of the deflections to the fringe orders:

$$\frac{\partial^2 \bar{w}}{\partial \bar{x}^2} = m_{xx} \; \frac{M_x^2}{2 \cdot \Delta x \cdot \nu_x \cdot z_o} \qquad (12)$$

$$\frac{\partial^2 \bar{w}}{\partial \bar{y}^2} = m_{yy} \; \frac{M_y^2}{2 \cdot \Delta y \cdot \nu_y \cdot z_o} \qquad (13)$$

$$\frac{\partial}{\partial \bar{y}} (\frac{\partial \bar{w}}{\partial \bar{x}}) = m_{xy} \; \frac{M_x M_y}{2 \cdot \Delta y \cdot \nu_x \cdot z_o} \qquad (14)$$

$$\frac{\partial}{\partial \bar{x}} (\frac{\partial \bar{w}}{\partial \bar{y}}) = m_{yx} \; \frac{M_y M_x}{2 \cdot \Delta x \cdot \nu_y \cdot z_o} \qquad (15)$$

In the above equations, the magnification M is subscripted, since the system may have different magnifications in the x and y directions (as for example, as a function of the reflection angle of the system).

Due to the different magnifications, the frequency of the RSI also changes. In this case, the general expression for FDT is:

$$(FDT)_x = \frac{\Delta x \cdot \nu_x}{M_x} \; \frac{f-z_o}{f} \qquad (16)$$

with similar expressions for $(FDT)_y$. For mixed derivatives, it is possible to use either a grating with lines parallel to the 'x' axis and a translation in the 'y' direction, or vice versa. The resolution R_c of the curvature and twist is defined as the value of the change of curvature or twist, which will cause a change of one fringe order. By use of eq. (12) the resolution with respect to 'x' is given by:

$$R_{c\bar{x}\bar{x}} = \frac{M_x^2}{2\,\Delta x \cdot \nu_x \cdot z_o} \qquad (17)$$

and the resolution of the other directions is being obtained in a similar manner.

For this recording procedure, Fig. 3 demonstrates the experimental results and contains four interferograms of the cantilever beam recorded at z_o = 54 mm, w_m = 0.60 mm. The value of $\Delta x'$, which defines the fringe contrast, is equal to 0.37 at the base of the beam. The insensitivity of the system is clearly seen. This system measures absolute curvatures. Table 1 describes the fringe values as calculated for interferograms, according to eq.(12) compared to those obtained in the experiment, and describes in Fig.3 (M=1).

TABLE 1   Fringe Values as Calculated by Eq. (12) for
Interferograms of Fig. 3 (M=1)

Δx	$\dfrac{M_x^2}{2\nu_x z_o \Delta x}$	m_{xx}	$\dfrac{d^2\bar{w}}{d\bar{x}^2}$
mm	mm^{-1}		mm^{-1}
1.215	$6.44 \cdot 10^{-4}$	1/2	$3.22 \cdot 10^{-4}$
1.82	$4.31 \cdot 10^{-4}$	1/2	$2.15 \cdot 10^{-4}$
2.43	$3.22 \cdot 10^{-4}$	1/2	$1.61 \cdot 10^{-4}$
2.43	$3.22 \cdot 10^{-4}$	1-1/2	$4.83 \cdot 10^{-4}$

Double Frequency Grating (DFG)

When using a DFG to obtain RSI, the same process is performed as for single frequency grating (SFG). In this case, the photographic plate is translated a distance $\Delta x'$ between exposures. This double exposed plate is developed, with the amplitude transmission of the processed plate written as the sum of the exposure intensities. The developed plate is spatially filtered, and the intensity in the output plane is given by:

$$I_0(x,y) = \frac{b^4}{8} (1+\cos[\ 2\pi\ \frac{f-z_0}{f}\ \Delta\nu\Delta x' + 2\pi\Delta\nu\Delta x' z_0\ \frac{\partial^2\bar{w}}{\partial\bar{x}^2}\ ])\ \ \ \ \ \ \ (18)$$

This expression is maximum, when the argument of 'cos' is m' , and m' = 0, ± 1, ± 2, ...

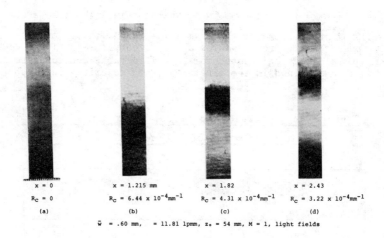

x = 0   x = 1.215 mm   x = 1.82   x = 2.43

$R_C = 0$ $R_C = 6.44 \times 10^{-4}mm^{-1}$ $R_C = 4.31 \times 10^{-4}mm^{-1}$ $R_C = 3.22 \times 10^{-4}mm^{-1}$

(a)   (b)   (c)   (d)

$\bar{w}$ = .60 mm,   = 11.81 lpmm, z_0 = 54 mm, M = 1, light fields

Fig. 3.   Interferograms of curvature recording for
          cantilever beam with a single frequency
          grating. Double exposure.

Similar equations are found for fringe values, where the only difference is the use of $\Delta\nu$ and $\Delta x'$ instead of ν and Δx. The resolution of the system is given by (for the x derivative):

$$R_{c\bar{x}\bar{x}} = \frac{M_{1x}M_{2x}}{2\Delta x'\Delta\nu_x z_0}\ \ \ \ \ \ \ (19)$$

and similar expressions for the other measurements. In spite of the similarity between the equations for fringe values, it can be seen that in DFG they include two different values of magnification. This is necessary because of the large diffraction angle, which the first order waves make with the zero order wave. X_0 is the size of the image due to zero order wave, and X' is the image size due to the first order wave. The magnifications M_1 and M_2 are defined as follows:

M_1 — magnification of the optical system = X_0/X

M_2 — magnification of the first order image = X'/X

The magnification also has effect on the FDT. For a translation in the x' direction, the FDT is given by:

$$(FDT)_{x'} = \frac{\Delta x' \Delta \nu_x}{M_{2x}} \frac{f - z_o}{f} \qquad (20)$$

Unfortunately, the large diffraction angle causes a variable magnification M_2 across the field. Since M_2 varies, then both the FDT and the fringe values vary across the field. Therefore for a correct evaluation of the output interferogram, both of these values must be calculated separately for each fringe.

In order to eliminate these distortions, it is possible to add an additional compensating grating between the DFG and the output plane. This makes the system also achromatic. Wyant [3] suggested that a blazed grating be used for this purpose, in order to keep the loss in light intensity to a minimum. Another possibility to eliminate these distortions is by use of a system based upon plane wave interference, rather than spherical waves. The experimental results and the theoretical comparisons using DFG are described in Fig. 4.

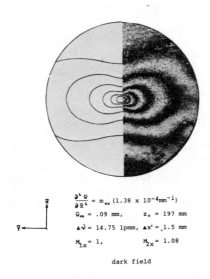

$$\frac{\partial^2 \overline{w}}{\partial x^2} = m_{xx} \ (1.38 \times 10^{-4} \text{mm}^{-1})$$

$\overline{w}_m = .09$ mm, $z_o = 197$ mm

$\Delta \nu = 14.75$ lpmm, $\Delta x' = 1.5$ mm

$M_{1x} = 1$, $M_{2x} = 1.08$

dark field

Fig. 4.   Theoretical and experimental curvature patterns
for circular plate (DFG, double exposure).

Hybrid Method for Measuring Curvature, Requiring a Single Exposure

Here a method will be presented to obtain an interferogram related to a component of the second derivative, which requires only a single exposure taken with the model under load. In this case for the spatial filtering procedure, the beam splitting element is placed in the Fourier plane.

In the previous chapter, the curvature measuring method can be considered as double shearing system, with the secondary shear being supplied by the

translation of the photographic plate or the object between the exposures.
However, it is obvious that the secondary shear can be provided by means
other than a translation of the plate. One of many possible arrangements
would be to place a grating (single or double frequency) at the filter plane,
which will shear one of the first order waves (the other waves are blocked).
However it is emphasized that the shear can be provided by any beam splitting
element, such as Wollaston prism, parallel plate, birefringent crystal, etc.
In addition, it is not required that the RSI be recorded with a grating
system. For this reason, the method is termed "hybrid".

We shall assume that both the primary and secondary shear are provided by
DFG. The value of the secondary shear is $\lambda \Delta \nu z_o$, where in general z_o is
equal to the focal length of the filtering system lens when the arrangement
is for unit magnification. In order to interpret the output interferograms,
relations can be obtained from eq. (12) to eq. (15) by substituting the
above value of secondary shear in place of $\Delta x'$ or $\Delta y'$. The resulting
expressions take the form:

$$\frac{\partial^2 \bar{w}}{\partial \bar{x}^2} = m_{xx} \frac{M_{1x} M_{2x} M_{3x}}{2(\Delta \nu_{x_1} z_{01})(\lambda_2 \Delta \nu_{x_2} z_{02})} \tag{21}$$

and a similar expression for the curvature in the y direction. For one of
the twist components:

$$\frac{\partial}{\partial \bar{y}} \left(\frac{\partial \bar{w}}{\partial \bar{x}} \right) = m_{xy} \frac{M_{1x} M_{2y} M_{3y}}{2(\Delta \nu_{x_1} z_{01})(\lambda_2 \Delta \nu_{y_2} z_{02})} \tag{22}$$

and a similar expression for $\partial/\partial \bar{x} \, (\partial \bar{w}/\partial \bar{y})$.

The subscript 1 refers to values during the recording step, and 2 — the
filtering step. An additional magnification M_3 is noted in each of the
above relations, which relates to the magnification of the filtering system.

M_2 and M_3 are associated with the secondary shear, and can be replaced by
a single value M ('lumped' magnification):

$$M_x = M_{x2} M_{x3} = \frac{x \text{ dimension of filtered image}}{x \text{ dimension of model}} \tag{23}$$

and a similar expression for M_y .

By inserting eq. (23) into eq. (21) and eq. (22), a simpler expression can
be found — from which the following general expression can be written for
the resolution:

$$R_c = \frac{M_1 M}{2(\Delta \nu_1 z_{01})(\lambda_2 \Delta \nu_2 z_{02})} \tag{24}$$

where the appropriate values of magnification and frequency are substituted,
according to which component is being measured. The experiments conducted,
using hybrid method, applied two beam splitting elements as DFG. The
grating used during the recording step had a value of $\Delta \nu_1 = 14.75$ lpmm, and

the grating used in the filtering step had a value $\Delta\nu_2 = 2.5$ lpmm. Quantitative results represented here are for the circular plate in Fig. 5. Similar results for a rectangular plate can be obtained.

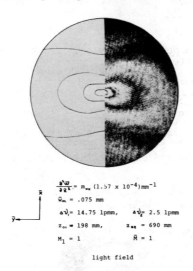

$$\frac{\partial^2 \overline{w}}{\partial \overline{x}^2} = m_{w\overline{x}} \, (1.57 \times 10^{-4}) \, mm^{-1}$$

$\overline{w}_m = .075$ mm

$\Delta\nu_1 = 14.75$ lpmm, $\Delta\nu_2 = 2.5$ lpmm

$z_{o_1} = 198$ mm, $z_{o_2} = 690$ mm

$M_1 = 1$ $\overline{M} = 1$

light field

Fig. 5. Theoretical and experimental curvature patterns
for circular plate (hybrid method).

Fourier Imaging Curvature Measurement

When the RSI is filtered, the intensity of the information term is sinusoidally dependent on the value of ψ' [5] according to:

$$I_o(x,y) = a^2 b^2 (1 + \cos^2(\pi\psi') + 2\cos(2\pi\nu z_o \frac{\partial\overline{w}}{\partial x}) \cos(\pi\psi')) \qquad (25)$$

In addition, ψ' is linearly dependent on the curvature, according to:

$$\psi' = \lambda\nu^2 z_o (1 - \frac{z_o}{f} + z_o \frac{\partial^2\overline{w}}{\partial\overline{x}^2}) \qquad (26)$$

Here we exploit this phenomenon as an effective means of curvature measurement. The set-up uses a Ronchi ruling, which is displaced from the focal plane. Only a single exposure is taken with the model under load. After exposure and developing, the intensity in the output plane (after spatial filtering) is found as:

$$I(x,y) = a^2 b^2 \cos^2(\pi\psi') \qquad (27)$$

as was expected. By eq. (26) and eq. (27), ψ' can be developed to:

$$\psi' = \lambda\nu^2 z_o (1 - \frac{z_o}{Mf} + \frac{2z_o}{M^2} \frac{\partial^2\overline{w}}{\partial\overline{x}^2}) \qquad (28)$$

if $\psi' = N(N = 0, \pm 1, \pm 2, ...)$, then the output intensity is zero. By the use of eq. (27), the system is set for one of the conditions of eq. (28), under the assumption that the curvature is zero. In this case it is the change of ψ' which is important. The relation for the fringe value for the x slope component becomes:

$$\frac{\partial^2 \bar{w}}{\partial \bar{x}^2} = m_{xx} \frac{M_{xx}}{2\lambda \nu_x^2 z_o^2} \tag{29}$$

and similarly for the y component. From eq. (29) a general expression for the resolution can be written:

$$R_c = \frac{M^2}{2\lambda \nu^2 z_o^2} \tag{30}$$

A significant feature of this technique, that it can yield results in real time, which is the only full field real time curvature display, for our knowledge. The theoretical and experimental results are compared and demonstrated in Fig. 6.

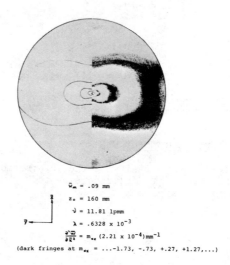

$\bar{w}_m = .09$ mm

$z_o = 160$ mm

$\nu = 11.81$ lpmm

$\lambda = .6328 \times 10^{-3}$

$\frac{\partial^2 \bar{w}}{\partial \bar{x}^2} = m_{xx} (2.21 \times 10^{-4}) \text{mm}^{-1}$

(dark fringes at $m_{xx} = ...-1.73, -.73, +.27, +1.27,...$)

Fig. 6. Theoretical and experimental curvature patterns for circular plate (Fourier imaging).

Summary and Comparison

The single frequency grating, such as Ronchi ruling, had such a low sensitivity, that it is good only for rough measurements. The double exposure method with double frequency grating, was capable of much higher resolution. However, correctly evaluate the interferograms (the curvature components), the fringe order due to translation must be known — which is difficult to determine with high accuracy. In addition, this method suffers from the distortion which the first order waves undergo, so that the accuracy is also reduced. Image distortions can be reduced by a compensating grating, or by the use of plane wave system.

The hybrid method was seen to produce the most accurate results of all the methods considered. It also has the advantages of requiring only a single exposure and of not having an FDT (Fringe order Due to Translation), as the first two methods. In addition, it can compensate (to some degree) for initial model curvature and varying magnification. This method has the disadvantage of needing a more complicated filtering operation, involving the careful setting of the beam splitting element.

The Fourier imaging technique has the advantage of not being a "double shear" technique. It has also the advantage of not having an FDT, which was seen to cause difficulties in the interpretation of the output interferograms. It is not distorting the output wavefront, due to the large diffraction angles and of a more efficient use of available light. Moreover this method can be used for "real time" analysis. It is the simplest method to apply, requiring only a single exposure followed by an ordinary spatial filtering step. The disadvantage of this method is that it can measure the two components of curvature, but not being able to measure the twist.

References

1. Timoshenko, S. and Woinovsky-Krieger, S., "Theory of Plates and Shells" 2nd Ed., McGraw-Hill, New-York, 1959.

2. Ronchi, V., "Forty years of history of a grating interferometer", Appl. Opt. 3, 437-451 (1964).

3. Wyant, J.C., "Double frequency grating lateral shear interferometer". Appl. Opt. 12, 2057-2060 (1973).

4. Heise, V., "A moire method for measuring plate curvatures", Exp. Mech. 7, 47-48 (1967).

5. Assa, A., "Some new optical methods for use in experimental stress analysis", D.Sc. Dissertation, Dept. of Aeronautical Eng., Technion I.I.T., Haifa, Israel (July, 1975).

SESSION 9.

SPECKLE PATTERNS

Chairman: R. DÄNDLIKER

PUPIL SIZE AND SPECKLE STATISTICS FOR A ROUGH METAL SURFACE

P. Hariharan and Z. S. Hegedus

National Measurement Laboratory, CSIRO, Sydney, Australia 2008

ABSTRACT

The cross correlation between two orthogonally polarized speckle images of a rough metal surface is studied. It is found to increase as the relative aperture of the imaging lens decreases. The theory of this effect and its influence on the visibility of the fringes in speckle interferometry are discussed.

INTRODUCTION

When a rough metal surface is illuminated with a polarized beam from a laser, the scattered light contains a depolarized component whose relative magnitude increases with the angle of incidence (1,2). This depolarized component can be observed quite easily, and its existence cannot be ignored in experimental studies of speckle.

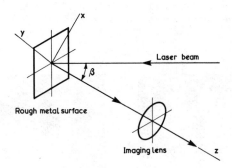

Fig. 1. Optical set-up

Consider the optical set-up shown in Fig. 1, in which the laser beam is polarized in the vertical plane, and the plane of incidence is horizontal. The illuminance in the speckled image of the surface formed by the lens is the sum of the illuminances of two component speckle patterns produced by scattered light polarized in two orthogonal directions, OX, OY. For symmetry, and also to equalize the average illuminances of the two component patterns, these directions can be taken at $+45°$ and $-45°$.

319

If the illuminances $I_1(x,y)$, $I_2(x,y)$ at corresponding points in the component speckle patterns are not identical, the illuminance distribution in the speckle image is no longer a negative exponential distribution. In general, the two component speckle patterns are only partially correlated, and the contrast of the speckle in the image decreases. Goodman (3) has shown that the resulting illuminance can be regarded as the sum of two uncorrelated speckle patterns whose average intensities are the eigenvalues of the coherency matrix. The statistical properties of the speckle image can then be derived in terms of these eigenvalues which are also the correlation coefficients of the component speckle illuminances. The latter can again be expressed in terms of the correlation coefficients of the amplitudes of the component fields.

EXPERIMENTAL RESULTS

The cross-correlation of the illuminances in the two component speckle images can be evaluated by a technique used by Chakraborty (4). It is well known that if two identical speckle patterns are recorded on the same plate with a small mutual displacement, the resulting transparency produces a system of Young's fringes in its Fourier plane. Any difference between the speckle patterns results in a reduction in the visibility of the fringes, which can therefore be looked upon as a measure of the correlation of the speckle patterns.

Experimental studies were carried out with a sandblasted aluminium surface (r.m.s. surface roughness $\sigma(h)$ = 2.7 μm) mounted on a micrometer slide so that it could be translated laterally in its own plane by known amounts between exposures. The surface was illuminated at an angle by the expanded beam from a He-Ne laser and viewed normally. The speckled image of the surface was recorded with a camera fitted with a 100 mm lens. The distance from the object to the camera lens was approximately 0.5 m. Kodak Holographic Film SO-253 was used, processed in Kodak X-ray Developer at 20 C for 5 min to a density of 0.7.

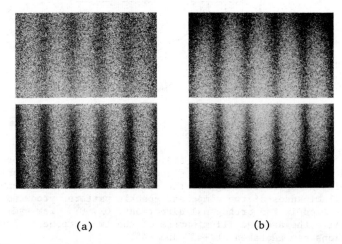

(a)   (b)

Fig. 2. Young's fringes obtained from speckle photographs
made at (a) f/4.5, (b) f/16.

Exposures were made through a polarizer in a rotating mount at relative apertures ranging from f/4.5 to f/22, and at angles of incidence ranging from 25°
to 45°. Figures 2(a) and 2(b) show the Young's fringes obtained in the Fourier
plane of an optical processor from photographs made at relative apertures of
f/4.5 and f/16 with the laser beam incident at an angle of 40°. The upper
half of each figure shows the Young's fringes obtained with exposures made
with the polarizer set first at +45° and then at -45°. As a control, the
lower half shows the fringes obtained in each case with no change in the polarizer setting between the exposures. There is a noticeable increase in the
visibility of the fringes, and hence in the cross correlation of the two component speckle patterns, when the relative aperture of the camera lens is
reduced. This is confirmed by microdensitometer records of the fringes which
show that the ratio of the visibilities of the two sets of fringes in the
upper and lower halves increased from only 0.16 in Fig. 2(a) to 0.64 in Fig.
2(b).

THEORETICAL ANALYSIS

While a rigorous analysis presents many difficulties, a useful approximation
for the cross correlation of the fields in the component speckle images can be
obtained if some simplifying assumptions are made. This analysis is an extension of a theoretical treatment of the effect of surface roughness on the
statistics of image speckle due to Goodman (3).

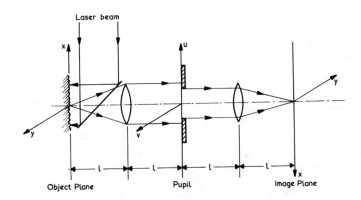

Fig. 3. Geometry for theoretical analysis

Consider the two-lens system shown in Fig. 3, in which the rough metal surface in the input plane is illuminated normally via a beam-splitter by a
laser beam polarized at 45°, and a speckled image is formed in the output
plane. The point spread function of the system is determined by the pupil
K(u,v) in the Fourier plane.

It is assumed that the rough metal surface can be described by a height function h(x,y) which is a stationary, random process with normal distribution
and zero mean. To evaluate the two components of the scattered field, it is
then necessary to take into account the fact that the complex reflection coefficients for the two polarizations at any point differ. Assuming that the
variations in the complex reflection coefficients affect mainly the phases of

the scattered waves (5), the complex fields immediately to the right of the surface can be written as

$$f_1(x,y) = \exp i[\theta(x,y) + \phi_1(x,y)]$$

$$f_2(x,y) = \exp i[\theta(x,y) + \phi_2(x,y)] \tag{1}$$

where $\theta(x,y) = (4\pi/\lambda)h(x,y)$ is the phase change due to the variations in surface height, and $\phi_1(x,y)$, $\phi_2(x,y)$ are the phase shifts for the two polarizations due to the local variations in the complex reflection coefficients.

The phase changes $\phi_1(x,y)$ and $\phi_2(x,y)$ are both stationary, random variables with normal distribution and zero mean. If they are linear functions of the partial derivatives of $h(x,y)$ they are independent of each other and of $\theta(x,y)$. In addition, they have the same variance, so that

$$\sigma^2(\phi_1) = \sigma^2(\phi_2) = \sigma^2(\phi). \tag{2}$$

If it is assumed that $\sigma^2(\theta)$, the variance of $\theta(x,y)$ is large enough that the component fields approximate to circular, complex, random variables, and the surface slopes are small enough that $\sigma^2(\phi) \ll 1$, the cross correlation of the two components of the scattered object field can be shown to be

$$R_{f_1f_2}(\Delta x,\Delta y) = \exp\{-\sigma^2(\theta)[1 - R_{hh}(\Delta x,\Delta y)]\}$$

$$\times \exp\{-\sigma^2(\phi)\} \tag{3}$$

where $R_{hh}(\Delta x,\Delta y)$ is the autocorrelation function of $h(x,y)$.

The two components of the image field, $g_1(x,y)$, $g_2(x,y)$ are obtained by convolving the object fields given by eq. (1) with $k(x,y)$ the point spread function of the imaging system. It can then be shown that $R_{g_1g_2}(0,0)$ the peak value of the cross-correlation of the component image fields is less than the peak value of their autocorrelation $R_{gg}(0,0)$ by an amount

$$R_{gg}(0,0) - R_{g_1g_2}(0,0) \simeq \sigma^2(\phi) \cdot R_{KK}(0,0) \cdot S_1 \tag{4}$$

where

$$S_1 = \iint R_{ff}(\Delta x,\Delta y)\, d\Delta x\, d\Delta y \tag{5}$$

can be defined as the correlation area of the scattered object wave, and $R_{KK}(0,0)$ is the area of the pupil.

For any given surface this difference becomes less and less as the area of the pupil decreases, so that the cross-correlation of the two component fields increases.

FRINGE VISIBILITY IN SPECKLE INTERFEROMETRY

The dependence of the cross correlation of the two component speckle patterns formed by a metal surface on the relative aperture of the camera lens also affects the visibility of the fringes in speckle interferometry techniques which use Fourier processing (6), since the two laterally displaced speckle images of the surface which are recorded are assumed to exhibit complete correlation at points corresponding to an optical path change $\Delta d = n\lambda$, while their correlation is a minimum at points where $\Delta d = (2n+1)\lambda/2$.

In the double exposure technique described by Archbold et al. (7), the resulting local variations in speckle contrast are converted into transmittance variations by a photographic material with a highly non-linear characteristic. The presence of even a relatively small depolarized component in the light scattered from the surface causes a reduction in the contrast of the individual speckle patterns and this, in turn, results in a much greater decrease in the visibility of the speckle fringes. With the sandblasted aluminium surface used in these experiments, the loss in visibility is so severe with an optical system of large relative aperture (f/4.5), that no speckle fringes are obtained without a polarizer. However, the loss in contrast of the individual speckle patterns is reduced to a level at which usable fringes are obtained, if the camera lens is stopped down to f/16.

REFERENCES

(1) Beckmann, P., The Depolarization of Electromagnetic Waves,
       Golem Press, Boulder, 1968.

(2) J. Renau, P.K. Cheo and H.G. Cooper, Depolarization of linearly polarized
       EM waves backscattered from rough metals and inhomogeneous dielec-
       trics, J. Opt. Soc. Am. 57, 459 (1967).

(3) J.W. Goodman in Laser Speckle and Related Phenomena, ed. Dainty, J.C.,
       Springer-Verlag, Berlin, 1975.

(4) A.K. Chakraborty, The effect of polarisation of the illuminating beam on
       the microstructure of speckles produced by a random diffuser,
       Opt. Commun. 8, 366 (1973).

(5) Born, M. and Wolf, E., Principles of Optics,
       Pergamon Press, Oxford, 1965.

(6) J. Butters and J. Leendertz, J. Phys. E 4, 277 (1971).

(7) E. Archbold, J. Burch and A. Ennos, Opt. Acta 17, 883 (1970).

TWO DECORRELATED DIFFUSERS-INTERFEROMETER

Yves Dzialowski and Marie May

Institut d'Optique et Laboratoire d'Optique, Université Paris VI,
4 Place Jussieu, 75230 Paris Cedex 05, France

ABSTRACT

We describe here an interferometer constituted of two axially shifted diffusers,illuminated by a laser beam.These diffusers, which are not correlated,give rise to an interference phenomenon in their Fourier plane.Each speckle of this plane is modulated by a ring system,the geometrical characteristics of which are depending on the geometry of the arrangement.The interference orders of the fringes are fixed by the random phase characterizing the considered speckle.As observation plane is Fourier plane,a lateral or a longitudinal shift of one of the diffusers causes only a change of the modulating fringes,the corresponding speckles remaining fixed and correlated.Consequently,a photographic recording of the irradiances of Fourier plane before and after the shift,exhibits,after processing,a Moiré fringe system,the contrast of which is maximum in the diffracted orders.

INTRODUCTION

The coherent addition of two speckle fields has been performed first by Leendertz (1) to study the deformations occuring to one of the fields.A Michelson interferometer,in which the mirrors are replaced by two scatterers and adjusted to the zero path difference,is illuminated by a parallel beam of laser light.The resulting intensity distribution is observed in image plane of the scatterers.The variation of this speckle pattern with an axial translation of one of the scatterers is studied.To study the deformations of an object,it is also possible to image it onto a photographic plate through a particular pupil (Refs. 1,2,3). The resulting speckle of image plane is characterized by the autocorrelation function of the pupil and the correlation fringes occuring between the two recorded irradiances of image plane (before and after the deformation) reveal the nature of the deformation.

We describe in this paper an interferometer constituted of two longitudinally shifted diffusers illuminated by a laser beam. The resulting intensity distribution is observed in the image plane of the laser source which is the Fourier plane of the diffusers.

INTERFERENCE PHENOMENON BETWEEN TWO DIFFUSERS

Let us consider the scheme of Fig. 1 in which D and D' are two

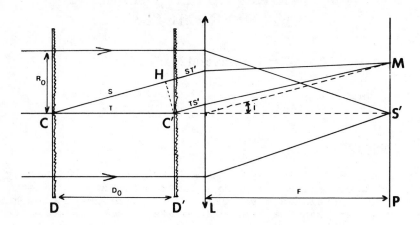

Fig. 1 Speckle pattern interferometer in parallel light

partially diffusing plates derived from the photographic recording
of two speckle fields.They are illuminated by a parallel beam of
laser light and the resulting phenomenon is studied in the image
plane of the source (i.e. the Fourier plane of the diffusers).
The incident beam is split into two beams by D:a directly transmitted beam t and a scattered beam s .Each of these two beams is
afterwards divided into two beams by D'.Finally,four beams are
reaching the focal plane of lens L.The (tt') beam,transmitted by
D and D',is converging at point S' and can easily be removed with
a stop.The beam (ss') scattered first by D and then by D' can also
be neglected due to its low irradiance.In any point M of plane P,
characterized by its rectangular coordinates (X,Y),there is the
amplitude superposition of two beams:the beam (st')scattered by
D and transmitted by D' and the beam (s't) transmitted by D and
scattered by D'.These two beams habe been scattered by D and D'
in a direction inclined at the angle $i=(X^2+Y^2)^{1/2}/f$ on the optical axis.They give rise to a speckle in point M.Their phase difference is the sum of two terms:a random term due to the diffusions by D and D' and a quadratic term due to the longitudinal
shift existing between D and D'.Let

$$D(x,y) = a + A(x,y) \tag{1}$$
$$D'(x',y') = a' + A'(x',y') \tag{2}$$

be the complex amplitude transmittances of D and D'.The terms a
and a' are constants and the terms $A(x,y)$ and $A'(x',y')$ are two
random complex functions.
The path difference between the (st') and (s't) beams is given by

$$\Delta = (CC')-(CH) = (CC')(1-\cos i) \simeq (CC')\frac{i^2}{2} \tag{3}$$

which can be rewritten as

$$\Delta = d_o \frac{(X^2+Y^2)}{2f^2} \qquad (4)$$

Under these conditions,the amplitude distribution at point M is proportional to

$$U(X,Y) = a'\hat{A}(X,Y) + a\hat{A}'(X,Y)\exp[j\frac{\pi d_o}{\lambda f^2}(X^2+Y^2)] \qquad (5)$$

where f is the focal length of lens L,d_o the distance between D

and D' and where $\hat{A}(X,Y)$ and $\hat{A}'(X,Y)$ are the respective Fourier transforms of A(x,y) and A'(x',y') corresponding to the spatial frequency$(X/\lambda f,Y/\lambda f)$.
The resulting irradiance at point M is given by

$$I(X,Y) = a'^2|\hat{A}(X,Y)|^2 + a^2|\hat{A}'(X,Y)|^2 + 2aa'|\hat{A}(X,Y)\|\hat{A}'(X,Y)|\times..$$

$$.. \cos[\frac{\pi d_o}{\lambda f^2}(X^2+Y^2) + \Phi_o'(X,Y)-\Phi_o(X,Y)] \qquad (6)$$

where $\Phi_o'(X,Y)$ and $\Phi_o(X,Y)$ are the respective random phases of $\hat{A}'(X,Y)$ and $\hat{A}(X,Y)$.
The speckle centered at point M is defined by the random phase difference $\Phi_o'(X,Y)-\Phi_o(X,Y)$ which remains constant over its area. Its mean size is given by the mean size of the speckles generated in plane P by D alone,on the one hand,and D' alone on the other hand.It corresponds to the diameter of the Fraunhofer diffraction pattern of the lens L and is given by $1.22 \lambda f/2r_o$,where $2r_o$ is the diameter of lens L.As seen in expression (6),this speckle is modulated by a ring system centered at point S'(X=Y=0),the radii of which are given by

$$r_K = f (\frac{\lambda}{d} (2K - \frac{\Phi_o'(X,Y)-\Phi_o(X,Y)}{\pi}))^{1/2} \qquad (7)$$

where K = 1,2 ..
Consequently,each speckle of Fourier plane is modulated by a ring system centered at S',the interference order of which depends on its random phase distribution.However,these fringes can be generated only if the (st') and the (ts') beams reaching the point M are laterally shifted.Then,they appear in the speckles situated ouside of the circle centered at S'(Fig.2),the radius of which is given by

$$(X_o^2+Y_o^2)^{1/2} = f\frac{2r_o}{d} \qquad (8)$$

For example,if f=300mm , $2r_o$ = 30mm and d_o = 500mm,the fringes appear in the speckles situated at least,at 18mm from the center of the field.

Let us note that the phenomena observed in plane P are the same whatever be the position of the diffusers D and D' between the point source and its image S'.Under the conditions when the diffusers are limited by the illuminating beam,the speckle size in plane P depends only on the numerical aperture of the lens L.

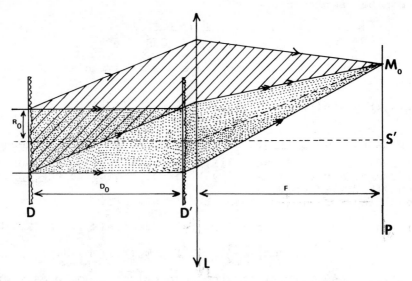

Fig. 2 Fringes appear in plane P outside the circle
the radius of which is S'M₀

If D and D' are illuminated by a convergent beam (Fig.3),the
intensity distribution in any point M(X,Y) of plane P is

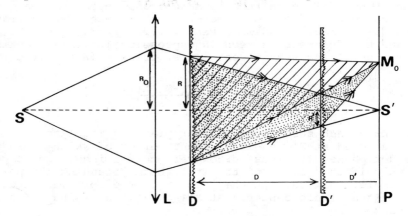

Fig. 3 Speckle pattern interferometer in convergent
light

proportional to

$$I'(X,Y) = a'^2|\hat{A}(\frac{X}{d+d'},\frac{Y}{d+d'})|^2 + a^2|\hat{A}'(\frac{X}{d'},\frac{Y}{d'})|^2 +$$

$$2aa'|\hat{A}(\frac{X}{d+d'},\frac{Y}{d+d'})\|\hat{A}'(\frac{X}{d'},\frac{Y}{d'})|\cos[\frac{\pi d(X^2+Y^2)}{\lambda d'(d+d')} + \Phi'(X,Y) - \Phi(X,Y)]$$

(9)

where d is the distance between D and D'and d' is the distance
between D' and the Fourier plane.

The radii of the rings modulating the speckle centered at point M
are given by

$$r'_K = (\lambda \frac{d'(d+d')}{d} (2K - \frac{\Phi'(X,Y) - \Phi(X,Y)}{\pi}))^{1/2} \tag{10}$$

The fringes appear only inside the speckles situated at least at

$$(X_o^2 + Y_o^2)^{1/2} = \frac{r'(d+d') + d'r}{d} = \frac{2rd'}{d} \tag{11}$$

from the point S', where r and r' are the respective radii of the
diffusers D and D'. For example, with 2r=20mm, d=300mm and d'=10mm,
the fringes appear in the speckles situated at 0.7mm from S'.
The irradiance of plane P illuminated in the way described above
(Fig. 1) is recorded onto a photographic plate H. After processing
H can be considered as a random Fresnel zone plate. The amplitude
Fourier spectrum of H is constituted of three terms: the central
order, lying in the Fourier plane of the photographic plate, is the
sum of the autocorrelation function of A and the autocorrelation
function of A'; the two diffracted orders (+1 and -1) are lying in
two planes on both sides of the Fourier plane and symetrical by
respect to it. Their amplitude distributions are complex conjugate
and given by the cross correlation function of A and A'. Let us
note that the three diffracted orders are in-line.

LATERAL TRANSLATION OF D

Let us suppose now that D is laterally shifted through x_o by respect to its position in Fig.1. This translation acts only on the
(st') beam and involves a linear phase change of the corresponding
amplitude in Fourier plane. The complex amplitude distribution at
point M(X,Y) can now be writen as

$$U_L(X,Y) = a'\hat{A}(X,Y)\exp(-j\frac{2\pi Xx_o}{\lambda f}) + a\hat{A}'(X,Y)\exp(j\frac{\pi d_o}{\lambda f^2}(X^2+Y^2)) \tag{12}$$

The corresponding intensity distribution is proportional to

$$I_L(X,Y) = a'^2|\hat{A}(X,Y)|^2 + a^2|\hat{A}'(X,Y)|^2 +$$

$$2aa'|\hat{A}(X,Y)| \| \hat{A}'(X,Y)|\cos(\frac{\pi d_o}{\lambda f^2}(X^2+Y^2)+\frac{2\pi}{\lambda f}Xx_o+\Phi'_o(X,Y)-\Phi_o(X,Y)) \tag{13}$$

The speckle centered at point M(X,Y) remains characterized by the
random phase $\Phi'_o(X,Y)-\Phi_o(X,Y)$, constant over its area. It is identical to the preceeding speckle but, it is now modulated by a ring
system centered at the point $(X=-x_o d_o/f, Y=0)$, the radii of which
are given by

$$r_{LK} = (r_K^2 + \frac{f^2 x_o^2}{d_o^2})^{1/2} \tag{14}$$

The lateral shift of D involves thus only a lateral translation
of the center of the ring systems and a constant increase of the
square of their radii. Save on the modulating fringes, the random
intensity distributions given by expressions (6) and (13) are
exactly superimposed and quite correlated.
The irradiance distributions lying in plane P (before and after

the lateral translation of D) are successively recorded on a photographic plate H'.The total intensity recorded by H' can be
written as

$$M(X,Y) = 2(a'^2|\hat{A}(X,Y)|^2 + a^2|\hat{A}'(X,Y)|^2) +$$

$$4aa'|\hat{A}(X,Y)\|\hat{A}'(X,Y)|\cos(\frac{\pi d_o}{\lambda f^2}(X^2+Y^2)+\Phi'_o(X,Y)-\Phi_o(X,Y)+\frac{\pi XX_o}{\lambda f}) \ ..$$

$$.. \cos(\frac{\pi}{\lambda f}XX_o) \tag{15}$$

After processing under the usual conditions of linearity,H' exhibits a system of rectilinear Moiré fringes,the fringe spacing of
which is given by

$$d_L = \frac{\lambda f}{X_o} \tag{16}$$

The contrast of these fringes is not maximum when they are seen
in the direction of the central order.If the central order is
removed by filtering the Fourier spectrum of the photographic
plate,the contrast of the Moiré fringes observed in the image
plane of H' is maximum since the two recorded speckle patterns
are identical.Let us note that the filtering operation is easier
if the diffusers D and D' are laterally shifted during the recording process.

LONGITUDINAL TRANSLATION OF D

Let us consider now the case when the diffuser D is axially shifted through z by respect to its position in Fig. 1 .The amplitude
distribution at point M(X,Y) of plane P is now given by

$$U_A(X,Y) = a'\hat{A}(X,Y)\exp(-j\frac{\pi z}{\lambda f^2}(X^2+Y^2))+a\hat{A}'(X,Y)\exp(j\frac{\pi d_o}{\lambda f^2}(X^2+Y^2)) \tag{17}$$

According to expression (17) it can be seen that the longitudinal
translation of D involves only a modulation of the (st') amplitude by a quadratic phase factor whatever be the value of z.This is
due to the fact that the diffuser D is illuminated by a parallel
beam of laser light and the observation plane is Fourier plane
(Ref. 4).Intensity distribution lying in plane P is proportional
to

$$I_A(X,Y) = a'^2|\hat{A}(X,Y)|^2 + a^2|\hat{A}'(X,Y)|^2 +$$

$$2aa'|\hat{A}(X,Y)\|\hat{A}'(X,Y)|\cos(\frac{\pi d_o}{\lambda f^2}(X^2+Y^2)(1+z/d_o)+\Phi'_o(X,Y)-\Phi_o(X,Y)) \tag{18}$$

As before,the longitudinal shift involves only a change of the
modulating fringes,each speckle of plane P remaining fixed and
defined by the same constant random phase.The rings modulating
the speckle centered at M(X,Y) are centered at (X=Y=0) and their
radii are given by

$$r_{AK} = r_K(1+z/d_o)^{-1/2} \tag{19}$$

The photographic recording of the irradiances lying in plane P
before and after the longitudinal shift of D,exhibits,after pro-

cessing,circular Moiré fringes.The contrast of these fringes is maximum if the zero order of the Fourier spectrum of the recording is removed by filtering.Under these conditions,the irradiance of the Moiré system is proportional to

$$M_A(X,Y) = 4a^2a'^2|\hat{A}(X,Y)|^2|\hat{A}'(X,Y)|^2\cos^2(\frac{\pi z}{2\lambda f^2}(X^2+Y^2)) \qquad (20)$$

The rings are centered at (X=Y=0) and their radii are given by

$$R_p = f(2\lambda p/z)^{1/2} \qquad (21)$$

where p = 1,2,..

Figure 4a represents the Moiré fringes obtained with a lateral translation of D through 60μm and Fig. 4b the circular Moiré fringes corresponding to an axial shift of D through 100μm. The focal length of lens L is 400 mm.

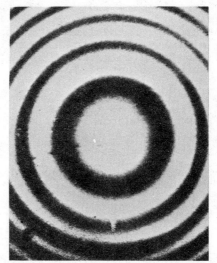

Fig. 4a Straight Moiré fringes obtained with $x_0=60\mu m$ and f=400mm.

Fig. 4b Circular Moiré fringes obtained with z=100μm and f=400mm.

These Moiré fringes are localised in the plane of the photographic plate.Consequently they are achromatic and they can be retrieved with a white light point source which has the advantage to remove the speckle appearance of the image of the photographic plate.

Let us note to end that the same phenomena occur with the scheme of Fig. 3.However in the case of the converging beam,a longitudinal translation of D involves in plane of S' both a quadratic change of phase of the corresponding amplitude and a similarity for the speckles (Refs. 4,5) in the ratio (1-z/d+d').If the radial shift cannot be neglected,the speckles are no more superimposed on their doubly exposed photographic recording.The contrast of the Moiré rings decreases from the center to the edge of the photographic plate.

CONCLUSION

In the speckle interferometer described above,a photographic plate records twice the irradiance lying in Fourier plane of two
axially shifted decorrelated diffusers.Between the two exposures
a translation (lateral or longitudinal) is given to one of the
diffusers.After processing,the photographic plate exhibits a system of Moiré fringes,the contrast of which is maximum in the diffracted orders.A lateral translation of the diffuser involves
rectilineat Moiré fringes in the plane of the plate.If the diffuser is axially shifted,the corresponding Moiré fringes are circular and centered at the image of the source.In both cases the
amount of the translation given to the diffuser can be obtained
from the spacing of the fringes or from the radii of the rings.
If the diffuser is laterally and axially translated,the resulting
Moiré fringes are circular but the center of the rings is laterally shifted by respect to S'.The value of this shift depends on
the amount of the lateral translation given to the diffuser.The
sensitivity of the interferometer being a function of the lengths
d_0 and f , displacements smaller than the speckle size in Fourier
plane can be detected.

REFERENCES

(1) J.A. Leendertz, Interferometric displacement measurement on
    scattering surfaces utilizing speckle effect,
    J.Phys.E (Sci. Instrum) 3,214(1970).

(2) D.E. Duffy, Moiré gauging of in-plane displacement using
    double aperture imaging, Appl. Opt. 11,1778 (1972).

(3) Y.Y. Hung,R.E. Rowlands,and I.M. Daniel, Speckle-shearing
    interferometric technique:a full-filled strain gauge,
    Appl. Opt. 14,618 (1975).

(4) Y. Dzialowski and M. May, Correlation of speckle patterns
    generated by laser point source-illuminated diffusers,
    Opt. Commun. 16,334 (1976).

(5) J.A. Mendez et M.L. Roblin, Relation entre les intensités
    lumineuses produites par un diffuseur dans deux plans
    parallèles, Opt. Commun. 11,245 (1974).

OBSERVATION AND EXPLANATION OF THE LOCALIZATION OF FRINGES FORMED WITH DOUBLE-EXPOSURE SPECKLE PHOTOGRAPH

Kazuo P. Miyake and Hiroshi Tamura

Institute for Optical Research, Kyoiku University, Shinjuku, Tokyo, 160 Japan

ABSTRACT

A system of interference fringes formed by diffraction at a double-exposure speckle photograph is recorded by a photographic camera and the localization of fringes is investigated. The interference fringes are located on a plane at twice the distance of the speckle photograph from the camera lens. The experimental fact is interpreted theoretically by calculation of the mean square deviation of phase difference between interfering rays. Physical significance of the calculation is derived from an integral formula for the visibility of interference fringes.

INTRODUCTION

The localization of interference fringes has been observed and discussed in holographic interferometry (1-11). The localized fringes are observed on the plane where phase difference between interfering rays remains stationary.

There are two separate theories to treat the problem of fringe localization. In the first theory, the optical path difference is differentiated with respect to variables such as propagation vectors to derive the condition for localization (2-8). The other is based on an elaborate concept of homologous rays (9-11). In this theory, it is realized that the fringe occurs at the location where corresponding homologous rays intersect or make the nearest approach.

Walles (9) proposed an analytical method for determining the visibility of interference fringes formed with double-exposure hologram, but he did not apply it to practical problems.

In this paper, we treat the localization of fringes formed with double-exposure speckle photograph. Speckle pattern photographs contain statistical randomness in intensity and position of speckles recorded. Therefore, there is no wonder if localized fringes similar to those observed in holographic interferometry are observed with speckle photographs. But there seems no description on the phenomena.

The authors observed an example of such fringe localization and

interpreted it by use of the method lying on the line of
Walles'.

OBSERVATION

Experimental setup used for observation of fringes is similar to
that used by Archbold and Ennos (12) to measure the surface
displacement in the direction of observation.

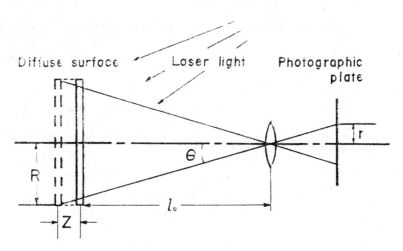

Fig. 1. Preparation of double-exposure photograph.

Figure 1 shows an experimental setup to obtain a double-
exposure photograph. A diffuse surface illuminated by a laser
beam is photographed by an imaging lens. Then, the surface is
displaced by an amount of Z along the optical axis of the lens.
After the displacement, the surface is photographed again onto
the same photographic plate. From Fig. 1, it is easily
understood that the displacement $\delta(r)$ of corresponding speckles
on the photographic plate is given by

$$\delta(r) = Zr/\ell_0 = \alpha r, \tag{1}$$

which is proportional to radial distance r of the speckle from
the optical axis.

Figure 2 shows an arrangement for observation of fringes. The
double-exposure photograph is illuminated from the back by a
collimated laser beam. Diffracted beam is collected to a
photographic plate by a lens. The interference takes place on
the photographic plate between coherent beams diffracted by
corresponding speckles. The condition for the n-th fringe to
appear is given by

$$\delta \sin\psi = \delta r/\ell = n\lambda. \tag{2}$$

From Eqs. (1) and (2), we obtain

$$r = \sqrt{n\ell\ell_0\lambda/Z}. \tag{3}$$

Equation (3) shows that fringes are concentric rings and look like either Haidinger fringes or Newton's rings. Therefore, we may suppose that the fringes are located in the former case at infinity and in the latter on the plane of the speckle photograph.

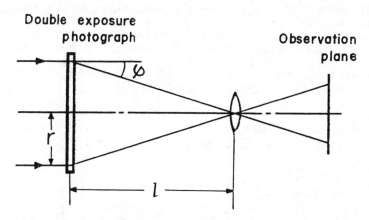

Fig. 2. Observation of interference fringes.

Although fringes of high contrast are observed by the eye through a pinhole, fringes are not recorded by use of a pinhole camera. To record the fringe system, an imaging lens must be used.

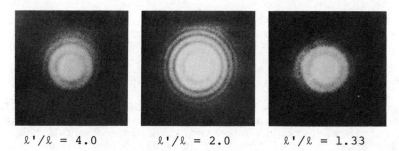

$\ell'/\ell = 4.0$ $\ell'/\ell = 2.0$ $\ell'/\ell = 1.33$

Fig. 3. Photographs of interference fringes.

Figure 3 shows photographs taken at different values of ℓ'/ℓ. It is seen that fringes of high contrast are located at the distance $\ell' = 2\ell$ from the lens.

The experimental conditions are as follows. For preparation of double-exposure photograph, $Z = 1.6mm$, $\ell_0 = -28cm$, and the imaging lens is of $f = 50mm$ and F/11. For observation of interference fringes, $\ell = -16cm$ and the lens is of $f = 50mm$ and F/8. The lens can be opened up to full aperture (F/4) without appreciable loss of contrast. However, the aperture is kept at F/8 to make sizable the central bright disc due to non-diffracted light.

THEORETICAL EXPLANATION

As stated before, fringes appear at the point where the path difference between interfering rays remains stationary. To explain the observed phenomena, we first calculate the function representing the optical path difference. The mean square deviation of this function is calculated by integration of the function over a bundle of rays which contributes to formation of fringes at a point on the photographic plate.

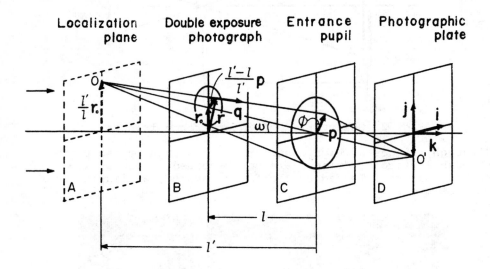

Fig. 4. Determination of localization plane.

In Fig. 4, the plane A is expected to be the localization plane; the plane B represents the double-exposure photograph through which the fringe system is observed; the plane C shows the plane of the entrance pupil of the lens; the plane D represents the photographic plate on which the fringe system is recorded. All planes are considered parallel to each other. Let the $\bar{k}$-axis be coincident with the optical axis of the lens. The $\bar{i}$- and $\bar{j}$-axes are taken as shown in Fig. 4. The distances of the planes B and A from the plane C are ℓ and ℓ' respectively. The planes A and D are assumed conjugate to each other with respect to the lens.

Let us consider a straight line connecting the center of the entrance pupil and a point on the plane B specified by a radius vector $\bar{r}_0$. No generality is lost by taking $\bar{r}_0$ parallel to the $\bar{j}$-axis. From Fig. 4, it is clear that the said straight line intersects the plane A at a point O with a radius vector $\bar{r}_0 \ell'/\ell$. We can take the straight line for the principal ray of a pencil of light rays starting from the point O. Let us consider an arbitrary ray belonging to the pencil and crossing the entrance pupil at a point with a radius vector $\bar{p}$. Assume that the ray intersects the plane B at a point with a radius vector $\vec{r}$. Then,

$$\vec{r} = \vec{r}_0 + \vec{p}(\ell' - \ell)/\ell'. \tag{4}$$

The ray which is diffracted at the double-exposure photograph and reaches the point O' conjugate to the point O, must belong to the pencil of light rays mentioned above. Its direction at the plane B is expressed by using a unit vector $\vec{q}$ as

$$\vec{q} = (-\vec{r} + \ell\vec{k} + \vec{p})/|-\vec{r} + \ell\vec{k} + \vec{p}|. \tag{5}$$

Interference can occur exclusively between components diffracted by corresponding speckle images before and after the displacement of the diffuse surface. If the displacement $\vec{d}(r)$ is small, the path difference F between correlated components diffracted at the plane B is given by

$$F = \vec{d}(r) \cdot \vec{q}(r). \tag{6}$$

By substitution of Eqs. (1), (2) and (5) into Eq. (6), F can be expressed as the function of only $\vec{p}$. On the assumption that the aperture of the lens is small, we expand F with respect to $\vec{p}$ and retain the terms up to the second order in $\vec{p}$. We obtain

$$F = a\{1 + b\cdot\cos\phi\cdot p + (c + d\cdot\cos^2\phi)p^2 + O(p^3)\}, \tag{7}$$

where (p, ϕ) are the polar coordinates on the plane C, and

$$a = -\alpha r_0^2/\sqrt{r_0^2 + \ell^2},$$

$$b = \{1 - \frac{\ell}{\ell'}(2 - \frac{r_0^2}{r_0^2 + \ell^2})\}/r_0,$$

$$c = \frac{-\ell}{2\ell'}\{2 - \frac{\ell}{\ell'}(2 - \frac{r_0^2}{r_0^2 + \ell^2})\}/r_0^2, \tag{8}$$

$$d = \frac{\ell}{2\ell'}\{2 - \frac{\ell}{\ell'}(4 - \frac{3r_0^2}{r_0^2 + \ell^2})\}/(r_0^2 + \ell^2).$$

Then, the squared deviation $(F - \bar{F})^2$ is integrated over a range of $\vec{p}$ corresponding to the aperture of the lens. Here $\bar{F}$ is the average value of F over the aperture. In performing the integration, we assume that the entrance pupil is a circle with a radius P.* Differentiation of the mean square deviation of F with respect to ℓ/ℓ' gives the condition required.

The mean square deviation is given as,

$$\iint_P (F - \bar{F})^2 p\,dp\,d\phi/\pi P^2$$

$$= \frac{\alpha^2}{4}\frac{r_0^2}{r_0^2 + \ell^2}\{1 - \frac{\ell}{\ell'}(2 - \frac{r_0^2}{r_0^2 + \ell^2})\}^2 P^2$$

$$+ \frac{a^2}{4}(\frac{c^2}{3} + \frac{cd}{3} + \frac{d^2}{4})P^4 + O(P^5), \tag{9}$$

where P attached to the integral sign means the integration over the aperture of radius P.

The condition for localization is derived from the P^2-term of

---

* In this work, the aperture of the imaging lens is assumed to be circular, but any form of aperture can be used as integration range.

Eq. (9) as

$$\frac{\ell'}{\ell} = 2 - \frac{r_0^2}{r_0^2 + \ell^2} = 2 - \sin^2 w, \tag{10}$$

where w denotes the field angle corresponding to the point
under observation. From Eq. (10) it is realized that the
localization surface is found to be somewhat curved. In our
experimental condition w is of the order of 3° and, Eq. (10)
gives $\ell'/\ell = 2$, which corresponds to the location of the highest
contrast observed experimentally.

DISCUSSION

The method of calculation given in the preceding section can be
derived from the concept of visibility.

Walles derived an integral formula for the visibility of
interference fringes formed with double-exposure holograms
(Ref 9, p.326). After his formula, the visibility is the
absolute value of a weighted mean of an exponential phase
function. On the assumption that the variation of phase
difference over the aperture is small, the exponential phase
function is expanded into power series up to the second order.
Then, a simplified formula convenient for calculation is
obtained.

Unlike diffuse surfaces in holographic interferometry, speckle
photographs contain rapid and random change in amplitude
transmission. Therefore, derivation of the integral formula for
the visibility needs a minor correction. It can be done using
conventional method in speckle interference.

The simplified formula for the visibility is given as,

$$V = 1 - (2\pi^2/\lambda^2) \iint_p (F - \overline{F})^2 pdpd\phi / \iint_p pdpd\phi, \tag{11}$$

where V means the visibility and λ the wavelength used.
Equation (11) shows that the decrease of visibility (1 - V) is
proportional to the mean square deviation of $(F - \overline{F})$. In this
degree of approximation, our calculation is equivalent to that
of the visibility.

Let us calculate the visibility in our case. In the
localization condition (10), the radius P of the aperture does
not appear. It means that the effect of aperture on the
visibility is not taken into account. To determine tolerable
extension of the aperture, the P^4-term of Eq. (9) must be
considered. The decrease of visibility on the localization
surface is obtained by substitution of Eq. (10) into Eqs. (8)
and (9). Tolerable limit of the visibility is taken as 0.8
according to a common practice. As we are concerned with an
approximate evaluation of tolerable value of P, it is reasonable
to assume that $\ell'/\ell = 2$ and $\ell \gg r_0$. Terms with d can be
neglected in comparison with that of only c in Eq. (9). Then,
the limit of P is given by

$$\frac{2\pi^2}{\lambda^2} \frac{a^2c^2}{12}P^2 \leq 0.2 \ . \tag{12}$$

From our experimental conditions and measured value of the diameter of interference fringes, the tolerable value of P is calculated as approximately 4.8mm and it corresponds to F/5.2. This value is compatible with the fact that the lens can be opened to F/4 without appreciable decrease of contrast.

REFERENCES

(1) W. H. Steel, Fringe localization and visibility in classical and hologram interferometers, Optica Acta 17, 873 (1970).

(2) K. A. Stetson, A rigorous treatment of the fringes of hologram interferometry, Optik 29, 386 (1969).

(3) K. A. Stetson, The argument of the fringe function in hologram interferometry of general deformations, Optik 31, 576 (1970).

(4) W. T. Welford, Fringe visibility and localization in hologram interferometry, Opt. Commun. 1, 123 (1969).

(5) W. T. Welford, Fringe visibility and localization in hologram interferometry with parallel displacement, Opt. Commun. 1, 311 (1970).

(6) J. Tsujiuchi, N. Takeya et K. Matsuda, Mesure de la deformation d'un objet par interferometrie holographique, Optica Acta 16, 709 (1969).

(7) I. Prikryl, Localization of interference fringes in holographic interferometry, Optica Acta 21, 675 (1974).

(8) M. Dubas et W. Schumann, Sur la determination holographique de l'etat de deformation a la surface d'un corps non-transparent, Optica Acta 21, 547 (1974).

(9) S. Walles, Visibility and localization of fringes in holographic interferometry of diffusely reflecting surfaces, Arkiv for Fysik 40, 299 (1969).

(10) S. Walles, On the concept of homologous rays in holographic interferometry of diffusely reflecting surfaces, Optica Acta 17, 899 (1970).

(11) T. Tsuruta, N. Shiotake and Y. Itoh, Formation and localization of holographically produced interference fringes, Optica Acta 16, 723 (1969).

(12) E. Archbold and A. E. Ennos, Displacement measurement from double-exposure laser photographs, Optca Acta 19, 253 (1972).

LASER SPECKLE REDUCTION IN
A TWO-STEP PROCESS

Leif A. Östlund and Klaus Biedermann

*Department of Physics II and Institute of Optical Research, The Royal Institute of
Technology S-100 44 Stockholm 70, Sweden*

ABSTRACT

*A new method for speckle reduction in coherent imaging is
presented. The method is based on a two-step process in which an
image is first recorded in coherent light and then spatially
filtered in incoherent light. This two-step process, which does
not use moving parts, is shown to be mathematically equivalent
to the well-known moving aperture technique.*

*Also a real-time version of the two-step process is presented,
where a liquid crystal cell provides the necessary coherent-to-
incoherent image conversion.*

*Finally, it is shown that the spatial filtering effect of a
single small aperture can also be achieved by a large-aperture
phase plate.*

1. INTRODUCTION

Images formed by coherent optical systems are often degraded by speckles. One
common situation where speckles arise is in imaging of objects with diffusely reflecting surfaces; an important technical case is hologram interferometry where closely spaced fringes may become buried in speckle noise [1].
The increasing application of coherent light has called forth a great many
papers on methods for speckle reduction [2] among which the moving aperture
technique [3], has received much attention.

We propose a new method for speckle reduction which, compared to the moving
aperture technique, offers some advantages. Our method is a two-step imaging
process where the first step is common imaging of the object in coherent
light while the second step, executed in a second imaging system, spatially
filters the image of the first step. We will show that our method is equivalent to the moving aperture method but in practice it offers at least two
advantages

  a) the exposure time in coherent light is considerably shorter

  b) the degree of speckle reduction can be tried out later,
     e.g. in the dark-room

341

Point b) is important because, if we assume that speckle reduction in some cases really may facilitate information retrieval, it is not clearly understood yet what character and degree of speckle reduction to choose, a choice which must be made beforehand in the existing methods.

On the other hand, sometimes a real-time speckle reduction system may be called for, especially when a useful speckle reduction characteristic already has been tried out. For that purpose we show an experiment that performs the two-step speckle reduction process in real time without moving parts. Essential components of this system are a liquid crystal cell, working as a coherent-to-incoherent image converter, and a spatial filtering aperture with a transfer function similar to that of a small circular aperture but with negligible light absorption. Analyses of the aperture, which consists of a random distribution of circular dielectric thin film discs on a glass plate, and experimental results to support the analyses are given.

We will use a slightly simplified notation in the analyses to that extent that double integrals are denoted by single integral signs, and, in some formulas the spatial coordinates are omitted.

2. THE TWO-STEP PROCESS

The principle of the two-step process is outlined in Fig. 1.

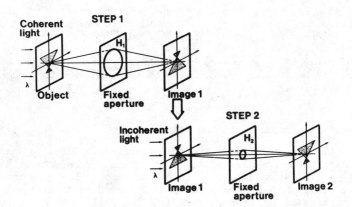

Fig. 1 The two-step process

In Step 1, an image of the coherently illuminated object is recorded by a linear process. This Image 1 is then used as an object and reimaged in incoherent light in Step 2. If we denote the intensity transmittance of Image 1 by i_1 and the intensity in the image plane of Step 2 by i_2, then i_2 is obtained from i_1 by the usual convolution formula for incoherent imaging systems as (e.g. [4])

$$i_2 = i_1 * |h_2|^2 \qquad (1)$$

where h_2 is the point spread function of Step 2. In Eq. (1), we have for simplicity assumed the magnification to be unity. To obtain an expression for i_1, we denote in Step 1 the object field by f and the field in the image

plane by g, then, by the convolution formula for coherent imaging systems
(e.g. [4]), i_1 is given by

$$i_1 = |g|^2 = |f' * h_1|^2 \qquad (2)$$

where f' is the complex amplitude of the ideal image expected from geometrical optics and h_1 is the point spread function of Step 1. If we combine Eq.
(1) and Eq. (2), the intensity in the final image plane i_2 can be expressed as

$$i_2 = |f' * h_1|^2 * |h_2|^2 \qquad (3)$$

3. THE MOVING APERTURE TECHNIQUE

Since the first paper on this method [3], several authors have investigated
the statistical character (variance, power spectrum etc.) of the speckles in
the resulting image [5, 6, 7]. We have taken another approach and shown that
the time average image in the moving aperture technique is equivalent to the
image recorded in the two-step process [8]. A necessary assumption is that
the moving aperture, during the exposure, visites uniformly all positions
that give any contribution to the image. This condition is plausible because,
in order to achieve maximum speckle reduction, the aperture shall take as
many independent positions as possible [6].

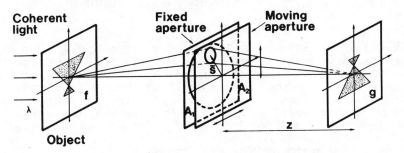

Fig. 2 The moving aperture technique

In Fig. 2, the principle of the moving aperture technique is outlined. The
notations are as before, i.e. f for the object field, f' for the complex amplitude of the ideal image expected from geometrical optics, and g for the
field in the image plane. The aperture of the system consists of two apertures; one fixed (usually the lens pupil) and one movable (usually a hole in
a movable opaque screen). If we denote the transmittance of the fixed aper-
ture by $A_1 (\bar{r})$ and that of the moving aperture by $A_2 (\bar{r} - \bar{s})$, then the transmittance of the combined aperture is

$$A(\bar{r}) = A_1(\bar{r}) \cdot A_2(\bar{r} - \bar{s}) \qquad (4)$$

The variable $\bar{s}$ is time-dependent and indicates the position of the moving
aperture. To simplify the analysis we will make the derivation in the spatial
frequency domain, and the Fourier transform of a function will be denoted by
the corresponding capital letter.

The spatial frequency spectrum I of the intensity in the image plane, for one
position of the moving aperture, is given by (e.g. [4])

$$I = (F' \cdot H) \circledast (F' \cdot H) \qquad (5)$$

where F' is the spectrum of the suitably scaled object field and H is the transfer function of the system. The symbol ⊛ denotes autocorrelation. By using the relation between the aperture and the transfer function in a coherent imaging system (e.g. [4]) we obtain from Eq. (4)

$$H(\overline{\nu}) = A_1(\lambda z \overline{\nu}) \cdot A_2(\lambda z \overline{\nu} - \overline{s}) = H_1(\overline{\nu}) \cdot H_2(\overline{\nu} - \overline{\mu}) \qquad (6)$$

where $\overline{\nu}$ is a two-dimensional spatial frequency vector, λ is the wavelength, z is the distance between the lens and the image plane and

$$H_1(\overline{\nu}) = A_1(\lambda z \overline{\nu})$$

$$H_2(\overline{\nu} - \overline{\mu}) = A_2(\lambda z \overline{\nu} - \overline{s}) \qquad (7)$$

$$\text{where } \overline{\mu} = \frac{\overline{s}}{\lambda z}$$

To obtain the Fourier transform of the total irradiance we have to add the contributions from all positions of the aperture. Because the motion is continuous and uniform this can be done by an integration of I over all values of $\overline{s}$. In Eq. (8) this integral is written down in full.

$$E(\overline{\nu}) = \int_{-\infty}^{\infty} F'(\overline{y}) \; H_1(\overline{y}) \; F'^{*}(\overline{y} - \overline{\nu}) \; H_1^{*}(\overline{y} - \overline{\nu}) \cdot \int_{-\infty}^{\infty} H_2(\overline{y} - \overline{\mu}) \; H_2^{*}(\overline{y} - \overline{\mu} - \overline{\nu}) \; d\overline{\mu} \; d\overline{y} \qquad (8)$$

The domain of the integration variable $\overline{\mu}$ is unlimited in Eq. (8). This is possible because we assumed that the moving aperture should take all positions that contribute to the image, and this includes positions where the moving aperture is between partly and just completely outside the fixed aperture. But when the moving aperture already is outside the fixed aperture, we can expand the integration domain to infinity, without altering the value of the integral.

If we, in Eq. (8), make the variable substitution

$$\xi = \overline{y} - \overline{\mu} \qquad (9)$$

then, because of the unlimited domain of $\overline{\mu}$, the inner integral becomes independent of $\overline{y}$ and the integrals can be separated. Hence, we can write the Fourier transform of the total irradiance as

$$E = [(F' \cdot H_1) \circledast (F' \cdot H_1)] \cdot [H_2 \circledast H_2] \qquad (10)$$

From Eq. (10), the total irradiance is obtained by a inverse Fourier transformation.

$$e = |f' * h_1|^2 * |h_2|^2 \qquad (11)$$

Comparison of Eqs. (3) and (11) shows that the two-step process and the moving aperture technique give identical images, and from Eq. (10) it is seen that the speckle reduction is obtained by incoherent spatial filtering of a common speckled image. For these conclusions to hold strictly it is necessary that aberrations are negligible and that the recording process in Step 1 is linear. In practical application, the assumption of linearity is only approximately correct when photographic recording material is used, but it is fulfilled in our real-time system, where, as we will show below, the image in Step 1 is not recorded but only converted from a coherent to an incoherent image by a liquid crystal cell.

4. REAL-TIME SPECKLE REDUCTION

The choice of the speckle reduction characteristic is usually made empirically, which partly is due to lack of a theory of speckle reduction and information retrieval. In the two-step process, this choice can be tried out in the dark-room with only one image of the object, but in the moving aperture method and the two-step real-time process every trial of a new filtering function requires a new exposure in coherent light, and possibly a repetition of the whole experiment. However, there are situations where a real-time process could be appropriate. For instance, if a sequence of images of similar objects has to be taken, labour is saved if the filtering function is tried out on one of the objects and then the rest of the objects is recorded directly in a real-time process.

We have made an experiment to develop the two-step process into a real-time system. The principle is outlined in Fig. 3.

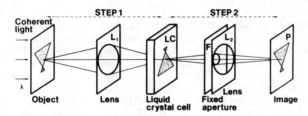

Fig. 3 Real-time speckle reduction in two steps

The object, which is illuminated by coherent light, is imaged by the lens L_1 to the plane of a liquid crystal layer in cell LC. LC acts as a coherent-to-incoherent image converter and separates Step 1 from Step 2, which consists of the lens L_2, the filtering aperture F and the image plane P.

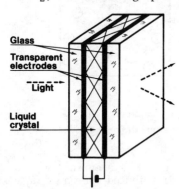

Fig. 4 The liquid crystal cell

The liquid crystal cell, outlined in Fig. 4, consists of two closely spaced glass plates, coated with transparent and conducting material on the sides facing each other. The space between the plates is filled with a nematic liquid crystal, and when a potential is applicated to the transparent electrodes, a turbulent motion is produced in the liquid crystal layer. This turbulence causes strong light scattering (dynamic scattering), and the cell starts acting like a moving ground glass [9, 10]. In this way, the spatial coherence is destroyed in the transmitted light, and Step 2 becomes an incoherent system. A condition for this last statement, which in our system is approximately fulfilled, is that the light must be scattered uniformly over the whole aperture F [11]. An experiment with this real-time system will be reported in Sect. 6.

The filtering aperture F is usually rather small. For instance, a reduction of the speckle standard deviation by a factor of 10 requires a numerical aperture for F of approximately 10 times the numerical aperture of Step 1. In this respect, our real-time system has the same drawback of an extended exposure time as the moving aperture technique. The light economy is further deteriorated in addition by the scattering in the liquid crystal cell. In the next section we try to eliminate this disadvantage by proposing a type of random aperture which gives a much better light yield and yet offers the same low spatial frequency cut-off.

5. RANDOM APERTURES

Apertures consisting of many small apertures in a random distribution can be classified with respect to different attributes. For instance, the small apertures can be of equal or different size, they can have equal or different orientation, or they can be distributed completely random or may be not allowed to overlap each other. In this paper, we will treat the simplest case only, i.e. when the apertures are of equal size, have the same orientation and are distributed completely random. To further simplify the analysis we will assume the total aperture to be "unlimited" in comparison to one of the small aperture.

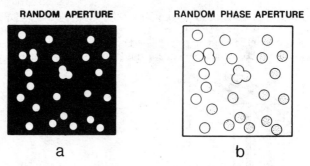

Fig. 5 Random apertures. a) random distribution of holes in an opaque screen, b) random distribution of dielectric thin film discs on a glass plate

The first type of random aperture we investigated is seen in Fig. 5 a. This aperture consists of circular holes random distributed over an opaque screen. The transmittance function, which only takes the values one and zero, is denoted by T. To obtain the transfer function, which is proportional to the autocorrelation of T, we will use a result from the theory of photographic film granularity.

In a model of grain structures, called the overlapping circular grain model, the grains are treated as circular, and the centers of the grains are distributed at random on a plate as independent events. The transmittance function, denoted by G, is zero for points covered by a grain, and one elsewhere. From this theory [12], the autocorrelation function of G, which we may denote GG, can be expressed as the transfer function which results when the grainy film is thought to be placed in the aperture of an incoherent imaging system. The "grains" are of equal size, and a single grain regarded as an aperture would give a certain modulation transfer function (MTF) mtf.

$$GG = \lim_{A \to \infty} \frac{1}{A} (G \circledast G) = \overline{G}^2 \exp(- \text{mtf} \cdot \ln(\overline{G})) \qquad (12)$$

where $\overline{G} = \lim_{A \to \infty} \frac{1}{A} \int_A G$ = average transmittance factor

$\text{mtf} = \frac{2}{\pi} (\arccos(\nu) - \nu\sqrt{1 - \nu^2})$ = modulation transfer function for one circular aperture

ν = normalized spatial frequency

A = integration domain

To use this result we assume the grains to be of the same size as the small apertures in the random aperture, then the function G is the complementary function to T, i.e.

$$T = 1 - G \qquad \text{and} \qquad \overline{T} = 1 - \overline{G} \qquad (13)$$

and the autocorrelation function of T is obtained as

$$TT = \lim_{A \to \infty} \frac{1}{A} (T \circledast T) = \lim_{A \to \infty} \frac{1}{A} (1 - G) \circledast (1 - G) =$$

$$= 1 - 2\overline{G} + GG$$

$$= (2\overline{T} - 1) + (1 - \overline{T})^2 \cdot \exp(- \text{mtf} \cdot \ln(1 - \overline{T})) \qquad (14)$$

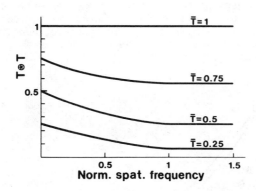

Fig. 6 Autocorrelation function of an aperture consisting of circular holes random distributed over an opaque screen

Eq. (14) and the plot in Fig. 6 show that TT for high spatial frequencies, that is where mtf = 0, approaches the value T^2. Hence, this kind of random aperture does not fulfill our requirements of a low cut-off frequency. (In practice, the transfer function will, of course, go to zero at some frequency, but this point is dependent on the maximum dimension of the total aperture and not on the small apertures).

The random aperture investigated above may well be used in speckle reduction, but we have found a second approach to be more advantageous for achieving low cut-off frequency combined with high light transfer: a random distribution of circular dielectric thin film discs on a glass plate, where the thickness of the dielectric film is λ/2 and the discs cover half the total area (c.f. [13]). To analyse this type of aperture, outlined in Fig. 5b, we express the transmittance function P as

$$P = G + (1 - G)\exp(i\varphi) \tag{15}$$

where G is the same function as before, and φ is the phase shift introduced by the dielectric layer. The autocorrelation function PP of P is then given by

$$PP = \lim_{A\to\infty} \frac{1}{A} (P \circledast P) = 1 - 2(1 - \cos\varphi)\,\overline{G} + 2(1 - \cos\varphi)\cdot GG \tag{16}$$

By substituting Eq. (12) into Eq. (16) we obtain the MTF of the phase plate, MTF_p,

$$MTF_p = 1 - 2(1 - \cos\varphi)\cdot(1 - \overline{T})\cdot\overline{T} -$$

$$- \quad 2(1 - \cos\varphi)\cdot(1 - \overline{T})^2\cdot(1 - \exp(-\,mtf\cdot\ln(1 - \overline{T}))) \tag{17}$$

To cover half the plate with dielectric discs of λ/2 means φ = π and $\overline{T}$ = 0.5 in Eq. (17), and we arrive at a simple form for this special MTF

$$MTF'_p = \exp(mtf\cdot\ln 2) - 1 \tag{18}$$

In Fig. 7, we have plotted MTF'_p, and, for comparison, also the modulation transfer function of a circular hole with the same size as one of the dielectric film discs. There is an obvious similarity between the two functions. Hence, if the random phase aperture is used in our real-time speckle reduction system, we achieve a speckle reduction characteristic very similar to that obtained by means of a single small hole-aperture, but with a gain of light given by the ratio of the area of the total aperture to the area of the single hole-aperture.

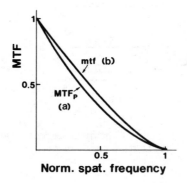

Fig. 7 Modulation transfer function of a) the random phase aperture and b) a single circular hole of the same size as the dielectric discs

6. EXPERIMENTS

To support the analysis a real-time speckle reduction experiment, was set up as sketched in Fig. 8.

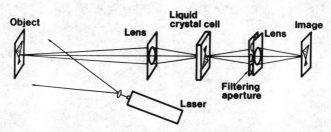

Fig. 8 Laboratory set-up for real-time speckle reduction

The object was illuminated by a 5 mW He-Ne laser. The lens in Step 1 had a f-number of 21. The liquid crystal used in the cell was Merck Nematic Phase 9 A. The layer had a thickness of 20 μm, and the voltage applied was 70 volts/50Hz.

Figs. 9 and 10 present some results.

a   b

Fig. 9

Fig. 9 shows two images of the same part of the object. The prints are mounted in mirror symmetry. Image a) was recorded with a lens of f-number 11 in Step 2 and no filtering aperture, while for image b), the lens was fitted

with a random phase aperture where the film discs had a diameter corresponding
to a f-number of 140. The exposure time was 5 sec for both images.

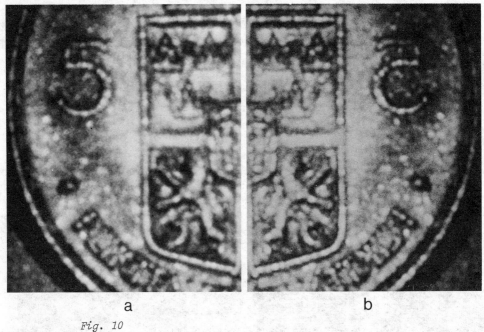

a   b

Fig. 10

Fig. 10 shows a comparison between the random phase aperture and a single
hole aperture. Image a)(identical to 9b) was recorded with the random phase
aperture as above, and image b) with a single-hole aperture of f-number 140.
The exposure time was 5 sec and 20 min respectively. As above, the lens used
in Step 2 had a f-number of 11.

7. DISCUSSION

We have shown that the moving aperture technique and the two-step process
can be thought of as speckle reduction by incoherent spatial filtering of a
speckled image. However, in experimentation with speckle reduction, the
choice of filter functions can be increased if coherent light is used in the
second step of the two-step process. Coherent spatial filtering could provide filter functions that cannot be obtained in the moving aperture technique, with the consequence that there might be a better chance to match the
filter characteristic to the object spectrum.

Speckle reduction by spatial filtering of the image implies that the decrease
in speckle noise is paid for by a decrease in bandwith. So far, a theory
seems not yet to exist that answers the question, under which conditions and
to which extent, speckle reduction really facilitates information retrieval.

REFERENCES

[1]   L.H. Tanner, A study of fringe clarity in laser interferometry and
      holography, J.Sci. Instr. 1, 517. (1968)

[2]   Laser Speckle and Related Phenomena, J.C. Dainty, Ed. (Topics in
      Applied Physics, Vol. 9) Springer-Verlag, 1975

[3]   J.C. Dainty, W.T. Welford, Reduction of speckle in image plane holo-
      gram reconstruction by moving pupils, Opt. Comm. 3, 289, (1971)

[4]   Goodman, J.W., Introduction to Fourier Optics, McGraw-Hill, (1968) Ch.6.

[5]   P. Hariharan, Z.S. Hegedus, Reduction of speckle in coherent imaging
      by spatial frequency sampling, Opt. Acta 21, 345, (1974)

[6]   T.S. McKechnie, Reduction of speckle by a moving aperture: theory and
      measurement, Optik 41, 34 (1974)

[7]   T.S. McKechnie, Reduction of speckle in an image by a moving aperture
      - second order statistics, Opt. Comm. 13, 29 (1975)

[8]   L.A. Östlund, K. Biedermann, Laser Speckle Reduction: Equivalence of
      the moving Aperture Method and Incoherent Spatial Filtering,
      Appl. Opt., (in printing)

[9]   H. Kiemle, U. Wolff, Application de cristaux liquides en holographie
      optique, Opt. Comm. 3, 26 (1971)

[10] F. Scudieri, M. Bertolotti, R. Bartolino, Light scattered by a liquid
      crystal: a new quasi-thermal source, Appl. Opt. 13, 181 (1974)

[11] H. Arsenault, S. Lowenthal, Partial coherence in the image of an
      object illuminated with laser light through a moving diffuser,
      Opt. Comm. 1, 451 (1970)

[12] O'Neill, E.L., Introduction to Statistical Optics, Addison-Wesley,
      1963, Ch. 7

[13] K. Sayanagi, Optical noise filter, Oyo Butsuri (J. Appl. Phys. (Japan)
      27, 623 (1958)

SESSION 10.

RECORDING MATERIALS

Chairman: K. BIEDERMANN

VOLUME HOLOGRAM STORAGE
IN IRON DOPED LiNbO$_3$ CRYSTALS

F. Micheron, J. P. Huignard and J. P. Herriau

Thomson-CSF, Laboratoire Central de Recherches B.P. 10 - 91401 Orsay (France)

ABSTRACT.

Fe doped LiNbO$_3$ Crystals, reduced in Argon atmosphere, are investigated as possible holographic storage media. Good image reconstruction quality can be achieved, even in the cases of grey scale images. The assymetric write-erase cycles allow multiple hologram superpositions ; their number is limited to some tenths by photoinduced scattering which occurs when recording at room temperature. Selective erasures in stacked holograms and logic operations between recorded images are performed, using a coherent substraction technique.

INTRODUCTION.

Holographic recording in LiNbO$_3$ is caused by a photoinduced change of refractive index, the so called "photorefractive effect". This effect involves the photoinduction of space charge fields which modulate the refractive index via the electrooptic effect. Space charge fields are attributed to charge transfers photoinduced in the visible range, between doping ions, or doping ions and host lattice ions. These transfers occur on microscopic and macroscopic scales :

- in the first case, the transfer occurs between neighbouring ions, and may be anisotropic if the crystal belongs to the pyroelectric class. Such transfer gives rise to the photovoltaïc effect (self biased photoconductivity), i.e. a permanent photocurrent exists without biasing field (1).

- in the second case, photocarriers are displaced over macroscopic distances by thermal diffusion or drift under the influence of an external applied field, before being trapped (2).

According to Young et al (3), the maximum photosensitivity is reached when the photocarriers displacement length is comparable with the holographic fringe spacing. This optimum situation is encountered only in K(Ta,Nb)O$_3$ (4), Bi$_{12}$ Si O$_{20}$ and Bi$_{12}$ Ge O$_{20}$ (5). In the other photorefractive crystals, the photosensitivity is in general a decreasing function of the fringe spacing.

LiNbO$_3$ is the most popular photorefractive crystal, for the following reasons :

- Crystal growth in large sizes and excellent optical quality, even when doped, is easily performed.

- The Curie temperature is very high (1100°C), therefore the crystals are stable in their single domain state.

A large number of doping experiments has been already performed, in order to improve the photosensibility and understand the origin of the photorefractive effect (6, 7). The best results are obtained by iron doping : $300mJ/cm^2$ at λ = 514.5nm is required for a 1% diffraction efficiency in a 1mm thick crystal. This photosensitivity is three orders of magnitude lower than photographic plate sensitivity, never theless, such Fe doped $LiNbO_3$ crystals show the following advantages :

- absence of developping process,
- large diffraction efficiencies, characteristic of three dimensional phase holograms,
- holograms superposition capability, therefore large storage capacity (10^8 - 10^9 bits/cm^3), due to large dynamic range of refractive index change, and assymetry of the recording-reading cycles,
- storage time of several months,
- reusable recording medium after optical or thermal erasure.

The aim of this paper is to demonstrate the reconstruction quality of grey scale images, the hologram superposition capability, and results of image processing, performed by coherent substractions and additions. The crystal used is a 3mm thick $LiNbO_3$ crystal, 0.015% iron doped, and reduced in Argon atmosphere, for a 50% absorption at λ = 514.5nm.

RECONSTRUCTIONS OF RECORDED IMAGES WITH GREY SCALE.

Photographs 1 and 2 are reconstructions of stored images recorded as out of focus Fourier holograms. The hologram diameter is $\emptyset$ = 4mm, and the fringe spacing Λ = 3/um. Reference beam to object beam ratio is 3 and diffraction efficiency reaches η = 2%. A 10^5 bits resolution is achieved, limited by speckle noise on the photographic plate.

SUPERIMPOSIONS OF THREE DIMENSIONAL HOLOGRAMS.

Ten holograms of different digital images have been recorded in the same volume in the geometrical conditions previously reported. The theoretical Bragg selectivity angle is $\Delta\theta$ = 10^{-3} radian, but changes in the beams angle between each hologram recording have been choosed 3 times $\Delta\theta$, in order to avoid any non linear crossover. Photograph 3 shows the reconstructions of the three first holograms, and the tenth. No difference can be observed in the signal to noise between these images. When increasing the superimposed holograms number (up to 20), a scattering effect occurs, which lowers the signal to noise ratio, due to cumulative recording of interferences between beams scattered by surface and bulk defects. Recording at higher temperature (200°C) than room temperature avoids this scattering effect, as demonstrated by Staebler et al (8).

COHERENT IMAGE SUBSTRACTIONS AND ADDITIONS.

We have demonstrated that in a $LiNbO_3$: Fe crystal, short circuited during holographic recording, the spatial average change of refractive index is zero, i.e., the continuous component of the hologram is not recorded : the reason is that, the spatial average of the photoinduced space charge field is zero, therefore, the linear electrooptic effect in $LiNbO_3$ can't give rise to any continuous component of the refractive index change. This property allows to increase the number of superimposed holograms by increasing the dynamic range of the refractive index change. This property is used to

FIGURE I

FIGURE II

FIGURE III

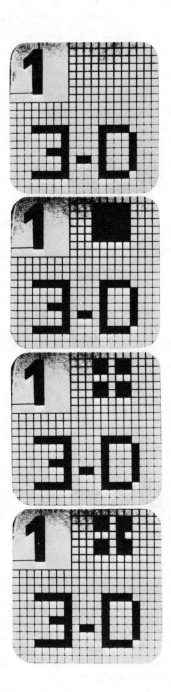

FIGURE IV

perform hologram substractions and additions, without changing the average value of the refractive index (9, 10, 11).

Coherent substractions between two images are performed by recording the first image, and then the second after having introduced a π shift in the reference beam. When this operation is performed with two digital images, the resulting reconstructed image corres ponds to the Exclusive OR operation between these two images. Coherent additions are obtained by recording the different holograms without changing the phase between the two beams. The corresponding result between two digital images is the OR operation.

Photograph 4 shows examples of such operations, summarized in the following table :

Recorded object	Reconstructed image
Square to be substracted	Substracted square
Cross to be added	Added cross
Point to be substracted	Substracted point

CONCLUSION.

Up to now, $LiNbO_3$: Fe is the best candidate as reusable volume storage medium. Practical applications would require a two orders of magnitude improvement in photosensitivity, and a complete absence of scattering after long exposures to coherent light.

BIBLIOGRAPHY.

- 1 - A.M. GLASS, D. VON DER LINDE and T.J. NEGRAN
        Appl. Phys. Letters 25, 233 (1974)

- 2 - G.A. ALPHONSE, R.C. ALIG, D.L. STAEBLER and W. PHILLIPS
        RCA Review 36, 213 (1975)

- 3 - L. YOUNG, W.K.Y. WONG, M.L. THEWALT and W.D. CORNISH
        Appl. Phys. Letters 24, 264 (1974)

- 4 - D. VON DER LINDE, A.M. GLASS and K.F. RODGERS
        Appl. Phys. Letters 25, 155 (1974)

- 5 - F. MICHERON, M. PELTIER and J.P. HUIGNARD
        This issue.

- 6 - W. PHILLIPS, J.J. AMODEI and D.L. STAEBLER
        RCA Review 33, 94 (1972)

- 7 - E. KRÄTZIG and H. KURZ
        Ferroelectrics 10, 159 (1976)

- 8 - D.L. STAEBLER, W.J. BURKE, W. PHILLIPS and J.J. AMODEI
        Appl. Phys. Letters 26, 182 (1975)

- 9 - J.P. HUIGNARD, J.P. HERRIAU and F. MICHERON
        Appl. Phys. Letters 26, 256 (1975)

- 10 - J.P. HUIGNARD, J.P. HERRIAU and F. MICHERON
         Rev. Phys. Appl. 10, 417 (1975)

- 11 - J.P. HUIGNARD, J.P. HERRIAU and F. MICHERON
         Ferroelectrics, 11, 393 (1976)

PHOTOREFRACTIVE RECORDING OF VOLUME HOLOGRAMS IN DOPED LiNbO₃

H. Kurz, V. Doormann and R. Kobs

Philips GmbH Forschungslaboratorium, Hamburg, 2000 Hamburg 54, BRD

ABSTRACT

The principles of the photorefractive recording process in doped LiNbO$_3$ are derived in a closed form expression containing the basic charge transport equations in pyroelectric materials.

The inspection of the obtained solutions allows the optimaliza- tion of the key parameters for multiple storage of volume holo- grams which increases the storage capacity of these crystals by two orders of magnitude.

INTRODUCTION

Holographic techniques and storage materials have found an in- creasing interest in the field of data storage and processing. The effort in research and development has been primarily on thin storage media. In an increasing number of applications thick storage media are preferable because of their capacity for high density storage of information. Using the angular selectivity of volume holograms recorded in thick storage media the theoretical limits of storage capacity lies in the order of $(1/\lambda)^3$. Photoin- duced refractive index changes as the principle storage effect are preferred, because of the high reconstruction efficiency of phase holograms required in optical storage systems.

The purpose of this contribution is to investigate the photo- refractive recording principles in LiNbO$_3$, one of the most inter- esting medium for optical storage. The key parameters of the storage process, like photorefractive sensitivity, range of line- arity, and saturation values are derived and related to photo- electrical quantities. The critical analysis of the derived re- lations and the comparison with experimental data aims to clari- fy the advantages of this storage material and also show some principal limitations of the storage performance.

PHOTOREFRACTIVE RECORDING PRINCIPLES

The holographic storage in photorefractive crystals is generally based on the generation of photocurrents by a suitable excita- tion process. The modulation of these currents by the varying light intensity of a hologram gives rise to space charge distri- butions which in turn create phase holograms via electrooptic effect of the storage medium. After removing the laser light

these photorefractive phase holograms decay with time constants
defined by the dielectric relaxation of the space charge fields.

A prominent representative in the group of photorefractive media
is pyroelectric $LiNbO_3$ showing unusual charge transport properties. Doped with transition metal concentrations in the order of
100-1000 ppm these crystals show characteristic optical absorption bands. Light absorbed by these bands generates steady state
photocurrents along the polar axis even in the absence of
applied fields. These zero field photocurrents are caused by a
bulk photovoltaic effect, which seems to be a unique property of
pyroelectric single domain crystals (1). Detailed analysis of
the current voltage relation in doped $LiNbO_3$ and corresponding
measurements of the photorefractive dynamics demonstrate clearly
the essential role of impurities like Fe, Cu and Mn as excitation and trap centers (1,2).

In addition the experimental results of these measurements indicate that the photorefractive storage process in doped $LiNbO_3$
is entirely governed by the photovoltaic effect while diffusion
plays a minor role.

Because of the anisotropic charge transport properties efficient
photorefractive recording is only possible if the photocurrents
are modulated along the polar axis. The fringes of the interference pattern have to be perpendicular to the c-axis of the
crystals. The laser beams are π-polarized to take the advantage
of the high electrooptic coefficient r_{33} in this geometry.

In a first order approximation the photorefractive dynamic can
be described by the space time development of the displacement
current $\varepsilon\varepsilon_o dE/dt(z,t)$ along the polar axis, determined by the
conduction currents $i_c = \kappa_o \alpha I + \sigma E$ as follows:

$$\varepsilon\varepsilon_o \frac{dE}{dt}(z,t) + i_c(z,t) = J(z,t). \qquad (1)$$

Here means I the light intensity
κ_o the photovoltaic constant,
α the optical absorption constant
$\sigma = \kappa_1 \alpha I + \sigma_d$ the conductivity
(photo- and dark conductivity)   and
$J(z,t)$ the total current density.

Under short circuit conditions the development of the fundamental component of the refractive index distribution
$\Delta n_f = \Delta n \cos Kz$ during sinusoidal illumination $I = I_o(1+m \cos Kz)$
with $m \leq 0.5$ can be described in a series expansion (3):

$$\Delta n_f = \frac{n^3 r_{33} \cdot \kappa_o m 2}{2\kappa_1} (1-e^{-\gamma} + \frac{3}{4}m^2 (1-e^{-\gamma}(1+\gamma+\frac{\gamma^2}{2})) + \ldots) \qquad (2)$$

where n and r_{33} are the appropriate refractive indices and electrooptic coefficients, $m_2 = m(1+\sigma_d/\kappa_1\alpha I_o)^{-1}$ a reduced value of the modulation index and $\gamma = (\kappa_1\alpha I_o+\sigma_d)t/\varepsilon\varepsilon_o$ the rate constant of the photorefractive storage process.

During the initial stage of illumination ($\gamma \leq 0.2$) the photorefractive sensitivity $s = d(\Delta n_f)/d(I_o t)$ is directly related to the photovoltaic constant κ_o:

$$s = \frac{n^3 r_{33}\kappa_o\alpha\, m}{2\varepsilon\varepsilon_o} , \tag{3}$$

and the photorefractive saturation value $\Delta n_f^s(\gamma\to\infty)$ to the κ_o/κ_1-ratio which corresponds to the photovoltaic saturation fields E_s measured under open circuit conditions:

$$\Delta n_f^s = \frac{n^3 r_{33}\kappa_o m_2}{2\kappa_1}\left(1 + \frac{3}{4}m_2^2 + \ldots\right) \tag{4}$$

Both the photorefractive saturation value Δn_f^s and the saturation fields E_s indicate the build up of space charge fields between 10^4 and 10^5 V/cm, depending on the impurity concentration (4).

To obtain comparable values by diffusion extremely high spatial frequencies between 10^5/cm $< K < 10^6$/cm have to be used. Thus for conventional holographic recording with laser light the photorefractive sensitivity should be independent of the spatial frequencies in crystals doped with sufficient photovoltaic centers. Only in cases where the average migration length of the charge carriers will be comparable with the grating spacing, the resulting phase shift between refractive index modulation and light intensity pattern leads to complex frequency dependence of the photorefractive sensitivity (5).

In the holographic storage technique particular attention is paid to the range of linear recording characteristic. According to Eq. (2) we find a linear recording range, defined by $\gamma \leq 0.2$, which is approximately:

$$\Delta n \cong \frac{0.1 n^3 r_{33}\kappa_o m_2}{\kappa_1} . \tag{5}$$

For example in highly doped LiNbO$_3$:Fe and LiNbO$_3$:Cu the linear recording characteristic holds up to refractive index changes in the order of 10^{-4}.

One of the interesting advantages of photorefractive storage materials is the full erase capability without any fatigue. By uniform illumination with coherent or incoherent light the refractive index pattern decays with a rate constant $\gamma_e = (\kappa_1\alpha I+\sigma_d)t/\varepsilon\varepsilon_o$ reestablising the origianl state of the crystal. An excellent agreement between measurements based on erasure constant γ_e and the photoconductivity $\sigma_o = \kappa_1\alpha I$ has been found in spectroscopic investigations (6). In the dark thermal erasure occurs with $\gamma_e = \sigma_d t_d/\varepsilon\varepsilon_o$ determining the dark storage time t_d of the phase holograms.

In conclusion the photorefractive storage parameters, defined
in Eq. (2)-(5), are mainly governed by the photovoltaic constant κ, the specific photosensitivity κ_1 and the dark conductivity σ_d. These three figures depend on the kind and valence
state of impurities, wavelength of excitation light and temperature as well as on the annealing treatments. Details on the
subject are beyond the scope of this contribution and will be
published elsewhere. However, to give an idea of the orders of
magnitude a set of typical data measured at λ = 514 nm is
shown in Table 1.

TABLE 1   Photorefractive parameters

Sample	Doping		κ_o	κ_1	t_d	Δn_f^s
	wt%		Acm/W	Acm^2/VW	days	
$LiNbO_3$:Fe	0.015	Fe_2O_3	$1 \cdot 10^{-9}$	$4 \cdot 10^{-14}$	10	$3 \cdot 10^{-4}$
$LiNbO_3$:Fe	0.1	Fe_2O_3	$2 \cdot 10^{-9}$	$1.15 \cdot 10^{-14}$	50	$3 \cdot 10^{-3}$
$LiNbO_3$:Cu	0.25	CuO	$0.3 \cdot 10^{-9}$	$3 \cdot 10^{-15}$	40	$1.8 \cdot 10^{-3}$

Photovoltaic constant κ_o, specific photosensitivity κ_1, dark
storage time t_d (estimated), and photorefractive saturation
value Δn_f^s.

Using the well known relation (7,8) between first order diffraction efficiency η_1 and amplitude of pure sinusoidal phase pattern
$\Delta n_f = \Delta n \cos Kz$ in the linear range of photorefractive dynamic
(see Eq. (2)), the optical exposure energy E(η=1%) for the recording of a phase hologram with 1% diffraction efficiency can
be calculated. Taking into account reflection losses R and the
attenuation of light at high optical densities in Eq. (1)-(5) we
obtain the following relation:

$$E(\eta=1\%) = \frac{1.5 \cdot 10^{-4} \lambda \cos \theta}{\kappa_o m_1 \pi (1-e^{-\alpha d})} \text{ arc sin } (\frac{0.01 e^{\alpha d/\cos \theta}}{R})^{1/2} \quad (6)$$

This energy figure of merit is inversely proportinal to the
photorefractive constant κ_o. The Eq. (6) is elucidated in Fig. 1,
where this exposure energy is displayed versus the αd-values of
the samples for defined parameters and π-polarized light.

For example with αd-values between $0.7 < \alpha d < 1.4$ still a relatively large exposure energy of 1.3 J/cm^2 has to be supplied.
The cross point between the arrow, defining the minimum exposure
energy with respect to αd and line at an energy level of
0.3 J/cm^2 in Fig. 1 represents the maximum of holographic recording sensitivity by choosing the optimum recording wavelength
at room temperature and using $LiNbO_3$ doped with Fe, which shows
the highest value of κ_o up to now. The photovoltaic constant
seems to be enhanced hardly by chemical treatments and dramatic
improvements cannot be expected for known theoretical reasons
(1,9).

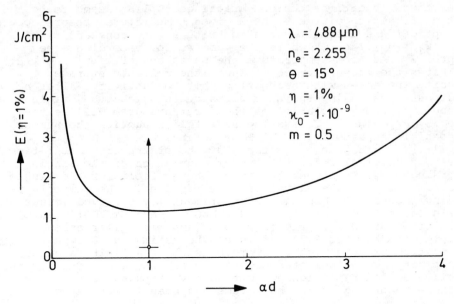

Fig. 1   Exposure energy, necessary to record
phase holograms with 1% diffraction
efficiency, versus αd-values.

MULTIPLE STORAGE OF VOLUME HOLOGRAMS

Multiple storage in $LiNbO_3$ proceeds by a time sequential recording of holograms under different angles of the reference beam,
using the angular selectivity of "thick" holograms. In our experiments we simulate this procedure by rotating the storage
crystal at fixed angle between signal and reference beam.

The basic conditions for multiple storage of phase holograms
in photorefractive crystals are following:

1. Optical decoupling of the individual holograms during recording and read-out. - This is provided by the angular selectivity
of the reconstruction process, derived by H. Kogelnik and modified by N. Uchida in the case of attenuated gratings (7,8).
According to these treatments the intrinsic angular selectivity
is maintained in the case $\alpha d \leq 1$ only.

2. Electrical decoupling of the single holograms during subsequent recordings. The space charge fields of previously recorded holograms should not affect the following recordings and
vice versa the erasure of already recorded holograms by the
following multiple recording should be a minimum. That means
that the asymmetry of recording and erasure sensitivity has to
be large.

Under these conditions the electrical decoupling seems to be the most difficult requirement. At a given temperature erasure occurs via optical and thermal excitation with a rate constant $\gamma_e = (\kappa_1\alpha I + \sigma_d)t/\varepsilon\varepsilon_0$ which is the same as for recording (see Chapter 2). At room temperature the photoconductivity $\sigma_0 = \kappa_1\alpha I$ dominates the erasure process. Thus, the multiple storage capacity increases with decreasing photoconductivity. After n superpositions, using the same exposure energies, the original refractive index modulation of the first hologram will be reduced by $\exp\gamma(n-1)$. Tolerating an e-fold reduction the number of possible superpositions will be defined by $n = (1+\gamma)/\gamma = \Delta n_f^s/\Delta n_f$, the ratio of photorefractive saturation value and original refractive index modulation of the single recordings Δn_f. The lower the space charge fields of the single hologram the less the electrical interaction in the multiple recording process. Regarding the experimentally observed values of Δn_f^s in Table 1 for heavily doped $LiNbO_3:Fe$ and $LiNbO_3:Cu$ an upper limit of 10^3 superimposed holograms with diffraction efficiencies between 2.7% and 1% can be estimated. But the full utilization of this figure is prevented by the finite angular selectivity. Using selected crystals whose photoconductivity is heavily reduced by enrichment with empty trapping centers the multiple recording of 100 holograms has been investigated.

In Fig. 2 the reconstructions of the first and 100th hologram recorded in a 5 mm thick crystal, containing $3 \cdot 10^{19}/cm^3$ Fe^{3+}-ions as trapping centers, are shown. The reconstruction efficiency of the first recorded holograms was measured to be $\eta_1 = 1\%$. The single holograms were spaced 0.3^0 apart.

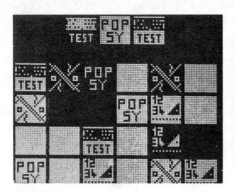 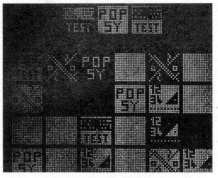

a)   b)

Fig. 2   Demonstration of the superposition of 100 holo-
          grams, each containing 10^4 bits, reconstruction
          of the 100th (a) and the first hologram (b).

A method for the elimination   of unwanted erasure during multiple
recording and read-out has been discovered by D.L. Staebler and
W. Phillips who suggested a thermal fixation of photorefractive
holograms. At temperatures above 100 °C the observable phase
pattern relaxes quickly due to the measured density of
thermally activated charge carriers. Information remains stored
in form of the thermally stable distribution of donors and traps.
Their space charge fields are compensated by thermally activated
ionic defects (11), thus avoiding any electrical interactions
for multiple recordings. In this stage of charge compensation
there appears a variation of $\Delta\alpha$ in the crystals corresponding to
the locally varying density of filled traps, forming a weak absorption hologram. The amplitude of these absorption holograms
is given by:

$$\Delta\alpha(z) = \frac{\alpha\kappa_o\sigma I_o tKm \sin Kz}{q} = A \sin Kz \qquad (7)$$

where q is the electronic charge unit and σ the absorption cross
section, in the order of $3\cdot10^{-18}$ cm^2 for $LiNbO_3:Fe^{2+}$ (12). There
is a $\pi/2$-phase shift between original phase- and absorption
hologram.

In Fig. 3 the recording procedure for the generation of thermally
fixed holograms is illustrated. By keeping the crystal at a
temperature of 140 °C the photorefractive recorded hologram
relaxes in a few seconds after removing the laser light. After
cooling down to room temperature the charge compensation mechanism is stopped and the original holograms can be recovered
by uniform illumination, preferably with incoherent light, due
to the forced modulation of the photocurrents according to the
pattern $\Delta\alpha(z)$: $i(z,t) = \kappa_o\Delta\alpha(z)I+\sigma(z)\cdot E(z,t)$.

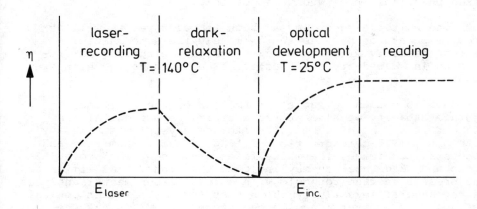

Fig. 3   Recording procedure of thermally fixed holograms.

Similar to the treatment of the photorefractive dynamic in
Eqs. (1)-(5) the recovery sensitivity and the saturation value
of the recovered phase pattern Δn_r^s can be calculated. A crude
estimation shows that Δn_r^s is approximately:

$$\Delta n_r^s = \frac{n^3 r_{33} \kappa_o A}{2 \kappa_1} = \frac{n^3 r_{33} \kappa_o^2 \sigma I_o t K m}{2 \kappa_1 q} \tag{8}$$

where $I_o t$ means the exposure energy invested for the recording
in the high temperature phase. With $\kappa_o = 1 \cdot 10^{-9}$, $\kappa_1 = 1 \cdot 10^{-14}$,
$\sigma = 3 \cdot 10^{-18}$ cm^2, $m = 0.5$ and $K = 10^4$ cm^{-1} we calculate an exposure energy of 30 J/cm^2 per single hologram to obtain re-
covered phase changes in the order of $\Delta n_r^s = 1 \cdot 10^{-5}$ after strong
exposure with incoherent light. These energy requirements represent a severe limitation for the introduction of thermal
fixing techniques in optical data storage systems. The principle
drawback of the thermal fixing technique, however, is the lack
of convenient erasure mechanism.

On the discussion of the multiple storage capacity obtainable
by the hot recording technique, we have to ask for the thermal
erasure of the filled traps reducing the amplitude of the absorption holograms. D.L. Staebler et al. have investigated the
time constant of this thermal erasure in LiNbO$_3$:Fe and explained by diffusion (13). At temperatures around 140 ^{o}C they
have found time constants in the order of 10^5 seconds, depending on the spatial frequency of the hologram and the doping
concentration.

Assuming a cycle time of 100 seconds for the recording and
ionic compensation of a single hologram, observed in storage
experiments, the time sequential superposition of 10^3 thermally
fixed phase holograms would be possible.

Both the multiple recording at high temperatures as well at
room temperatures provide approximately the same multiple
storage capacity. The improved electrical decoupling for recording in heated crystals has to be balanced to the high exposure energies required.

In order to obtain reliable information on the practical
storage capacity we have stored a binary object containing
10^4 bit in a cube of 1 mm^3. The storage crystal was LiNbO$_3$
with optimum Fe-concentration. In Fig. 4 the transmitted and
reconstructed image, thermally fixed are shown. The poor SNR-
ratio and image quality may indicate the fundamental limits of
the present storage technology. A careful experimental analysis
of the angular selectivity shows that for a sufficient optical
decoupling the crystal has to be rotated by 0.3^o. Thus the
multiple storage technique is limited to around 100 holograms
by optical reasons representing a total storage capacity of
10^6 bit/mm^3.

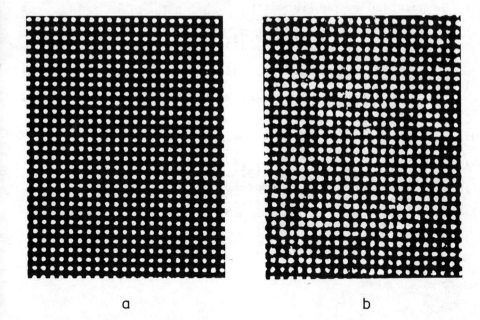

a   b

Fig. 4   Transmitted (a) and reconstructed image (b) of a
         binary object containing 10^4 bit. The thermally
         fixed hologram is stored in a volume of 1 mm^3.

In conclusion we find doped LiNbO$_3$ to be one of the most attractive
candidate for holographic storage at ultrahigh capacities.
Deduced from the fundamental material parameters the theoretical
number of superpositions is somewhat limited for optical reasons.
Also, the high exposure energy required for the recording of
phase holograms in photorefractive LiNbO$_3$, especially in the case
of thermally fixed holograms restrict the likely applications
to archival or DRAW (direct read after write) storage systems,
with low recording data rate.

ACKNOWLEDGEMENTS

The authors thank J. Reck and F. Welz for technical assistance,
G. Jentschke for helpful electronic engineering and H. Dammann
for valuable discussions. The work was supported partly by the
German government.

REFERENCES

(1)   A.M. Glass, D. van der Linde, and T.J. Negram,
      Appl. Phys. Letters 25, 233 (1974).
(2)   W. Phillips, J.A. Amodei, and D.L. Staebler,
      RCA Review 33, 94 (1972).
(3)   G.A. Alphons, R.C. Alig, D.L. Staebler, and W. Phillips,
      RCA Review 36, 213 (1975).

(4)   H. Kurz, to be published in
Ferroelectrics 12 (1976).

(5)   L. Young, W.K.Y. Wong, M.L.W. Thewalt, and W.D. Cornish,
Appl. Phys. Letters 24, 264 (1975).

(6)   H. Kurz and E. Kraetzig,
Appl. Phys. Letters 26, 635 (1975).

(7)   H. Kogelnik,
Bell System Techn. J. 48, 2909 (1969).

(8)   N. Uchida,
J. Opt. Soc. Am. 63, 280 (1973).

(9)   E. Kraetzig and H. Kurz to be published in
J. Electrochem. Soc.

(10)   D.L. Staebler and W. Phillips,
Appl. Optics 13, 788 (1974).

(11)   D.L. Staebler and J.J. Amodei,
Ferroelectrics 3, 107 (1972).

(12)   M.G. Clark, F.J. Di Salvo, A.M. Glass, and G.E. Peterson,
J. Chem. Phys. 59, 6209 (1973).

(13)   D.L. Staebler, W.J. Burke, W. Phillips and J.J. Amodei,
Appl. Phys. Letters 26, 182 (1975).

HOLOGRAM FORMATION MECHANISM IN SOME PHOTODIELECTRIC MATERIALS

S. Reich, A. A. Friesem and Z. Rav-Noy

Weizmann Institute of Science, Rehovot, Israel

ABSTRACT

The holographic storage mechanism in a novel, self developing, photodielectric polymer system of poly (alkyl cyano acrylate) sensitized with parabenzoquinone is described. In this volume phase material, the recording is achieved by means of small changes in the real part of the complex dielectric constant resulting in changes in the index of refraction; these changes are due to density modulations caused by photocrosslinking reaction in the polymer bulk. Experimental results reveal, that as in the case of other macromolecular media, certain bulk properties, such as the time-dependent elastic behavior and the cooperative noise generation, impose special demands on the parameters of the photodielectrics.

INTRODUCTION

Photodielectric recording materials have been particularly interesting as holographic recording media because of their in-situ self development capabilities, high resolution, high diffraction efficiencies, and excellent noise characteristics. The best known examples are the polymethylmethacrylate (PMMA) in which the recording may be due to either photoinduced crosslinking or photodegradation [1-5], and photopolymers in which the recording is due to polymerization of monomers [6-9]. These materials, unfortunately have a short shelf life and require relatively complex preparation procedure in order to achieve high optical quality recording area. Generally, the preparation involves the use of a solvent to dissolve the polymeric host material before the necessary visible light sensitisers can be introduced; the solvent must then be removed by means of slow and careful evaporation resulting in volumetric changes which affect the quality of the surface of the recording media.

Recently we reported on the holographic response of a newly developed polymeric material which is relatively easy to prepare, with no need for solvents, possesses long shelf-life, and yet exhibits recording and readout characteristics comparable to the other materials [10]. For example, the diffraction efficiency for simple holographic gratings exceeded 90 percent, the resolution is greater then 3000 lines/mm, and the signal-to-noise ratio was about 200 for complex input signals. In this paper we expand on our earlier work emphasizing the holographic storage mechanism in these new materials.

371

PREPARATION OF THE MATERIALS

The first step of the preparation employs depolymerization techniques to obtain methyl α -cyanoacrylate monomer [11]. Alternatively, commercially available monomer such as Eastman 910 can be used directly, although it contains about 10 percent by weight of unwanted additives. A purified parabenzoquinone sensitizer to obtain a spectral response in the blue-green region of the visible spectrum is then introduced into the monomer, and the resulting solution is polymerized between glass plates or between other equivalent substrate materials. Thicknesses ranging from a few microns to several millimeters can be obtained by merely placing appropriate spacers between the glass plates.

The above preparation method offers several advantages over preparation techniques which are used for other materials. The first important advantage is that no solvent is needed to introduce the sensitizers into the polymer. Thus the slow evaporation usually necessary with other materials is avoided, resulting in excellent optical quality. Furthermore, by eliminating the solvent, its plasticizing effect on the polymer and its reaction with the sensitizer are avoided, thereby improving both the dimensional stability and chemical properties of the recording media. This improvement leads to increased shelf life; in our experiments, high quality holograms were recorded in samples which had been prepared six months earlier.

A second advantage is the high reactivity of the methyl α -cyanoacrylate monomer, allowing rapid preparation of the recording plates. For example, plates of about 50 microns thickness and areas greater than 1 cm^2 can be prepared in several minutes, whereas plates of 200 microns thickness can be prepared in about one hour. The high reactivity results from the fact that in this material polymerization is caused not only by the free radical initiation but also by anionic initiation. While the free radical path is probably blocked by the introduction of the quinone sensitizer, the anionic initiation is dominant so that polymerization occurs rapidly. The chemical process may be described by the follwing reaction:

$$
\underset{\delta+ \ \ \delta-}{CH_2 = \overset{\overset{\displaystyle CN}{|}}{C} \ COOR} \ \rightleftharpoons \ CH_2 \ \overset{\overset{\displaystyle CN}{|}}{\underset{}{C}} \ \overset{\overset{\displaystyle A^-}{}}{COOR} \ \rightarrow \ A - CH_2 \ \overset{\overset{\displaystyle CN}{|}}{\underset{\ominus}{C}} \ COOR \ \ \xrightarrow{\ \ CH_2 = \overset{\overset{\displaystyle CN}{|}}{C}COOR \ \ }
$$

$$
A - CH_2 \ \overset{\overset{\displaystyle CN}{|}}{\underset{|}{C}} \ CH_2 \ \overset{\overset{\displaystyle CN}{|}}{\underset{\ominus}{C}}COOR \ \longrightarrow \ polymer
$$
$$
\ \ \ \ \ \ \ \ \ \ \ \ \ \ COOR
$$

The presence of two electron acceptor groups in the monomer, -CN and CO, at a multiply bonded carbon, contributes toward polymerization even under the influence of such a weak base as water. This polymerization occurs at room temperature and does not involve a solvent or an added catalyst. For example the bonding action observed when the cyanoacrylate is placed between glass plates is probably catalyzed by minute amounts of water or other weak bases present on the glass surfaces.

Another advantage is the excellent bonding and filling properties of the cyanoacrylate when polymerization occurs. This is, in part, due to the material's unusual property the volume of the material remains constant during the polymerization process [12]. This allows for the formation of an exact replica of the glass substrates, so that high optical quality can be achieved.

RECORDING MECHANISM

The dielectric constant of a photodielectric material is described by the Clausius-Mosotti relation given by

$$\varepsilon(\nu_i, \alpha_i) = \frac{1 + 2 \frac{4\pi}{3} \sum_i \nu_i \alpha_i}{1 - \frac{4\pi}{3} \sum_i \nu_i \alpha_i} , \tag{1}$$

where ν_i is the molecular density of the ith component of the material, and α_i is the molecular polarizability. Therefore, the change of the dielectric constant, $\Delta\varepsilon$, may be described by the relation,

$$\Delta\varepsilon = \sum_i \frac{\partial\varepsilon}{\partial\alpha_i} \Delta\alpha_i + \sum_i \frac{\partial\varepsilon}{\partial\nu_i} \Delta\nu_i . \tag{2}$$

It is through the exposure dependent changes of α_i and ν_i that volume hologram formation is made possible.

The holographic information then is stored by means of small changes in the dielectric constant -$\Delta\varepsilon$; in pure phase polymeric materials only the real part of the dielectric constant is affected. This change is induced during exposure to light either by the modulation of the molecular polarizability, $\Delta\alpha$, or by the change in density resulting from a change in the number of molecules per unit volume, $\Delta\nu$, or by both mechanisms simultaneously. We conducted a number of experiments to identify which of these two mechanisms is the major contribution to the change of the dielectric constant, and thereby to the change of index of refraction.

In one experiment we recorded simple reflection holograms (two plane wave with λ = 488 nm) in a photodielectric layer, of 130 microns thickness, and monitored the diffracted intensity during recording with λ = 632.8 nm. We observed that during recording, the readout beam incident angle had to be constantly increased for maximizing the diffracted intensity. We attribute this behavior to a continual shrinkage (creep) of the recording material which reduces the spacing, d, between the recorded fringe surfaces. As a consequence the readout angle, θ, must be correspondingly changed to satisfy the Bragg condition, $2d \sin\theta = \lambda$.

The change in fringe spacing as a function of readout angular change is obtained from the Bragg relation to yield

$$\Delta d = -d \cot \Delta\theta \qquad\qquad (3)$$

In the experiment the reference and signal beams were oriented at 45° with respect to the normal of the hologram plane, so that Eq. (3) reduces to

$$\Delta d = -d \ \Delta\theta \qquad\qquad (4)$$

The recorded fringe surfaces are parallel to the front and back surfaces of the recording medium; thus the relative change of $\Delta d/d$ corresponds to a relative change of the thickness of the medium $\Delta t/t$, where t is the thickness. The measurements revealed that, as the exposure levels increased, the required change, $\Delta\theta$, to maximize the diffracted intensity ranged from 0.003 to 0.018 radians, implying that the percentage of density increase ranged from 0.3 percent to 1.8 percent.

The effect of this "creep" behavior is negligible in an unslanted transmission holograms, where the boundary conditions are different. In these holograms the planes of the fringes (corresponding to higher densities) will buckle upon creeping, as illustrated in the upper part of part of Fig. 1, but the average distance between the planes will remain the same. For the reflection holograms, however, the creep process leads to an average decrease in the period of the grating, as illustrated in the lower part of Fig. 1.

Transmission Hologram

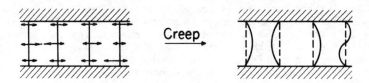

Reflection Hologram

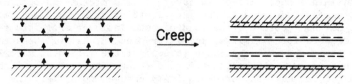

Fig. 1 - An illustration of the creep mechanism in simple transmission and reflection holograms.

In a second experiment a strip of gold was evaporated on an optically flat
glass plate, which served as a substrate for a 60 microns layer of photodielectric material. Adjacent to the gold strip we recorded a simple holographic grating with a certain exposure level to give a diffraction
efficiency in excess of 90 percent. For this high diffraction efficiency
and for a small incidence angle θ, the predicted value of optical path
change due to a change of index of refraction should be approximately
$\Delta n \cdot t = \lambda/2$; this was derived from the basic thick transmission hologram

diffraction efficiency equation,   D.E. $= \sin^2 [\frac{\pi \Delta n \cdot t}{\lambda \cos \theta}]$

An area of the material under the gold strip was then uniformly irradiated
(image of a knife edge) through the transparent substrate to the same
exposure level; because of the reflective gold layer, the predicted overall
path change should be λ. When examined microscopically with a Nomarski
polarization interferometer, however, we determined that the change in
optical path was only $\lambda/4$ as shown in Fig. 2; the distance between adjacent
fringes correspond to $\lambda/2$. Since the overall optical path change
$\Delta n \cdot t + \Delta t \cdot n$, the discrepancy between the expected and experimental results
is due to the decrease of the term $\Delta t \cdot n$, i.e. a reduction of thickness t.
From these results we calculate that the percentage of density increase is
about 0.6 percent.

Fig. 2 - Microinterferogram from a Nomarski interferometer

In a third interference experiment a stack of ten parallel photodielectric layers, separated by glass plates, was placed in one arm of an interferometer as shown in Fig. 3. The lens magnifies the simple interference pattern which is displayed on some viewing screen. A stack rather than a single 1.3 mm thick layer was used in order to avoid the excessive noise which would be generated in the thick layer for the high exposure level necessary to conveniently detect a change in the material. Due to the good index match between the glass plates and the photodielectric layers, the absorption of the overall stack differs only slightly from that of a 1.3 mm sample.

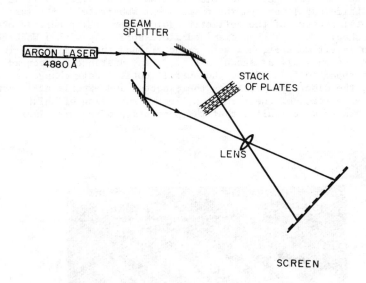

Fig. 3 - Interferometric arrangement for measuring optical path change of stacked photodielectric layers.

The stack was irradiated to an exposure level that should have caused a 5 fringe displacement in the observed interference pattern if the refractive index change were only due to molecular polarizability changes. The actual shift, however, was 1/2 fringe, implying a shrinkage of the photodielectric material. This again confirms that the major contribution to the optical path change stems from local density increase, which in this experiment was about 0.1 percent.

DISCUSSION

The results of the experiments described in the preceding section provide
evidence that the index of refraction changes are primarily caused by local
increases of density in the photodielectric bulk, and should be independent
of absolute thickness of the sample. Indeed, we experimentally measured the
diffraction efficiency of simple holographic gratings and converted this data
to index of refraction modulation, Δn, as a function of exposure. As
expected, the results, shown in Fig. 4, follow a master curve regardless of
the thickness of the recording samples.

The chemical process responsible for the changes in density is photocrosslinking of the dielectric material. This process can be represented by the
following reaction:

When irradiated with actinic radiation the parabenzoquinone is excited to the
(n, π^*) triplet state which is responsible for the hydrogen abstraction [13];
double abstraction forms hydroquinone and -R-R crosslinks. Thus, the
parabenzoquinone is photochemically branching and subsequently crosslinking
the photodielectric material. This process was corroborated by results from
a recent study dealing with sensitized PMMA [14], and is also in agreement
with the crosslinking mechanism suggested for UV sensitive PMMA [1].

When the crosslinking density exceeds a certain value, local mechanical
stresses develop. Stresses in the vicinity of one cluster of crosslinks
affects adjacent clusters causing a small relative displacement. This
displacement will perturb the initial crosslinking pattern leading to

cooperative noise generation which reduces the signal-to-noise ratio of the reconstructed imagery; this is particularly severe with thick samples.   In addition, a time-dependent creep effect between adjacent planes of high density crosslinks, will alter the spacing of the holographic interference pattern, so as to exclude recording of reflection holograms.   We therefore, conclude that extensive crosslinking should be avoided when recording holograms in these materials.

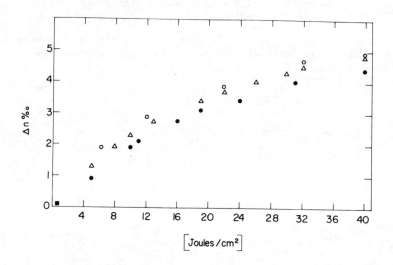

Fig. 4 - Index of refraction modulation as a function of exposure for different sample thicknesses. △ - 20μm thick; ● -30μm thick; ○ - 130μm thick, ■ - 1.3mm thick.

ACKNOWLEDGMENT

The authors express their appreciation to D. Peri for helpful discussion. This work was supported in part by the U.S. Air Force Flight Dynamics Laboratory under Grant AFOSR 76-3004.

REFERENCES

(1) W.J. Tomlinson, I.P. Kaminow, E.A. Chandross, R.L. Fork, and W.T. Silfvast, Appl. Phys. Lett. 16, 486 (1970).

(2) J.M. Moran and I.P. Kaminow, Appl. Opt. 12, 1964 (1973).

(3) M.J. Bowden, E.A. Chandross, and I.P. Kaminow, Appl. Opt. 13, 112 (1974).

(4) F.P. Laming, Society of Plastics Engrs. Tech. Meeting, Oct. 1970; and Pol. Eng. Sci. 11, 421 (1972).

(5) R.G. Zech, J. Opt. Soc. Am. 62, 1396A (1972).

(6) J.A. Jenney, J. Opt. Soc. Am. 60, 1155 (1970).

(7) W.S. Colburn and K.A. Haines, Appl. Opt. 10, 1636 (1971).

(8) B.L. Booth, Appl. Opt. 14, 593 (1975).

(9) W.J. Tomlinson, E.A. Chandross, H.P. Weber, and G.D. Aumiller, Appl. Opt. 15, 534 (1976).

(10) A.A. Friesem, Z. Rav Noy and S. Reich, Appl. Opt. 15, (1976) [in press].

(11) F. Leonard, R.R. Kalkarni, G. Brandes, J. Nelson and J.J. Cameron, J. Appl. Pol. Sci. 10, 259 (1969).

(12) H.W. Coover, Jr., F.B. Joynes, N.H. Shearer, Jr., and T.H. Wicker, Jr. SPE Journal, 413 (1959).

(13) Z. Malcom Bruce, Photochemistry of Quinones, The Chemistry of Quinoid Compounds, Part 1, Chapter 9, J. Wiley and Sons, (1974).

(14) Y.L. Freilich, M. Levy, and S. Reich - J. of Polymer Sci. (Chem.), [in press].

HIGHEST SENSITIVITY MATERIALS FOR VOLUME
HOLOGRAPHY: $Bi_{12} Si O_{20}$ and $Bi_{12}Ge O_{20}$

F. Micheron, M. Peltier and J. P. Huignard

Thomson-CSF-Laboratoire Central de Recherches, B.P. 10 - 91401 Orsay

ABSTRACT.

$Bi_{12} Si O_{20}$ and $Bi_{12} Ge O_{20}$ present the best known photorefractive photosensitivity for read write volume holographic storage ($S^{-1} = 300/uJ/cm^2$) combined with high quality image reconstruction. Recording processes by photocarriers diffusion (Zero Field applied) and by photocarriers drift are identified ; the high photosensitivity is attributed to photocarriers displacements comparable with fringe spacings or larger. Saturation diffraction efficiency at light power densities larger than $600/uW/cm^2$ at $\lambda = 514,5$ nm occurs from complete photocarriers trap filling.

INTRODUCTION.

In the field of data storage and processing using optical techniques, the main difficulty arises from the active materials in which the informations are optically written. For instance, silver halide photographic plates are photosensitive enough, but do not allow real time processing and cannot be erased ; photochromic crystals with colour centers require too large light energy density, but are erasable ; photopolymers are quite photosensitive, but in general not reversible. At the end of the last decade a new class of erasable photosensitive materials has been introduced which are the photosensitive electrooptic crystals such as $LiNbO_3$. In principle these materials allow real time processing and erasure without any fatigue but unfortunately, the photosensitivity remains very low (typically, three order of magnitude below photographic plate photosensitivity). This paper is the first report of electrooptic materials with a photosensitivity comparable to the photographic plate photosensitivity, which allows three-dimensional recording of phase holograms with an excellent image reconstruction quality. The crystals are BSO ($Bi_{12} Si O_{20}$) and isomorphous BGO ($Bi_{12} Ge O_{20}$), already used in the Pockel Read Out Optical Modulator (1), and also for surface piezoelectric devices.

This paper deals with the experimental results of holographic recording in BSO and BGO, and the physical process underlying the photoinduced change of refractive index.

PHOTOSENSITIVE ELECTROOPTIC MATERIALS AND THE PHOTOREFRACTIVE EFFECT.

In Photosensitive electrooptic materials, the photoinduced change of refractive index (PCI) occurs via the electrooptic effect. That is to say, PCI Δ n is related to a photoinduced change of electric field Δ E inside the material

$$\Delta n = R(\Delta E)$$

where $R(\Delta E)$ describes the electrooptic effect. $R(\Delta E)$ may be a quadratic function, such as in non polar ferroelectric materials (Pb, La) (Zr, Ti)O_3 (2) or a linear function, for instance in LiNbO$_3$, BSO and BGO :

$$R(\Delta E) = R.\Delta E$$

where R is the electrooptic coefficient; the photoinduced change of electric field arises in general from a photoinduced space charge, due to photocarriers drift or diffusion, which are trapped in centers spatially different from the primary donnors centers.

The photosensitivity of this process is measured by the energy W required for obtaining a given Δn ; this definition is commonly applied to the case of a holographic diffraction efficiency of 10^{-2}, in a one millimeter thick crystal. In oxygen octaedra ferroelectrics such as LiNbO$_3$, Glass (3) has evaluated the minimum required energy, assuming a unity quantum efficiency, and a photocarrier displacement length L equal to the holographic fringe spacing Λ. This minimum energy is found equal to $20\,\mu J/cm^{-2}$ at wavelength $\lambda = 0.5/um$, but has never been measured in any photosensitive electrooptic crystal. KTN (4) and SBN (5) reach respectively $100\,\mu J/cm^2$ and $10mJ/cm^2$, with bad optical quality, and LiNbO$_3$ (6) 300 mJ/cm^2 with excellent optical quality. The reason is that quantum efficiency β is probably smaller than unity, and displacement length L is much smaller than the fringe spacing Λ.

In BSO crystals, we have measured $W = 300\,\mu W/cm^2$, which constitutes the best performance, since the optical quality can be compared to the LiNbO$_3$ optical quality.

CHARGE TRANSFER PROCESS IN BSO AND BGO CRYSTAL

The simplified energy band diagram of BSO and BGO are shown in Fig.I.1 ; the center B, responsible for extrinsic absorption is situated at 2.60ev under the conduction band, with a population density $N_B = 10^{19}cm^{-3}$, and is believed to be a silicon (or germanium) vacancy (7) ; the center A is the electron trapping center with population density $N_A = 10^{16}cm^{-3}$, and its nature is not already known (8). The quantum efficiency β has been found (9) equal to $\beta = 0.7$ at $\lambda = 514.5$ nm, which means that, after absorbing N_A/β photons, the crystal shows a saturation of the B centers. Since the absorption coefficient is $\alpha = 2.3cm^{-1}$, this saturation occurs for incident light energy densities equal or larger than

$$W_S = \frac{N_a\,h\nu}{\alpha\,\beta} = 3mJ/cm^2$$

For light energy densities larger than W_S, the band diagram is shown in Fig.I.2, with A centers fully occupied, and trapped holes at the center B.

Now, let us consider the photocurrent voltage characteristics of a BSO crystal shown in Fig.II. These characteristics are linear versus illumination and applied voltage, no photocurrent is found at zero voltage applied. These characteristics can be described by the relation

$$i - id = \mu\tau g \quad q E a \quad , \quad g = \frac{\alpha\beta I}{h\nu}$$

Where id is the dark current (neglectable versus photocurrent for $I > 10\,\mu W/cm^2$) and g the photoelectron generation rate. This linear relation

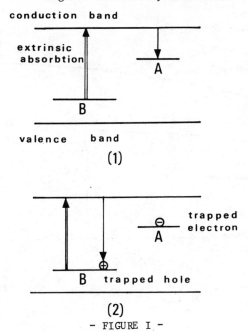

conduction band

extrinsic absorbtion

A

B

valence band

(1)

trapped electron
A

B trapped hole

(2)

- FIGURE I -

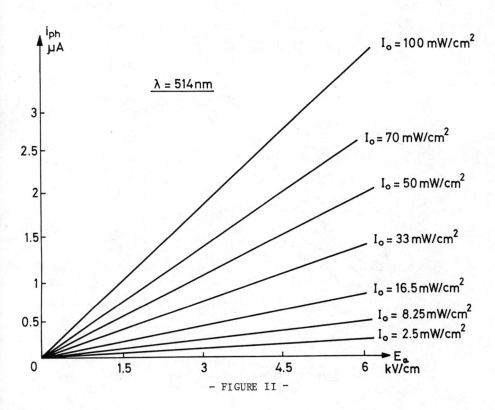

- FIGURE II -

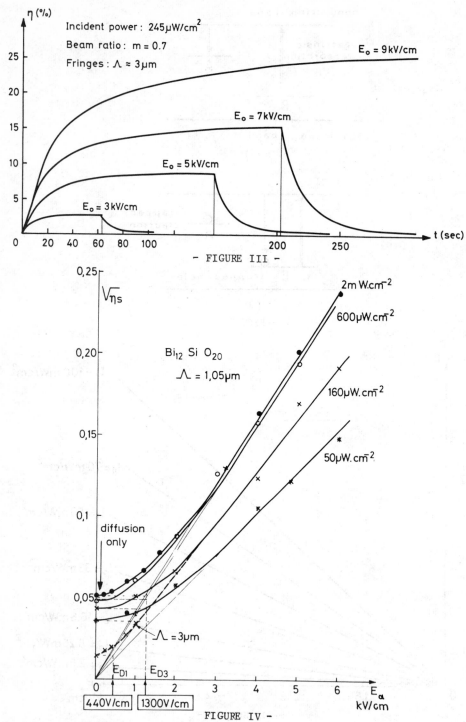

$\eta\,(\%)$

Incident power : 245 μW/cm^2
Beam ratio : m = 0.7
Fringes : $\Lambda \approx 3\,\mu$m

$E_o = 9\,$kV/cm

$E_o = 7\,$kV/cm

$E_o = 5\,$kV/cm

$E_o = 3\,$kV/cm

t (sec)

− FIGURE III −

$\sqrt{\eta_s}$

2m W.cm^{-2}

600μW.cm^{-2}

$Bi_{12}\,Si\,O_{20}$

$\Lambda = 1{,}05\,\mu$m

160μW.cm^{-2}

50μW.cm^{-2}

diffusion
only

$\Lambda = 3\,\mu$m

E_{D1} E_{D3}

440V/cm   1300V/cm

E_α
kV/cm

FIGURE IV −

means that both the generation rate g and photocarriers lifetime τ are constant values, that is to say, the relative change of N_B under uniform illumination can be neglected.

The photoelectron drift length L = /uτ Ea and diffusion length L' = $(D\tau)^{1/2}$ = $(kT/u\tau/q)^{1/2}$ are :

$$\text{BSO} \quad L = 8.4/um \ (Ea = 6kV/cm) \quad ; \quad L' = 0.6/um$$
$$\text{BGO} \quad L = 7.2/um \ (Ea = 6kV/cm) \quad ; \quad L' = 0.56/um$$

These results show that the characteristic displacement lengths L and L' are three or four orders of magnitude larger than estimated values in other photorefractive materials with equivalent optical quality such as LiNbO$_3$. According to the model of Young et al (10), such large length compared to holographic fringe spacings do not limit the spatial resolution, and the ultimate photorefractive photosensitivity for volume hologram storage can be reached.

VOLUME HOLOGRAM RECORDING

Hologram grating recording is performed with Argon Laser line λ = 514 nm (fringe modulation $m = 1$) and diffraction efficiency is continously monitored with a low power He-Ne Laser (reading power density : several /uW/cm^2) under Bragg incidence. Hologram erasure is achieved by illumination with one of the recording beams providing uniform space charge field relaxation. Due to large optical activity of these crystals and inhomogeneous light density in the thickness, diffraction efficiency η is not easily related to photoinduced space charge field ΔE : a linearly polarised recording beam is diffracted into an elliptical one. Nevertheless we will assume that for low diffraction efficiencies η is proportionnal to the square of the photoinduced field amplitude, and space charge fields at saturation is of the order of Ea. Fig.III shows the time dependance of diffraction efficiency, up to saturation, and erasure for different applied field values. These cycles are nearly symetrical and the initial slope becomes independant of applied field for Ea = 7.10^3V.cm^{-1}. This means that photoelectrons traps A are uniformely filled, the hologram space charge field being developped by non uniform distribution of trapped holes on centers B under sinusoïdal illumination I = Io (1 + mcos Kx). For such a situation, the space charge distribution at initial stage of hologram recording is ρ (x,t) = $\rho_0$mt cosKx, with $\rho_0 = \dfrac{\alpha Io \beta q}{h\nu}$; the space charge field derived from Poisson equation is :

$$\Delta E = \frac{m \rho_0 t}{\varepsilon K} \sin Kx$$

For Io = 245/uW cm^{-2} , K = $\dfrac{2\pi}{\Lambda}$, Λ = 3/um, the rate of change of photoinduced field is :

$$R = \frac{\partial(\Delta E)}{\partial t} = 10^3 \ Vcm^{-1}s^{-1}$$

and the corresponding recording time constant for a linear behavior would be $\tau_s = Ea/R$ i.e τ_s = one second per KV cm^{-1}. The experimental recording time constants in Fig.III are found larger due to non linear time evolution of space charge field and beam coupling with a corresponding non sinusoïdal field saturation as discussed by KIM et al in reference (11) :

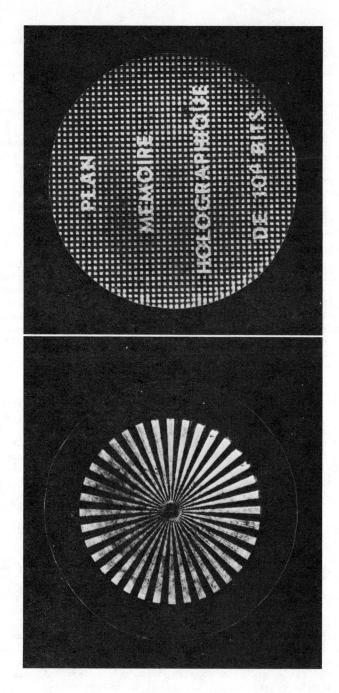

FIGURE V

- The holographic grating shows $\pm$ $\pi/2$ shift versus illumination pattern at initial stage of hologram recording and no shift at saturation,

- At saturation, the second and third harmonics of fundamental spatial frequency give rise to large diffraction efficiencies : $\eta_{s2} = 0.25\,\eta_s$ $\eta_{s3} = 0.1\,\eta_s$ (η_s : saturation efficiency at spatial frequency K).

Fig. IV shows the square root of saturation diffraction efficiencies, η_s, versus applied field, for different light power densities. The representative curves η_s (Ea) are asymptotic to straight lines, passing through the origin ($\eta_s = 0$, Ea = 0). This confirms the assumptions $\eta_s \simeq E_s^2$, and $E_s \neq Ea$. The equivalent fields E_D deduced from these curves corresponding to saturation efficiency η_{so} for zero applied field : Ea = 0 are $E_{D1} = 1400$ Vcm^{-1} at $\Lambda = 1/$um and $E_{D3} = 440$ Vcm^{-1} at $\Lambda = 3/$um.

These values are in accordance with space charge fields developped by diffusion only :

$$E_{SD} = \frac{RT}{q}\,K \qquad \begin{aligned} E_{SD} &= 1600 \text{ V.cm}^{-1} \ (\Lambda = 1\mu m) \\ E_{SD} &= 540 \text{ V.cm}^{-1} \ (\Lambda = 3\mu m) \end{aligned}$$

For applied fields Ea $\gg E_{sD}$, the photoelectron diffusion process is neglectable versus the drift process. The photoinduced field ΔE becomes proportionnal to the fringe spacing Λ , therefore, the photosensitivity must be an increasing function of the fringe spacing. This is confirmed by measurements of the light energy densities required for obtaining 10^{-2} diffraction efficiency in a 1mm thick crystal, for different fringe spacings, and applied field Ea = 6KV/cm :

Λ =	5 /um	1 /um	0.6 /um
BSO	300 /uJ/cm^2	2mJ/cm^2	4mJ/cm^2
BGO	1.25mJ/cm^2	8mJ/cm^2	16mJ/cm^2

Differences in photosensitivity between BSO and BGO are due to the lower electrooptic coefficient of BGO. Erasure times, under uniform illumination or in the complete dark are found experimentally larger than the corresponding dielectric relaxation times.

For instance, erasure time in the dark is $\tau_e = 10^5$s for BSO, and the dielectric relaxation time is $\tau_o \simeq 3.10^3$s. This effect is attributed to the non uniform population of absorption centers B (12).

CONCLUSION

Volume hologram recording characteristics of BSO and BGO crystals are governed by two main properties ; the quantum efficiency is nearly unity at $\lambda = 514.5$nm, and the photoelectron drift lengths are comparable with the holographic fringe spacings. This efficient charge transfer process allows to reach a photosensitivity nearly equal to the photographic plate photosensitivity with excellent image reconstruction quality (Fig. V). No fatigue effect have been observed after multiple recording-erasure cycles, and spatial resolution is diffraction limited. These crystals are therefore good candidates as active optical media in fast real-time holographic processors.

BIBLIOGRAPHY.

- 1 - S.L. LOU and D.S. OLIVER
      Appl. Phys. Letters 18, 325 (1971)

- 2 - F. MICHERON, J.M. ROUCHON and M. VERGNOLLE
      Appl. Phys. Letters 24, 605 (1974)

- 3 - A.M. GLASS - PHOTONICS
      ed. M. Balkanski and P. Lallemand ; Gauthiers Villars (1973)

- 4 - D. VON DER LINDE, A.M. GLASS and K.F. RODGERS
      Appl. Phys. Letters 26, 22 (1975)

- 5 - J.B. THAXTER and M. KESTIGIAN
      Appl. Optics 12, 1675 (1974)

- 6 - J.P. HUIGNARD, J.P. HERRIAU and F. MICHERON
      Ferroelectrics 11, 393 (1976)

- 7 - S.C. ABRAHAMS, P.B. JAMESON and J.L. BERNSTEIN
      J. Chem. Phys. 47, 4034 (1967)

- 8 - R.B. LAUER
      J. Appl. Phys. 42, 2147 (1971)

- 9 - R.A. SPRAGUE
      J. Appl. Phys. 46, 1673 (1975)

- 10 - L. YOUNG, W.K.Y. WONG, M.L.W. THEWALT and W.D. CORNISH
       Appl. Phys. Letters 24, 264 (1974)

- 11 - D.M. KIM, K.R. SHAH, T.A. RABSON and F.K. TITTEL
       Appl. Phys. Letters 28, 338 (1976)

- 12 - M. PELTIER and F. MICHERON
       To be published J. Appl. Phys.

A SIMPLE MATHEMATICAL MODEL FOR NEGATIVE PHOTORESIST BEHAVIOUR

Jean J. Clair and Jaime Frejlich

Institut d'Optique - Université Paris VI, lab. Françon 4, Place Jussieu,
75005 Paris, France

ABSTRACT

A simple mathematical model is proposed to account for experimental behaviour of a negative commercial photoresist:Microresist 747 of Kodak.Influence of variation of the three parameters caracterizing the resist can be roughly predicted refering to the computed theoretical curves presented.An estimated value for the overall coefficient involving quantum efficiency and kinetics of reaction is found.Brief discussion of basic hypothesis involved is presented,and attention is called for the fact that the whole of the reaction mechanism and not only the photoabsorption process must be considered in order to find basic equations governing the phenomenon.Experimental and theoretical results are presented.

INTRODUCTION

Negative photoresists are substances that get polimerized under the action of light,usually in the near U.V.range;there are then changes of physical and chemical properties,especially the refractive index and solubility.
Development consists in dissolving the non-exposed,soluble material.Photoresist response is defined as the curve showing the remaining resist film relative thickness after development vs light energy received by unit surface.
In this way,mapping of a light energy spatial distribution,into a transparent phase object relief,is possible.

A MATHEMATICAL MODEL

Basic Hypothesis and Definitions

Hypothesis 1.We suppose a constant and uniform non-exposed initial chemical composition and mass concentration.We designate as monomer and polymer,respectively the non-exposed and totally exposed product.Because of assumptions of no molecular diffusion and experimental evidence of no volume changes in the film before development,we can suppose an invariable total mass concentration at each point in the film during exposure.Because of symmetry, only the x-axis direction in the film (Fig. 1) will be considered.

389

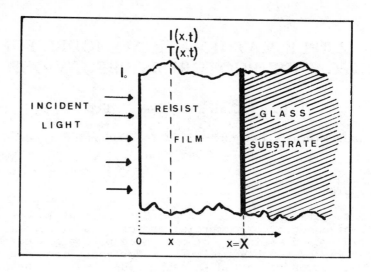

Fig. 1. Description

Hypothesis 2.We suppose our fundamental law to be

$$\partial T/\partial t = K.I.(1-T)^2 \tag{1}$$

$$T = \left[C(z,0) - C(z,t) \right] / C(z,0) \quad \text{and} \quad z = x/X$$

$T(z,t)$, $C(z,t)$ and $I(z,t)$ being the polymerization rate, monomer mass concentration, and irradiance, respectively, at normalized position $\underline{z}$ in the film thickness, and time $\underline{t}$; $\underline{K}$ is a constant related to kinetics of reaction and quantum efficiency of photoreaction. Conclusions using eq.(1) fit better our experimental results than those using the usual law:

$$\partial T/\partial t = K.I.(1-T) \tag{2}$$

as will be shown in Fig. 5.
We do not provide definite theoretical support for eq.(1) because of lack of knowledge on the complex reaction mechanism involved. A possible mechanism that would account for this law is represented in the following scheme:

$$M + h\nu \xrightarrow{k_0} M^* \qquad \text{photoabsorption}$$

$$M^* \xrightarrow{k_1} M \qquad \text{spontaneous relaxation}$$

$$M^* + M \xrightarrow{k_2} P \qquad \text{polymerization}$$

$$\partial [M^*]/\partial t = k_0.[M].I - k_1.[M^*] - k_2.[M^*].[M]$$

and at equilibrium: $\partial [M^*]/\partial t = 0$, so $[M^*] = k_0.[M].I/(k_1+k_2.[M])$

and : $\partial [P]/\partial t = k_2.[M^*].[M]$

with $\underline{\nu}$ the time frequency of light, $\underline{M}$ the photosensitive monomer,

$\underline{M^*}$the optically excited monomer,$\underline{P}$ the polymer, $[\underline{M}]$ the molar concentration of species $\underline{M}$, and $\bar{k}_o, k_1, k_2$ are the corresponding reaction rate constants,the rate of polymer formation being:

$$\partial[P]/\partial t = k_2 . k_o . [M]^2 . I/(k_1+k_2 . [M])$$

If spontaneous relaxation is relatively slow,that is if $k_1 \ll k_2 . [M]$ we find an equation similar to eq.(2): $\partial[P]/\partial t = k_o . [M] . I$.
In the opposite situation, $k_1 \gg k_2 . [M]$,we find an equation similar to eq.(1),the one used here: $\partial[P]/\partial t = (k_2 . k_o/k_1) . [M]^2 . I$

As can be seen,the law to be used depends on the whole reaction mechanism and not only on the photoabsorption step.The reaction mechanism here is just one possibility among others,but it shows why the whole process should be considered.No isolation of any sensitizer from the bulk resist was reached through repetitive solution and precipitation procedures on the bulk material,so we may think the sensitizer to be chemically linked to the <u>monomer</u> molecule forming a chemical unit: the photosensible monomer <u>M</u>.

<u>Hypothesis 3</u>.We suppose the optical absorption coefficient <u>g(z,t)</u> at point <u>z</u> and time <u>t</u>,to be linearly dependant on polymerization rate <u>T</u>,so as to write:

$$g(z,t) = a . [1-T(z,t)] + b . T(z,t) \qquad (3)$$

<u>a</u> and <u>b</u> being the decadic absorption coefficients of monomer and polymer respectively.

<u>Hypothesis 4</u>.We assume the resist film to follow Lambert-Beer's absorption law,whatever the polymerization rate presented.The law has been tested to be valid for pure monomer and pure polymer,at the 365 nm wavelength utilized in photoreaction (Fig. 2)

<u>Hypothesis 5</u>.Molecular diffusion is supposed negligible during operation.This can be easily accepted when phenomenon on a transverse to the x-axis direction is considered (Fig. 1),because wide surfaces are involved and because of the high viscosity of the film(85 to 115 cp in solution,solvent evaporation during the film lying process increases further the viscosity).This is less evident when diffusion in the x-axis direction is considered, because of the small film thickness ($X \cong 1 \mu m$)and the relatively long exposure times (several minutes).

<u>Hypothesis 6</u>.Our model being concerned only with photopolymeriza-tion reactions,we drop off the <u>threshold energy</u> concept,as no solvent action is involved;this concept is related to solubility properties of polymer:a threshold energy is associated to the formation of a certain number of molecular bonds so as gel formation and begining of insolubilization is reached (Ref. 1). Even if our model is not concerned with solubilization process itself,practical realization of phase profiles must deal with it; so a mathematical model accounting on solubilization process should be developped.Anyhow,we have realized that we can modify the <u>resist response</u> curve,so as to fit fairly well the <u>photopolymerization rate</u> curve,just by slightly modifying the development conditions,as can be seen in Fig. 3.We should insist however in the fact that there is not necessarily any close relationship between both curves.

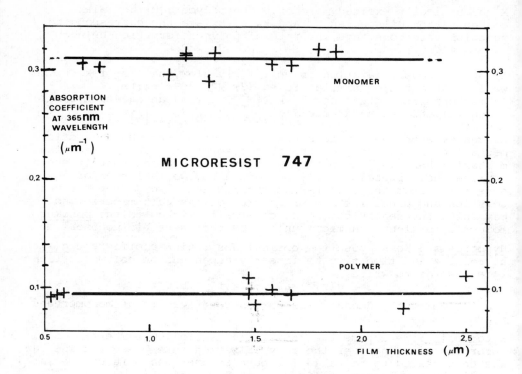

Fig. 2. Lambert-Beer's Law verification for
pure Monomer and Polymer.

Derivation of Equations

Integration of eq.(1) with $\partial E/\partial t=I$, E being the light energy
received by unit film surface at z and time t, with boundary
conditions $T(z,0)=0$ and $E(z,0)=0$, gives:

$$T = K.E/(1+K.E) \tag{4}$$

From Lambert-Beer's law and eq.(3), we find

$$\partial I/\partial z = -I.\left[A.(1-T) + B.T\right]\ln(10) \tag{5}$$

with $E=E(z,t)$, $T=T(z,t)$, $I=I(z,t)$, $A=a.X$ and $B=b.X$.
From eqs.(4) and (5):

$$(\partial/\partial t)(\partial E/\partial z) =-(\partial/\partial t)\left[(A-B)/K .\ln(1+K.E) + B.E\right]\ln(10)$$

so, after integration in time t, with boundary conditions $\partial E/\partial z=0$
for $E=0$, and accounting on eq.(4), we find:

$$\partial T/\partial z= \left[(1-h).(1-T)^2.\ln(1-T) - h.T.(1-T)\right].A.\ln(10) \tag{6}$$

with $h=b/a$. Equations (6) and (4) must be solved simultaneously
with boundary conditions: $T(0,t)=K.E_o/(1+K.E_o)$ with
$E_o=E(0,t)=I_o.t$. By numerical methods eq.(6) can be computed to

find $T(z,t)=f(z,K.E_o,h,A)$.
From a practical point of view we are interested on the average
polymerization rate over the whole film thickness $\underline{<T>}$:

$$<T> = \int_{z=0}^{z=1} T(z,t).dz = j(K.E_o , h , A) \tag{7}$$

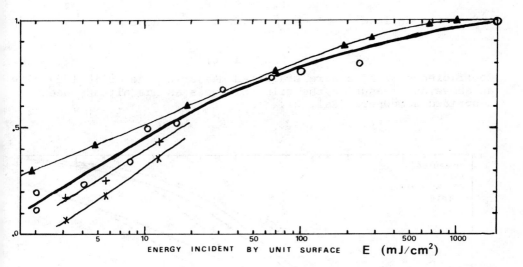

Fig. 3. Resist Response Curves and Development Conditions.
<u>Photopolymerization rate</u>:experimental data ($\mathbf{O}$).
<u>Resist Response</u>:experimental data with varying conditions of
development:2 min in Developer solvent and 1 min in Rinse
solvent ($\blacktriangle$);then 1 min in Develop.and 1 min Rinse ($+$);then
5 min in Develop. and 5 min in Rinse ($\times$).

Display of Computed Results

As seen in eq.(7),photoresist film properties are well described
by parameters $\underline{h}$ and $\underline{A}$,the former caracteristic of the ratio of
absorption coefficients of polymer and monomer,and the latter
depending on the total absorption of initial monomer film.
Exposure,kinetics and quantum efficiency may be advantageously
described by a single product: $\underline{K.E_o}$.Computed results for eq.(7)
are seen in Fig. 4.

EXPERIMENTAL DATA

Experimental work has been done on <u>Microresist 747</u>,a Kodak
negative photoresist.Light was provided by a high pressure
mercury lamp (Osram HBO 500 W/2) and it was mainly that of 365
nm wavelength.

The average experimental polymerization rate $<T>$ was found by measuring the optical density $\underline{D}$ of the film,using a Cary spectrophotometer,and using basic hypothesis included in eq.(3):

$$D = X.\int_{z=0}^{z=1} g.dz = X.(b-a).\int_{z=0}^{z=1} T.dz + X.\int_{z=0}^{z=1} a.dz$$

and as was defined: $<T> = \int_{z=0}^{z=1} T.dz$, so we have

$$<T> = (a- D/X)/(a- b) \tag{8}$$

Coefficients $\underline{a}$ and $\underline{b}$ were evaluated measuring the optical density as above,and measuring the film thickness as has already been described somewhere (Ref. 2).

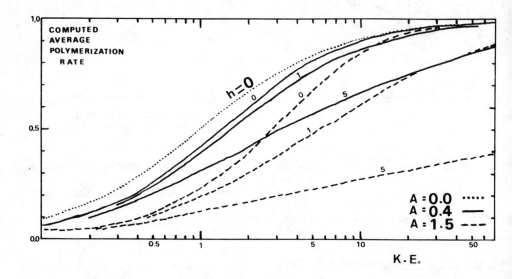

Fig. 4. Computed Average Polymerization Rate
for h=0,1,5 and A= 0,0.4,1.5 .

In Fig. 5 we compare experimental and computed values for the average polymerization rate,using eq.(1) and eq.(2),all others hypothesis and conditions being the same.As can be seen there, computed curve using eq.(1) fits better the experimental data,and the shift on horizontal axis required(experimental values were shifted so as to fit computed curves),allows a primary very rough estimation of the constant $\underline{K}$;it is found to be approximately 0.1 cm^2/mJ. A better estimation of this value follows.

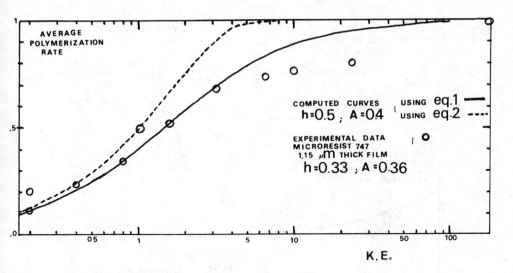

Fig. 5.Experimental and Computed Average Polymerization Rate. Experimental data being:A=0.36 and h=0.33(using average values at 365 nm Wavelength),for $a=0.31\mu m^{-1}$ and $b=0.1\ \mu m^{-1}$ and $X=1.15\mu m \pm 10\%$.Parameter values used for computation are A=0.4 and h=0.5.

Estimation of the Constant K

As was similarly calculated somewhere for a positive resist(Ref.3) and using eq.(3):

$$D=X. \int_{z=0}^{z=1} g.dz = X.\int_{z=0}^{z=1}\left[a.(1-T)+b.T\right].dz , \quad then$$

$$\partial\langle g\rangle/\partial t = \int_{z=0}^{z=1} (\partial T/\partial t).(b-a).dz \quad ,being \ \langle g\rangle = D/X, \ and \ using \ eq.(1),$$

dividing both members by $I_o=E_o/t$, using Lambert-Beer's law, rearrenging terms and taking limits:

$$\lim_{E_o\to 0}\left[\partial\langle g\rangle/\partial E_o\right] = (b-a).K.\lim_{E_o\to 0}\left[\int_{z=0}^{z=1} 10^{-g.X.z}.dz\right] \quad and \ then$$

$$\left[\partial\langle g\rangle/\partial E_o\right]_{E_o=0} = (b-a).K.(1-10^{-a.X})/(a.X.ln(10)) \quad so \ finally:$$

$$K=\left[\partial\langle g\rangle/\partial E_o\right]_{E_o=0} .a.X.\ln(10)/\left[(1-10^{-a.X}).(b-a)\right] \qquad (9)$$

So, using eq.(9) and calculating graphically the limit of the derivative above, in Fig.6 where we plot some values of $\langle g\rangle$ vs E_o, we find the value of $\underline{K}$:

$$\left[\partial\langle g\rangle/\partial E_o\right]_{E_o=0} =-0.0251 \text{ cm}^2/(\text{mJ}.\mu\text{m}) \text{ and } \underline{b-a}=-0.21 \text{ m}^{-1}(\text{average})$$

and $\underline{a.X}=0.35$ so as we find the coefficient $K = 0.17 \text{ cm}^2/\text{mJ}$

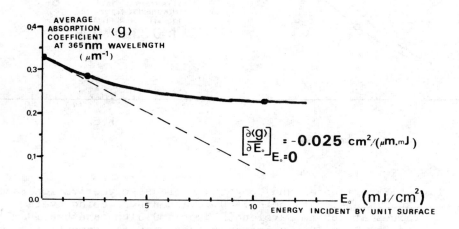

Fig. 6. Absorption Coefficient Curve for Estimation of
the Photoresist's Coefficient $\underline{K}$.

Further experimental results on $\underline{\text{Microresist 747}}$ and details on experimental techniques can be found in Ref.2.

Error Estimation and Corrections

Film thickness measurements are made with a precision of about $0.05\mu\text{m}$, and film thickness variation over the central portion of the sample (about 2cm x 4cm) is usually less than 10%.
Optical densities of the resist film were measured accounting on the glass substrate presence, so as the corrected value is:

$(\langle g\rangle.X) = D+e$, $\underline{e}$ being a correcting term, and $D=D_s-D_v$

$$e=(r-R)/(2.3(1-R)) \simeq (R_r-R.\left[1+ (10^{-2.D_s})/(1-R)^2\right])/(2.3)$$

$$R \simeq R_v/\left[1+ (10^{-2.D_v})/(1-R_v/2)^2\right]$$

Ds:measured apparent optical density of the resist-glass substrate,composite sample;Dv:measured apparent optical density of glass substrate;($<g>$.X):actual optical density of the resist film;r and R:air-resist and glass-air interface reflexion coefficients;Rr and Rv:overall measured reflexion coefficients of the resist-substrate composite sample,and substrate plate, respectively.

The above formulas were obtained on the assumptions that $e \ll 1$, and $R \simeq r \ll 1$,thus neglecting reflexions at the resist-substrate interface.With $R \simeq 0.075$ and $Rv \simeq 0.10$,the correcting term is found to be $-0.007 \leqslant e \leqslant +0.007$,depending on the actual value of Ds,and representing a less than 4% correction term in the most disadvantageable situation we were concerned with,so we conclude that it is not relevant compared to other sources of experimental error.

Energy incident on entrance surface of the film,E_o,should be eventually corrected($\simeq -5\%$) accounting on the $\underline{r}$ coefficient.

CONCLUSION

Though a mathematical model for negative photoresist has already been proposed,it was developped for quite restrictive operation conditions(Ref. 4).Other models have been recently develo ped,but for positive resist(Ref. 3).

The mathematical model we describe here represents a good approximation to experimental behaviour of this negative Microresist 747,within wide varying experimental conditions. Influence of variation of its parameters can be roughly predicted within wide limits,refering to Fig.4.We have found a rough approximative value for the constant $\underline{K}$ caracterizing chemical kinetics and quantum efficiency,both of them remaining basically unknown.Attention is called for the fact that the whole of the reaction mechanism and not only photon absorption,is to be considered in order to find basic laws governing the phenomenon. Our model is to be considered just as an oversimplified though fairly suited representation of reality.

REFERENCES

(1) Barret Broyde,J.Electrochem.Soc.,71,1241,(1969).

(2) J.J.Clair,J.Frejlich,J.M.Jonathan,J.M.Van Haecke,J.Rosiu, Proceedings of the Electro-optics/Laser International'76 UK Conference,IPC Science and Technology Press,Englend 1976(in press)

(3) Alain Brochet,Thèse Docteur 3e Cycle,Université Paris-Sud (Orsay),France,June 18th 1976.

(4) J.P.Kirk and G.L.Fillmore,Model of Negative Working Photoresist as Continuous Tone Photographic Material, Appl.Opt. 11,2347,(1972).

SESSION 11.

HOLOGRAPHIC DEVICES & TECHNIQUES

Chairman: E. SPITZ

WHITE-LIGHT TRANSMISSION/REFLECTION HOLOGRAPHIC IMAGING

Stephen A. Benton

Research Laboratories, Polaroid Corporation,
750 Main St-LA, Cambridge, Mass. 02139 U.S.A.

Holography has provided a spectacular advance toward the idealized three-dimensional "window view upon reality", as Lippmann put it, that is the goal of autostereoscopic imaging research. But despite the fantastic images that one hardly dreamed of a few years ago, holography's acceptance in commercial and educational display applications has fallen far short of all predictions. This reluctance has been due to many factors, not least among them being cost, low image luminance, and the general need for special illuminators, such as lasers or arc lamps, for hologram viewing.

Here we describe the principles and techniques of producing a recently developed type of transmission hologram that reconstructs colorful, bright, three-dimensional images using commonly available white-light sources, such as high-intensity desk lamps and the sun (1). Of course, Denisyuk-van Heerden volume reflection holograms can also be viewed with such sources, but the limitations imposed on their image depth and luminance by the reflected spectral bandwidth recommend them for separate applications. White-light transmission holograms, on the other hand, can offer images that are several meters deep, in large format, and bright enough to be viewed in average room-light conditions, and they offer the hope of very low cost replication by the embossing of plastic surfaces for viewing in reflection as well as transmission.

Horizontal-Parallax-Only Imaging

The reduction of hologram illumination temporal coherence is made possible by a reduction of the information content of the image. Specifically, all vertical parallax is sacrificed, so that one can no longer look over and under objects in the scene, but full horizontal parallax is retained in order to provide the play of right to left perspective views necessary for binocular stereopsis and motion parallax. Thus virtually all of the three-dimensional impact of the scene is preserved.

In horizontal-parallax-only imaging, information about an out-of-plane object point is not distributed over the entire image surface, but over only a thin horizontal blur-like strip corresponding to the parallax between the point and the surface as the eye moves from right to left. In holography, information is encoded within that strip by a system of interference fringes to present a sharp image to each eye, as detailed later. Because the image is sharply focused in the vertical direction, its recording represents an interesting hybrid of photographic and holographic concepts.

Because of the missing information, all horizontal-parallax-only image types, including white-light-transmission holograms and parallax panoramagrams, share certain features. For example, they have a unique "stigmatic surface", usually the information bearing surface, so called because only image points at that surface depth give rise to extended ray bundles with a common center (effectively spherical wavefronts). Ray bundles from points at any other depth are horizontally centered at that depth in order to provide parallax, but are always vertically centered on the stigmatic surface, and so become highly astigmatic at extreme image depths, setting a limit on the useful depth range in addition to any due to particular optical designs of the image types. Fortunately the eye seems quite tolerant of moderate astigmatism in uncritical viewing, perhaps as much as one-half diopter, so that image distances of 0.64 to 7.5 meters from the observer can be reproduced by a stigmatic surface (e.g. a hologram) 1.5 meters away.

Another feature shared by horizontal-parallax-only images is a particular intended viewing distance. At that distance, objects at all depths appear normally, but as the viewer moves nearer or farther, a height-to-width distortion appears for objects away from the stigmatic surface that increases with separation from that surface. For example, a square object that is apparently at infinity will have a horizontal angular subtense that is independent of viewer distance, as in normal experience, but its vertical subtense will be determined by the ratio of the fixed image height on the stigmatic surface and the viewer distance, and so will increase as the viewer approaches the hologram. Again, modest amounts of distortion are visually acceptable, and viewing ranges of several meters are readily attained if the image depth is not extreme. This range is extended even further for white-light transmission holograms by their variation of designed viewing distance with image color.

We see that the enormous informational economies of parallax limiting, and their concommitant practical advantages, have imposed a loose structure of constraints on image depth and viewer location, within which we are able to find a wide variety of impressive three-dimensional displays. We go on to describe the principles of a particular technique for exploiting the informational economies by holography.

Real Image Projection

White-light transmission holograms can be made in a single step by means of lenses, mirrors, and other techniques, but the use of a holographic intermediate step offers the clearest conceptual explanation, and is often the method of choice in actual practice. All of these techniques depend on the holographic recording of interference phase relationships, and so require spatially and temporally coherent laser radiation in all but the last step, viewing with white-light.

We begin by recording a first, or master, hologram (H1) in the usual Leith-Upatnieks way by exposure of a photographic plate to wavefronts reflected from the object scene and to a mutually coherent off-axis reference wavefront, perhaps diverging from a point source above, as illustrated in Fig. 1.

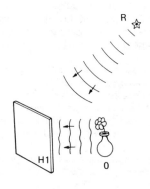

Fig. 1   Recording of the master hologram, H1, by interference of waves from
the object, 0, and from a coherent reference point source, R.

After exposure and processing, H1 may be returned to its position and illuminated by the reference beam to reconstruct the virtual image of the scene. Alternately, it may be illuminated from the opposite side and below by wavefronts **conjugate** to those of the reference beam. That is, the wavefronts are homologous to those of the reference beam (ignoring extinction by the plate) but moving in the opposite direction, as though a stop-action motion picture of the wavefronts were played in reversed time (2). In Fig. 2 the conjugate illumination is a spherical wave converging on the earlier location of the reference source. One of the wavefronts diffracted by the hologram will, in turn, be the conjugate of the object wavefront (note that the interference pattern that would be formed by the two conjugate wavefronts is the same as that formed by the original wavefronts), so that a spherical wavelet will converge at each object point location, creating a precise and undistorted real image of the object scene in space, as Fig. 2 illustrates. This image has an unfamiliar "outside-in" or "pseudoscopic" appearance when viewed from the right, but it is well suited to be an object for a further holographic recording.

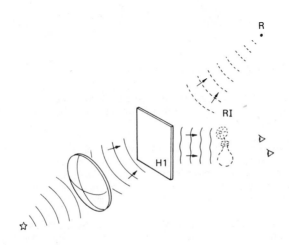

Fig. 2   Illumination of H1 with waves converging to R (reference beam conjugate) produces the object wave conjugate, projecting a three-dimensional real image, RI, in space. When viewed from the right, RI appears pseudoscopic.

Reduced-Information Real Image Projection

Each small area of H1 can be considered as projecting into space its particular perspective view of the object, which happens to overlap in register with the others at the object location, but could be projected onto a screen or plate at any distance. If only a narrow horizontal strip of H1 is used, perhaps by aperturing the conjugate illumination as in Fig. 3, then only those views differing in right-to-left parallax will be projected, producing a reduced-information real image that can also serve as a holographic object wavefront for recording. A reference wavefront of essentially vertical alignment is used, and we show a converging wavefront incident from below, anticipating a conjugate reconstruction source located at the focus.

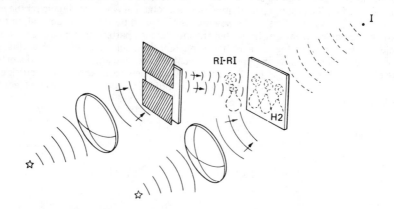

Fig. 3   Illumination of only a narrow horizontal strip of H1 projects a reduced-information real image, RI-RI, and a horizontal-parallax-only image onto the second plate, H2, where it is holographically recorded by interference with a vertically inclined reference wave converging to the intended illumination location, I.

After exposure and processing, illumination of the second hologram (H2) with the wavefront conjugate to its reference wave produces a real image of the strip of hologram H1 in space as well as a reduced-information image of the object, as illustrated in Fig. 4. Because the object image is a pseudoscopic reconstruction of the previous pseudoscopic real image, the doubly-depth-reversed image appears orthoscopically, or correctly arranged in space, but is entirely visible only when the eyes are placed within the real image of the hologram strip. One might consider H2 as projecting back to the location of each elemental area of H1 just the perspective view it recorded of the object scene, and presenting that view to an eye if it is placed at that location. Moving the eye right to left within the strip image then presents the play of parallax creating the three-dimensional image.

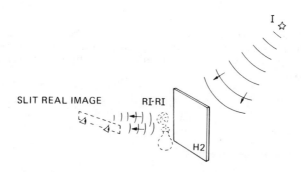

Fig. 4   Illumination of H2 from I produces a real image of the H1 slit. Eyes placed in that real image view the three-dimensional horizontal-parallax-only image RI-RI.

White Light Illumination

If a point source of white light is used for illumination at the same location, as in Fig. 5, a continuum of H1 strip real images is produced as a vertically dispersed spectrum array. As the eyes move vertically, they see the same variety of perspective views in the corresponding spectral band, having a bandwidth determined by the H1 strip height and the eye pupil diameter. Because the dispersion is vertical, there can be a negligible overlap of perspectives from differing viewpoints, so that an unblurred image is seen in a single hue from every viewpoint. The red end of the spectrum is tipped toward the hologram, so that the distance for single-color and distortionless viewing changes with viewer height. When viewed from very far away, the rainbow-like spectrum array can be seen suspended before the hologram. In general, the viewer may move throughout a substantial volume while enjoying a sharp and deep three-dimensional image in varying hues.

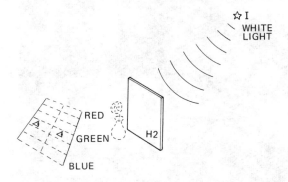

Fig. 5   Illumination of H2 with a white-light point source at I produces a spectral continuum of slit real images, so that the eyes may move up and down and back and forth to view the three-dimensional image.

Fig. 6   Two views of a white-light transmission hologram image, showing variation of horizontal parallax. Vertical viewer motion produces the same views, but in changing hues. Hologram designed by Fritz Goro.

Image Brightness

Because the entire spectral content of the illuminating source is diffracted into a relatively small solid angle for viewing, the hologram images can be very bright. For example, a 3000°K tungsten lamp subtending 10 x 30 minutes (vertical source extent is relatively unimportant) at 45° to a hologram of only 20% diffraction efficiency and a 30° wide viewing zone produces an image-plane "white area" luminance at 550 nm of about 70 cd/m^2 (20 ft-L), bright enough for viewing against most ambient room conditions.

Fig. 7   A white-light hologram displayed in an ordinary laboratory environment, showing the high visibility of the image. The illumination source is a microscope illuminator above and behind the hologram. Hologram designed by S.A. Benton, J.L. Benton, and R. Silberman.

Some Practical Considerations

In the real image projection of Fig. 3, image points at all distances ought to be vertically focused at the plane of H2 to minimize blurring in the final image. Unfortunately this can be true only for points imaged at the plane of H2 (or some other plane if an astigmatic reconstruction wave is used), and out-of-plane points are blurred by an amount proportional to the H1 strip height. However, decreasing the strip height increases the visibility of speckle in the final image by increasing its size and decreasing the spectral waveband over which it is averaged during white-light viewing, and ultimately degrades the image resolution by diffraction. The strip height is therefore chosen to compromise between speckle visibility and depth of field in each case, and generally ranges between one and six millimeters.

Few laboratories are equipped with the large-diameter collimator-type optics necessary for ideal conjugate-wavefront projections, but good results can be obtained by using fairly long radii wavefronts in each case to approximate plane waves, which are self-conjugate. Off-axis holograms made in this way will exhibit considerable astigmatism so that, for example, the distortionless viewing surface for H2 will no longer coincide with the spectral focal plane, and some variation of hue may be apparent over the image. It is the horizontal focusing and magnification relationships that are central to distortion-free imaging, and they are adequately described by "off-axis paraxial" forms such as:

recording: $\dfrac{1}{D_{object}} - \dfrac{1}{D_{reference}} = \dfrac{\lambda\,record}{F_{hologram}}$

projection: $\dfrac{1}{D_{real\,image}} = \dfrac{\lambda\,project}{F_{hologram}} - \dfrac{1}{D_{illumination}}$

$$\text{Magnification} = \dfrac{D_{real\,image}}{D_{object}}$$

Extra-axial effects give rise to a cylindrical space distortion that seems to curl the image backwards.

Thick and Thin Holograms

Volume diffraction effects can markedly enhance the brightness of the hologram, but Bragg-angle selection in even 6 μm thick emulsions noticeably lowers the brightness of the red and blue reconstructions when maximized in the green, and causes brightness variations along the height of a hologram if imperfect conjugates are used. Because very high diffraction efficiencies are often not required, surface-relief phase-modulating structures can be very useful for white-light transmission holograms. Such structures are readily replicated on plastic surfaces by casting, embossing, and so forth, to produce very inexpensive holograms. If the surface is metallized, an image is also easily visible in reflection (3).

Fig. 8   A reflection analog of a white-light transmission hologram. This piece was mass-produced by embossing and metallizing plastic for decorative use. Designed by Selwyn Lissack for Holovision International, Inc.

Holographic Stereograms

The economies of horizontal-parallax-only imaging were quickly appreciated in the synthesis of holographic images from large numbers of conventional photographs, electron micrographs, radiographs, computer generated perspectives, and so forth (4). Such techniques are readily adapted to allow white-light illumination, but image brightness can be kept needlessly low by incoherently overlapping many exposures (5). The recent commercial development of a cylindrical stereogram technique for white-light illurnination avoids exposure overlap to produce bright three-dimensional images that float and move within a cylinder as it turns (6).

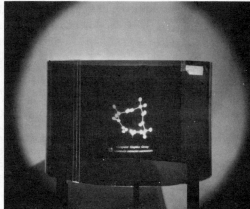

Fig. 9   Two views of a cylindrical holographic stereogram illuminated by an incandescent lamp. The computer generated three-dimensional image of a tetraglycine molecule evolves from a ball-and-stick model to a space-filling model as the stereogram turns. Images created by S. Levine and D.L. Vickers at the Computer Graphics Center, Lawrence Livermore Laboratories.

Conclusions

Within the framework of horizontal-parallax-only autostereoscopic imaging, white-light transmission holography, and its reflection analog, offers high luminance imagery of remarkable depth and clarity, while requiring only simple incandescent light sources for viewing. These features have already fostered a wider acceptance of holograms in museums and galleries, where they are set up and maintained by non-technical personnel. The two-step generating process invokes most of the concepts of simple holographic physics, and its complexity greatly discourages most artists. But as the techniques become more simplified and refined, and more interesting images are designed, holography will continue to define its domain of excellence as an imaging medium.

The author gratefully acknowledges the assistance of H.S. Mingace Jr. in the development of these techniques, and the efforts of H. Casdin-Silver in several fine-arts imaging collaborations. This work has benefited from significant contributions by many groups within the Polaroid Research Laboratories, and especially from many discussions with Dr. E. H. Land.

References:

1. S.A. Benton, "Hologram Reconstructions with Extended Light Sources" **J. Opt. Soc. Amer. 59,** 1545A (1969).

2. R.J. Collier, C.B. Burkhardt, and L.H. Lin, **Optical Holography** (Academic Press, New York, 1971) p. 25

3. A prominent developer of this technology in the U.S.A. is Michael S. Foster, 3530 Big Cottonwood Road, Salt Lake City, Utah 84126, U.S.A.

4. For a bibliography on holographic stereograms see: S.A. Benton, "Holographic Displays-A Review" **Optical Eng'g. 14,** 402-407 (Sept/Oct 1975).

5. N.D. Haig, "Three-Dimensional Holograms by Rotational Multiplexing of Two-Dimensional Films" **Applied Optics 12,** 419L (1973).

6. L. Cross, Multiplex Co. 454 Shotwell Ave., San Francisco, California 94110, U.S.A.

USE OF TRIHEDRAL REFLECTORS IN
HOLOGRAPHY WITH A LOCAL REFERENCE BEAM[+]

Max L. A. Gassend and Wolfgang M. Boerner

Laser Optical Research Laboratory, Electrical Engineering Dept. University of Manitoba
Winnipeg, Manitoba, Canada R3T 2N2

ABSTRACT

Although holographic methods using local optically processed reference beams have been known for many years, no real applications of these methods have so far been disclosed in the literature. The lack of interest for these methods primarily stems
from the fact that the holographic recording system which processes a local reference beam must be accurately aligned with the
illuminating beam from which both reference and object waves are
derived.

It is the aim of this paper to present a new method that considerably enhances the practicality of holography with a local reference beam.

In this method a retroreflector, preferably a trihedral reflector,
provides stability of alignment of the holographic recording
system with the illuminating beam, since it permits location of
the laser source near the element which filters the local reference beam. The position of the retroreflector is not critical,
and it does not affect the phase difference between the two interfering beams.

INTRODUCTION

In essence, a hologram is an interferogram, the fringe pattern of
which is often very complex. It is well known that the visibility of fringes is a function of the degree of coherence of the
interfering waves (Ref. 1). Thus, the brightness of the images
reconstructed from the hologram is a direct function of the
degree of coherence. If both interfering waves (reference and
object waves) are derived from the same beam generated by a laser
source, and if the difference between the optical paths of the
two interfering beams is time-invariant and much smaller than the

---

[+] This work forms part of a dissertation to be submitted by
M.L.A. Gassend in partial fulfillment of the requirements for
the Ph.D. degree.

coherence length of the source, then the degree of coherence is
very high indeed. It is readily seen that if the optical path
difference existing between the two beams is time-varying, the
Doppler frequency shift reduces the initial degree of coherence
and, as a result, the fringe visibility may be reduced to zero.
In order to alleviate changes in optical paths, it is important
to reduce the dimensions of the hologram recording system (the
interferometric arrangement). Methods of reducing the size of
the recording system have been proposed (Ref. 2-6). In these
methods the reference wave is locally processed and derived from
the beam propagating through a transilluminated scene, and is
spatially filtered by means of a pinhole aperature placed in the
focal plane of a Fourier transform lens (by analogy with photographic cameras, a holographic system which generates a local
reference wave may be called a holocamera). These methods do not
require any reference wave that bypasses the illuminated scene.
However, alignment of the optical components of the holocamera
with the illuminating beam is very critical (this is due to the
fact that spatial filtering of the local reference wave neces-
sitates that all spatial frequency components except the central
ordinate be blocked).

It is the object of the present paper to reveal a method in
which the alignment requirement of the holocamera is met by using
a retroreflector. The use of retroreflectors in optical interferometry was suggested for the first time by Peck (7) in a
paper entitled: "A New Principle in Interferometric Design". This
paper was concerned with the description of a modified Michelson
interferometer in which each one of the two mirrors of a conventional Michelson interferometer was replaced either by a two-
mirror or a three-mirror reflector. The new properties of the
scheme were: firstly, to reduce the critical adjustment inherent
in the conventional Michelson interferometer and, secondly, to
control angular motion by interferometric measurement of the cord
length when one of the retroreflector pair is carried by a rotating arm. In 1967, Goodman, et al. (Ref. 8) used a tetrahedral prism reflector in order to obtain a strong reflected
signal from the scene in long distance holography (this strong
signal was used for the reference wave). Finally, very recently
a holographic configuration embodying a three-mirror reflector,
a rotating mirror, a laser and a holocamera generating a local
reference beam was proposed (Ref. 9); this configuration could be
used for measurements of hypervelocity particle fields.

THE CONFIGURATION

The basic beam-folded configuration of the method is shown in
Fig. 1. A collimated beam generated either by a cw or by a pulse
laser is folded back by a trihedral reflector and the three-
dimensional scene scatters the illuminating beam before it reaches
the holocamera. In fact, the illuminating beam passes twice
through the scene, namely: when propagating toward the retroreflector and when propagating toward the holocamera. Two interfering waves are obtained from the light gathered by the

holocamera: (i) the object bearing wave and (ii) the reference
wave which is optically processed. Amplitude/wavefront division methods (Ref. 2-4) as well as the method of spatial
frequency component separation (Ref. 5,6) may be used for processing the reference wave. Mechanical vibrations, air turbulence and thermal drift have negligible influence upon the stability of the interference fringes in the recording plane, since
both the object and the reference waves propagate from the laser
source to the holocamera through a common optical path. It is
seen that the alignment of the holocamera and the optical path
matching between the object and reference waves are not affected
by the locations and/or the angular positions of the retroreflector. With proper design of the holocamera, the coherence
requirements of the laser source may be reduced to those of in-
line holography (Ref. 5).

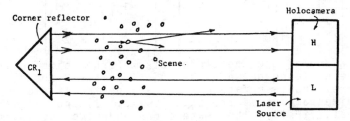

Fig. 1. Setup illustrating the method. The holocamera and the
laser are rigidly assembled, while the retroreflector can be
either assembled on a light structure or remotely located.

THEORETICAL ANALYSIS

Alignment of the Holocamera with the Illuminating Beam

A trihedral reflector is composed of three front-surface mirrors
perpendicular to each other. The law of reflection for each
front surface mirror may be sequentially applied in order to
find the image of an object point. It can be shown (Ref. 9)
that if M is a point of the object space (see Fig. 2) with coordinates x_m, y_m and z_m, the coordinates x_m', y_m' and z_m' of the

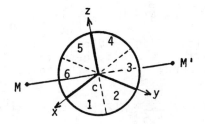

Fig. 2. The axes of the rectangular system of coordinates x,y
and z coincide with the lines of intersection of the three
reflecting surfaces; the backward virtual reflections of these
lines are shown in dotted lines.

image point M' are expressed in terms of the matrix equation:

$$
\begin{bmatrix} x'_m \\ y'_m \\ z'_m \end{bmatrix} = \begin{bmatrix} 1 & 0 & 0 \\ 0 & 1 & 0 \\ 0 & 0 & 1 \end{bmatrix} \begin{bmatrix} -x_m \\ -y_m \\ -z_m \end{bmatrix} , \qquad (1)
$$

where the axes of the rectangular system of coordinates x, y and z coincide with the lines of intersection of the three front-surface mirrors. In addition, due to the linearity properties of the matrix form, an object vector passing through the object points M and N is transformed by the trihedral reflector into a vector having the following direction cosines:

$$
\begin{bmatrix} c_x \\ c_y \\ c_z \end{bmatrix} = \begin{bmatrix} 1 & 0 & 0 \\ 0 & 1 & 0 \\ 0 & 0 & 1 \end{bmatrix} \begin{bmatrix} -g^{-1}(x_n - x_m) \\ -g^{-1}(y_n - y_m) \\ -g^{-1}(z_n - z_m) \end{bmatrix} , \qquad (2)
$$

where x_n, y_n and z_n are the coordinates of the object point N, and g is the distance between the object points M and N. Assuming that the three front-surface mirrors are perfectly flat and that the three dihedral angles are exactly $\pi/2$, (1) and (2) merely indicate that, firstly, the image of an object point is sym-metrically located with respect to the corner point of the trihedral reflector and, secondly, the object and image vectors are antiparallel.

In the absence of the scene, the illuminating beam impinging upon the holocamera aperature is exactly antiparallel to the collimated beam emerging from the laser source; however, according to (1) the retroreflected beam is transversely shifted.

Effect of the Retroreflector Motion on the Fringe Visibility

Let the object and reference waves at a point P in the hologram plane be represented respectively by the functions:

$$
a(P, t) = A(P, t) \exp [i\, 2\pi\, \nu_o\, t] \qquad (3)
$$

and

$$
b(P, t) = B(P, t) \exp [i\, 2\pi\, \nu_o\, t] ,
$$

where ν_o is the optical frequency of the laser source. Fig. 3 schematically shows the recording geometry. If T is the exposure time, the total exposure, $E_t(P)$ at the point P can be written as follows:

$$
E_t(P) = \int_{-T/2}^{T/2} |A(P,t) + B(P,t)|^2 dt , \qquad (4)
$$

and the term of the exposure responsible for the reconstruction of the virtual image is:

$$E_v(P) = \int_{-T/2}^{T/2} A(P,t)\, B^*(P,t)\,dt. \tag{5}$$

According to the analysis given by Goodman (10), the hologram may be regarded as a temporal filter, and (5) can be rewritten in the frequency domain as:

$$E_v(P) = \int_{-\infty}^{\infty} \mathrm{sinc}(\pi T\nu)\, F[A(P,t)\, B^*(P,t)]\,d\nu, \tag{6}$$

where F is the Fourier transform operator. Eq. (6) has been obtained by using the general form of Parseval's formula.

Since both the object and the reference waves are derived from the same illuminating beam, any axial motion of the corner point of the retroreflector induces an identical phase shift into the two interfering waves. Thus, the phasors $A(P,t)$ and $B(P,t)$ can be expressed as follows:

$$A(P,t) = A_0(P)\, \exp i\,[-\phi_0(P) - \phi_i(t)] \tag{7}$$

and

$$B(P,t) = B_0(P)\, \exp i\,[-\phi_r(P) - \phi_i(t)], \tag{8}$$

where $\phi_0(P)$, $\phi_r(P)$ are the phases of the object and reference waves at $t=0$, and $\phi_i(t)$ represents the phase shift created by the retroreflector motion. Using expressions (7) and (8), (6) yields:

$$E_v(P) = T\, A_0(P)\, B_0^*(P) \tag{9}$$

Eq. (9) indicates that the fringe visibility is maximum and, therefore, the brightness of the reconstructed virtual image is unaffected by the motion of the corner point. It is immediately seen that the retroreflector can be held by a very light structure or can even be remotely located, since its relative position and motion with respect to the ensemble laser/holocamera are not critical. In case of rapid axial motion the optical frequency of the illuminating beam is Doppler shifted. As a result, the temporal frequencies of both the reference and the object waves are shifted by a similar amount and, thus, the mutual coherence of the two waves is preserved. However, the Doppler frequency shift affects the effective aperature of the hologram when the axial motion is non-linear, because the frequency shift becomes a time-varying function and the fringe spacing in the recording plane changes during the exposure time T.

Effects of the Scene Motion

Figure 3 gives the recording geometry for a holocamera in which the scene is not imaged by any lens prior to recording. If the hologram records an aerial image of the scene, the object motion is transformed into an image motion by means of the imaging system; in this case, the image motion must be considered in the analysis.

416   M. L. A. Gassend, W. M. Boerner

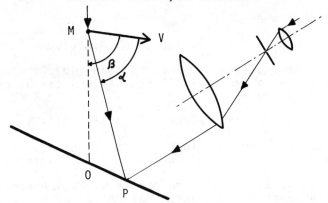

Fig. 3. Recording geometry.

Consider a ray of the reference wave and a ray scattered by a
point M of the scene which interfere in a point P of the Fig. 3
recording plane. By limiting the analysis to two-dimensions,
the directions of the illuminating beam and of the scattered ray
MP with respect to the velocity vector of the point M are simply
determined by the angles α and β (see Fig. 3). If the motion of
the point M is linear with velocity v, the phasors $A(P,t)$ and
$B(P,t)$ can be expressed as follows:

$$A(P,t) = A_0(P) \exp i \ [- (2\pi vt/\lambda)(\cos \beta - \cos \alpha) - \phi_0(P)] \quad (10)$$

and

$$B(P,t) = B_0(P) \exp i [-\phi_r(P)] \quad (11)$$

Using the expressions of $A(P,t)$ and $B(P,t)$, (6) yields:

$$E_v(P) = T B_0(P) A_0(P) \exp i [\phi_r(P) - \phi_0(P)]$$
$$\int_{-\infty}^{\infty} \text{sinc} (\pi T\nu) F[\exp i(2\pi vt/\lambda)(\cos \alpha - \cos \beta)]d\nu. \quad (12)$$

The above result is transformed, using the expression of the
Dirac delta function, to obtain:

$$E_v(P) = \text{sinc}(b)T B_0(P) A_0(P) \cdot \exp i [\phi_r(P) - \phi_0(P)], \quad (13)$$

with

$$b = (\pi Tv/\lambda)(\cos \alpha - \cos \beta). \quad (14)$$

The fringe contrast is maximum for sinc (b) = 1, that is for b=0.
Discarding the trivial value v=0 (no motion), the fringe contrast
is maximum when (cos α - cos β) = 0. A case of interest is when
the point on the hologram is exactly in the projection of the
unscattered ray passing through the point M. Thus for the point
0 the fringe visibility is maximum for any direction of the
velocity vectors. However, as the point of observation in the
recording plane is moved away from the point 0, the fringe
contrast decreases until it reaches the value zero (first zero of

the sinc function). It is readily seen that the effect of scene
motion on the reconstructed image is to reduce the diffraction-
limited image resolution by reducing the effective aperture of
the recorded hologram. As pointed out by Neumann (11), the
spatial modulation of the fringe contrast over the hologram may
be considered as an apodization of the image field recorded by
a hologram when the scene is steady.

Another effect of scene motion on the holographic process is the
influence of linear motion on the sharpness of the reconstructed
image. Neumann (11) showed that the reconstructed image of a
point of the scene is a line of length (v T). Thus, a streaked
image is reconstructed and the loss in image sharpness is the
same as the loss occurring in normal photography with similar
object motion and exposure time.

EXPERIMENTAL INVESTIGATION

The above theoretical study has treated the case of a trihedral
reflector made of three front surface mirrors, as the mathematical analysis is facilitated when the retroreflector is purely
reflective. However, three-mirror reflectors for interferometric work are not readily available and, for this reason,
tetrahedral prism reflectors are used in the experimental investigation. A tetrahedral prism reflector is simply a corner cut
off from a cubical piece of glass, in which the three mutually
perpendicular faces of the cube are the reflecting surfaces of
the prism, and the plane cut surface is the face through which
light enters and emerges from the prism. The optical properties
of such retroreflectors have been extensively studied by Peck
(12,13). A tetrahedral prism reflector has the advantage over
a trihedral reflector in that it can be easily produced to
tolerances suitable for interferometric work. However, the
corner point no longer coincides with the optical center of the
retroreflector (Ref. 12,14); in addition, since there is total
reflection on the reflecting surfaces, the emerging rays have a
state of polarization different from that of the incident rays,
and the polarization phase shift is a function of the angle of
incidence of the rays upon the reflecting surfaces. In order to
maintain a high fringe contrast, it becomes necessary to have a
ray to ray correspondence between the interfering rays in the
hologram plane (i.e., a diffracted ray must interfere with the
ray that illuminates the diffracting point). In practice, if
the collimated beam generated by the laser source is linearly
polarized, all the rays emerging from the same sextant of the
retroreflector (see Fig. 2) have the same state of polarization.
Hence, the interfering rays derived from the same extant will
give maximum fringe contrast.

A general view of the experimental arrangement is shown in Fig.
4. The holocamera used generated a reference beam by amplitude/
wavefront division.

Fig. 4. General view of the experimental arrangement. During
recording of the holograms, the retroreflector was placed at
∿ 12 m from the holocamera. The equipment is identified as
follows: "BC" beam collimator, "CR" retroreflector, "o" scene,
"P" prism, "M" mirror, "SF" spatial filter and "H" hologram
holder.

The property of the retroreflector to maintain rigorous alignment of the holocamera was clearly demonstrated by placing the
retroreflector at approximately 12 m from the holocamera and by
translating and rotating it during the recording of the hologram. The effect of the retroreflector motion on the reconstructed images is shown in Fig. 5(a)-(c): in (a) for a transverse displacement of 10 mm, in (b) for an axial displacement
of 1.5 m and in (c) for a rotation of $\pm$ 15° angle. It is seen
that the retroreflector motion did not affect the quality of
the reconstructed image.

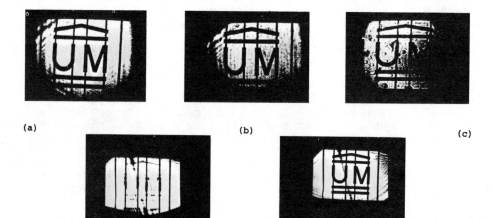

Fig. 5.   Reconstructed images.

The effect of scene motion was also experimentally investigated.
Figure 5(d) and (e) show the reconstructed images when the scene
had a transverse and an axial displacement of 3 mm respectively.
As expected, the reconstructed images are similar to photographic images.

MODIFIED CONFIGURATION

The basic beam-folded configuration of Fig. 1 can be modified as
shown in Fig. 6. Such a configuration embodies two trihedral
reflectors and allows compensation for scene motion when the
motion of the retroreflector placed near the holocamera is synchronized with that of the scene. The coherence requirements
are those of the first method and, again, providing that the
mirrors M_1 and M_2, the laser and the holocamera are rigidly
assembled, the alignment of the holocamera and the optical path
matching of object and reference waves are not affected by the
locations and/or the angular positions of the retroreflectors.

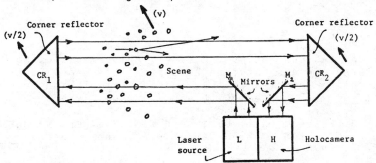

Fig. 6. Modified beam-folded configuration. This configuration
is well suited to holographic imaging of moving scenes. The
scene motion is compensated for by synchronizing the motion of
the retroreflectors with that of the scene.

CONCLUSIONS

It has been shown that the alignment of holocameras (using local
reference beams) with the illuminating beams is no longer
critical when laser sources and holocameras are placed side by
side, the illuminating beam being folded back by means of retroreflectors.

The method presented can be adapted to a great variety of applications, such as measurement of particle fields. The coherence
requirements are very low, since any change in optical path of
the illuminating beam induces similar changes in the reference
and object waves. In addition, scene motion does not necessarily
jeopardize the holographic process because the scene is back
illuminated.

ACKNOWLEDGEMENT

The authors wish to thank Ms. B. Kruse for typing the manuscript.

This work has been supported by the National Research Council of
Canada under Grant No. A-7240 and by the Department of External
Affairs (Award No. 759193 administered by the Canada Council).

REFERENCES

(1)   Born, M. and Wolf, E., _Principles of Optics_, Pergamon Press,
       Oxford, 1976.

(2)   W.T. Cathey, Jr., Local Reference Beam Generation, _U.S._
       Patent No. 3, 415, 387, Dec. 10, 1968.

(3)   H.J. Caulfield, J.L. Harris, H.W. Hemstreet, Jr. and J.G.
       Cobb, Local Reference Beam Generation in Holography, _IEEE_
       Proc. 55, 1758, 1967.

(4)   S.C. Som and R.A. Lessard, Holography with Off-Axis
       Reference Beam Derived from the Object Beam, _Appl. Phys._
       Lett. 24, 600, 1974.

(5)   M.L.A. Gassend and W.M. Boerner, Holocameras Using Local
       Reference Beams Obtained by Spatial Frequency Separation,
       Invention Reported to the _Canadian Patents and Development_
       Limited, CPDL case no. 265-6087-1, July 1975.

(6)   M.L.A. Gassend, K.I. Davis and W.M. Boerner, Holographic
       Recording of Magnified Dark-Field Images, _Opt. Soc. Am._ 66,
       384 A, 1976.

(7)   E.R. Peck, A New Principle in Interferometer Design, _Opt._
       Soc. Am. 38, 66, 1948.

(8)   J.W. Goodman, D.W. Jackson, M. Lehman and J. Knotto, Ex-
       periments in Long-Distance Holographic Imagery, _Appl. Opt._
       8, 1581, 1969.

(9)   M.L.A. Gassend and W.M. Boerner, Hypervelocity Particle
       Measurements by a Holographic Method, _SPIE Proc._ 97, 1976.

(10)   G.W. Goodman, Temporal Filtering Properties of Holograms,
       Appl. Opt. 6, 857, 1967.

(11)   D.B. Neumann, Holography of Moving Scenes, _J. Opt. Soc. Am._
       58, 447, 1968.

(12)   E.R. Peck, Theory of the Corner-Cube Interferometer, J.
       Opt. Soc.. Am. 38, 1015, 1948.

(13)   E.R. Peck, Polarization Properties of Corner Reflectors and
       Cavities, J. Opt. Soc. Am. 52, 253, 1962.

(14)   D.C. Hogg, Optical Centre of a Glass Corner Cube: its
       Measurement, Appl. Opt. 15, 304, 1976.

KANJI CHARACTER GENERATION BY A HOLOGRAPHIC MEMORY SYSTEM

Makoto Kato, Isao Sato, Katsuyuki Fujito and Yoshikazu Nakayama

Central Research Laboratories, Matsushita Electric Industrial Co., Ltd.
Moriguchi , Osaka, 571 Japan

ABSTRACT

A selfconsistent holographic memory system for generating high
quality Kanji ideograms is presented. Attainment of storage
capacity corresponding to the order of 10^7 bits was possible
by the use of a pseudorandom diffuser and Fourier transform
lens systems of high performances. Random access to specified
characters is effected through (i) a page address by means of
driving a X-Y deflector, and (ii) a character address on a
camera tube. Accurate image reconstruction of 8×8 characters
from respective 44 pages were gained on the basis of exact
Fourier transform optics, whereby the degrees of freedom in
designing the total system were considerably increased and the
implementation of the Kanji generator could be facilitated.

INTRODUCTION

So far potential capabilities of the holographic storage of
information as a part of a computer memory system has been
extensively studied, in which digital data up to 10^8 bits could
be stored in the form of the page-oriented memories (Ref 1, 2).
The essential features of the holographic memories are also
attractive in view of the analogue data processings. The high
storage density, high redundancy, and the remarkable simplicity
of the information retrieval optics seem to be the promising
characteristics of the holographic memories. The approaches
to a variety of the practical applications by the holographic
methods, however, have met the difficulty that those fundamental
features of the holograms are not always consistent with each
other (Ref 2). The basic concept of the page-oriented
holographic memories involves a certain limitation that the
same wavelength and the same reference beam should be used for
recording the holograms and reconstructing the information in
order to achieve the abberation free imaging: it is generally
advantageous to record holograms by a comparatively high power
laser of the short wavelength, while in reconstruction a small
source with stability and low cost is preferred especially in

the case of the read-only holographic memories. Another
difficulty in developing the holographic storage system, in
particular, with analogue information is that the storage
density has to be sacrificed so as to reduce the spatial noise
inherent in the coherent imagings (Ref 3 - 7). The recent
development of Fourier holography with pseudorandom diffusers
(Ref 8 - 9) motivated an alternative approach to the page-
oriented holographic memories.

The paper sketches a holographic memory system which consists
of a hologram recorder and a Kanji generator suitable for
storing the Kanji characters (Chinese ideograms) of over 5000
kinds. Information capacity required to store the Kanji
characters generally used in Japanese documents amounts to the
order of 10^7 bits, assuming 50×50 bits are assigned per one
character. Since those characters are detected themselves by
human perception, the styles and forms of the characters, e.g.
"Gothic bold-faced", are greatly valued in the choice of the
Kanji generators. It is the subject of this investigation to
demonstrate that the holographic memories can indeed play a
unique role in the field of processing analogue data of
moderate amount with a considerable access time. The outline
and some notable performances of the proposed system will be
described with experimental results.

PROPOSED SYSTEM AND ITS PERFORMANCES

Hologram Recorder

Figure 1 illustrates a hologram recorder, wherein a Fourier
hologram of a diffuser with a pseudorandom phase code (Ref 8,
9) is imaged by a double diffraction optical system onto a
recording plate almost in contact with an aperture of the size
$D_h \times D_h$. The diffuser is a phase plate composed of an array of
square areas each of which impart a phase shift of one of the
four levels (0, $\pi/2$, π, $3/2 \pi$) to the light incident on it, in
which phase steps of $\pi/2$ or $-\pi/2$ are repeated at random.

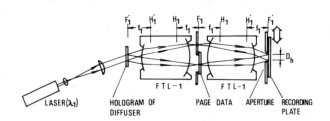

Fig. 1. Hologram recorder

A pair of Fourier transform
lens of the focal length f_1 was
employed in the same manner as
demonstrated before (Ref 8).
The reconstructed image of the
diffuser illuminates page data,
which are composed of N × N
characters and are inserted
sequentially in the back focal
plane F_1 of the left hand lens
FTL-1. Recording of a hologram
matrix is completed after M
times of the exposure and the
parallel shift of the recording
plate. Figure 2 shows an
example of the configuration
when forty four subholograms
were arranged as is the case
in our experiment. It is to

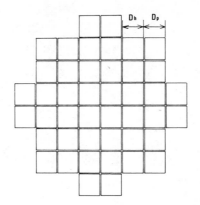

Fig. 2. Hologram matrix.

be noted that the recording system is much more stable to
mechanical vibrations as well as to fluctuations of the source
coherence than conventional interferometric systems (Ref 10).
Hence the practical feasibility of making a large number of
holograms with high reliability.

Kanji Character Generator

Figure 3 shows a block diagram of the Kanji character generator,
which consists of an image reconstruction system and a character
display system with electronic signal processor. Both the
subsystems are optically joined on the face plate of the camera
tube, the performance of which specifies the size and the
numbers of the characters per unit page. The effective size
of the image on the tube was approximately 10 mm × 10 mm since a
1-inch Newvicon* was used in our set-up. The possible number
of the raster scan and the quality of the final output were
considered, whereby the optimum number of the characters was
determined to be 8 × 8 for a unit page.

The image reconstruction system is characterized by a Fourier
transform lens FTL-2 of large apertures at both focal planes
F_2, F_2', which was developed by S. Ishiyama and T. Kojima after
specific discussions on the holographic memories based on exact
Fourier transform (Ref 8). The maximum size of the aperture at
the hologram plane F_2' was safely expanded to 18mm in dia for
the focal length $f_2 = 28$mm, while that at the image plane F_2
could also be as wide as 18mm in dia beyond the required size
of $10\sqrt{2}$mm.
The successful development of the Fourier transform lens of
high performance presents an essential advantage that the
reconstruction system of the holographic memories can be

---

* Trademark of Matsushita Electronic Co., Ltd.

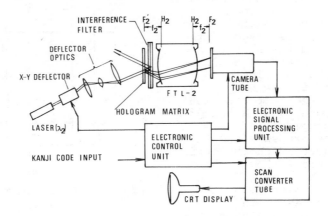

Fig. 3. Kanji Character Generator.

constructed independent of the recording system. The Fourier
transform planes at both systems are simply matched only if
the shift invariant property of the Fourier transform optics
is taken into account.

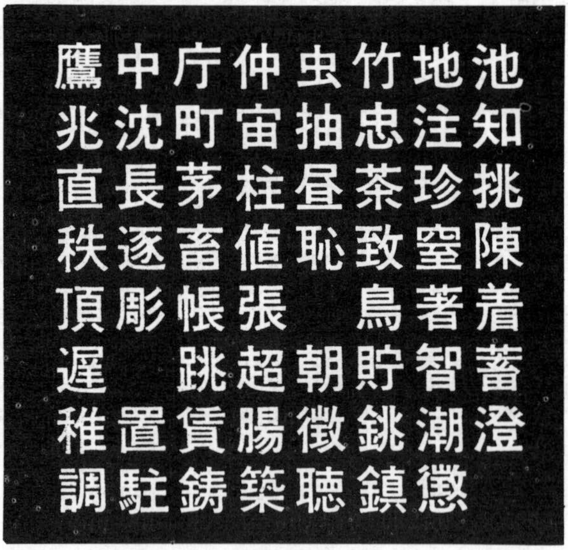

Fig. 4. Reconstruction of a page pattern.

Now such hologram matrix as proposed is accessed by a parallel
beam of the wavelength λ_2 different from that of λ_1 in the
recording stage. From a viewpoint of cost-performance a 2-mW
He-Ne laser and a mirror deflector system were arranged to
effect the optical access as shown in Fig. 3. Several thousands
of the Kanji ideograms including a group of symbols were
associated with given codes. A page code addresses the X-Y
deflecter, by which a specified subhologram out of the 44 pages
is illuminated. Figure 4 is an example of the image
reconstructed from a subhologram, in which the aperture was
$D_h = 1.9mm$, the pitch $D_h = 2mm$, the wavelength $\lambda_1 = 4880\text{Å}$, and the
lens focal length $f_1 = 70mm$; the number of sampling points in
the diffuser were 800×800, which corresponded to 100×100
points per one character. Abberations in the whole images
reconstructed from the hologram matrix were not appreciable,
and the positional errors in the images were less than half the
sampling spacing of the diffuser. Addressing to a character is
then attained by controlling the electronic deflection in the
camera tube. The retrieved characters are stored in a scan
converter tube to compose one frame, and are displayed on a
cathode-ray tube. An example of a CRT display is shown in Fig.
5. The access speed of 164 characters/sec was obtained for
raster scan of 64 lines/character.

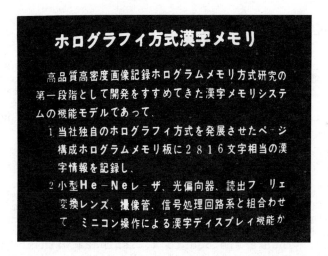

Fig. 5. Example of a CRT display.

One of the serious problems expected before implementing the
system was the effect of the lens flare which would be produced
if an intense zero-order diffracted beam from the hologram
matrix passed through the Fourier transform lens system.   The

trouble was finally solved by introducing an interference
filter in front of the lens*. The corresponding attenuation in
the images was compensated by using an inverse filter in the
stage of making the hologram of the diffuser.

Storage Capacity

The storage capacity obtained in our experimental set-up was
the 44 pages of subholograms in which 2816 characters were
recorded. Apart from some ambiguity in the amount of the
analogue information equivalent capacity of 2.8×10^7 bits could
be stored in a unit hologram matrix since a page data
corresponded to 800×800 sampling points. The total number of
the characters in the proposed system is further increased
naturally by improving the packing density. It is hopeful
enough to reduce the hologram aperture to $D_h = 1.4mm$ and also
the spacing constant to $D_h = 1.5mm$, which gives the capacity of
7000 characters. An interesting feature of the proposed systems
is the possibility to increase the storage capacity simply by
recording as much pages as can be accessed, if the corresponding
dimensions of the image sensors could be enlarged. Additional
change in the specific design is to use the Fourier transform
lens FTL-2 of the same F number but with the increased focal
length. Then storage capacity in excess of 10^8 bits would be
possible in the form of the page oriented holographic memories.

CONCLUSION

A new type of the holographic memory system was proposed on the
basis of Fourier transform holography, wherein intrinsic
advantages of the holographic memories can be well matched to
the conventional arts of the mechanical and the electronic
technologies. As an example of such applications the hologram
recorder and the character generator were developed and
implemented.

ACKNOWLEDGEMENTS

We thank Y. Wakaki of Japan Science Engineering Co., Ltd. who
assisted us to design the hologram recorder with precision
mechanism. The cooperation of Konishiroku Photo Industrial Co.,
Ltd. in the development of Fourier transform lens is
acknowledged. Thanks are due to K. Ishihara with the production
of the phase plates. We also thank H. Matsuda, M. Ohnishi, and
S. Hasegawa for their elaborate construction of the electronic
circuit of the character display system. Helpful support of
Y. Miyazaki in the mechanical design aspect, and finally the
continuous encouragement of Professor Tatsuro Suzuki are
gratefully acknowledged.

---

* After a suggestion by T. Kojima et al.

REFERENCES

(1) F.M. Smits and L.E. Gallenger, Design considerations for a semipermanet optical memory, Bell Syst. Tech. J. 46, 1267 (1967).

(2) Di Chen and J.David Zook, An overview of optical data storage technology, Proc. IEEE, 63, 1207 (1975).

(3) H.J. Gerritsen, W.J. Hannan, and E.G. Ramberg, Elimination of speckle noise in holograms with redundancy, Appl. Opt. 7, 2301 (1968).

(4) E.N. Leith and J. Upatnieks, Imagery with pseudo-randomly diffused coherent illumination, Appl. Opt. 7, 2085 (1968).

(5) D. Gabor, Laser speckle and its elimination, IBM J. Res. Dev. 509 (1970).

(6) Y. Tsunoda and Y. Takeda, High density image storage holograms by sampling and random phase shifter method, J. Appl. Physics, 44, 2422 (1973).

(7) M. Minami and K. Yamada, Artificial diffuser for the holographic picture memory, Procc. Int. Conf. E.O. March 19-21, 1974 (England).

(8) Makoto Kato, Yoshikazu Nakayama, and Tatsuro Suzuki, Speckle reduction in holography with a spatially incoherent source, Appl. Opt. 14, 1093 (1975).

(9) Isao Sato and Makoto Kato, Speckle-noise simulation of Fourier-transform holography with random phase sequences, J. Opt. Soc. Am. 65, 856 (1975).

(10) Makoto Kato, Speckle suppressed holography with spatially incoherent source, J. Opt. Soc. Am. 64, 1507 (1974).

A REAL TIME TWO COLOR INCOHERENT TO COHERENT PLZT IMAGE CONVERTER

N. Bar-Chaim,* A. Seidman* and E. Weiner-Avnear**

*Dept. of Electronics, School of Engineering, Tel-Aviv University, Tel-Aviv, Israel
**Dept. of Physics, Ben-Gurion University, Beer-Sheva, Israel

ABSTRACT

The use of dual stage PLZT color cell for a two color real time incoherent
to coherent image converter is presented. The cell consists of two stages:
a color selector and a display unit, operated in tandem through a photoconductive switch. The operation of both stages is based on the electroopic
effect due to induced effective birefringence in PLZT 9-65/35 ferrolectric
ceramic by an external electric field. Thus, a three state incoherent input:
ON (λ_1)-ON(λ_2)-OFF, is converted to a three state coherent output: ON(λ_3)-ON
(λ_4)-OFF. Contrast ratio of 10^3 was measured for red (6328 Å) and blue
(4765 Å) laser lights and a distinction of 1:500 between the colors was
obtained. Switching time of the ferroelectric material (<10μS) was limited by
the CdS photoconductive switch.

INTRODUCTION

A real time incoherent to coherent image converter (ICC) is a vital stage in
optical data processing systems. In many cases the input of the optical
system is an incoherent one. This requires utilization of an ICC to produce
the Fourier transform of the input at the back focal plane of a lens. A real
time operation is of primary importance for a rapid processing. A schematic
optical processing system is described in Fig. 1. The incoherent input data
is converted to coherent one and further processed.

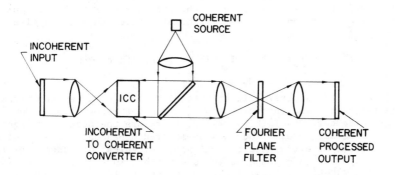

Fig. 1 A schematic optical data processing system.

429

The following techniques were described to obtain incoherent to coherent
conversion: Ruticon (1), Thermoplastic devices (2), Titus (3), Prom (4), and
liquid crystal (hybrid field effect) (5). All these devices can be used only
for monochromatic data processing.

Ferroelectric ceramics exhibit several electrooptic effects which can serve as
a basis for optical devices. The outstanding effects which were investigated
lately are (6): electrically controlled birefringence, electrically
controlled scattering and surface deformation effects. We have utilized the
electrically controlled birefringence due to the transverse quadratic electrooptic effect in slim loop Lanthanum modified Lead Titanate - Lead Zirconate
ceramics (abbreviated PLZT) as a two color, real time ICC. The basic cell
consists of two stages: a color selector and a display unit, each containing
a PLZT plate placed between crossed polorizers. The display unit is gated by
the color selector through a photoconductive switch. In this paper we
describe the structure and the operation of the device which enables a real
time processing of a two color input data. The device is distinguished by
its compactness, low power and relatively low voltages. A matrix configuration of the basic ICC cells leads to an incoherent to coherent image
conversion in two colors. However, the resolution of the matrix, which is
determined by its structure, is limited by the grain size of the crystallites
(several microns).

ICC CELL - PRINCIPLE OF OPERATION

Slim loop PLZT 9-65/35 ceramics exhibit a quadratic dependence of the effective birefringence - $\overline{\Delta n}$, vs. the applied electric field. In the transverse
electrooptic effect, $\overline{\Delta n}$ can be expressed as (7)

$$\overline{\Delta n} = - \frac{1}{2} n_1^3 RE_3^2$$

where n_1 is the index of refraction, measured normally to the applied field -
E_3 and R is the transverse quadratic electrooptic coefficient (of the order
of magnitude of $9 \cdot 10^{-6}$ cm^2/kV^2). At room temperature this material has a
cubic symmetry. By applying an external electric field a ferrolectric
tetragonal phase is induced, causing the material to become birefringent.
When the ceramic plate is placed between two crossed polarizers with its
optic axis oriented at 45° relative to the polarizers axis, the output
intensity - I_{out}, for monochromatic light of wavelength - λ, is given by

$$I_{out}(E,\lambda) = I_s(E,\lambda) + [I_{in}T(\lambda) - I_s(E,\lambda)]\sin^2[\frac{\pi \Gamma(E,\lambda)}{\lambda}]$$

where $I_s(E,\lambda)$ is the scattered light intensity, $T(\lambda)$ - the system transmission
I_{in} - the incident light intensity and $\Gamma(E,\lambda)$ - the retardation. Usually, for
small grain size ceramics $I_s(E,\lambda) \ll I_{in}$, therefore the scattered light
intensity can be neglected. Thus a sine-squared dependence of I_{out} on the
induced effective birefringence is achieved. A typical dependence of I_{out} on
E in PLZT 9-65/35 at room temperature is described in Fig. 2 for red (6328Å)
He-Ne and blue (4765Å) Ar laser lights. The first minimum value of I_{out}

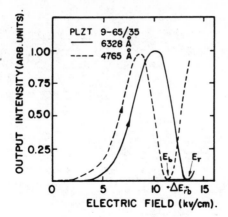

Fig. 2. A typical I_{out}-E dependence in PLZT 9-65/35 for red (6328Å)
and blue (4765Å) laser lights.

is obtained when the electric field magnitude is E_r = 13.1 kV/cm for the red
light and E_b = 10.8 kV/cm for the blue light, providing a working region
ΔE_{rb} = 2.3 kV/cm. The typical contrast ratio for each color is 10^3 and a
good distinction ratio can be achieved through the visible region. Similar
results were obtained for two incoherent filtered light sources ($\Delta\lambda$ = 200Å)
λ_1 = 5780Å and λ_2 = 5090Å, however, the contrast ratio is obviously decreased
(~100).

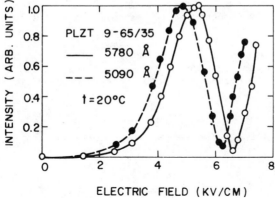

Fig. 3. A typical I_{out}-E dependence in PLZT 9-65/35 for 5780Å and
5090Å incoherent lights.

In a recent paper the operation of a two color memory cell has been described
(8). The ICC cell has a similar structure to that of the memory cell,
however, it consists of two slim loop PLZT in tandem, through a photoconductive switch. This is described in Fig. 4 for transmission and reflection modes of operation.

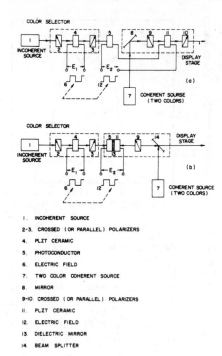

I.   INCOHERENT SOURCE
2-3. CROSSED (OR PARALLEL) POLARIZERS
4.   PLZT CERAMIC
5.   PHOTOCONDUCTOR
6.   ELECTRIC FIELD
7.   TWO COLOR COHERENT SOURCE
8.   MIRROR
9-10. CROSSED (OR PARALLEL) POLARIZERS
II.   PLZT CERAMIC
12.   ELECTRIC FIELD
13.   DIELECTRIC MIRROR
14.   BEAM SPLITTER

Fig. 4. ICC structure: (a) transmission mode (b) reflection mode.

Typical dimensions of the ceramic plates are: thickness – 0.2 mm; electrodes
gap – 0.8 mm. The bias square wave electric field – E(t), applied simultaneously to both stages is described in Fig. 5 for red and blue lights
operation. During the periods T_1 both stages are opened to the red light

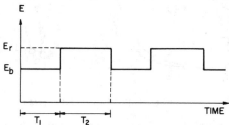

Fig. 5. Electric field – E, applied to the ICC.

and during the periods T_2 – to the blue light. When a red incoherent light
is incident on the color selector it drives the photoconductive switch only
at time intervals T_1 and gates the display unit, enabling the transmission
(reflection) of coherent red light. Similar process holds for the blue light
at the periods T_2. We have used a CdS photoconductor as the coupling element
due to its simultaneous sensitivity in the red and blue lights.   The response
time of the ferroelectric material was measured to be less than 10μS.
However, the response time of the complete cell is severely increased due to
the relatively slow response of the photoconductor.

A color conversion is achieved by varying the applied fields E_r and E_b at the display unit to a different pair of values, corresponding to the transmission of another pair of colors. Thus, $ON(\lambda_1)-ON(\lambda_2)$ incoherent states are converted to $ON(\lambda_3)-ON(\lambda_4)$ coherent states. Since in the absence of electric field the effective birefringence is zero, the OFF state (dark state) is achieved for all visible spectrum.

IMAGE CONVERTER - ICC CELLS MATRIX CONFIGURATION

A matrix configuration of the previously described ICC cells which allow an incoherent to coherent image conversion in two colors is evaluated. The resolution of the matrix, which is determined by its structure, is limited by the grain size of the crystallites - and the electrodes network for operating the matrix reduces the system simplicity. One can overcome this difficulty by utilizing the ceramic in the ferroelectric longitudinal effect with transparent electrodes on photoconductor - ferroelectric sandwich configuration. However, at the present there are two limiting factors for the utilization of the longitudinal effect: the absence of a true birefringent effect and the existence of a preturbing scattering effect (9). The former arises from the cylindrical symmetry axis (∞mm), parallel to the field direction and can be partially overcome by oblique field electrooptic effect. The latter effect (scattering) is basically a drawback for coherent light operation and should be minimized to assure to ICC operation.

In both cases, if the electrodes gap is reduced to 50 microns or less, the average voltage, needed to operate the device is 70 volts and the operating power is very low due to the minimal leakage current.

CONCLUSION

The electrically controlled birefringent effect in slim-loop PLZT ceramics is utilized for incoherent to coherent two color converter in real time. This device offers a three state: $ON(\lambda_1)-ON(\lambda_2)-OFF$ distinction of the input and a color conversion to $ON(\lambda_3)-ON(\lambda_4)-OFF$ output. An image conversion is feasible in the transversal electrooptic mode with a matrix configuration of typical cells. The application of this device in real time optical data processing systems can extend the present monochromatic techniques.

REFERENCES

(1) N.K. Sheridon, The Ruticon family of erasable image recording devices, IEEE Trans. Elect. Dev. ED-19, 1003 (1972).

(2) T.L. Credelle and F.W. Spong, Thermoplastic media for holographic recordings, RCA Rev. 33, 206 (1972).

(3) M. Grenot et.al., New electrooptic light valve device for image storage and processing, Appl. Phys. Lett. 21, 83 (1972).

(4) P.Vohl, P.Nisenson and D.S. Oliver, Real time incoherent-to-coherent optical converter, IEEE Trans. Elect. Dev. ED-20, 1032 (1973).

(5) J. Grinberg et.al., J. Soc. Photo-Opt. Inst. Engn. 14, 217 (1975).

(6) C.E. Land, P.D. Thacher and G.H. Haertling, Electrooptic ceramics, Appl. Sol. State Sci., ed. by R. Wolfe, 4, 137 (1974).

(7) G.H. Haertling and C.E. Land, Hot-presses $(Pb,La)(Zr,Ti)O_3$ ferroelectric ceramics for electrooptic applications, J. Am. Ceram. Soc. 54, 1 (1971).

(8) N. Bar-Chaim, A. Seidman and E. Wiener-Anvear, A color memory mode based on the variable birefringence in PLZT ferroelectric ceramics, Ferroelectrics 11, 385 (1976).

(9) G.H. Haertling and C.B. McCampbell, A new longitudinal display mode for ceramic electrooptic devices, Proc. IEEE 60, 450 (1972).

SESSION 12.

DEVICES & TECHNIQUES

Chairman: A. LOHMANN

NETWORK METHODS FOR
INTEGRATED OPTICS DEVICES

Theodor Tamir and Song-Tsuen Peng

Dept. of Electrical Engineering and Electrophysics
Polytechnic Institute of New York, Brooklyn, New York 11201

ABSTRACT

Network methods using equivalent electrical circuits are shown to provide a
systematic approach which facilitates the analysis and design of integrated-
optics devices. This approach is illustrated by examples which include dielectric gratings for beam couplers and other applications.

INTRODUCTION

The similarities between guided-wave components in the two areas of integrated
optics and microwave engineering have been well recognized. However, while
equivalent networks involving lumped and distributed elements have served as a
powerful tool in treating a wide range of microwave problems, the use of network methods in integrated optics has been so far very limited in scope. The
aim of this paper is to show the effectiveness of equivalent networks in integrated optics by presenting their novel application to dielectric gratings,
which play a dominant role in beam couplers, filters, distributed-feedback
lasers and other devices that incorporate periodic structures.

THE NETWORK APPROACH

Network terminology has already been employed in the early microphotolitho-
graphic work at infra-red (Ref 1), while equivalent networks have been subsequently used mostly in transverse-resonance analysis of dispersion curves for
thin-film waveguides of the strip (Ref 2) or planar varieties (Ref 3). In
that context, the usual interpretation of propagation along optical waveguides
is in terms of a plane wave following a zig-zag path which is produced by
total reflections at the two boundaries. As shown in Fig. 1(a), a waveguide
mode is then supported by the thin-film waveguide if the phase-shift $k_z^{(f)}$ inside the film satisfies a consistency condition in terms of the reflection
coefficients r_0 and r_f at $z = 0$ and $z = -t_f$, respectively.

In the network approach, the configuration of Fig. 1(a) is viewed transversely along z and each one of its regions is described by a transmission line
of appropriate length, as shown in Fig. 1(b). These transmission lines are
associated with characteristic admittances

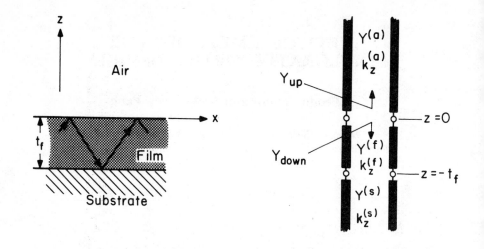

$$| - r_0 \exp(ik_z^{(f)} t_f) r_f \exp(ik_z^{(f)} t_f) = 0 \qquad\qquad Y_{up} + Y_{down} = 0$$

$$(a) \qquad\qquad\qquad\qquad\qquad (b)$$

Fig. 1. Surface-wave modes in optical waveguide:
(a) zig-zag plane wave description;
(b) equivalent transverse network representation.

$$Y^{(u)} = \frac{1}{Z^{(u)}} = \left\{ \begin{array}{ll} k_z^{(u)}/\omega\mu_o, & \text{for TE modes,} \\ \omega\varepsilon_o\varepsilon_u/k_z^{(u)}, & \text{for TM modes,} \end{array} \right. \qquad (1)$$

and propagation factors

$$k_z^{(u)} = (k_o^2 \varepsilon_u - \beta^2)^{\frac{1}{2}}, \qquad (2)$$

where $k_o = \omega(\mu_o\varepsilon_o)^{\frac{1}{2}}$, β is the modal propagation factor along the longitudinal x direction, ε_u is the dielectric constant in medium u and the index $u = a, f$ or s denotes the air, film or substrate regions, respectively.

In this network approach, the modes are found by employing the transverse-resonance condition

$$Y_{up} + Y_{down} = 0, \qquad (3)$$

where Y_{up} and Y_{down} are input admittances looking up and down, respectively, at any one cross-section of the transmission-line network. In Fig. 1(b), condition (3) is applied at $z = 0$. It may easily be shown that the

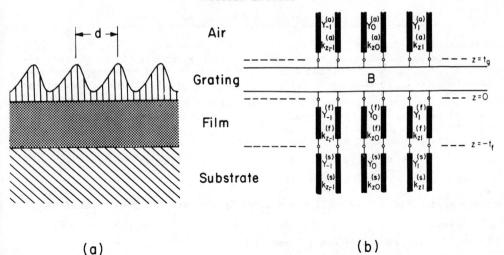

Fig. 2.   Network approach to dielectric gratings:
          (a) actual grating configuration;
          (b) equivalent transverse network representation.

transverse-resonance relation is identical to the consistency condition dictated by the bouncing plane wave in Fig. 1(a).

While it thus leads to the same result as that given by the zig-zag plane wave, the network approach provides a systematic procedure for deriving mode solutions in situations that are more complicated than that described in Fig. 1. In integrated optics, such a situation occurs in the case of dielectric gratings of the type shown in Fig. 2(a), which can be described (Ref 4) in terms of the equivalent transverse network displayed in Fig. 2(b).

For the grating configuration, the various media in the regions $z < 0$ and $z > t_g$ are represented by a multiplicity of transmission lines, each one of which identifies a particular diffraction order via a subscript n. Thus $Y_n^{(u)}$ and $k_{zn}^{(u)}$ are analogous to the quantities given in Eqs. (1) and (2), provided that now β is changed to

$$\beta_n = \beta_0 + (2n\pi/d), \quad \text{with } n = 0, \pm 1, \pm 2 \ldots, \tag{4}$$

where d is the period and β_0 is a fundamental phase factor. The grating region ($0 < z < t_g$) is represented by a lossless network, denoted by B in Fig. 2(b), which couples all of the grating orders. Whereas the complete network is rather complicated if a large number of grating orders must be accounted for, it nevertheless leads to systematic computational procedures for structures with a larger number of layers and/or for gratings having complex profiles.

SIMPLIFIED NETWORK REPRESENTATION OF THE GRATING

A great simplification in the equivalent network of Fig. 2 can be achieved by adopting a perturbation approach (Ref 5), which assumes that the grating region appears as a modification of a uniform layer having an average refractive index ϵ_g. In that case, the complicated coupling network B disappears and, instead, each one of the diffraction orders reduces to a simple and uncoupled (independent) transmission-line circuit, as shown in Fig. 3. The effect of coupling due to the network B is now replaced by equivalent voltage $v_n(z)$ and current $j_n(z)$ sources, which are distributed inside the grating region $(0 < z < t_g)$ only. These sources are prescribed by the field incident on the grating, as discussed below.

The periodic grating layer $(0 < z < t_g)$ can be described by a relative permittivity

$$\epsilon(x,z) = \epsilon_g + \sum_n \epsilon_n(z) \exp(i\frac{2n\pi}{d} x), \tag{5}$$

where ϵ_g is the average value inside the grating layer. If all ϵ_n are zero in Eq. (5), the configuration of Fig. 3 reduces to a multi-layered structure of the type shown in Fig. 1(a). Let now a TE or TM field of the form exp $(i\beta_0 x - i\omega t)$ be incident on this structure. This will produce a scattered field which can be found by relatively simple considerations, so that the total electromagnetic field solution in the grating region can be phrased in terms of the following transverse components

TE fields	TM fields	
$H_x = -I_g(z) \exp(i\beta_0 x)$	$E_x = V_g(z) \exp(i\beta_0 x)$,	(6)
$E_y = V_g(z) \exp(i\beta_0 x)$	$H_y = I_g(z) \exp(i\beta_0 x)$,	(7)

where the time dependence $\exp(-i\omega t)$ has been suppressed.

The sources (generators) in the grating region in Fig. 3 are then given by

$$v_n(z) = \begin{cases} 0 \text{ , for TE modes } , \\ -i(\beta_n\beta_0/\omega\epsilon_o\epsilon_g^2) \ \epsilon_n(z) \ I_g(z) \text{ , for TM modes } , \end{cases} \tag{8}$$

$$j_n(z) = i\omega\epsilon_o \ \epsilon_n(z) \ V_g(z) \text{ , for both TE and TM modes.} \tag{9}$$

Recalling that $V_g(z)$ and $I_g(z)$ are known solutions to specific incident fields, the sources v_n and j_n are well specified and they reduce the complex grating problem to a simpler problem in terms of source-excited transmission lines. The incident field may be a plane wave impinging obliquely on the grating (for scattering phenomena), a surface wave travelling longitudinally (for beam-coupling or filtering applications), a standing wave along the grating (for distributed-feedback lasers), or any other suitable field form. Each one of these situations is distinguished from the other only by the fact

that the voltage and current sources are different.
However, these sources are known in every case, and
they are simply prescribed by the given incident
field via Eqs. (8) and (9). Because of their relative simplicity, equivalent networks such as that
in Fig. 3 are extremely helpful in providing both
physical insight and accurate quantitative evalua-
tions for practical design problems. Two specific
applications are discussed below.

APPLICATION TO BEAM COUPLERS

An important application of dielectric gratings is
the grating coupler, which converts an incident
surface-wave mode into outgoing beams or performs
the reciprocal operation, i. e., it converts an incident beam into a surface wave. The former function is shown in Fig. 4, where the geometry conforms to the situations shown in Fig. 1(a) and 2(a).
The surface-wave-to-beam conversion in this case is
based on the fact that the surface wave, which
varies as $\exp(i\beta_{sw}x)$, is scattered by the grating

and transformed into a leaky wave, which varies as
$\exp(i\beta_0 x - \alpha x)$, where usually $\beta_0 \simeq \beta_{sw}$. Here α

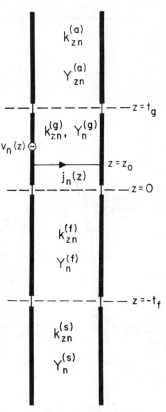

Fig. 3. Simplified
network
description
for the n-th
diffraction
order on a
dielectric
grating.

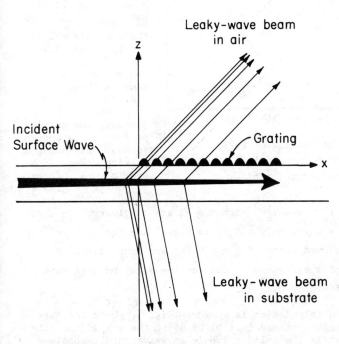

Fig. 4. Conversion of a surface wave
into leaky-wave beams by a
grating coupler.

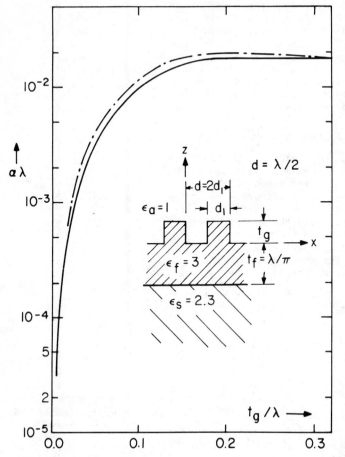

denotes the attenua-
tion due to energy
that radiates out
from the grating re-
gion in the form of
leaky-wave beams.

The leakage factor α
is the single most
important parameter
for designing effi-
cient beam couplers
(Ref 6). This factor
is given by

$$\alpha = p_{rad}/2\,P_{inc}, \quad (10)$$

where p_{rad} is the
power radiated per
unit length along the
grating, whereas P_{inc}
is the total power of
the incident surface
wave. The latter
quantity is known,
and p_{rad} is given by
the sum over all the
energy radiated into
the air and substrate
regions by the
sources shown in
Fig. 3.

Fig. 5. Variation of leakage $\alpha\lambda$ versus
the grating thickness t_g/λ for a
typical dielectric grating.

Usually, the grating
period d is chosen
so that only one
grating order
(for $n = -1$) propa-
gates, i.e., $Y_{-1}^{(s)}$ and
$Y_{-1}^{(a)}$ are real whereas all $Y_n^{(s)}$ and $Y_n^{(a)}$ with $n \neq -1$ are imaginary. In that
case, p_{rad} is calculated by using only $n = -1$ in Fig. 3 and Eqs. (8) and (9).
For that purpose, it is assumed that $I_g(z)$ and $V_g(z)$ are proportional to the
surface-wave variation $\exp(-a_g z)$, where a_g denotes the decay of that wave
away from the film waveguide.

A typical result for such a calculation is shown in Fig. 5, where the dot-
dashed curve shows the result obtained (Ref 5) by using the simplified network approach of Fig. 3, while the solid curve is an exact result obtained by
a rigorous method (Ref 4). The high accuracy of the simpler technique is
evident. We also note that, as a function of the grating height t_g, α

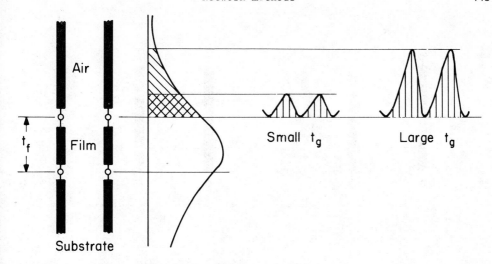

Fig. 6.   Explanation of the behavior of $\alpha\lambda$ in Fig. 5
by means of network and field considerations.

increases first as t_g^2 but then reaches a saturation value. This behavior
can also be understood in terms of the network shown in Fig. 3, as described
below.

From Eq. (10), we note that α is proportional to p_{rad}, which is itself proportional to the power delivered by the voltage and current sources $v_{-1}(z)$
and $j_{-1}(z)$. However, the amplitudes of these sources are proportional to the
magnitudes of $V_g(z)$ and $I_g(z)$, both of which vary as the amplitude of the
incident surface wave. The profile of this wave is shown to the right of the
network in Fig. 6, and we note that this profile conforms to the exponential
variation $\exp(-a_g z)$ that occurs in the grating region. As α is given by the
totality of sources in that region, it follows that α is proportional to the
square of the area under the profile within the grating region, i.e., the area
between $z = 0$ and $z = t_g$. Thus, as t_g increases, this area increases linearly at first but, as t_g becomes large, the area tends to approach a limit
value, as suggested by the shaded areas in Fig. 6. Hence α increases as t_g^2
for small t_g and approaches a constant value for large t_g, as described in
Fig. 5.

The above illustration shows that the network approach provides a simple procedure for calculating important parameters and also brings a considerable
amount of physical insight into the functional variation of such parameters.
Another example of the insight achieved by these network techniques is discussed below.

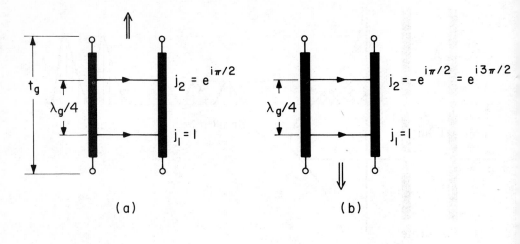

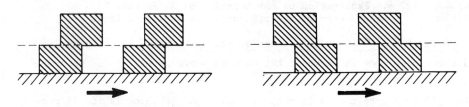

Fig. 7. Directional effects produced by special
         grating profiles and their network interpretation.

SURFACE-WAVE BLAZING

Consider two sources j_1 and j_2 spaced $\lambda_g/4$ apart on a transmission line as
shown in Fig. 7, where λ_g is the wavelength along the line. If these sources
are in phase quadrature and if reflections are neglected at the ends of the
transmission lines, all of the power will be radiated into the upper or lower
regions at the ends of the transmission line, as shown by the vertical arrows
in Fig. 7(a) or 7(b).

The above directional effect on the radiated power can be realized on a dielectric grating by regarding j_1 and j_2 as equivalent sources for generators
which are distributed over two distinct regions inside a dielectric grating
layer. Such a situation could be achieved by the grating profiles shown at
the bottom of Fig. 7. We note that j_1 and j_2 then represent the effects due
to two separate rectangular regions. The appropriate $\pi/2$ phase shift is then
obtained by placing these rectangular shapes in proper position with respect
to the incident surface wave which propagates longitudinally as shown by the
horizontal arrows.

The scheme described in Fig. 7 suggests that, by proper shaping of the grating profile, it is possible to eliminate one of the two outgoing beams in the

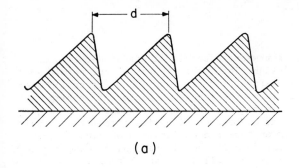

(a)

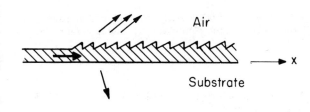

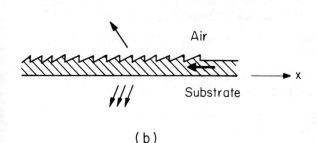

(b)

Fig. 8.   Blazing of surface waves:
(a) Asymmetric grating profile
for surface-wave blazing;

(b) A properly shaped grating
leaks a surface wave incident
from the left mostly into the
air region, whereas a surface
wave incident from the right
is leaked mostly into the
substrate.

coupler shown in Fig. 4.
This amounts to a "blaz-
ing" effect produced by a
properly shaped grating on
an incident surface wave;
the dielectric grating
thus yields a directional
behavior analogous to that
produced by a blazed me-
tallic grating which can
selectively suppress an
unwanted diffraction order
due to an incident plane
wave. In the beam coupler
of Fig. 4, in particular,
the elimination of one of
the two beams is important
because the coupling
efficiency can then be
substantially increased.

Obviously, the double rectangular profile suggested
in Fig. 7 does not lend
itself to practical implementation. However, it
has been shown by a rigorous calculation (Ref 7)
that the ideas provided by
the simple network considerations discussed above
are in perfect agreement
with the behavior expected
from gratings having more
realistic profiles. As an
example, Fig. 8 shows how
to achieve the desired directional blazing of sur-
face waves by means of a
grating having a simple
asymmetric triangular profile. We note that this
grating directs most of
the incoming surface-wave
energy into either the air
or the substrate regions,
depending on whether incidence is from the left or
right, respectively.

It is pertinent to note
that, while the above blazing effects due to gratings have been verified by
a recent experimental study (Ref 8), the design and systematic fabrication
of dielectric gratings for beam couplers or for other applications is still
an open research area. The considerations presented here indicate that network methods should play a major role in all of the design and implementation

stages of such a research effort.

CONCLUSIONS

To summarize, equivalent-network representations for integrated-optics devices provide a large amount of insight into their operation and permit a direct evaluation of performance characteristics. Thus, the dispersion properties of thin-film optical waveguides, the leakage behavior of grating couplers and the directional discrimination achieved by asymmetric triangular gratings can all be phrased in terms of transmission-line circuits and their associated incident fields. By extension, the application of such network methods to a wider class of integrated-optics configurations is expected to provide a powerful analytic technique, as well as a useful tool in the design of optical devices using guided waves.

ACKNOWLEDGMENT

This work was supported by the Office of Naval Research, under Contract No. N00014-75-C-0421, and by the National Science Foundation, under Grant No. ENG74-23908.

REFERENCES

1. D. B. Anderson, Applications of microphotolithography to millimeter and infrared devices, Proc. IEEE, 54, 657 (1966).

2. E. A. J. Marcatili, Dielectric rectangular waveguide and directional coupler for integrated optics, Bell System Tech. J., 48, 2071 (1969).

3. T. Tamir, Inhomogeneous wave types at planar interfaces: II - Surface waves, Optik, 37, 204 (1973).

4. S. T. Peng, T. Tamir and H. L. Bertoni, Theory of periodic dielectric waveguides, IEEE Trans. Microwave Theory Tech., MTT-23, 123 (1975).

5. S. T. Peng and T. Tamir, TM-mode perturbation analysis of dielectric gratings, Appl. Physics, 6, 35 (1975).

6. T. Tamir, Integrated Optics, Springer-Verlag, New York, 1975, Sec. 3, p. 84.

7. S. T. Peng and T. Tamir, Directional blazing of waves guided by asymmetrical dielectric gratings, Optics Commun., 11, 405 (1974).

8. T. Aoyagi, Y. Aoyagi and S. Namba, High efficient blazed grating couplers, J. Opt. Soc. Am., 66, 292 (1976).

OPTICAL COMMUNICATIONS USING SURFACE ACOUSTIC WAVES*

P. Das, D. Schumer and H. Estrada-Vazquez

*Electrical and Systems Engineering Dept., Rensselaer Polytechnic Institute,
Troy, New York 12181 U.S.A.*

ABSTRACT

After a brief review of signal to noise considerations in a general communications link, including matched filtering and correlation techniques, an optical communication system is described which employs a surface wave acousto-optic modulator at the transmitter end and a surface wave acousto-optic correlator at the receiver end. Some experimental results are reported for the transmission and reception of Barker encoded data. Lithium niobate surface wave acousto-optic modulators operated at 100 MHz center frequency and an argon ion laser ($5145°$A) were used in the system.

INTRODUCTION

Since the discovery of the laser, research efforts in the areas of coherent optical processing and optical lightwave communications have been considerable. The application of linear system theory and communication theory, which previously had proven so fruitful in the domain of electrical signal processing, to optical spatial filtering and holography, must be counted among the most impressive successes of this research (1). In the optical communications area, recent success in fabricating low loss fiber-optic-waveguides, the demonstration of high performance electro-optic and acousto-optic devices, and the availability of inexpensive and efficient photodiodes for light detection give promise that a practical optical communications system is a realistic goal (2). Whether such a system will be competitive with the existing technology remains to be seen.

THEORY

In its simplest form, a communication system is characterized by a transmitter end where information is encoded and imposed upon some carrier, a transmission medium, and a receiver end where the information is abstracted from the received signal. A major consideration in all real communication systems is the signal to noise ratio at the output of the receiver. From a practical point of view, signal/noise considerations determine the number of repeaters required along the transmission path, and thus have a direct bearing on the cost of the system.

It is well known (3,4) that the signal to noise ratio with respect to power at the output of a receiver is given by

$$\rho(t) = \frac{1/4 \pi^2 \left| \int_{-\infty}^{\infty} S(\omega) H(\omega) e^{j\omega t} d\omega \right|^2}{N/2\pi \int_{-\infty}^{\infty} \left| H(\omega) \right|^2 d\omega} \tag{1}$$

where ω represents angular frequency, t is time, $S(\omega)$ is the frequency spectrum of the received signal, $H(\omega)$ is the frequency characteristic of the receiver which is assumed linear and time invariant, and N is the noise spectrum power density, which for simplicity has been assumed flat. For a given signal, this ratio can be maximized by taking the receiver frequency characteristic to be of the form

$$H(\omega) = \alpha S^*(\omega) e^{-j\omega \tau} \tag{2}$$

where $S^*(\omega)$ is the complex conjugate of the signal spectrum, and α and τ are arbitrary real constants. This approach to matched filtering has of course, found widespread use in electrical and optical processing. The output of such a matched receiver to a signal of frequency spectrum $S(\omega)$ is given by

$$V_o = 1/2\pi \int_{-\infty}^{\infty} S(\omega) S^*(\omega) e^{-j\omega t} d\omega \tag{3}$$

where for simplicity, we have taken, $\alpha = 1$, $\tau = 0$. It is easily shown that this output can be equivalently written as an autocorrelation function of the signal $s(t)$ itself, where

$$s(t) = 1/2\pi \int_{-\infty}^{\infty} S(\omega) e^{j\omega t} d\omega \tag{4a}$$

$$S(\omega) = \int_{-\infty}^{\infty} s(t) e^{-j\omega t} dt \tag{4b}$$

That is

$$V_o = \int_{-\infty}^{\infty} s(t') s(t' - t) dt' \tag{5}$$

For the receiver output in a real communication system $s(t')$ in Eqn. (5) would be replaced by $s(t') + n(t')$, where $n(t')$ represents the noise. It is apparent from Eqns. (3) and (5) that best reception requires that either the receiver be matched in its frequency response to the incoming signal, or that the receiver be capable of signal correlation. The former method requires that the receiver have a different frequency response for each signal anticipated; the latter method requires a means for real-time correlation. These mathematically equivalent methods have very different realizations, as depicted in Fig. 1.

Regardless of which method of reception is implemented, the question remains as to how the information should be encoded and modulated. For an optical communication system carrying serial information on fibers, a practical scheme would probably involve the digital transfer of information, employing possibly two states, such as zero and one. To provide a high immunity to interference, associated with each state a particular encoded digital sequence would be transmitted. One code which has found widespread use, especially in multiple-access time slot systems, is the Barker code. This code is especially attractive since the autocorrelation of an N-bit Barker code has a correlation peak which is N times the magnitude of the side lobes. Barker sequences $\{q\}$, defined through the relation

(a) Matched Filtering

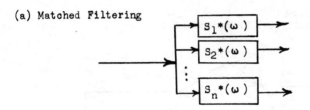

(b) Correlation

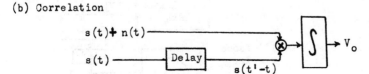

Fig. 1. Signal/Noise Enhancement Techniques

$$\sum_{i=0}^{N-k-1} q_i \, q_{i+k} = \begin{array}{l} 0, \text{ N-k even with } k \neq 0 \\ \pm 1, \text{ N-k odd,} \end{array} \qquad (6)$$

have been found empirically for N = 2, 3, 4, 5, 7, 11, 13. The individual elements may take on values +1 or -1. For N = 13, no Barker codes exist (3,4). It is to be mentioned that other important codes do exist and have been of use. Of special interest are the pseudo-random noise code and linear fm chirp.

It was the realization depicted in Fig. 1(b) which aroused the initial interest of our group in optical communications. Real-time signal processing including Fourier transformation, convolution, ambiguity function generation, and correlation had been demonstrated using the acousto-optic interaction by us and others (5). Most experiments involved the general configuration shown in Fig. 2.

Signals $s_1(t)$ and $s_2(t)$ are used to amplitude modulate surface acoustic waves at frequency ω_a which are launched from opposite ends of a delay line fabricated on $LiNbO_3$. Associated with the propagating acoustic wave is a periodic variation in the crystal refractive index which serves as a phase grating to diffract incident laser light into several spatially separated diffraction orders. The grating modulation depth depends upon the amplitude of the surface acoustic wave. For time-limited signals $s_1(t)$ and $s_2(t)$, a photodiode positioned to collect the light from one of the diffraction orders will contain in its output a signal at twice the acoustic frequency, $2\omega_a$, which can be isolated by heterodyning and has an envelope given by

$$s_1 * s_2 = \int_{-\infty}^{\infty} s_1(t') \, s_2(t - t') \, dt' \qquad (7)$$

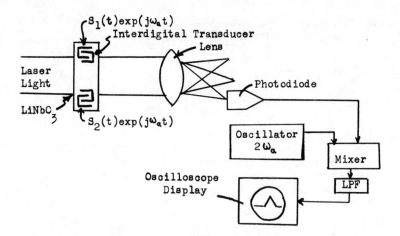

Fig. 2. Acousto-Optic Convolver

This is just the convolution of the two signals. The multiplication and integration of the two functions is provided by the photodiode, while the inversion and shift of one with respect to the other is provided by the propagation of the acoustic signals from opposite ends of the delay line. By time-inverting one of the signals, the convolver can be used to obtain correlation.

To use such a convolver/correlator in an optical communication system it is necessary to receive the transmitted data at the receiver, convert this to an electrical signal, use this signal as one input to the acousto-optic correlator and supply the desired correlation code to the other input.

Far more desirable would be to use the transmitted light directly as one of the correlator inputs. That this is possible is suggested by the interaction configuration shown in Fig. 3 and reported earlier (6).

The acoustic tracks are nonoverlapping. The incident light is modulated first by one acoustic wave, then by the second after a short optical propagation path. The convolution of the two signals is contained in the photodiode output, as was the case shown in Fig. 2. A natural extension of the parallel track configuration would be to separate the two tracks completely, using one for light modulation at the transmitter end of a communication system, and using the second acoustic track for correlation of the received light signal at the receiver end (7). It is a requirement of this particular scheme that the transmission medium preserve the phase front of the transmitted light.

Such a system is applicable where the optical transmission line can be modeled as a series of lenses, as indicated in Fig. 4(a). Under this assumption, Fig. 4(b) depicts the communication system implementation.

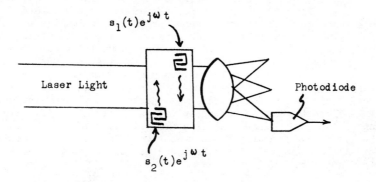

$$s_1(t)e^{j\omega t}$$

Laser Light

Photodiode

$$s_2(t)e^{j\omega t}$$

Fig. 3.   Parallel Track Configuration

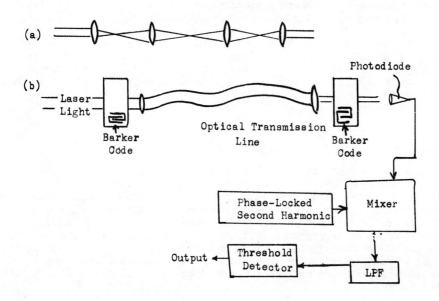

(a)

(b)

Laser Light

Barker Code

Optical Transmission Line

Barker Code

Photodiode

Phase-Locked Second Harmonic

Mixer

Output

Threshold Detector

LPF

Fig. 4.   (a) Lens Model of Optical Waveguide
          (b) Optical Communication System Implementation

EXPERIMENT

The scheme depicted in Fig. 2 was employed for obtaining the autoconvolution of Barker codes. YZ $LiNbO_3$ was used for the acousto-optic modulators using transducers tuned to $50\,\Omega^3$ at the center frequency, which was 100 MHz. Green light (5145 Å) from an argon laser was use, and detection was by means of a photodiode. The Barker code was impressed upon the acoustic signal by introducing a phase shift of π in the electrical cw signal used to launch the acoustic wave. In this way, +1 was distinguished from -1. Figures 5(a-h) show the input signal and detected convolution for a single rf pulse, a 3 bit Barker code, 7 bit Barker code, and 13 bit Barker code. The disproportionate side-lobe level in the last case was due to an electronic difficulty in the generation of the code. Figures 5(i-k) demonstrate time compression capability of the system. A linearly fm chirped signal introduced at one input of the convolver was convolved with a down chirped signal to provide a time compression factor of 20.

The arrangement shown in Fig. 6 was used to approximate an actual communication system. A uniform laser beam was brought to a line focus in the transmitter acousto-optic modulator. Lens L1, placed intermediately between the two $LiNbO_3$ acousto-optic modulators and twice its focal length from both produced a real inverted image with unit magnification of the transmitter acousto-optic modulator in the plane of the receiver acousto-optic correlator. L2 then provided the Fourier transformation of this spatial light distribution producing distinct spots in its focal plane corresponding to the various diffraction orders. A photodiode collected light in the first order diffracted light, and electronic processing of the photodiode signal yielded the oscilloscope display of the desired convolution.

An alternative arrangement is shown in Fig. 7. In this configuration only the first order diffracted light from the first acousto-optic modulator is permitted to interact with the second acousto-optic modulator. The photodiode output in this case provides the desired convolution without further electronic processing. Shown in Fig. 8 are experimental results using the latter arrangement.

CONCLUSIONS

Correlation techniques in an optical communication link were discussed and experimental results given. The system is capable of handling Barker encoded serial data. While light wave communications of the near future will probably employ 0's and 1's, sent as synchronized light pulses, demands for higher bit rates and longer repeater distances may ultimately require the transmission of Barker encoded data. A system such as that discussed will prove useful at that time. Finally, bulk acoustic waves could have been used for the acousto-optic devices. Surface wave devices, however, such as those reported here, are more efficient and will be more compatible with integrated optics systems.

ACKNOWLEDGMENTS

The authors wish to acknowledge the technical assistance and helpful suggestions of Messrs. Colin Lanzl, F. M. Mohammed Ayub, and Richard T. Webster. Discussions with Prof. L. Milstein concerning aspects of communication theory have been most helpful and are greatly appreciated.

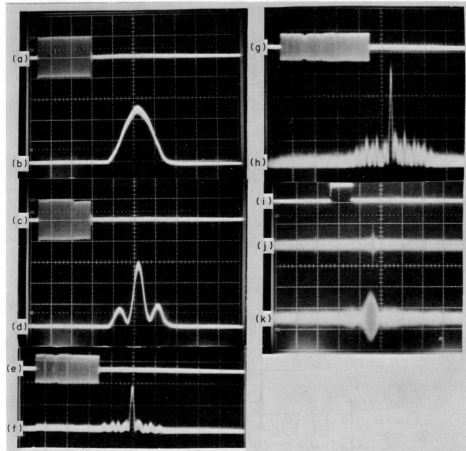

Fig. 5 Convolution of 100 MHz signals. (a) Input rf pulse and (b) output autoconvolution (1 μsec/div). (c) Input 3 bit Barker code and (d) Output autoconvolution (1 μsec/div). (e) Input 7 bit Barker code and (f) Output autoconvolution (2 μsec/div). (g) Input 13 bit Barker code and (h) Output autoconvolution (2 μsec/div). (i) Envelope of linear fm chirp input and (j) time-compressed output (2 μsec/div). (k) Expanded scale of time-compressed output (0. 1 μsec/div).

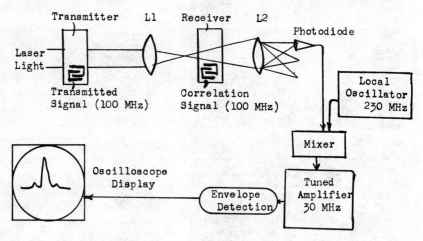

Fig. 6. Experimental Arrangement

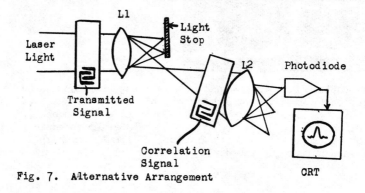

Fig. 7. Alternative Arrangement

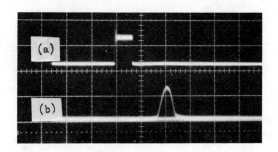

Fig. 8 Convolution using separated LiNbO$_3$ samples. (a) Envelope of 100 MHz
rf input pulse. (b) Autoconvolution output of photodiode (2 μsec/div).

REFERENCES

1. A. Vander Lugt, Proc. IEEE, 62, 1300 (1974).

2. S. J. Bucksbaum, Physics Today, 29, 23 (1976).

3. H. L. Van Trees, Detection, Estimation and Modulation Theory, John Wiley
      and Sons, New York (1971).

4. D. Vakman, Sophisticated Signals and the Uncertainty Principle in Radar,
      Edited by E. Jacobs, translated by K. N. Trirogoff, Springer Verlag,
      New York (1968).

5. a) C. J. Kramer, M. N. Araghi and P. Das, Appl. Phys. Letters, 24, 180
          (1974).
    b) K. Preston, Coherent Optical Computers, Chapter IX, McGraw-Hill, Inc.,
          (1972).
    c) G. I. Stegeman, IEEE Trans. on Sonics and Ultrasonics, SU-23, 33
          (1976).

6. P. Das and S. Schumer, Ferroelectrics, 10, 77 (1976).

7. R. Normandin and G. I. Stegeman, 1974 Ultrasonics Symposium Proceedings,
      74CHO 896-ISU, IEEE, 95 (1974).

*Partially supported by ONR Contract No. N00014-75-C-0772.

WAVEGUIDES AND GRATINGS
BY OPTICAL METHODS

Jean Jacques Clair and Michel Henry

Institut d'Optique 5 Bd Pasteur Paris (France)

ABSTRACT

We present methods for realization of phase profiles,mainly gratings
and waveguides.These methods are intended to eliminate projection
and associated diffraction limitations.Some devices are described
and results indicated.The photosensitive material used is mainly
photoresist.A simple mathematical model of polymerisation rate ver
sus incident energy is given,leading to the analogous of a H & D
response curve.

INTRODUCTION

The growing volume of information exchange between men urged scien
tists and engineers to improve ways for transmitting information.
A long recognized one is use of higher and higher frequencies for
carriers.One can say that a major part of telecommunications techni
ques is struggle for mastery of high frequencies.
Light,whose frequency is about 10^{14} Hz can carry a tremendous quan
tity of information,up to gigabits per second,orders of magnitude
more than VHF or UHF radio waves.
Since the advent of intense and coherent light sources,the principal
problem,coding light for information transport,appears to be solved.
Simultaneously,technology of light carriers,optical fibers,has grown
and allows realization of high quality fibers,with losses as low as
2 dB/km.However,fibers are passive elements.More important are act
ive ones:emitters,modulators,...which link information generator
and receiver to transmitting fibers.A wide variety of studies on
them appears since a few years,usually gathered under the vocable
of integrated optics.
Daughter of electronical microcircuitry,the commonest technique is
the planar one,where a plane structure supports emitters,modulators
and couplers.In this planar structure,light must travel near the
surface of a plane,thus allowing its emission,modulation and detec
tion.We are interested,in the laboratory,in possibilities offered
by optics for realization of such elements.Up to now,we restrict
ourselves to gratings and waveguides.

REALIZATION OF INTEGRATED OPTICS ELEMENTS

Quite a lot of processes have been set up,but the wideliest used
one is masking a convenient substrate and etching or sputtering it
(1).More than one operation may be needed,but all of them are based
upon the same technique.
The mask,realized by electronical means may be quite good.However,
etching or sputtering requires contact between mask and substrate,
either direct or by photoprojection.This contact can never be quite
good,which results in a minor quality of the elements.A major limi
tation comes from diffraction,which undulates edges of guides.These
undesired undulations entail light losses,and can even lead to loss
of carried information.Even with very sophisticated (and expensive)
objectives,one cannot hope to get,in visible light,details sharper
than 0.5 µm or so.
In order to avoid these limitations,and to get rid of the complex
apparatus required by electronics,we study realization of phase pro
files by optical means,using photosensitive materials like photogra
phic plates or photoresist.We shall now describe some techniques set
up recently.

Slit Translation (2)
A first idea to get a profile is to image a slit on a photosensitive
plate,then translate the plate while varying intensity of the light
Adjustment of plate translation and light intensity variation allow
us to get a wide variety of profiles;e.g. uniform translation and
sawtooth modulation give a blazed grating.
Diffraction and modulation transfer function (MTF) of photosensitive
plate restrict the method to somewhat low spatial frequencies,about
500 mm^{-1} to 1,000 mm^{-1}.
The apparatus used is quite simple and cheap (Fig.1)

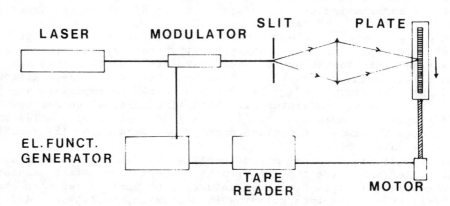

Fig.1 Slit translation apparatus

The light source is a laser for convenience,either visible or ultra
violet according to the photosensitive material used.
A cylindrical lens images a very fine slit on the plate.
The laser beam goes through an electrooptic modulator,driven by an

electronic functions generator.
The plate is driven by a motor,either continuously or step by step.
A program must be established to coordinate translation speed and
light intensity modulation amplitude and frequency.This program is
punched on paper tape,and read out by an optical tape reader which
delivers information to the motor and the modulator.Once this pro
gram is established,operations are fully automatized .

Interferences.

To get sinusoidal gratings,a quite simple way is registering of si
nusoidal interference fringes,e.g.two waves interferences.Pace of
the fringes can be modified according to the particular type of in
terferometer used.This method seems very promising as diffraction
does not intervene.
A very rugged apparatus allowed us to realise fringes up to 2,000 mm^{-1}
and little improvement will certainly allow us to get 5,000 mm^{-1}
or better(Fig. 3).
We postpone apparatus description until after discussion of Fourier
synthesis,which is a mere extension of the preceding method.

Fourier Synthesis

Let $f(x)$ be any one variable function.It can be resolved in sinusoi
dal components according to the well known relations:

$$f(x) = \int A(\nu)\exp(2i\kappa\nu x)\ d\nu \tag{1}$$

$$A(\nu) = \int f(x)\exp(-2i\kappa\nu x)\ dx \tag{2}$$

If $f(x)$ is periodical,one can write:

$$f(x) = \sum a_n \cos(2n\kappa\nu x + \varphi_n) \tag{3}$$

Well known examples are:

$$\sum \frac{1}{2n+1} \sin\left[(2n+1)2\kappa\nu x\right] = \text{square wave} \tag{4}$$

$$\sum \frac{(-1)^n}{n} \sin(2n\kappa\nu x) = \text{sawtooth wave} \tag{5}$$

Physically,it means than one can realise any,or nearly any,required
profile by superposition of two waves interference fringes.Practical
realisation,however,rises difficulties we shall now investigate.
First,we do not register amplitudes,but rather intensities,i.e.func
tions of the type

$$I = I_o \cos^2(2\kappa\nu x) \tag{6}$$

or $\qquad\qquad I = I_o' \left[1 + \cos(4\kappa\nu x)\right] \tag{7}$

So,a constant term is added to the variable ones,creating a conti
nuous background.On account of that background,it is preferable to
use a photoresist,rather than a photographic plate.
Then,some profiles require negative amplitudes a_n.This can be easi
ly done by a convenient phase shifting.The phase shifter itself de
pends on the particular apparatus used.

Next,the frequency range cannot be infinite,so a truncation is una
voidable.The profile is then convoluted with the Fourier transform
of the truncation function.If this truncation function is a square
one,its Fourier transform is a sinc function.Undulations appear on
the profile (Fig.2).However,these undulations can be strongly redu
ced by taking more terms of the Fourier expansion,and by use of con
venient apodizing functions (3).

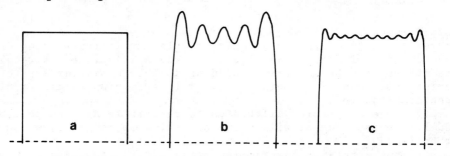

Fig.2 Square wave with limited Fourier expansion.
a)Theoretical   b) 5 terms   c) 40 terms

Next,fringes must be exactly superposed,which means that the centra
fringe of each system must be fixed relative to the plate.It is the
preferable to use interferometers which realize that fixity by thei
principle.We shall come back later on that condition.
Next,fringe contrast,i.e.component amplitude depends on the polari
zation state of interfering light (4) (5).We study that dependance
for our apparatus.It may be necessary to adjust exposure time accor
ding to polarization state,or to lay on well defined polarization
states for incident light.
Least,most interferometers are restricted to rather low spatial fre
quencies.It may be convenient,even necessary,to use high spatial
frequencies.Evanescent waves perhaps bring a solution (6).
Let us take the simplest interferometer where two plane waves inter
fere.(Fig.3)

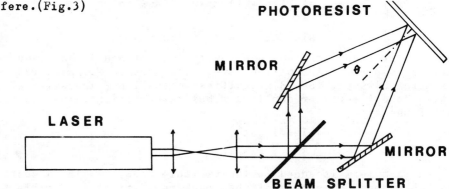

Fig.3 Two waves interferometer

The pace of the fringes is given by $a = \lambda/2\sin\theta$ and the spatial fre
quency by $\nu = 2\sin\theta/\lambda$ .Obviously,the maximum value of ν is ν_{max}
such as $\nu_{max} = 2/\lambda$.

Evanescent waves can be used,to get $v > v_{max}$.However,as pointed out
by NASSENSTEIN (6),the interference figure with evanescent waves
does only appear in a thin surface layer.We also study usefulness
of such fringes for integrated optics.
Of course,MTF of the photosensitive material must allow use of eva
nescent waves interference.This is a new reason to use photoresist
rather than photographic plate.

Apparatus
To get sinusoidal fringes,quite a lot of interferometers can be used
it least in principle.We already made mention of a very simple and
rugged apparatus(Fig.3).For Fourier synthesis,the interferometer
must allow exact superposition of the central fringe of the various
fringe systems,although step by step process allows control and re
positionning of a fringe system,if necessary.
So far,we designed two systems:a two mirrors interferometer and a
birefringent plates interferometer.
Two mirrors interferometer (7)(Fig.4).The light beam out of the sou
rce – a laser for convenience – is divided by a beam splitter.The
two beams are so reflected by a V-shaped mirror prism that they join
back symmetrically on the photosensitive plate.

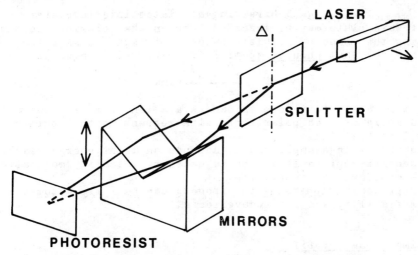

Fig.4 Two mirrors interferometer
Rotation of the source and translation of the mirror prism allow
modification of the spatial frequency of fringes,by variation of
the incidence angle of the beams on the photosensitive plate.This
arrangement lets beam splitter and plate motionless,and provides
easiness for control and adjustment.A birefringent phase shifter
may be inserted in one of the arms to provide phase shifting.
Birefringent plates interferometer.(8)(Fig.5) It relies upon well
known properties of Savart polariscopes.If two of these polariscopes
are illuminated by an extended light source,one gets nearly straight
fringes at infinity,the angular spatial frequency of which depends

on the angle of the polariscopes.So,relative rotation allows varia
tion of fringe pace,the central fringe remaining fixed.
Care must be taken not to tilt the polariscopes,which would alter
the central fringe direction,and to rotate them symmetrically,as
failure in symmetry results in a rotation of the whole fringe system

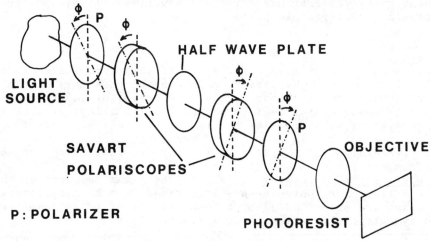

Fig. 5 Birefringent plates interferometer
A half wave plate may be inserted between the polariscopes,thus
shifting the phase by π .This provides for negative amplitudes.One
can show (8) that <u>angular</u> spatial frequency of fringes is

$$v = \frac{2t}{\lambda} \ \frac{a^2 - b^2}{a^2 + b^2} \ \sin\phi \tag{8}$$

where t is one polariscope thickness,a = $1/n_e$,b = $1/n_o$ and ϕ the ro
tation angle of each polariscope with respect to the position giving
$v = 0$,i.e. flat tinge.
As concerns fringe amplitude,it depends on light energy reaching the
photosensitive plate.It can be adjusted by intensity modulation and
– or – exposure time variation.
As for preceding apparatus,the process can be fully automatized,for
better fiability and more convenience.

<u>Two Dimensional Profiles</u>
The above mentioned methods give one dimensional profiles,i.e. str
aight grating lines or waveguides.Integrated optics often requires
curved profiles,so it would be quite interesting to get two dimen
sonal profiles by these techniques or similar ones,thus avoiding
masking as told before.
Some years ago,we set up a technique for realization of two dimen
sional profiles,in the present case kinoforms (9).We shall restrict
ourselves to a brief account,referring to the quoted reference for
further detail.
A point source is imaged by an objective on a photosensitive plate.
(Fig.6).A prism deviates the light beam.Translation of the prism

parallelly to the objective axis gives straight displacement of the
image;rotation of the prism leads to circular displacement.Combina
tion of translation and rotation would allow drawing of curves,e.g.
waveguides.The resulting profile is however diffraction modulated
by the objective and the prism.We are now studying extensions of the
method to get rid of diffraction.At all events,this process as it is
remains simple and well suited for somewhat rugged designs like two
dimensional gratings of low pace.

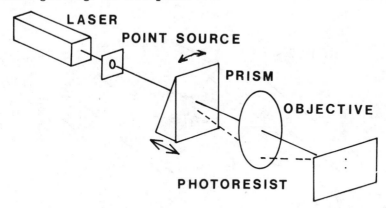

Fig.6 Apparatus for two dimensional profiles

PHOTORESISTS

Until now,we did not insist on the photosensitive material used.We
shall give here some details.One can make use,essentially,of either
photographic plates or photoresists.
Photographic plates are not well suited for two major reasons.
First,grain size (or MTF if one prefers) limits us to low spatial
frequencies,even when using holographic high resolution plates.Pho
toresist,on that point of view,appears to put no limit on spatial
frequencies.This is probably related to the molecular nature of the
photosensitive process.
Next,superposition of fringes gives a troublesome uniform background
as explained in Fourier synthesis study.On photographic plates,it
appears as a darkening of the whole field,while in photoresists,it
only results in a uniform variation of thickness,according to the
resist type used.
Besides,photographic plates give intensity profiles and we are more
interested in phase ones.Bleaching can do,but we find direct use of
photoresist more convenient.
We use a negative photoresist sold by KODAK under the trade name
"Microresist 747".It is sold as a rather viscous solution which has
to be laid out on a convenient substrate,here an optically flat
glass plate.
A spinning machine was first used for coating,but gave inconsistent
results,so we devised another technique and apparatus(10).
The essential parts of the apparatus are an inclined plane suppor

ting the glass plate and a cylindrical roller driven along the pla
te by a little motor.First,the roller is climbed up the inclined
plane and a drop ($\sim$ 3 cm^3) of liquid is deposited between the plate
and the roller,where it stays by capillarity.Then,the roller is dri
ven down and the resist spreads on the plate (Fig.7).This process
allows us to get optically flat coatings,about 1 - 2 μ m thick.After
baking in oven at moderate temperature for some tens of minutes and
cooling,the resist plate is ready for use.

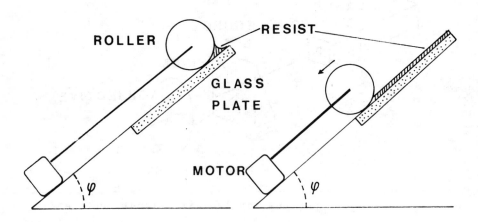

Fig.7 Coating machine

This resist is only sensitive to ultra violet light,so one can ope
rate every handling and experiment in red light,which is more conve
nient than total darkness required by photographic plates.
Photoresist exposure characteristics must be taken into account for
exposure time calculations.We studied polymerisation rate as a func
tion of incident light energy,and we were able to find a simple,if
not very accurate,model fitting experimental results(11).
The basic equations of this model are a chemical kinetics equation
for one part,and Beer's absorption law for the other part.

$$\frac{dC}{dt} = - KC^2 I \qquad (9)$$

$$\frac{dI}{dy} = - \alpha(k_M CI + k_P(1 - C)I) \qquad (10)$$

where C is monomer concentration,I light intensity,k_M monomer absor
ption coefficient,k_P polymer absorption coefficient,K and α constan
t time and y normalized resist thickness.Mathematical computation,
omitted here leads to:

$$\tau(y,t) = 1 - \frac{1}{KE + 1} \qquad (11)$$

$$\frac{dE}{dy} = ALn(KE + 1) - BE \qquad (12)$$

where E is light energy,$A = \frac{\alpha}{K} k_M(\frac{k_P}{k_M} - 1)$,$B = \alpha k_P C_o$,$\tau(y,t) = 1 - C$

and C_o the initial monomer concentration.Differential equation 12
has to be solved numerically.Its solution gives the required poly
merization rate vs. light energy relations.
Experimental values of the polymerization rate are computed from
optical density of exposed resist,the latter being measured by means
of a CARY spectrophotometer.
Agreement between this model and experimental values is good (Fig. 8)
We obtain that way the equivalent of the classical H & D curves.
Improvement of the model is in progress,by physico-chemical study
of the polymerization reaction essentially.

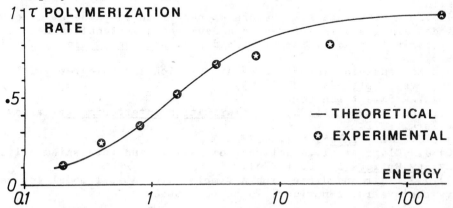

Fig. 8 Polymerization rate vs. light energy.
Besides,we do study modulation of photoresist under periodical illu
mination.We realize exposures through binary amplitude grids set in
contact with the resist.The resulting profile is studied at low spa
tial frequencies,up to 100 mm^{-1} or so by means of an interferometer
microscope.For higher frequencies,beyond microscope resolution,we
try to deduce the profile from the diffracted spectrum of the gra
ting,according to a theory set up by professor PETIT of Marseille
University.Furthermore,we hope to get high frequency gratings with
selective efficiency better than usual blazed amplitude gratings.

CONCLUSION

The main trend of our work is realization of phase profiles using
photoresist,especially high frequency gratings and integrated optics
elements such as waveguides.We try to set up processes and methods
easily feasible in any optics laboratory.Over all,our aim is to eli
minate,or at least strongly reduce diffraction initiated defects in
profiles.First results appear very promising,but much work is still
necessary to improve our methods and get consistent results.

REFERENCES

(1) Tamir,T,editor, <u>Integrated Optics</u> Springer Verlag,Berlin,1975
(2) J.J.Clair et al.,A new device for realization of gratings and modulation grids,<u>Optics Comm</u>.13, 183 (1975).
(3) J.J.Clair,A.Landrault, Phase profile synthesis for integrated optics, <u>Optics Comm</u>. 17, 115 (1976)
(4) Françon,M., <u>Handbuch der physik</u> (Ed.S.Flügge) Springer Verlag Berlin,1976
(5) S.Franck,K.Leonhardt, Mesure et calcul de la phase et du contraste des franges d'interférences à deux ondes en fonction de l'état de polarisation de la lumière incidente, <u>Nouv.Rev.Optique</u>, 4, 257 (1973)
(6) H.Nassenstein, Interference,diffraction and holography with sur face waves, <u>Optik</u> 29, 597 (1969)
(7) ANVAR patent pending.
(8)Françon,M.and Mallick,S., <u>Polarization Interferometers</u> Wiley New York, 1971.
(9) J.J.Clair, <u>Thèse</u>, Paris 1972
(10) J.J.Clair et al.,Negative photoresists and integrated optics, <u>Nouv.Rev.Optique</u> 6, 303 (1975).
(11) J.J.Clair and J.Frejlich,A simple mathematical model for nega tive photoresist behaviour, To be published.

IMAGING BY SAMPLING

I. Glaser and A. A. Friesem

Dept. Electronics, The Weizmann Institute of Science, Rehovot, Israel

ABSTRACT

An extremely compact device consisting of a lenslet
array and a periodic sampling structure is used for
image formation. It is especially useful with
diffusion transfer ("instant developing") cameras,
which can be made much smaller with no loss in picture
size. For example, an imaging device as thin as 1 cm
(front surface to image plane), can have an effective
focal length of 10 cm, image diagonal of 15 cm and 5
lines/mm resolution.

We determine the image size, resolution and image
illuminance as a function of the imaging device
parameters and object brightness, and derive
expressions for the effective focal length and
f-number. Experimental results demonstrating the
feasibility of the technique are presented.

INTRODUCTION

The human eye, as well as that of all mammals, behaves as a lens
with a single aperture to collect light from the entire scene and
to distribute it over the retina, forming an image. The complex
eyes of insects, however, contain an array of directional light
sensitive nerve endings, each having an independent aperture.
Each element of the array detects light from a single resolution
element of the external scene to form an image. Figure 1
compares the human eye (A) with the complex eye of an insect (B).

In this work we report on an implementation of the multiple
apertures approach to imaging devices. In this implementation we
use an array of miniature lenses each forming its own image. A
periodic sampling structure then derives from this array of

467

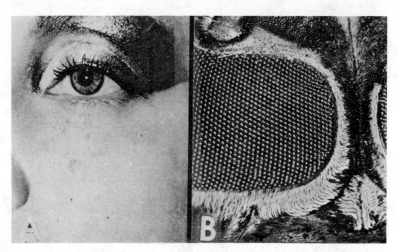

Fig. 1. Comparison between the human eye (A) and the eye of
an insect (B).

images a single magnified image. Such a composite device may
have an effective focal length and image size much larger than
that which is possible with conventional lenses of comparable
thickness. We have named this new imaging device "Bee's Eye
Lens" (B.E.L.) for the complex eye of the bee, which is the most
advanced found among insects (Ref. 1).

ANALYSIS

The principle of the operation of the Bee's Eye Lens (B.E.L.) is
illustrated in Fig. 2. The lenticular array, L, forms a series of
images on a single plane where the pinhole array, P, selects one
point from each image to form a single image, which is then
recorded on the film, F. To ensure continuity of the image, a
more pleasant display and better utilization of the the recording
film dynamic range, plane F is displaced from the pinhole array,
P, to effect some blurring of the sampling structure. In the
following, we determine the magnification, resolution and
photometric behavior of the B.E.L. device as a function of the
geometry of the arrays.

Magnification
We first consider the simplified case where the object is at
infinity and the size of the pinholes is infinitesimally small (a
δ function describes their transmittance). At the plane of the

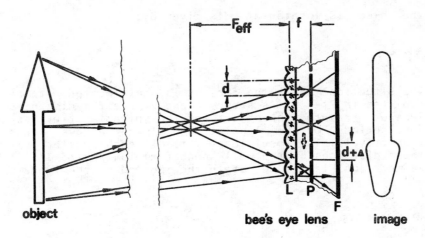

Fig. 2. Geometry of the B.E.L. device

pinhole array the incident illuminance is:

$$g(x, y)= \sum_{i_L=-n_x}^{n_x} \sum_{j_L=-n_y}^{n_y} \{e(x-i_L d, y-j_L d)\}, \tag{1}$$

where d is the period of the lenslet array and e(x,y) is the illuminance of a single image. We assume that images do not overlap so e(x,y)=0 for |x|>d/2 or |y|>d/2; this condition can be readily satisfied with proper baffeling structure. We select the period of the pinhole array to be d+Δ, where Δ<<d, so we write for the transmittance function of the pinhole array, T:

$$T(x, y)= \sum_{i_p=-n_x}^{n_x} \sum_{j_p=-n_y}^{n_y} \{\delta[x-i_p(d+\Delta), y-j_p(d+\Delta)]\}. \tag{2}$$

The illuminance transmitted by the pinhole array is given by p=gT, which after expansion becomes:

$$p(x, y)= \sum_{i_p=-n_x}^{n_x} \sum_{j_p=-n_y}^{n_y} \{e(x-i_p\Delta, y-j_p\Delta) \cdot \delta[x-i_p(d+\Delta), y-j_p(d+\Delta)]\}= \tag{3}$$

$$= e[x(\Delta/d), y(\Delta/d)] \cdot T(x, y).$$

Equation (3) indicates that the final image is a magnified version of an elemental image. The magnification factor, M_∞, when the object is located at ∞ is given by:

$$M_\infty = \frac{d}{\Delta}.$$

(4)

The situation where the object is not located at infinity is depicted in Fig. 3. As shown, S and S' are the distances from the lenslet array to the object and the pinhole array respectively. The periods of the lenslet and image arrays are d and d' respectively. The focal length, f, of each lenslet is related to these distances by the well known formula $1/f = 1/S + 1/S'$. From the geometry of Fig. 3, we obtain the period of the image array, d', as:

$$d' = d(S+S')/S,$$

(5)

Using similar derivation as for Eq. (4) we obtain that the magnification of the sampling process, M_S, may be written as:

$$M_S = d'/[\Delta - (d'-d)].$$

(6)

Now the magnification of each lenslet is $M_1 = f/(S-f)$, so that the overall B.E.L. magnification is $M_T = M_1 M_S$. For f<<S we have S'≃ f and we obtain:

$$M_T = [f(d/\Delta)]/\{S - [f(d/\Delta)]\},$$

(7)

We may now define an effective focal length for the B.E.L. as:

$$F_{eff} = f(d/\Delta) = f M_\infty.$$

(8)

After substituting Eq. (8) into Eq. (7) we obtain a familiar form which is identical with that for a conventional lens. The geometrical significance of F_{eff} is evident from Fig. 2. This was also demonstrated in our experimentation with the B.E.L. - we entered illuminated the rear side of the B.E.L. with a collimated light beam, forming a single spot of light at a distance F_{eff} in front of it.

Resolution
The resolution of the image produced by the B.E.L. is limited by several interrelated factors: a) The spatial frequency of the sampling: the period of the lenslet array; b) The size of individual pinholes; c) Diffraction caused by the aperture of individual lenslets; and d) Aberrations by individual lenslets.

There are obvious tradeoffs between some of these factors: for example, a higher spatial sampling rate results in smaller lenslets, with more pronounced diffraction; image illuminance decreases when the size of the pinholes is reduced in order to

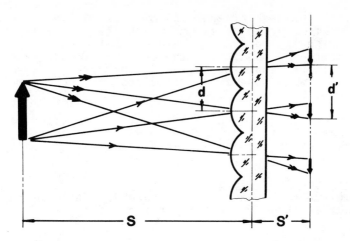

Fig. 3. Geometry of the B.E.L. for finite object distance.

obtain sharper imagery.

Since the size of a resolution element cannot be smaller than the sampling rate, there is a minimum pinhole size below which there is no improvement in resolution. for a circular pinhole of diameter D tne resolution limit, $1.2/|M_\infty|D$, is equated with the resolution limit, $1/d$, due to the sampling period to obtain the minimum useful diameter:

$$D \approx 1.2d/|M_\infty| = 1.2|\Delta|,\qquad\qquad (9)$$

To determine the effect of diffraction on the B.E.L. image we first consider the resolution of each diffraction limited lenslet (focal length f and a square aperture of size d) in incoherent light (Ref. 2),

$$r = d/(\lambda f),\qquad\qquad (10)$$

so that the resolution of the entire diffraction limited B.E.L. can be written as:

$$R = [d/(\lambda f)]/|M_\infty| = d/(\lambda|F_{eff}|)\qquad\qquad (11)$$

The form of Eq. (11) is identical to the resolution of a diffraction limited conventional lens with a focal length F_{eff} and aperture d. Because sampling limits the resolution of the B.E.L. to $1/d$, we have a situation similar to that of a pinhole camera. Increasing d degrades the image because of geometrical considirations whereas decreasing d increases diffraction. Thus there is an optimum value of d that provides the best possible resolution, R_{opt}, given by:

$$d_{opt} = \sqrt{\lambda|F_{eff}|} \;;\quad \text{or}\quad R_{opt} = (\lambda|F_{eff}|)^{-1/2}\qquad\qquad (12)$$

For typical values of F_{eff} =100 mm and λ=500 nanometer, the resolution is R_{opt}=5 lines/mm which is adequate for direct viewing. The performance of the B.E.L. can be improved by overlapping the lenslets in the array - the aperture of each lenslet is larger than the period of the array. Such a scheme may be achieved when the array consists of holographic optical elements.

Finally, aberrations may also affect the resolution of the B.E.L. However these aberrations may be adequately controlled by use of established design techniques (Refs. 3 and 4).

Photometric Considerations

When a B.E.L. forms an image of an object of brightness L, the illuminance at the pinhole array plane is:

$$E_p = d^2 L / f^2,$$

(13)

To obtain the flux through a single pinhole, we multiply Eq. (13) by the area of a single pinhole, to get:

$$\phi = \frac{1}{4}(\pi D^2 E_p) \approx 1.1(\Delta^2 d^2 L)/f^2,$$

(14)

This is distributed over an area d^2 in the film plane, to give the illuminance, E_i, at the film:

$$E_i = \phi/d^2 \approx 1.1\Delta^2/f^2,$$

(15)

If we may choose the effective f-number of the B.E.L. as:

$$f^\#_{eff} = 0.85 \, f/|\Delta| = 0.85 \, |F_{eff}|/d,$$

(16)

and substitute this result into Eq. (15) we obtain:

$$E_i \approx [\pi/(4 \, f^{\#2}_{eff})] \, L$$

(17)

Equation (17) is identical to that for the illuminance for conventional lenses having an f-number $f\#=F/A$, where F is the focal length and A is the diameter of the lens aperture. It is interesting to note that the photometric performance of the B.E.L. is comparable to a pinhole camera. However, its compactness makes the the B.E.L. much more attractive. The low light efficiency which might have limited the usefulness of the B.E.L. devices in the past is being overcome as the film sensitivity of modern photographic emulsions increases. Today, with diffusion transfer ("instant development") emulsions, the low light efficiency is no longer prohibitive. For example with commercially available diffusion transfer film having exposure index (ASA) of 3000, the exposure in daylight, with a typical B.E.L. camera having $f\#_{eff}$ =200, would be about 1/50 of second, adequate for hand held photography.

Distortions and Tolerances

When the periodicity of either the pinhole or the lenslet array
is slightly disturbed, the final imagery from the B.E.L. will be
distorted. Larger deviations from precise array periodicity would
cause image rotation and multiple image effects, described in the
former sections.   In this section we deal with two related
topics: (A) the creation of intentional small distortions in the
B.E.L. image, and (B) the precision required in the periodicity
and the relative alignment of the two arrays.

The location of each pinhole relative to the optical axis of the
corresponding lenslet determines the image point sampled by it.
For example, for an object at infinity the pinhole (m,n) will
sample image point (x-md,y-nd) so that point (x,y) will get an
illuminance proportional to $e(x-m\Delta,y-n\Delta)$, given by Eq.   (1), and
for $[x,y]=[m(d+\Delta),n(d+\Delta)]$, the magnification will not distort the
image.   If,   however,   (x,y)   slightly   deviates   from
$(m(d+\Delta),n(d+\Delta))$, the final image will be distorted; The factors
$x/(mM_\infty\Delta)$ and $y/(nM_\infty\Delta)$ will give the variable local magnifications
in the x and y directions respectively.   Thus any intentional
distortions can be readily introduced.

An example where controlled distortions are desired is in certain
recording systems (such as some direct print materials and
copying processes) affecting a mirror like image reversal;
another reversal is required (by the optical system) to have a
final unrevesed image. By using pinhole periods of d+ ' and d- '
in the B.E.L., such reversal is readily implemented. We obtain
different effective focal lengths in the x and y directions so
that $F_{eff,x} = -F_{eff,y}$.

Now the tolerances for the location of the lenslets and pinholes
in their arrays is dictated by how much distortions (as defined
by the variable magnifiction terms above), can be tolerated.   The
tolerance on the relative distance and parallelism (in the x and
y axes) between the lenslet and pinhole arrays is the depth of
focus of the individual lenslet images.

To determine the angular tolerance in the z axis, for the
alignment of the two arrays, two missregistration effects must be
taken into account: (a) rotation of the B.E.L. image, and (b)
improper magnification, leading to multiple images. To estimate
these effects, we consider the images from one row of lenslets,
sampled by one row of pinholes. Because of an angular alignment
error the two rows are not parallel but have a small angle
between them.   This introduces a height difference $(d+\Delta)\sin \alpha \approx d\alpha$
between adjacent pinholes, and changes the effective period of
the pinhole array from d+Δ to d+ Δ', where:

$$(d+\Delta')= (d+\Delta) \cos\alpha \approx (d+\Delta)(1-\tfrac{1}{2} \alpha^2). \qquad (18)$$

Now, the sampling magnifies the height difference between adjacent pinholes by approximately M_∞, causing an overall angular image rotation of β; thus, the B.F.L. image misalignment tolerance may be written as:

$$\alpha \lesssim \beta / |M_\infty| \, , \tag{19}$$

where β is now the permitted final image rotation. Because of the effective change in the period of the pinhole array the sampling increment Δ' given in Eq. (18) now becomes $\Delta' \simeq \Delta - \alpha^2 d/2$, and the sampling magnification becomes M_∞', where:

$$1/M'_\infty \simeq \Delta'/d \simeq 1/M_\infty - \alpha^2/2 . \tag{20}$$

Note that image rotation is related to the first order of α, whereas the magnifiction change depends on α^2. Thus, the rotation effect is more pronounced for small angular errors, so practical tolerance estimtions should be based on Eq. (19).

EXPERIMENTAL DEMONSTRATION

To illustrate the principle of the B.E.L. device we used a commercially available lenslet array having a period, d, of about 1 mm, focal length , f, of about 4 mm and a matching pinhole array. In our experiment the sampling increment, Δ, was about 0.05 mm, giving a sampling magnification, M, of about 20. Thus, while the overall thickness of the B.E.L. was about 10 mm, its effective focal length, F_{eff}, was 80 mm.

To fabricate the pinhole array we used the optical arrangement shown in Fig. 4. The lenslet array was illuminated with a diverging light beam, and a photographic plate (we used Kodak H.R.P.) was placed at the image plane. After development, the photographic plate contained the desired pinhole array which matched the corresponding lenslet array - compensating for any geometrical irregularities. The two arrays were then aligned together to form a B.E.L. device.

By proper choice of the distance between the divergence center of the light and the lenslet array, it is possible to obtain different pinhole array periods; consequently image magnification can be readily controlled.

Finally, Fig. 5 shows a representative result, using the above B.E.L. device. It is an image of an incoherently illuminated transparency containing a simple pattern. Because of the relatively large lenslet array period, the individual sampling

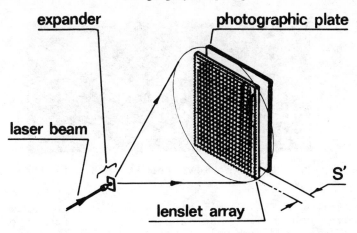

Fig. 4. Photoproduction of the pinhole array.

elements can be readily observed.

ALTERNATIVE CONFIGURATIONS

In the following we describe two B.E.L. configurations suitable
for specialized applications.

The Telecentric B.E.L.

By modifying the B.E.L. device so that the sampling increment, Δ,
is zero, we obtain $F_{eff} = \infty$. With conventional lenses, an infinite
F is meaningless. However, a B.E.L. with infinite F_{eff} behaves
as a telecentric system giving 1 to 1 image magnification
regardless of object location; we name such devices as
"Telecentric B.E.L.". To analyze the behavior of the telecentric
B.E.L., we first assume that $S' \simeq fS/(S-f)$. Thus, each image formed
by the individual lenslets would be limited mainly by
diffraction. For an object distance S and lenslet period d,
diffraction would limit the object and image to spatial frequency
of $d/\lambda S$. Since the sampling limits resolution to spatial
frequency of $1/d$ the optimum period d_0 is given by:

$$d_{opt} = \sqrt{\lambda S} \; . \tag{21}$$

For example, with s=5 cm we obtain $d_0 \simeq 1/7$ mm or resolution of
about 7 lines/mm, which is greater than the 5 lines/mm necessary
for office copying machine applications.

Unlike in the case of conventional lenses, the depth of field of
the Telecentric B.E.L. is limited by diffraction, rather than by
defocussing. Consequently, if the optimum resolution were to be
halved, then the object could be placed at any location from 0 to
2S.

Fig. 5. Representative result: a B.E.L. image.

The telecentric B.E.L. devices, therefore, offer several advantages over conventional copying lenses: (a) compactness, (b) greater depth of field and (c) telecentric imaging (magnification independent of object location).

Temporal Sampling

An example of a system with sequential sampling in one dimension is shown in Fig. 6. It is composed of a linear lenslet array, L, linear pinhole array, P, screen, S, drive mechanism, D, and recording film, F. The lenslet array forms a line of images which is sampled by the pinhole array and recorded on the film while the screen blocks other light from entering the film anywhere except through the piholes. The drive mechanism moves the lenslet array and pinhole array in a direction perpendicular to their linear direction at velocities V_L and V_p respectively.

The imaging properties of such array in the y direction are similar to those of the simultaneous sampling B.E.L., disucssed earlier . We will consider lenslets that may not have the same focal lengths in the y and x directions, denoted by f_y and f_x respectively (Practically, such lenslets may be realized by use of cylindrical or thoroidal surfaces). For undistorted imagery, we will require that the effective focal length of the B.E.L. would be the same in both x and y directions.

Following the derivation of eq. (8) we obtain for the effective focal length of the sequential sampling B.E.L. (S.S.B.E.L.):

$$F_{eff} = f_x v_L / (v_p - v_L) = f_y d / \Delta ,$$

(22)

To calculate the illuminance and effective f-number we first consider the flux transmitted through an individual pinhole:

$$\Phi = L \ [W \ d \ D_x D_y / (f_x f_y)] ,$$

(23)

where W is the width of the lenslet array, D_x and D_y the

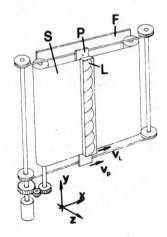

Fig. 6. Sequential sampling B.E.L.

dimensions of individual pinhole (we assume a rectangular pinhole), and L is the brightness of the object. We require that the resolution will be the same in the x and y directions, so we have:

$$\begin{cases} D_y = \Delta \\ D_x = \Delta \cdot f_x/f_y \, , \end{cases} \qquad (24)$$

Since the resolution requirement dictates that the spot, on the film, formed through a single pinhole would have the same size, d in both dimensions, the width of the lenslet array must be:

$$W = d \, f_x/f_y \, , \qquad (25)$$

Substituting Eqs. (25) and (24) into Eq. (23) and using $E = /d^2$, we obtain the illuminance at the film as:

$$E = L \, [(\Delta/f_y)^2 \cdot f_x/f_y \,] \, , \qquad (26)$$

Using the procedure for deriving Eq. (16) we may derive two different effective f-numbers for the S.S.B.E.L. - the instantaneous and the averaged. The instantaneous effective f-number is derived directly from Eq. (26), and is given by:

$$f^{\#}_{inst} = 1.1 \, (f/|\Delta|) \, \sqrt{d/W} \, . \qquad (27)$$

When including the time necessary for scanning and assuming that the scan distance is equal to the width of the lenslet array (it is usually not practical to make it shorter), we get for the time

I. Glaser, A. A. Friesem

averaged illuminance at the film:

$$E_{av} = L \, (\Delta/f_y)^2.$$
                                                                    (28)

from which we may derive the averaged effective f-number:

$$f^{\#}_{av} \simeq 1.1 \, (|F_{eff}|/d) = 1.1 \, f/|\Delta| .$$
                                                                    (29)

For some applications Eq. (27) would apply, whereas for others we
must use Eq. (29). For example, when using continuous light
(daylight) Eq. (27) should be used, while for pulsed illumination
(flash) the averaged illuminance and Eq. (29) are valid. also,
we note that the instantaneous exposure time, rather than
scanning time, determines the "motion freezing" ability.

Finally, the configuration described above is just one of many
possible sequential sampling configurations. Another
configuration, for example, similar to the track finish line
camera, may involve the movement of object and film while both
arrays remain stationary.

ACKNOWLEDGMENTS

We thank D. Treves, C. Barissac and S. Blit, for their helpful
comments.

REFERENCES

1) R. P. Feynman, R. B. Leighton and M. Sand, The Feynmann
   Lectures on Physics, vol. 1, section 36-4, (Addison-Wesley,
   Mass. 1963).
2) G. W. Goodman, Introduction to Fourier Optics, p. 113,
   (McGraw-Hill, San-Francisco, 1968).
3) A. E. Cornady, Applied Optics and Optical Design, part II,
   p. 777, (Dover, N.Y., 1960).
4) W. J. Smith, Modern Optical Engineering, p. 58,
   (McGraw-Hill, San Francisco, 1966).

LIQUID CRYSTAL LIGHT VALVE FOR COHERENT OPTICAL DATA PROCESSING

Jan Grinberg, Alex Jacobson, William Bleha,
Leroy Miller, Lewis Fraas, Donald Boswell
and Gary Myer

*Hughes Research Laboratories, 3011 Malibu, Canyon Road,
Malibu, California 90265*

ABSTRACT

A new, high performance device has been developed for application to real-time coherent optical data processing. The new device embodies a CdS photoconductor, a CdTe light-absorbent layer, a dielectric mirror and a liquid crystal layer sandwiched between indium-tin-oxide transparent electrodes deposited on optical quality glass flats. The noncoherent image is directed onto the photoconductor; this reduces the impedance of the photoconductor, thereby switching the ac voltage that is impressed across the electrodes onto the liquid crystal to activate the device. The liquid crystal is operated in a hybrid field effect mode. It utilizes the twisted nematic effect to create a dark off-state (voltage off the liquid crystal) and the optical birefringence effect to create the bright on-state. The liquid crystal modulates the polarization of the coherent read-out light so an analyzer must be used to create an intensity modulated output beam. Performance figures for the device include: Resolution >100 lines/mm; Input Sensitivity 160 μW/cm^2 at 525 nm; Time Response on: 10 msec; off: 15 msec; Contrast >100:1; Aperture 1 in. by 1 in.

The new device is a special adaptation of the ac photoactivated liquid crystal light valve reported elsewhere (1, 2). It represents a development of an earlier dc device that suffered from limited lifetimes (3). Basically, the device consists of a sandwich of thin films that electrically control the optical birefringence of a thin ($\sim$ 2 μm) liquid crystal layer. The device has high resolution (>100 lines/mm, limiting resolution), high contrast (>100:1), high speed (10 msec, on; 15 msec, off) and high input sensitivity ($\approx$ 0.3 ergs/cm^2 at threshold). Moreover, it has several practical advantages. It is compact (solid state), low power (several milliwatts), inexpensive to manufacture (thin film technology), and operates from a single, low voltage (5 to 10 V r.m.s.) power supply. In this paper we describe the design, operation, and structure of the device. We discuss in some detail the novel liquid crystal electro-optic mechanism that we use that enables us to reduce the thickness of the liquid crystal layer without sacrificing image quality. Finally, we discuss the performance of the device and show the results of several application experiments. The discussion throughout is primarily qualitative and descriptive in nature. Due

to the breadth of the   topic we were unable to include the quantitative
analysis as well. However, we do plan to publish this material, and where
appropriate we have referenced such publications.

DEVICE DESCRIPTION

The general configuration of the device is shown in Fig. 1. The ac light
valve consists of a number of thin film layers sandwiched between two
glass substrates. A low voltage (4 to 6 V$_{r.m.s}$) audio frequency power
supply is connected to the two outer, thin film indium-tin-oxide (ITO)
transparent electrodes (thus it is connected across the entire thin film
indium-tin-oxide (ITO) transparent electrodes (thus it is connected across
the entire thin film sandwich). The photoconductor (cadmium sulfide) and
the light blocking layer (cadium telluride) combine to create a rectifying
heterojunction. The dielectric mirror and the blocking layer separate
the photoconductor from the readout light beam. This is a major design
feature of the ac light valve. It enables simultaneous writing and reading
of the device without regard to the spectral composition of the two light
beams. Furthermore, the dielectric mirror prevents the flow of the device.
Finally, the mirror can be "tuned" to reflect any portion of the visible
spectrum, thereby optimizing the separation efficacy of the mirror and,
at the same time, maximizing the reflectivity of the device. This, together
with the chemically inert insulating layer, SiO_2, that bounds the liquid
crystal layer ensures a very long lifetime for the device. The liquid crystal that we use in this device is typically a biphenyl nematic material.

In order to understand the operation of the device, let us look at the
equivalent circuit of an ideal light valve substrate (Fig. 2). The diode
represents the CdS/CdTe heterojunction diode, and the capacitor represents
the capacitance of the dielectric mirror. The circuit is drawn for the case
of no input illumination light (to the CdS). If an ac voltage power supply
is connected to such a circuit, the capacitor will be charged to the negative peak voltage ($-V_p$) of the power supply during the first cycle. This
voltage will then serve as a back bias voltage on the diode for all values
of the sinusoidally varying power supply voltage. Assuming infinite back
resistance for the diode, the steady-state current flow in this circuit
will be zero, independent of the frequency, waveform (providing it is
periodic), and amplitude of the power supply voltage. Thus, there will
be no current flow in the nonilluminated resolution element of the ideal
ACLV.

3616-8

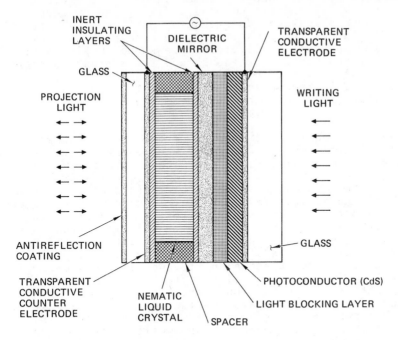

INERT
INSULATING
LAYERS

DIELECTRIC
MIRROR

TRANSPARENT
CONDUCTIVE
ELECTRODE

GLASS

PROJECTION
LIGHT

WRITING
LIGHT

ANTIREFLECTION
COATING

GLASS

TRANSPARENT
CONDUCTIVE
COUNTER
ELECTRODE

NEMATIC
LIQUID
CRYSTAL

SPACER

PHOTOCONDUCTOR (CdS)

LIGHT BLOCKING LAYER

Fig. 1.   Schematic of the ac liquid crystal light
valve.

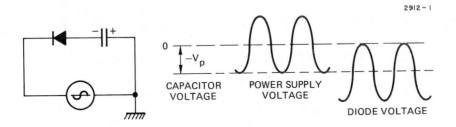

2912 - 1

$-V_p$

CAPACITOR
VOLTAGE

POWER SUPPLY
VOLTAGE

DIODE VOLTAGE

Fig. 2.   Equivalent circuit and voltage waveforms of an
ideal ac light valve substrate in the absnece
of illumination.

Now, let us consider what happens in an illuminated element. In the case of the idealized circuit (Fig. 3(a)), the incident photons introduce a leakage resistance across the diode; this resistance discharges the capacitor during the back bias cycle of the diode. The approximate waveforms are shown in Fig. 3(b). If the operating frequency is low enough and the current is high enough, the liquid crystal in the illuminated element will be driven above its electro-optic threshold by the field developed across it. This effect can then be read out by the projection beam.

The more realistic equivalent circuit of the ACLV is shown in Fig. 4 (4). In this circuit R_1 and C_1 represent the resistance and capacitance, respectively, of the liquid crystal, C_2 the capacitance of the mirror, R_3, $R_3 + R_4$ and C_3 represent, respectively, the forward resistance, back resistance and the capacitance of the diode. In deriving this circuit, we assume that the leakage resistance of the mirror, R_2, is very high, which means that $R_2 C_2 \omega \gg 1$, where ω is the basic frequency of the power supply, so that we can neglect the influence of R_2 on the circuit. Unlike the idealized circuit, a substrate that is represented by the circuit shown in Fig. 4, will pass current when the photoconductor is not illuminated as well as when it is. In this circuit, the illumination decreases the values of R_3 and R_4, and increases the capacitance C_3 (due to a photocapacitance effect in the junction). As a result, the current flow in the illuminated element is larger than the nonilluminated element. It is this difference in the current flow that provides us with the means for controlling the liquid crystal electro-optic effect with the photoconductor. So, the substrate has to be designed in such a fashion that the current flow in the nonilluminated element is less than the liquid crystal threshold level, and in the illuminated element is above this threshold by the desired amount. Generally speaking, our goal is to maximize this current ratio, which we call the switching ratio of the device, since this ratio determines the photoelectric efficacy of the light valve. Table 1 lists the switching ration and the substrate time response as a function of the input light intensity.

TABLE 1   Sputtered CdS Film (12 μm Thick), 2 kHz

Input Light Intensity ($\mu W/cm^2$)	Switching Ratio	Time Response	
		Excitation (msec)	Decay (msec)
10	1.1	250	30
40	1.3	50	30
100	1.5	20	50
400	1.75	10	30
sat	2.1	5	15

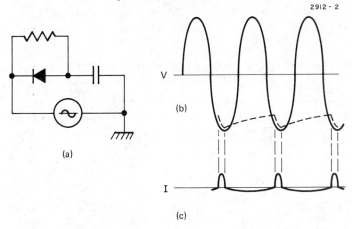

2912-2

(a)

(b)

(c)

Fig. 3. Equivalent circuit voltage and current waveforms of an ideal ac light valve substrate in the presence of illumination.

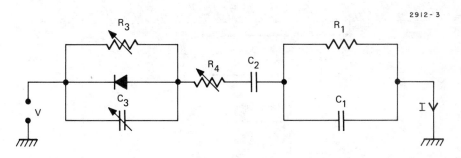

2912-3

Fig. 4. Equivalent circuit of actual ac light valve.

LIQUID CRYSTAL CONSIDERATIONS — THE HYBRID FIELD EFFECT

The ac light valve is inherently a multipurpose device (5, 6). Its different functions are realized by changing the manner in which the lqiuid crystal is applied to and used in the device. Liquid crystals exhibit several different electro-optic effects: Dynamic scattering (15), and two field effects — optical birefringence (16) and the twisted nematic effect. For a variety of reasons, discussed elsewhere, none of these effects, by itself, is directly suited for this application of the ac light valve. To obtain the performance characteristics that we describe in the next section we have developed a hybrid field effect mode — one that uses the conventional twisted nematic effect in the off-state (no voltage on the liquid crystal) and the pure optical birefringence effect of the liquid crystal in the on-state (voltage on the liquid crystal).

To implement this hybrid electro-optic effect, we fabricate the
liquid crystal layer in a twisted alignment configuration; the
liquid crystal molecules at the electrodes are aligned with their
long axes parallel to the electrode surfaces. In addition, they
are aligned to lie parallel to each other along a preferred direction that is fabricated into the device. The twisted alignment configuration is obtained by orienting the two electrodes so
that the direction of liquid crystal alignment on the two electrode surfaces make an acute (or right) angle with respect to
each other. As a result, as shown in Fig. 5, molecules in the
bulk of the liquid crystal layer rotate through this angle in
traversing the space between the electrodes. This twisted alignment configuration, combined with the intrinsic optical birefringence of the liquid crystal, causes the polarization direction of linearly polarized incident light to rotate exactly
through the twist angle. This is the so-called twisted nematic
effect. In conventional twisted nematic devices the twist angle
is 90°. As described below, in the device under discussion here,
we twist the molecules through 45°.

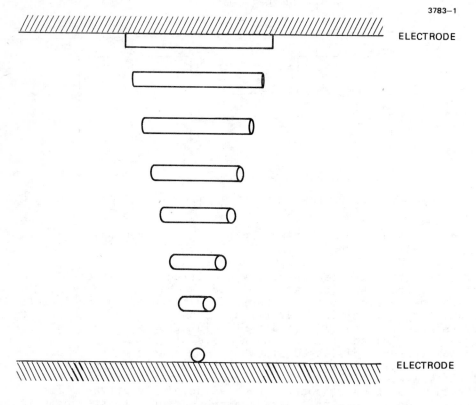

Fig. 5.   Schematic of the twisted nematic configuration of
          liquid crystal alignment.

To understand the operation of the hybrid field effect mode,
first   consider the off-state. As shown in Fig. 6(a), we
place a crossed polarizer/analyzer pair between the light valve
and the read-out light source. The polarizer is placed in the
incident beam and the analyzer is placed in the reflected beam.
This provides a dark off-state, because after its first pass
through the liquid crystal layer the direction of polarization
of the linearly polarized incident light is rotated through 45°.
But upon reflection from the dielectric mirror, the light passes
a second time through the liquid crystal and its polarization is
rotated back to the direction of the incident light, where it is
blocked by the crossed analyzer. Thus the off-state of the device is determined entirely by the twisted nematic effect.

The on-state is more complicated. If we apply voltage and rotate
the molecules to the homeotropic alignment,* the polarization of
the light would be unaffected by the liquid crystal and we would
have a dark on-state as well. This would be of no value. Closer
scrutiny of the process whereby the molecules untwist, however,
shows that between full "off" and full "on" there exists a voltage regime where the device will transmit light. As the voltage
is applied to the liquid crystal the molecules begin to tilt
toward the homeotropic alignment (see Fig. 6(b)). In this orientation of the molecules, between horizontal and perpendicular,
the optical birefringence of the molecules can affect the polarization of the light. As a result, at these intermediate voltages
the light that emerges from the device after reflection from the
mirror is no longer linearly polarized, so that some transmission
can occur. The question is — How much? To answer the question,
let us consider the orientation of the molecules as a function of
position across the layer, with voltage applied to the device.
Figure 7 shows calculated values (19) for the twist angle and for
the tilt angle of the molecules as a function of position within
the liquid crystal layer for a twisted alignment configuration
device. As shown in Fig. 7(a), the effect of the voltage is to
destroy the twist spiral. In the ideal case, with the voltage
on, half the molecules in the layer adopt the preferred alignment direction associated with one electrode, and the other half
adopt the alignment direction associated with the other electrode.

There is a realizable voltage regime in which the practical twist
angle distribution of a twisted nematic device is close to this
ideal distribution. The physical explanation for this behavior is
as follows. The twist of the molecules is transmitted from layer
to layer by means of "long" range intermolecular alignment forces
that are inherent in the liquid crystal. Generally speaking, as
the tilt angle of the molecules grows, (towards the perpendicular) the transmittance of the twist, from layer to layer, becomes
less effective. If any layer has molecules aligned perpendicular
to the electrodes, the transmittance of the twist by that layer
goes to zero. This has the effect of cutting the entire twist

*Alignment in which the long axis of the molecules is oriented
perpendicular to the electrode surfaces.

J. Grinberg *et al.*

3783—4 R2

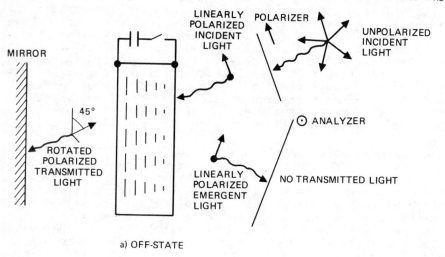

a) OFF-STATE

3783—5

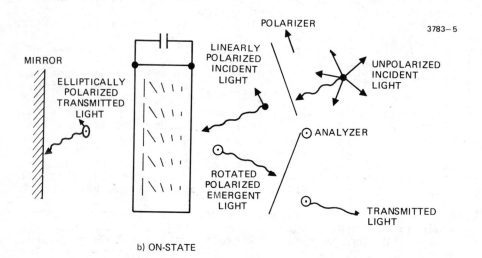

b) ON-STATE

Fig. 6.   Operation of the hybrid field effect device:
(a) The off-state;   (b) The on-state.

spiral into two separate parts. When this happens the molecules
snap into an alignment orientation that is determined by the
closest electrode. This in turn causes the twist angle distribution to transform to the ideal one (Fig. 7(a)). The foregoing
describes the nature of the twist mechanism. Next, consider the
effect of the voltage on the twist of an actual device.

The calculated tilt angle (θ) as function of position along the
cell thickness is shown in Fig. 7(b). Close to the electrodes
the tilt angle is small; but at the center of the layer it is
large, because there the influence of the electrodes on the
alignment of the molecules is at a minimum. For voltages that are
just twice the threshold voltage, the tile angle at the center of
the cell is already 80°. Thus with relatively low voltages
switched to the liquid crystal, the spiral can be snapped and the
distribution of twist angle will be close to the ideal shown in
Fig. 7(a). Moreover, in this near-ideal state the average tilt
angle is much less than 90°. The device takes advantage of the
birefringence of this state in the following manner.

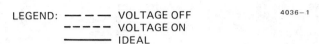

LEGEND: — — — VOLTAGE OFF   4036–1
        – – – – VOLTAGE ON
        ———— IDEAL

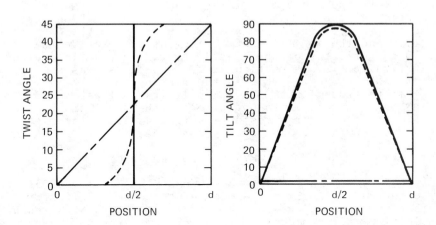

Fig. 7. Calculated values (19) of molecule orienta-
         tion as a function of position across the
         liquid crystal layer: (a) The twist angle;
         (b) The tilt angle.

The polarization of the light entering the device must be aligned
along the preferred alignment direction of the entrance electrode
in order to make the twisted nematic off-state work. Thus when
the molecules untwist, the polarization of the light would be
either at $0°$ or at $90°$ (in a $90°$ twist cell) with respect to the
majority of the liquid crystal molecules and the net birefringence would be very small. Given this picture, clearly the way to
maximize the birefringence (at least to first order) is to orient
the preferred directions of the two electrodes at an angle of $45°$
with respect to each other. In this way the polarization of the
light will make an angle of $45°$ with respect to the extraordinary
axis of the liquid crystal throughout half of the layer. This
optimizes the transmission of the device.

In Fig. 8 we show a plot of transmission versus voltage for the
hybrid field effect. The two curves correspond to twist angles
of $90°$ and $45°$, respectively, between the preferred with 2 μm
thick, reflection mode cells filled with an ester nematic liquid
crystal. The polarizer was oriented parallel to the liquid crystal optical axis on the front electrode and the analyzer was
oriented perpendicular to the polarizer. The read-out beam was a
HeNe laser. As expected from the above reasoning, the birefringence of the $45°$ cell is considerably stronger than that of the
$90°$ cell so that the maximum transmission for the $45°$ cell is
much larger than for the $90°$ cell.

Consider now the characteristic curve of the $45°$ device shown in
Fig. 8. The curve is characterized by low operating voltage
(below 4 V), by the steep and linear change of the transmission
as a function of the applied voltage, and by the high on-state
transmission (86%). These features, combined with the fast response time and the -ow off-state transmission, provide the
unique characteristics of the liquid crystal hybrid field effect
mode light valve. This particular cell thickness was not optimized for HeNe read-out;[*] this is the reason for the relatively
high off-state transmission (2% which limits contrast to 43:1)
shown in Fig. 8. The need to optimize the thickness of the liquid
crystal layer results from residual birefringence effects that
create a cross-polarized component in the off-state.

DEVICE PERFORMANCE, EVALUATION AND RESULTS

In order to generate a complete picture of the device performance, we tested it in several different systems. The test facil-
ities that we used include two main systems: (1) a 50 W xenon
arc reflection mode projection system (Fig. 9) for use where high
intensity projection light levels are needed to improve signal-
to-noise, thereby, to assure accurate measurements; (2) an 18
mW helium-neon laser optical data processing system and test
bench (Fig. 10). Both systems are used for observing optical
quality and general performance of the light valve. We report

---

[*] For a discussion of this off-state optimization process,
see reference 18.

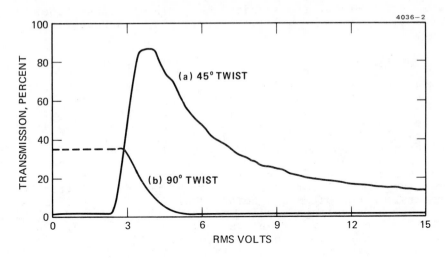

Fig. 8.   Experimental curve that characterizes the hybrid
          field effect liquid crystal device. Curve (a):
          45° twist angle; Curve (b): 90° twist angle.

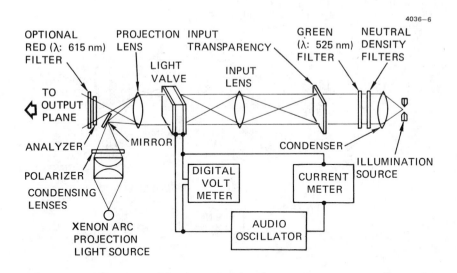

Fig. 9.   Schematic of the high intensity arc lamp test
          facility.

here measurements of the sensitometry, MTF, response time,
optical flatness and Fourier Transform plane characteristics of
the device.

Sensitometry

The sensitometry curve shown in Fig. 11 is that of a typical
light valve designed for application to coherent optical data
processing. The xenon projector system was used to record these
data because of the large dynamic range of illumination light
available in that system. The illumination light was filtered to
simulate a P-1 phosphor; it had a central wavelength of 527.5 nm
and a 50% bandwidth of 23.3 nm. The read-out (projection) light
was filtered to a narrow spectral band centered at 615 nm to
approximate He-Ne laser emission for the purposes of these measurements. A conventional photodiode radiometric detector was
used. The bias voltage applied to the light valve was 6 $V_{r.m.s.}$
at 20 kHz. The r.m.s. current was measured to be 5 mA. To obtain
data, shown in Fig. 11, we fixed and measured the light incident
on the CdS photoconductor and then we measured the readout light
transmitted from the device to the screen. These data represent one point on the curve in Fig. 11.

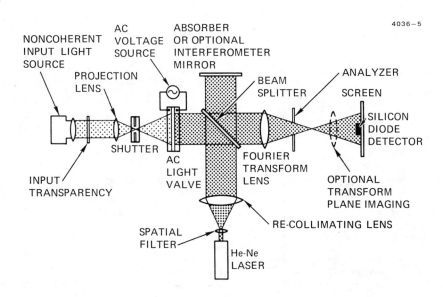

Fig. 10.   Schematic of the coherent optical data proces-
           sing system facility.

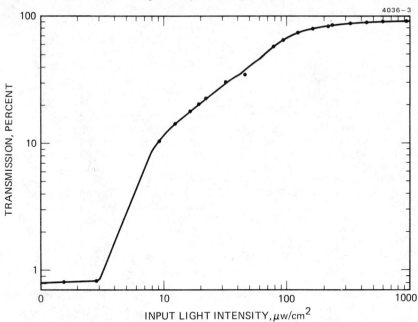

Fig. 11.   Sensitometry curve for the hybrid field effect
light valve: graph of percent transmission
versus input light intensity.

As it can be seen from the figure, the threshold sensitivity
occurs at about 3.3 μW/cm^2. If we use 10 msec for the excitation time response (see below) we find that the threshold exposure for the device is 0.33 ergs/cm^2. We attribute this low input light requirement at threshold to the design of the CdS
photoconductor that permits the main part of the input light to
be absorbed in the vicinity of the CdS/CdTe interface. Typically
we operate the device at voltages such that 100 μW/cm^2 is the
peak input intensity. This corresponds to a maximum contrast of
90:1 for the measured device and a sensitivity of 10 ergs/cm^2 at
maximum contrast. These data were taken with a device whose
thickness had not been tuned. We believe the contrast of a tuned
device will be considerably in excess of 100:1. However, one
was not available at the time of this publication.

The gray scale content of the output can also be derived from
this curve. If we use the standard sensitometric definition of
a gray scale step, 0.15 N.D. per step, and mark off these steps
along the input axis of this curve, then upon cross plotting the
points of intersection to the output axis, we find that the output contains nine shades of gray. A photograph that demonstrates
the gray scale performance of the device is shown in Fig. 17(a).

The MTF Measuring System

The MTF measurements are also carried out using the xenon projector assembly. A Sayce Resolution Chart transparency is used to input a continuously varying spatial frequency line pattern (10 lines/mm to 200 lines/mm) to the light valve. The light valve and projection optics project the line pattern to the screen with magnification of 50X to 70X. The projected line pattern is then mechanically scanned with a photodiode detector that has a 500 μm diameter aperture. The output current from the photodetector is recorded on an x-y recorder as a function of detector position on the line pattern. The limiting spatial frequency response of this system when used with 70X magnification and the 500 μm diameter detector of 140 lines/mm at the light valve. The data obtained in this manner shows photoresponse as a function of resolution for the light valve. Using the definition

$$\frac{I_L - I_S}{I_L + I_S} \times 100 = \text{percent modulation}$$

where I_L is the intensity maximum and I_S is the intesntiy minimum at a given spatial frequency, the percent modulation for the light valve at each spatial frequency can be calculated. A typical MTF curve for the light valve, obtained in this manner, is shown in Fig. 12. If contrast ratio is defined to be the ratio I_L/I_S, then 50% modulation corresponds to a contrast ratio of 3:1. As can be seen from the figure, the resolution at 50% MTF is 60 lines/mm. This is the so-called TV resolution capability of the light valve. Thus a device with a one inch aperture can resolve 1500 TV lines. Referring again to the figure we see that limiting resolution is in excess of 100 lines/mm.

To demonstrate this resolution capability pictorially, we show in Fig. 13, an Air Force resolution chart image that is projected from the light valve to a screen and is then photographed from the screen. Figure 13(a) shows the uniformity of the contrast and the overall sharpness of the image. Figure 13(b) shows a magnified view of the central section of the resolution chart. From this latter photo one can read the limiting resolution to be the group 5 to 6 which corresponds to 114 lines/mm. The curve in Fig. 12 was taken without the use of baseline subtraction. Chaging the power supply voltage, it is possible to trade transmission efficiency for additional resolution (20). Although this device should demonstrate this effect, we have not yet evaluated it in this regard.

Response Time Measurements

The laser optical data processing bench (Fig. 10) was used for the response time measurements. The input illumination beam was controlled with an Ilex Universal No. 4 shutter. The exposure intensity was controlled by the voltage applied to the tungsten lamp illumination source. We used a standard Air Force Resolution Chart for the input transparency. Projection optics were

used in the output laser beam. These optics were chosen to
project the coherent read-out image onto a screen with approximately 20X magnification. A silicon photodiode that has a time
response in excess of 10 kHz was placed in a large, uniformly
illuminated rectangle of the projected image to detect the output light variation as the input light was switched on and off.
The calibrated sweep of an oscilloscope was used as the time
reference and the shutter exposures were calibrated to this
reference. The exposure intensity was calibrated to a United
Detector Technology Model 11a Radiometer. The 10 kHz bias voltage that was applied to the light valve for these time response
measurements was set just below the threshold for liquid crystal excitation, unless otherwise indicated. The image was observed on the output screen during the measurements to verify
that high resolution images were being projected. Typical response measurements are presented in Fig. 14.

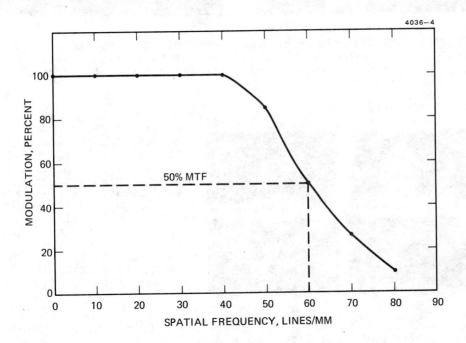

Fig. 12. MTF curve for the hybrid field effect light
          valve: Graph of percent modulation versus spa-
          tial frequency.

M10523   M10524

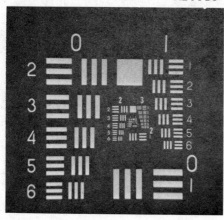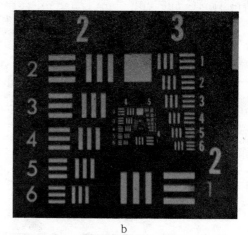

a   b

Fig. 13.   Photographs of Air Force resolution chart images read-
           out from the light valve. (a) Complete image and
           (b) magnified view of central portion of the chart.

4076-31   Fig. 14.

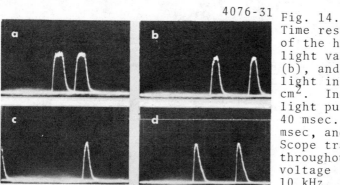

Time response measurements
of the hybrid field effect
light valve. In Fig. (a),
(b), and (c) the input
light intensity is 350 μW/
cm^2. In (a) and (d) the
light pulse duration is
40 msec. In (b) it is 23
msec, and in (c) 18 msec.
Scope trace if 50 msec/cm,
throughout. Light valve
voltage is 5.4 $V_{r.m.s}$ at
10 kHz.

In Fig. 14(a), (b), and (c), we show scope traces obtained for
three different input light pulse durations. As can be seen
from the figure, the rise time is approximately 10 msec and the
decay time is about 15 msec. These times are unusually fast for
a low voltage liquid crystal device. We attribute them to the
thinness of the liquid crystal layer that we use in the device.
Note also that these response times are essentially independent
of the duration of the light pulse. It can be seen in Fig.
14(d), that the fast response time is preserved, even at very
low light levels. In fact, Fig. 14(d) demonstrates that the
liquid crystal light valve is simultaneously capable of fast
response time and high sensitivity. A full contrast response is
achieved with an excitation of 24 erg/cm^2 (60 μW/cm^2 x 40 msec)
and a rise time of about 10 msec.

Coherent Wavefront Distortion Tests

Tests for phase distortion of the laser beam wavefront were
made on the optical data processing test bench. The optional
mirror (Fig. 10) was used to make the system into a Michelson
Interferometer.   This provided a reference beam with which to
interfere the read-out beam that is reflected from light valve
mirror.   The output plane of the system was then observed and
photographed for fringe curvature to determine the amount of
bowing in the substrate mirror. The resulting photograph is
shown in Fig. 15. By counting the numter of intereference
fringes (n) displaced across the aperture and using the equation
$\phi = (2n - 1)\ \pi$ radians, we can calculate the total phase shift
of the light valve reflection with respect to the reference beam.
Then $\phi/2\pi = N_\lambda$, the number of wavelengths in the path difference
(2π). The actual bow of the mirror is found using the relationship $\Delta\ell = N_\lambda\ \lambda/2$. Using this line of reasoning the device shown
in the figure is found to have a curvature of 1-3/4 wavelengths.
We plan to improve this characteristic of the device to better
than $\lambda/4$ by improving the polishing of the substrate.   We have
been able to produce mirrors in the laboratory that are flat to
within $\lambda/10$.

Fourier Plane

In Fig. 16 we present a photograph of the spatila Fourier transform generated by imaging a Ronchi Ruling of 24 lines/mm onto
the light valve and reading-out the light valve with 6238 Å HeNe
laser light.   In this photo, the very low Fourier Plane noise
and the high resolution capability of the light valve are clearly
demonstrated. The irregularity of the diffracted spots was
caused in part from the poor quality optics that are usec, and in
part by the 1-3/4 wavelength curvature of the light valve surface. A major thrust of the continuing work on this device is to
improve its flatness.

Summary of Performance

In the present configuration and at its present level of development, the HYFEM light valve has the following performance
characteristics:

Aperture Size	1 in.^{2}
Sensitivity (Full Contrast)	160 $\mu W/cm^2$ at 525 nm
Resolution	60 lines/mm at 50% MTF
Contrast	>100:1
Grayscales	9
Speed	
Excitation (0 to 90%)	10 msec
Extinction (100% to 10%)	15 msec
Projection Light Throughout	>100 mW/cm^2
Reflectivity	>90%

CONCLUSIONS

We have described a new device that can be used to input noncoherently illuminated images into a coherent optical data processing system in real time. The device has impressive performance characteristics and because of its compact, solid-state construction and very low power and voltage requirements, it is exceptionally easy to use. It simply requires a single, low voltage audio power supply. Furthermore, due to its high sensitivity to input light it can readily be driven from the conventional CRT of a television monitor.

At present, the device is still undergo ng development. The main areas of development are cosmetic quality and optical flatness. The cosmetic quality suffers principally from point defects caused by inclusions in the thin films that underlay the mirror. The optical flatness of the device is less than optimum because internal stresses in the thin films bow the substrate glass. Both of these development areas appear to be tractable. Hence we expect to improve the quality of the device substantially to complete its development. Hopefully, the fully developed device will exert sufficient impact on CODP systems to accelerate their application to practical, operational problems.

REFERENCES

1. T. D. Beard, W. P. Bleha, S. Y. Wong, Appl. Phys. Lett. 22, 90 (1973).

2. W. P. Bleha, J. Grinberg, A. D. Jacobson, 1973 SID International Symposium Digest of Technical Papers, Vol. IV, (N.Y.C., N.Y.) Pg. 42 (1973).

3. A. D. Jacobson, et al., Pattern Recognition 5, 13 (1973).

4. L. M. Fraas, J. Grinberg, W. P. Bleha, and A. D. Jacobson, J. Appl. Phys. 47, 2 576-583 (1976).

5. J. Grinberg, W. P. Bleha, A. D. Jacobson, A. M. Lackner, G. D. Myer, L. J. Miller, J. D. Margerum, L. M. Fraas, and D. D. Boswell. IEEE Trans. on Electron Devices, ED-22. 9, 775-783 (1975).

6. A. D. Jacobson, et al., to be published in Proc. of the SID Conference, April 22-25 1975.

7. G. H. Heilmeier, L. A. Zansoni and L. A. Barton, 56, 1162-1171, July 1968 and IEEE Trans. on Electron Devices, ED-17, 22-26, Jan. 1970.

8. M. F. Schiekel and K. Fahrenschon, Appl. Phys. Lett., 19, 391 (1971). F. J. Kan, Appl. Phys. Lett., 20, 199 (1972). R. A. Soref and M. J. Rafuse, J. Appl. Phys., 43, 2029.

SESSION 13.

OPTICAL ELEMENTS

Chairman: J. TSUJIUCHI

APPLICATIONS OF HOLOGRAPHIC STRUCTURES AS OPTICAL ELEMENTS - X-RAY MICROSCOPY

D. Rudolph, G. Schmahl and B. Niemann

Universitats-Sternwarte, Gottingen, FRG

ABSTRACT

Using holographically made zone plates and gratings we have built an x-ray microscope. It is possible to microscopize biological objects in vito without damaging.

The wavelength region between 1 and 10 nm is suitable for X-ray microscopy. In this region the image formation is dominated by photoelectric absorption. A monochromatic beam of soft X-rays is attenuated when passing through material according to the general law

$$I = I_o \exp\left(- \frac{\mu}{\rho}\rho d\right) \tag{1}$$

where I_o = Intensity of the incident X-rays

I = Intensity of the transmitted X-rays

$\frac{\mu}{\rho}$ = Mass absorption coefficient in $cm^2 g^{-1}$

d = Thickness of the material

ρ = Density of material

Between the absorption edges the mass absorption coefficient in this wavelength region is given by

$$\frac{\mu}{\rho} = k\lambda^n z^m \tag{2}$$

where $n \approx 3$ and $m \approx 4$. Because of the above mentioned law and the absorption edges the photoelectric absorption depends critically on the wavelength used and the density and chemical composition of the microscopic object. A great advantage of X-ray microscopy is that biological objects can be examined directly in a living state without severe radiation damaging. For example living cells can be examined in a chamber filled with air or water which is impossible in electron microscopy. In principle the resolution of a microscope using soft X-rays is lower than that of an electron microscope but higher than that of a microscope for visible light.

Optical Elements for Soft X-Rays

It is well known, that the refraction index of all materials in the soft X-ray region is slightly less than unity. Therefore, it is impossible to use normal refractive optics. There are, however, two other possibilities for imaging in this spectral region, namely, reflectance optics with grazing incidence and diffraction optics. Grazing incidence optics with rotational symmetry accord-

499

ing to designs of Wolter (1,2) are used widely in X-ray astronomy. Furthermore, there are intentions to use such systems for X-ray microscopy (3,4).

Zone Plates

The purpose of this paper is to discuss the application of zone plates - a special case of diffraction optics, namely circular gratings with radial increasing line density - as imaging elements for X-ray microscopy. Zone plates with large zone numbers can be realized holographically by superposition of two spherical waves or one spherical wave and one plane wave. This interference pattern is produced by using the radiation of an Ar^+ (457.9 nm) or a Kr^+ (350.7 nm) ion laser. By subsequent preparation the interference pattern is converted into a zone plate consisting of opaque gold rings on a thin organic layer transparent to soft X-rays.

The optical path difference between rays from P to Q via two subsequent zones with the radii r_n and r_{n-1} is $\lambda/2$, this means

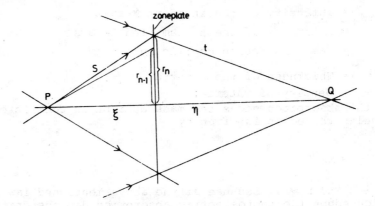

Fig. 1. Principle of zone plate construction

$$\xi + \eta = s + t - n\lambda/2, \tag{3}$$

or

$$\xi + \eta + n\lambda/2 = (\xi^2 + r_n^2)^{1/2} + (\eta^2 + r_n^2)^{1/2}. \tag{4}$$

The expansion of (4) gives

$$n\lambda = r_n^2\left(\frac{1}{\xi} + \frac{1}{\eta}\right) - \frac{1}{4}r_n^4\left(\frac{1}{\xi^3} + \frac{1}{\eta^3}\right) \pm \cdots . \tag{5}$$

If one is only interested in abberations up to the third order, one can use in the second term of (5) the approximation $r_n^2 = n\lambda f$

with $f^{-1} = \xi^{-1} + \eta^{-1}$. With $V = \xi/\eta$ one obtains

$$r_n^2 = n\lambda f + \frac{n^2\lambda^2}{4}\left\{\frac{V^3+1}{(V+1)^3}\right\}.$$ (6)

Up to now only zone plates with f-numbers of 50 to 100 in the soft X-ray region ($\lambda \simeq 5$ nm) have been made holographically. At this point of development the third-order aberrations can be neglected. The only aberration of practical importance at this point of development is the spherical aberration which occurs by using zone plates with wavelengths different from the wavelength used for construction. This aberration can be corrected by using wavefronts with spherical aberration for constructing the zone plate. In the case that a zone plate is made with a spherical wave and a plane wave ($\eta \to \infty$) a variation of ξ has to be used in the following form (5)

$$\Delta\xi_n = \xi_1\left[1 - \left(1 - \frac{r_n^2}{\xi_1^2}\right)^{1/2}\right]$$ (7)

ξ_1 refers to the first zone, ξ_n to the nth zone with $\Delta\xi_n = \xi_1 - \xi_n$.

The proper value of $\Delta\xi_n$ can be obtained by using r_n of eq. (6).

Thus it is possible to make zone plates holographically for a desired wavelength, e.g., 4.4 nm, and a desired V.

For X-ray zone plates which are made with visible light the correction for example can be obtained with good approximation by using a plano-parallel plate P as shown in Fig. 2.

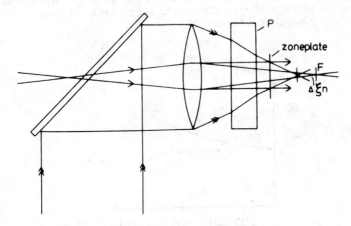

Fig. 2. Optical arrangement for the construction of zone plates with corrected spherical aberration.

A zone plate with high zone numbers behaves as a lens and resolves, according to the Rayleigh criterion, two pinholes with a separation of $s = \dfrac{1.22\lambda \cdot f}{D} = 1.22 \cdot \delta$ with δ = width of the outermost zone.

The efficiency η of a zone is defined as the ratio of the diffracted flux in a given order m for the incident flux. In the case of an amplitude zone plate with a groove to land ratio of 1 the efficiency is given by

$$\eta = \frac{1}{m^2 \pi^2}(\tau_2 - \tau_1)^2 \qquad (8)$$

with τ_2, τ_1 amplitude transmission of the open rings resp. gold rings. In the first order the maximum efficiency which can be obtained is 10.1 %.

For the prototype of a X-ray microscope the following zone plates were made holographically in our laboratory:

1.) Micro zone plates with r_N = 0.5 mm, r_1 = 17.4 µm
    ($f_{4.5 \text{ nm}}$ = 68 mm), N = 850 and r_N = 0.9 mm, r_1 = 18.6 µm
    ($f_{4.5 \text{ nm}}$ = 77 mm), N = 2340

2.) Condensor zone plate: r_N = 3.5 mm, r_1 = 50.5 µm
    ($f_{4.5 \text{ nm}}$ = 567 mm), N = 4800.

The zone plates are initially recorded in photoresist. Subsequent preparation yields a ring system in gold on a thin organic layer transparent to soft X-rays. To test such zone plates a strong X-ray source has been developed (5).

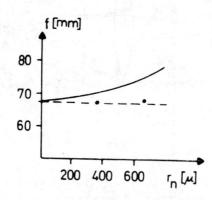

Fig. 3.

Fig. 3 shows the focal length of the zone plate with r_1 = 17.4μm
for the X-ray wavelength of CK_α (4.5 nm) as a function of r_n.
The solid curve corresponds to the uncorrected zone plate if made
with 457.9 nm. The points are measured values and show that this
zone plate is well corrected concerning spherical aberration.

The up to now obtained efficiency values are between 5 % and 6 %.

X-Ray Microscopy with Synchrotron Radiation

Synchrotrons and electron storage rings emit a higher intensity
into a small solid angle than any other X-ray source. Therefore,
experiments were made at the "Deutsches Elektronensynchrotron
(DESY)" in Hamburg.

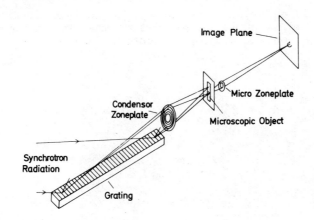

Fig. 4. Principle of the experimental arrangement.

Fig. 4 shows the experimental arrangement. The polychromatic,
slightly divergent radiation is dispersed by a 100 mm holographic
laminar grating with 600 grooves per millimeter used in grazing
incidence. The restriction of the bandwidth of the continuous
synchrotron radiation is necessary because of wavelength dependence of the focal length of zone plates. A holographically made
zone plate with a diameter D = 7 mm, 4800 zones and a focal
length of $f_{4.5 \text{ nm}}$ = 567 mm generates a reduced monochromatic
image of the synchrotron source in the object plane. A magnified
image of the object is generated by a microzone plate.

The registration of the image can be done either by using a photographic plate or an image converter (channel plate). In the
soft X-ray region photographic plates have detective quantum efficiencies up to 14 % (Ref. 7), channel plates have comparable
values.

The photograph of Fig. 5 shows the experimental setup at DESY.

Fig. 5. The experimental setup at DESY (Hamburg).

Fig. 6 shows an enlarged image of a 3 μm section of Eremosphaera viridis, stained with osemium acid and embedded in epoxy resin. Fig. 7 shows an enlarged image of cotton fibres.

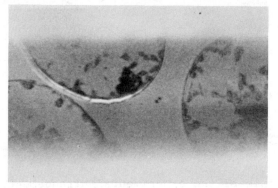

Fig. 6.A 3 μm section of Eremosphaera viridis
        X-ray magnification 15 x, total 330 x, λ = 4.6 nm.

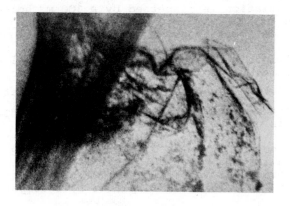

Fig. 7. Cotton fibres, X-ray magnification 15 x, total 470 x,
        λ = 4.6 nm.

The photographs have been made with the above mentioned zone plate with r_1 = 17.4 µm when used in the first order. The best resolution obtained with such zone plates is 0.5 µm.

Further Improvements and Prospects

The resolution limit of biological objects with soft X-radiation is determined by the absorption necessary to provide a detectable contrast for image formation. To demonstrate the contrast which can be obtained in microstructures of living cells Fig. 8 shows the transmission of a 50 nm thick layer of protein resp. water as a function of wavelength.

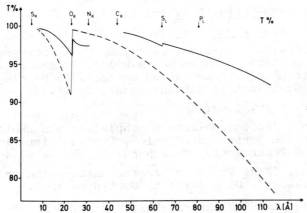

Fig. 8. Transmission of protein (solid curve) and water (dashed curve) - each 50 nm thick - as function of wavelength.

The values are based on the tables of Henke and Ebisu (8). This figure shows that protein structures of several ten nanometers can be discriminated against water. Therefore, it is useful to improve the resolution of zone plates to these values which will be achieved by a new generation of zone plates under construction at the Göttingen laboratory. Furthermore, it will be possible to get very short exposure times (in the order of seconds compared with minutes necessary for exposure of Figures 6 and 7) by using the synchrotron radiation of the storage ring DORIS, Hamburg.

On the other hand it should be mentioned that the monochromatism of the characteristic radiation of normal laboratory sources is sufficient for medium resolution microscopy (resolution of about 0.1 µm) provided that zone plates of about 100 µm diameter are used.

Holographically made zone plates with large zone numbers are also useful to obtain monochromatic synchrotron radiation for diffraction experiments in biology.

It should be considered, too, to use zone plate cameras for analysing hot plasmas in fusion experiments.

Furthermore, zone plates of this type are necessary optical elements for X-ray holography. Experiments in this field using storage ring synchrotron sources and high power zone plates are
under consideration. In particular, X-ray holography experiments
will become interesting, if coherent X-ray sources will be available. Such a development would be extremely relevant for X-ray
microscopy, too.

REFERENCES

(1) H. Wolter, Ann. Phys. Lpz. 10, 94 (1952).

(2) H. Wolter, Ann. Phys. Lpz. 10, 286 (1952).

(3) M. Bystricky, Perkin-Elmer Corp., Norwalk, Connecticut
    "Abbildende optische Systeme im Wellenlängenbereich der
    Röntgenstrahlung", Paper given on the 1975 Meeting of the
    Deutsche Gesellschaft für Angewandte Optik.

(4) H. Wolter, priv. comm.

(5) D. Rudolph, "Holographische Zonenplatten als abbildende Sy-
    steme für die Röntgenastronomie", Bundesministerium für For-
    schung und Technologie, Forschungsbericht FB W 74-07 (1974).

(6) B. Niemann, "Eine Röntgenröhre für ultraweiche charakteristi-
    sche Strahlung", Diplomarbeit, Göttingen 1971.

(7) B. Niemann, Space Science Instr., 1976, in press.

(8) B.L. Henke and E.S. Ebisu, "Low energy X-ray and electron
    absorption within solids", AFOSR Report 72-2174, Univ. of
    Hawai, Aug. 1973.

COMPUTER-GENERATED HOLOGRAMS AS SPACE-VARIANT OPTICAL ELEMENTS

Olof Bryngdahl

Xerox Research Center, Palo Alto, Calif. 94304, U.S.A.

(1975/76: Institut d'Optique, 91405 Orsay, France)

ABSTRACT

Computer-generated holograms are suitable for realization of generalized optical elements. The synthetic production of these filters allows incorporation of peculiar and useful features.

INTRODUCTION

A possible extension of conventional coherent optical information processing systems is introduction of space-variant techniques. In particular, I like here to show how it is possible to manufacture optical diffractive elements for the realization of schemes necessary to this end.[1,2]

With space-variant systems, it is possible to perform general types of phase as well as geometrical transformations on optical light distributions. Functions that can be achieved with this tool are, for example, correction of image distortion, optical map transformations, imaging on curved surfaces, shaping of reference waves in interferometry, and formation of laser scanning systems.

COMPUTER-GENERATED FILTERS

There are several kinds of elements that can be used to implement a desired phase variation, e.g. to focus and/or deflect light. They can be either refractive, reflective, and/or diffractive. A surface deformation or thickness variation is required of a refractive element. This makes it three-dimensional. From a fabricational standpoint, diffractive-type filters are less complicated. They can be formed as generalized gratings and are two-dimensional. Hybrid types, like blazed gratings, are other possibilities.

Computer-generated filters in shape of generalized grid structures represent functions that are determined by the relative lateral positions of their spatial configurations. This characteristic makes it possible to realize unusual optical components and thus transform wavefronts to almost any shape.

Wavefronts in optics are usually expressed as complex numbers. In computer-generated holograms, these numbers are manipulated into suitable codes which can be recorded on a material. These computer-generated structures will, when illuminated, reproduce certain complex wavefronts. Several types of coding schemes have been developed for different applications and to accomodate the different hardware available for producing computer-generated holograms. The type of computer-generated hologram described here resembles interferograms.

In fact, the holographic concept implies that when a structure formed by interference between two waves is illuminated with one of these waves, the diffracted light will constitute the other. Thus, in a strict sense, there is not much difference between a computer-generated hologram and a conventional hologram. Both can be considered as interferograms, an interpretation which simplifies both the explanation of the phenomena and the construction of the hologram when computer-generated filters are to serve as optical components. Actually, we can generate an interferogram with a computer and plotter from a formulation of a particular wavefront.

When dealing with optical components, we are primarily interested in phase information. The fringe pattern in the hologram (interferogram) plane can then be expressed by a grating equation

$$x/T + \phi(x,y)/2\pi = n, \tag{1}$$

where T is the average fringe spacing and n the index for the different fringes. This equation, which determines the fringe position, must be solved to produce the hologram. If we calculate and plot the positions of the fringes we will have a computer-generated hologram of the wavefront $\exp\{i\phi(x,y)\}$.

The synthetic way described here to produce holograms allows introduction of different types of carrier. The filter expressed by Eq. (1) has a linear carrier with a frequency proportional to $1/T$. The corresponding recording with a circular carrier frequency $1/r_c$ is

$$r/r_c + \phi(r,\theta)/2\pi = n. \tag{2}$$

The computer-generated hologram made according to this technique has an amplitude transmission function

$$t(x,y) \propto \sum_{m=-\infty}^{\infty} \exp\{im[2\pi x/T + \phi(x,y)]\}. \tag{3}$$

Due to the binary nature of the hologram, many diffracted orders indicated by different m values will be reconstructed. The first diffraction order can be separated out from the rest in a frequency plane as long as the slope of $\phi(x,y)$ does not exceed $2\pi/3T$.

SPACE-VARIANT ELEMENTS

As mentioned in the introduction, there exist situations where space-variant elements are required. Fig. 1 illustrates two such cases. The local phase variation of the space-variant filter, E, may be thought of as a combination of deflecting (prisms) and focusing (lenses) components. If the phase function introduced by the optical element, E, is $\phi(x,y)$, the normal to the wavefront leaving E has the direction

$$k^{-1} \partial\phi/\partial x , \quad k^{-1} \partial\phi/\partial y . \tag{4}$$

The coordinates in the front focal plane of the lens L with focal length f_L are x,y and in the back one x',y' and $k = 2\pi/\lambda$. Thus, in the paraxial region,

$$x' = \frac{f_L}{k} \frac{\partial\phi(x,y)}{\partial x} , \quad y' = \frac{f_L}{k} \frac{\partial\phi(x,y)}{\partial y} \tag{5}$$

hold. The light is focused a distance

$$z' = - f_L^2/f_E(x,y) \tag{6}$$

from the back focal plane of L. $f_E(x,y)$ represents the focal length of the filter E.

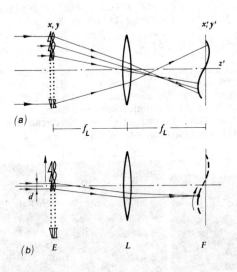

Fig. 1. Illustration of space-variant optical systems.

In Fig. 1(a) is illustrated the principle of using a computer-generated optical element in combination with a lens to transform a light wave to any
desired pattern.[1] This means that we can display any light distribution on a
surface of general shape onto another arbitrarily shaped surface and incorporate such modifications as local stretch, rotation, and translation.

How can a conventional imaging system with coherent light illumination be
converted to a space-variant one using this principle ? A solution is to incorporate two elements.[2] One of these elements, which can be a conventional
lens, is placed close to the object so that light from different parts of the
object falls on laterally separate locations in the frequency plane. The
other element is placed in the frequency plane and can now modify the information from various portions of the object in different ways.

Another principle of using computer-generated space-variant optical elements
is illustrated in Fig. 1(b). This is the scanning equivalent to the map transformation system of Fig. 1(a). A laser beam enters along the optical axis of
L and the optical element E is moved across the beam. Around the back focal
plane of L, a pattern will now be formed by a scanning mode.[3,4]

EXAMPLES

To illustrate the potential for fabricating and using generalized optical
components in the form of computer-generated filters, I like to show two

examples here.

<u>Geometrical transformation</u>. Suppose we want to optically perform the confor-
mal mapping

$$iw^* = z^2 \ ,$$

(7)

where z represents the x,y-plane, and w the u,v(x',y')-plane.
Introduction of Eq. (7) in Eq. (5) results in

$$\phi(x,y) = \frac{k}{f_L}(x^2y - y^3/3) \ .$$

(8)

This phase function was incorporated in a computer-generated hologram and a
set-up similar to the one in Fig. 1(a) was used. The object distribution to
be transformed was imaged onto the element E and the light distribution in
the back focal plane of L was recorded. Some examples are shown in Fig. 2.
The object transparencies that illuminated the hologram are on the left and
the corresponding transformations on the right of Fig. 2.

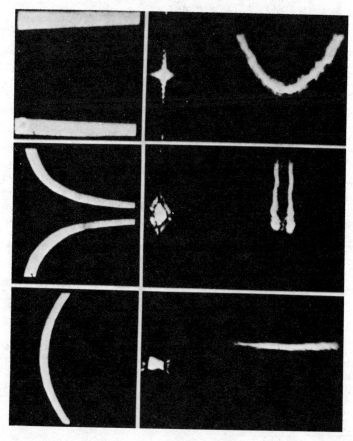

Fig. 2. Example of conformal mapping showing how straight lines
are transformed into hyperbolas and vice versa.

Laser beam scanner. The advantage of these space-variant elements for display of optical information in a scanning mode over more conventional mirror or prism systems is their flexibility. For example, elements for line scanners with corrections incorporated for both linearity and field curvature can be fabricated. Furthermore, complicated scan patterns can easily be achieved.

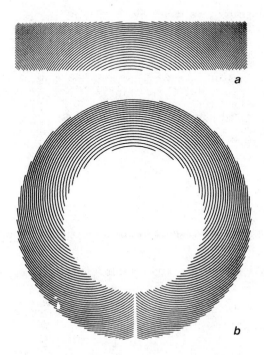

a

b

Fig. 3. Computer-generated zone-plate elements for a line scanner.

Examples of elements that can be used for line scanner realizations are shown in Fig. 3. Figure 3(a) illustrates an element intended for translational movement. It is an off-axis portion of a zone-plate pattern with the phase variation

$$\phi(x,y) = 2\pi y/T + k(x^2 + y^2)/2f. \tag{9}$$

This filter is able to deflect as well as focus a light beam. If this element is translated along a straight path across a laser beam, a focused spot will move in the same direction. A further advantage with the synthetic production of the filter is that its geometrical shape can be arbitrarily selected, e.g. independent of the desired scan pattern configuration.[3,4] An example of this is shown in Fig. 3(b). Here the zone-plate pattern of Fig. 3(a) has been bent to an annular band. This pattern

$$\phi(r,\theta) = 2\pi r/r_c + \pi(r^2 + r_a^2\theta^2)/2f \tag{10}$$

is a geometrical transformation of the one in Fig. 3(a). r_a is the average radius of the hologram. A proper sized version of this pattern was mounted as a disc on the axis of an electric motor and a collimated laser beam was directed towards the annular pattern. The achieved scan pattern remained the same as in case of the translated pattern of Fig. 3(a).

COMMENTS

In conventional holography, we must have a copy of the reconstructed wavefront to form the hologram. This requirement is hard to meet when dealing with general shapes and, in this respect, the techniques of forming computer-generated holograms offer new possibilities. As pointed out here, computer-generated holograms have similarities with conventional holograms and can be considered as recordings of interference patterns. Thus, the grating equation can be used to compute and plot their fringes.

The examples demonstrated here show the possibility of producing unusual kinds of optical space-variant components as diffractive elements. Because of their diffractive nature, the problem to construct them consists of formation of a generalized grating structure.

REFERENCES

1. O. Bryngdahl, Optical map transformations, Opt. Commun. 10, 164 (1974).

2. O. Bryngdahl, Geometrical transformations in Optics, J. Opt. Soc. Am. 64, 1092 (1974).

3. O. Bryngdahl, Optical scanner-light deflection using computer-generated diffractive elements, Opt. Commun. 15, 237 (1975).

4. O. Bryngdahl and W.-H. Lee, Laser beam scanning using computer-generated holograms, Appl. Opt. 15, 183 (1976).

INFRARED SPECTROPHOTOMETRIC STUDY OF NEGATIVE PHOTORESISTS AND THEIR USE AS PHASE PROFILES MATERIAL FOR THE NIR

C. I. Abitbol and J. J. Clair

Institut d'Optique - Université de Paris VI 4, place Jussieu -
Tour 13 - 3 ème étage 75231 Paris Cedex 05

ABSTRACT

In the search for new infrared optical materials, we show here that photoresists exhibits very interesting optical properties when illuminated in the 2 - 15 microns spectral radiation.
We found namely that the transmittance function reaches 80% over a large range centered at 1000 cm^{-1} (10 microns) and is strongly related to the percentage of solvent left in the resist after processing.
The study of the absorption spectrum of the resist and its solvent separately shows interesting characteristics. These lead us to make certain conclusions about the transmitting behavior of a photoresist film and its improvement, when knowing the thickness and the amount of solvent present in the layer.

The purpose of this work is to realize good, very light and inexpensive phase filters for the nir by using the optical synthesis method developed by us recently.
We report here the spectrophotometric study on Kodak micro-resist 747, from commercial stock and discuss the principal results.

INTRODUCTION

In the search for new, low-weight and low-cost materials capable of making good phase filters for the nir, we found that Kodak micro-resist 747, available from commercial stock, exhibits very interesting transmitting properties when illuminated over the 2 - 15 microns spectral radiation.
Observation of the infrared spectrum of photoresist layers of different thicknesses spread on KBr substrates reveals the presence of typical absorption bands of the photosensitive material compounds. We noticed namely that these bands change in intensity when the rate of evaporation of the solvent increases. This behavior indicates the importance of this compound while studying the transmittance capability of the resist in the I.R.

Previous publications from our laboratory reported the physico-

chemical properties of negative photoresists (1, 2, 3), and we
illustrate already the obvious advantages in using this material
by producing good optical components (4,5). Phase profiles such
as Schmidt plate and lenses, kinoform or not, have been realized
by using the simple optical synthesis method developed by us.
Figure 1 illustrates the conception of a phase profile modulated
from 0 to 2 π to correct the sperical aberration of a f/2.7
concave mirror. Figures 2 and 3 show the equal fringes obtained
by a Michelson interferometer corresponding to the phase filter
produced on photoresist.

The purpose of this work is to realize very light and inexpensive
phase filters for the nir. This paper deals with the infrared
spectrophotometric analysis of negative photoresist. Transmittance
function and absorption coefficient curves of the sensitive
material are reported just as a discussion on the absorption
bands distribution while varying certain parameters in the nir.

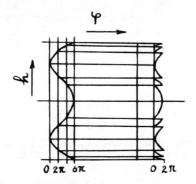

Fig. 1 : Profile of the phase plate and its equivalent
         kinoform representation.

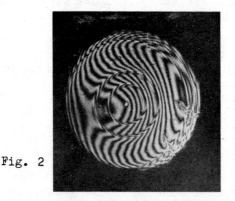

Fig. 2   Fig. 3

Fig. 2 and 3 : Equal thickness fringes obtained by a Michelson
               interferometer for the corresponding phase
               filter produced on photoresist.

Experimental procedure

To study the transmittance variation of the material we spread
the photoresist onto KBr (Potassium bromide) substrates whose
transmittance is previously determined. Inexpensive samples have
been easily obtained showing an average transmission value of 90%
from 4000 to 200 cm^{-1}.

Layer and multi-layers samples have been obtained by different
coating processes developed in our laboratory. The centrifugal
technique used to obtain thicknesses of about one to 10 microns
gives good surface precision ($\lambda/5$) when the samples are correctly
placed on the wheel. For thicker layers (above 20 microns) used
to observe the weak absorption bands, we drop the resist from a
graduated pipette and obtain generally a flat central area of
about 1 cm^2 after evaporation of the solvent. The observed
precision is a few microns for a 50 micron layer.

We use a Perkin-Elmer spectrophotometer model n°225 to record the
transmission spectrum from 4000 cm^{-1} to 200 cm^{-1}. Samples are
studied in normal incidence.

RESULTS

Infrared optical properties of photoresist film

Infrared spectrum of photoresist, recorded just after polymerization, plotted linearly in wavelength (microns) against percentage
transmission is given in fig. 4 for 2 different thicknesses.

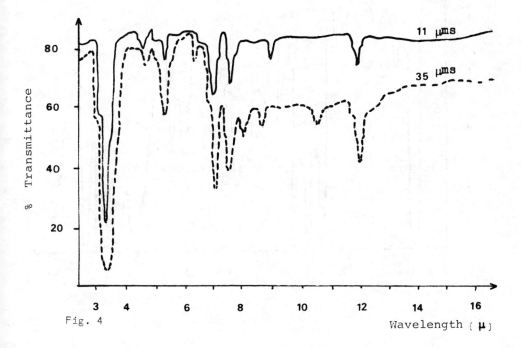

Fig. 4

For an 11 micron layer, the transmission factor reaches 0.8 over
a large range of the spectrum with which we are concerned (3,5
and 15 microns). For a layer about 3 times thicker, this factor
is still 60% its maximum value.
Samples have been exposed to a high pressure mercury lamp HBO
Philips 500 W to provide a polymerization of the coat.
Observation of this spectrum reveals the presence of a broad band
at 3.3 microns (3000 cm^{-1}), characteristic of the aromatic bound
of the polymer. Other bands of various intensity but less important than the later, belong to the different chemical compounds
present in the resist. We notice that some bands are missing
(bands at 950 cm^{-1} and 1200 cm^{-1}). To determine the influence of
the thickness on the different processes involved inside the layer,
we recorded the spectrum of the same sample some days later and
found valuable changes in intensity of many absorption bands. We
thaught the solvent should be of particular importance in this
behavior. To verify this, we recorded the absorption spectrum of
the solvent distilled from the resit solution. This spectrum is
given in Fig. 5. We used a cell of solvent of fifty microns in
depth.

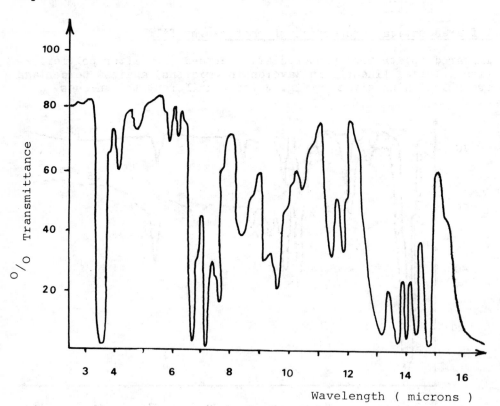

Fig. 5

Inspection of this curve shows the great analogy with the absorption spectrum of the resist given in Fig. 1, especially the bands
centered at 3000 cm^{-1}, around 1400 cm^{-1} and 900 cm^{-1}. We notice
that these bands particularly affect the transmittance in certain
domains of the I.R. spectrum of the resist. Some bands of the
solvent spectrum (between 650 and 750 cm^{-1}) are very strong but
belong to a highly volatile compound of the resist and do not
appear in the resist spectrum given in Fig. 4.

From a spectrophotometric study we determined that our solvent is
composed of benzene with combination of toluene and chlorobenzene.
We plotted in Fig. 6, the curve of percentage transmission versus
time of evaporation of solvent for a specified wavelength (ν = 2.9
microns). Transmission increases of 10% after 5 hours of evaporation at room temperature.

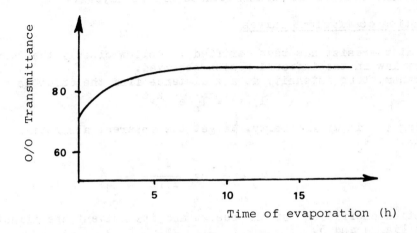

Fig. 6 : Change in percentage transmission versus time of
         evaporation of the solvent for ν = 2.9 microns.
         Sample has been placed in a drying-room at 85°C.

From the infrared spectrum of a photoresist sample it is
possible to relate the intensity of some typical absorbing bands
to the amount of solvent in the layer.

Solvent action

The role of the solvent has been taken up in previous reports
from our laboratory. As a term "solvent" generally refers to a
large group of chemicals that are liquids at room temperature

and that are employed to make solids or viscous liquids fluid
enough for practical use. They are also used for developing and
related processing. Solvents do not react chemically with the
material they dissolve and they evaporate to allow the formation
of a film from the dissolved non-volatiles. Beside that, solvents
hinder the realization of superimposed layers since depositing
successive coats partially dissolves the dry layer just underneath.
A sample of 10 microns thick, for exemple, will show an average
depth of 14 or 15 microns after depositing 3 coats of the same
amount of liquid resist.
From an optical point of view, slow evaporating solvent seems to
have rather a pernicious part in the processing and in the use of
the material.
We think solvents with a greater evaporation rate than the one
we use here (e.g. in combination with a thinner), provided that
accurate precautions are taken, will facilitate our multi-coating
process and improve the transmittance of the layers.

Absorption coefficient curves

Kodak micro-resist has been verified to follow closely the Beer-
Lambert law in the infrared region observed.
The transmitting intensity at a z distance from the input is :

$$I_t(z) = I_o \, e^{- \, k_\nu z}$$

I_o being the input intensity. We get the apparent absorption
coefficient k_ν

$$k_\nu = \frac{1}{z} \, \text{Log}_e \, \frac{I_o}{I_t(z)}$$

Absorption coefficient of the resist and its solvent are illustrated in Fig. 4 and 5.
We studied a resist layer of 11 micron thick and a cell of solvent
having 50 microns in depth. As we were expecting, we observed a
big similarity between the absorption coefficient curves of the
resist and its solvent, (e.g. bands at 3000 cm^{-1}, 1600 cm^{-1},
1400 cm^{-1}, 1200 cm^{-1} and 1000 cm^{-1}). Provided we know the amount
of solvent still present in the layer after processing and baking,
it is possible to determine the transmittance capability and its
eventual improvement of a sample of resist since we know exactly
the value of k_ν each compound (6).

CONCLUSION

The study we have reported here, with a special reference to the
transmittance, shows the particular interest in using photoresist
as near infrared material to realize good and inexpensive phase
optical filters. Though we have not conducted a complete comparison with the infrared materials actually commercialized, it seems
however, photoresist, because of availability, malleability, uni-

formity and cost can be a very suitable medium for phase i.r. filters production.

In particular it would be very interesting to investigate the possible improvement of the transmittance capability of our films since the solvent can be easily modified.

We are however concerned with the study of the physical behavior of the material, such as temporal adherance of the resist with the substrate and the thermal resistance of the coat.

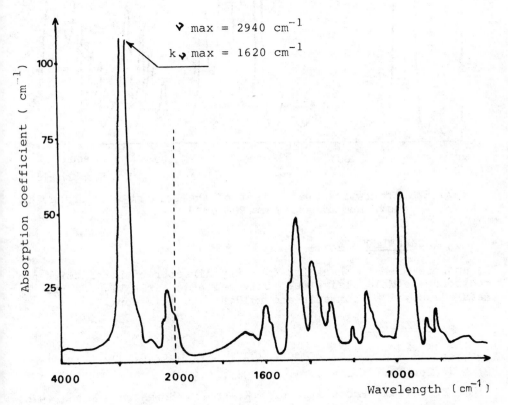

Fig. 7 : Absorption coefficient of the Kodak micro-resist with change in wave number scale at 200 cm^{-1}.

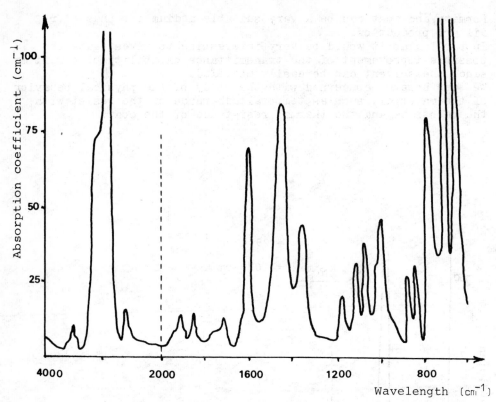

Fig. 8 : Absorption coefficient of the solvent with change in wave number scale at 200 cm^{-1}.

ACKNOWLEDGMENT

The authors wish to thank Dr. L. Henry director of Laboratoire de Spectroscopie Moléculaire de l'Université de Paris VI, for his useful discussion, comments and guidance in this work.

REFERENCES

(1) J.J. Clair and L.H. Torrès. Nouv. Rev. Opt. 4, 353, (1974).

(2) L.H. Torrès. Optica Acta, 22, 963, (1975).

(3) J. Freijlich. Proceeding of U.K. Laser. Brighton - March 1976.

(4) J.J. Clair. Appl. Opt. 11, 480, (1972).

(5) C.I. Abitbol and J.J. Clair. Optica Acta, 22, 144, (1975).

(6) A. Weisberger. Technic of Organic Chemistry. Vol. 9
Interscience - London 1956.

HOLOGRAPHIC DIFFRACTION GRATINGS
WITH ASYMMETRICAL GROOVE PROFILES

Stefan Johansson, Lars-Erik Nilsson, Klaus Biedermann
and Kent Leveby

Department of Physics II and Institute of Optical Research, The Royal Institute of
Technology, S-100 44 Stockholm 70, Sweden

ABSTRACT

Gratings with a sinusoidal profile, as they are obtained in photo-resist by holographic interference techniques, can provide very high efficiency values for wavelengths of about the same dimension as the groove spacing. When the high dispersion following from the high groove density is undesirable and broadband efficiency is required, asymmetrical groove profiles become necessary. We report on a method of Fourier synthesizing an asymmetrical profile by super-imposing two sinusoidal exposure distributions of spatial frequency ratio 1:2. Amplitudes and the relative phase of the two interference exposures are obtained from a calculation optimizing diffraction efficiency by optimizing the combined action of the diffracted fields. The technical realisation of the method, comprising, e.g., exact phase control of the two individual exposures, is described in detail. As an example, experimental results are reported for a grating of 600 grooves/mm blazed for λ = 340 nm The diffraction efficiency obtained in the +1 order is 67 per cent, and this value is higher by a factor of 9.5 than that of the −1 order.

1. INTRODUCTION

Diffraction gratings made by means of the holographic technique of exposing two-beam interference distributions into photoresist normally come out with a symmetrical and more or less sinusoidal groove profile. Gratings produced by the traditional ruling technique usually are given a blazed profile for increasing efficiency in a certain diffraction order. Therefore, a common question from users of gratings is: "Can also blazed gratings be made by holographic techniques?" The answer is "yes"; in fact, several methods have been suggested in the literature.

To begin with, however, one should point out that it is not absolutely neces-sary to blaze a holographic grating in order to achieve high efficiency. Holographic gratings can be made with very high groove densities and it has been derived that extremely high efficiency values can be obtained with a sinusoidal profile in a configuration that sustains only zero and one first diffraction order, i.e.,if the ratio of wavelength λ and groove spacing d is $0.7 \lesssim \lambda/d \lesssim 1.2$ [1]. This condition entails rather high groove densities with a spectral range too limited or a dispersion too large for some applications as a result. Holographic gratings with comparatively low groove densities, resulting in several diffraction orders, have to be blazed in order to make them competitive with ruled gratings also with respect to efficiency.

2. BLAZE OF HOLOGRAPHIC GRATINGS

Experiments to achieve asymmetrical groove profiles [2] rely either on a sinusoidal exposure of the photoresist followed by special processing (2.1), or Fourier synthesis of an asymmetrical exposure distribution and linear resist development (2.2).

2.1 Blaze from Sinusoidal Exposures

In 1968, Sheridon [3] proposed to expose skew fringes in the depth of the photoresist layer, which may lead to a sawtooth-like groove shape. This method has been used to produce gratings for the UV-region [4]. Successfull application of the method, however, depends strongly on the developing procedure. Moreover, one of the interfering fields has to enter the photoresist layer through the substrate. Therefore, a plane-parallel blank of extreme optical quality has to be used, and blazed gratings cannot be generated on curved surfaces by this method.

Another possible method is to use a grating in photoresist as a mask in a sputtering process [5]. With a skew angle of the ion beam, blazed grating profiles have been etched into GaAs single crystals [6].

2.2 Blaze by Fourier Synthesis

In theory, any exposure profile can be Fourier analysed and accordingly synthesized by incoherent addition of a series of sinusoidal exposure distributions generated by two-beam interference. As an example, the first two components of the Fourier expansion of a sawtooth profile are (apart from mean exposure and constant common factors)

$$\Delta E \cdot \sin 2\pi \cdot \nu_0 \cdot x \quad \text{and}$$

$$\frac{1}{2} \cdot \Delta E \cdot \sin 2\pi \cdot 2\nu_0 \cdot x,$$

where ΔE is the amplitude of the exposure variation, ν_0 is the fundamental spatial frequency, and x a coordinate in the direction of the grating vector. If a grating is to be blazed for the wavelength λ_b, the Fourier components have to be chosen in such a way, that they, after transfer from exposure to surface height variations, result in peak-to-peak phase excursions of $0.32 \cdot \lambda_b$ and $0.16 \cdot \lambda_b$, respectively.

The possibility of synthesizing grating profiles from sinusoidal components has been discussed [7]. The technical problems, however, seem enormous, since the method requires creating fringe patterns of exact multiple spatial frequencies and superimposing them in correct position both with regard to direction and relative phase. To produce a grating of 10 mm diameter with 600 grooves per mm from two superimposed gratings, the relative frequency deviation must be less than 10^{-5}, and the grooves of the two gratings must be parallel within 0.02 mrad. Also the groove position must be controlled within 1/10 of the grating period. One suggestion aimed at solving the problem of mutual adjustment was made by Schmahl and Rudolph [8]. They used first and second order diffracted fields from different parts of a large high-quality master grating to expose a small substrate in the region where these fields overlap and interfere.

3. SUPERIMPOSED INDIVIDUAL SINUSOIDAL GRATINGS WITH INTERFEROMETRIC ADJUSTMENT

In the method we want to report on in this paper, individually generated sinusoidal interference patterns are superimposed and their mutual phase relations are adjusted and controlled interferometrically by means of a reference grating. Another special feature is that not the common sawtooth profile is regarded as given, rather the desirable values for amplitude and relative phase of the Fourier components are determined from a calculation maximizing diffraction efficiency obtained in a certain order from the interaction of the contributing fields.

4. MATHEMATICS

In this section, we want to derive the design parameters of a grating with maximized diffraction efficiency in the first diffraction order. The desired grating may be built up from two superimposed (sinusoidal) phase gratings with equally oriented grating vectors and with spatial frequencies related as 1:2. The design parameters are then the modulations of the two gratings and their relative position. Let the grating vectors define the direction of the x-coordinate. Then the phase delay $\phi_1(x)$ impressed onto an incident field as a function of the x-coordinate of the fundamental grating with spatial frequency ν_0 is

$$\phi_1(x) = \frac{m_1}{2} \sin 2\pi \nu_0 x$$

and the phase delay $\phi_2(x)$ from a grating with exactly twice the spatial frequency and with a relative displacement of β is

$$\phi_2(x) = \frac{m_2}{2} \sin (2\pi 2\nu_0 x + \beta).$$

The parameters m_1 and m_2 describe the peak-to-peak excursion of the phase delay. As an illustration, Fig. 1 shows that an asymmetrical profile can be obtained as the sum of two symmetrical functions $\phi_1(x)$ and $\phi_2(x)$.

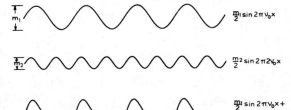

Fig. 1 *Illustrative example demonstrating how an asymmetric function is obtained from adding two sinusoidal functions together.*

We may describe the superposition of the two gratings by a complex amplitude transmittance

$$\tau(x) = e^{i\phi_1(x)} \cdot e^{i\phi_2(x)}$$

If this grating is irradiated by a normally incident plane wave U_0, scalar electromagnetic theory yields for the field immediately behind the grating (e.g. [9])

$$U(x) = U_0 \cdot \tau(x)$$

This theory is applicable for low spatial frequencies only, which is the case when holographic gratings have to be blazed.

Using the identity

$$e^{i\frac{m_1}{2} \cdot \sin 2\pi\nu_0 x} = \sum_{p=-\infty}^{\infty} J_p\left(\frac{m_1}{2}\right) \cdot e^{i2\pi p\nu_0 x}$$

for both gratings, where J_p is a Bessel function of the first kind order p, the analysis can be simplified.

$$U = U_0 \cdot \sum_{p=-\infty}^{\infty} J_p\left(\frac{m_1}{2}\right) \cdot e^{i2\pi p\nu_0 x} \cdot$$

$$\sum_{q=-\infty}^{\infty} J_q\left(\frac{m_2}{2}\right) \cdot e^{i2\pi q 2\nu_0 x} \cdot e^{iq\beta} =$$

$$= U_0 \cdot \sum_{p=-\infty}^{\infty} \sum_{q=-\infty}^{\infty} J_p\left(\frac{m_1}{2}\right) \cdot J_q\left(\frac{m_2}{2}\right) \cdot e^{iq\beta} \cdot e^{i2\pi(p + 2q)\nu_0 x}$$

The integer $p + 2q$ may be interpreted as the diffraction order of the grating. Therefore, $p + 2q = 1$ or $p = 1 - 2q$ inserted into the expression gives the amplitude of the field diffracted into the first order,

$$U_{+1} = U_0 \cdot \sum_{q=-\infty}^{\infty} J_{1-2q}\left(\frac{m_1}{2}\right) \cdot J_q\left(\frac{m_2}{2}\right) \cdot e^{iq\beta} \cdot e^{i2\pi\nu_0 x}$$

The corresponding intensity is

$$I_{+1} = U_{+1} \cdot U_{+1}^* =$$

$$= I_0 \cdot \sum_{q=-\infty}^{\infty} \sum_{l=-\infty}^{\infty} J_{1-2q}\left(\frac{m_1}{2}\right) \cdot J_{1-2l}\left(\frac{m_1}{2}\right) \cdot J_q\left(\frac{m_2}{2}\right) J_l\left(\frac{m_2}{2}\right) \cdot e^{i(q-l)\beta}$$

where l is a summation index introduced for the second summation, and $I_0 = |U_0|^2$. The infinite number of terms can be decreased considerably if we, for simplicity, consider zero and first order Bessel functions only

$$I_{+1} = I_0 \cdot J_1^2\left(\frac{m_1}{2}\right)\left[J_0^2\left(\frac{m_2}{2}\right) + J_1^2\left(\frac{m_2}{2}\right) + 2 J_1\left(\frac{m_2}{2}\right) \cdot J_0\left(\frac{m_2}{2}\right) \cos\beta\right]$$

The task of designing an efficient blazed grating requires to find that set of values for β, m_1 and m_2 which maximizes diffraction efficiency $\eta = I_{+1} / I_0$. We see that the expression for η consists of two factors, the first one $J_1^2(m_1/2)$, describes the diffraction efficiency of the grating of the fundamental frequency alone, while the second one, in square brackets, can be interpreted as the amplification, resulting from coherent addition of fields diffracted at the second grating (this factor is not necessarily greater than one). We see at once that $\beta=0$ is one condition for maximum η.

In Fig. 2, the two factors of the equation for η are plotted as a function of $m_1/2$ and $m_2/2$, respectively. Numerical calculation yields $m_1/2 = 1.8$ and $m_2/2 = 0.8$ for maximum η.

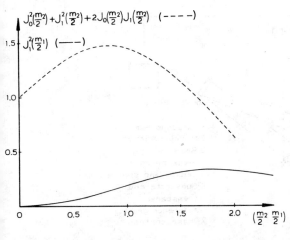

Fig. 2 *Plot of the two functions whose product determines diffraction efficiency of a composite grating.* $J_1^2(m_1/2)$ *(full line) is the well-known function describing diffraction efficiency of a sinusoidal grating of phase delay amplitude* $m_1/2$.

Expressed in phase shift for the blaze wavelength λ_b, the exposure modulations have to be adjusted to give peak-to-peak phase excursions in the final grating components of $0.29 \cdot \lambda_b$ and $0.13 \cdot \lambda_b$, respectively. (These values come close to those mentioned in Sect. 2.2 for the straight-forward approximation of a sawtooth profile by its two first Fourier components, namely $0.32 \cdot \lambda_b$ and $0.16 \cdot \lambda_b$). If we look at the contributions of the two terms to diffraction efficiency, we read from Fig. 2 the well-known value of $\eta = 33.9$ per cent for the maximum efficiency of a sinusoidal grating at $m_1/2 = 1.83$ and an amplification factor of 1.48 at $m_2/2 = 0.8$. The result of superimposing the two matched gratings is an increase of maximum diffraction efficiency to 50 per cent. The asymmetry of the compound grating is also well illustrated when we calculate the efficiency in the -1 order, which is 7.7% yielding the ratio $\eta_{+1} / \eta_{-1} = 6.5$.

5. EXPERIMENTAL REALIZATION

The chief problem is, whether it is technically possible at all to generate and adjust high spatial frequency patterns with the accuracy discussed in Sect. 2.2. In a theoretical proposal [7], the idea of aligning the wavefields in the holographic exposure set-up by some type of autocollimation arrangement had been put forward. We built on previous experience with a microdensitometer principle, where high-frequency periodic structures are scanned with interference patterns of exactly the same or a multiple frequency [10], and the practical realization in an instrument [11]. For the grating synthesis, the interferometric principle of checking, adjusting and locking one wavefield in relation to another is implemented by means of a reference grating serving simultaneously as a spatial frequency standard and a beam splitter. The adjustment and the phase-locking are done by observing and monitoring the interferogram resulting from the superposition of two different diffraction orders derived from the two incident fields. The technical realization will be described in detail in the following subsections.

5.1 The Holographic Set-up

The arrangement of the principal components is schematically shown in Fig. 3.

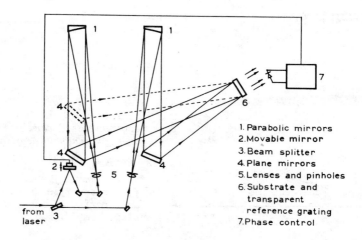

1. Parabolic mirrors
2. Movable mirror
3. Beam splitter
4. Plane mirrors
5. Lenses and pinholes
6. Substrate and transparent reference grating
7. Phase control

Fig. 3 Arrangement for exposing holographic gratings

The beam from an argon ion laser (Spectra Physics 170) is split up into two separate beams by a continuously variable beamsplitter (3). One of the two beams is reflected from a mirror (2) which can be moved in order to change the relative pathlength between the two beams. The beams are expanded by microscope lenses and spatially filtered by pin holes (5), which are located at the foci of parabolic mirrors (1) in the usual way. In either of the collimated beams plane mirrors (4) are placed with micrometer screws for adjustments. The fields reflected at the mirrors (4) intersect each other at an angle which can be changed easily. At the intersection, the substrate and reference grating (6) are placed. Behind the reference grating, a photodetector senses an interference fringe which indicates the relative phase between the grooves of the grating and the two-beam interference pattern. The output from the detector controls, via a specially designed feedback system (7), the position of the movable mirror (2) and, in this way, the phase is held constant during exposure.

5.2 The Phase Control System

The signal from the photodiode behind the reference grating (6) is amplified and electronically divided by a reference signal from another photodiode, monitoring the laser output. In this way, variations in laser output level are prevented from being interpreted as fringe motion due to phase variations. The normalized signal is subtracted from a preset reference value, which corresponds to the desired phase. The difference signal is used to drive the membrane of a telephone loudspeaker, to which the phase controlling movable mirror (2) is attached.

5.3 The Adjustment

As a reference for phase control during exposure and as a gauge for precision adjustment of spatial frequencies and orientation of the exposing interference patterns, first only an outer ring of the coated blank is exposed to an interference pattern of the fundamental frequency of the grating to be made. To have this reference grating generated on the very substrate for the future blazed grating is an essential precondition for perfect alignment. Mirrors (4) are adjusted until the zero order of one of the fields incident on the reference grating coincides with the first order diffraction of the other field. Fig. 4 shows a photograph of an interferogram indicating need for further adjustment of spatial frequency and orientation.

Fig. 4 Photograph of the interferogram obtained at the reference grating demonstrating non-perfect adjustment of both spatial frequency and angular orientation of the irradiance distribution.

When zero fringe adjustment has been achieved in the interferometer, the fringe density of the interference pattern coincides with the groove density of the reference grating (an adjustment comparable to the formation of a moiré-fringe of a very large period). In the same way, we make use of interference between zero and second order diffraction fields to adjust the exposing interference pattern of exactly twice the fundamental grating frequency.

The extent of the zero fringe is a measure for the accuracy of the matching

of the interference pattern to the reference grating. The level of the
irradiance in the zero fringe relative to its maximum irradiance is a measure
for the phase of the interference pattern relative to the reference grating.
Therefore, the level of the zero fringe is monitored to lock the exposing
interference pattern into a predetermined position by means of the phase
control system.

5.4 Experimental Procedure

Glass blanks of optical quality are spin coated with positive photoresist
Shipley AZ - 1350 to a thickness of approximately 0.5 μm. The coated blank is
mounted in the set-up, shielded for the interfering beams with the exception
of a collar which is exposed to the fundamental frequency to yield the reference grating. After a first development, the blank is put back into its
holder. The unexposed central part, which remained unaffected by the developer,
continues to be covered while zero fringe and phase are adjusted at the ring-
shaped reference grating. Thereafter, the phase is locked and the central part
of the blank receives an interference exposure with the data ensuring maximum
diffraction efficiency for a single grating. The central part is covered again,
the mirrors are adjusted for twice the fundamental frequency, the phase is adjusted and locked again, and the central part is exposed according to the data
providing maximum amplification factor. By now, the reference grating is
heavily overexposed, hence, it is dissolved completely in the following development, while the blazed grating appears on the substrate.

We found that development for 30 sec in developer AZ - 303 diluted with ten
parts of distilled water provides a sufficiently linear depth vs. exposure
relationship. The influence of development parameters on the linearity of the
resist have been studied in [12] and [13].

6. EXPERIMENTAL RESULTS

With the method described, so far, a few experimental gratings with 600
grooves/mm and a diameter of 25 mm have been made. A limitation in grating
size is given mainly by the area over which plane wavefronts can be generated,
and maintained stable during exposure. This is a question of size and quality
of the mirrors, and of stability of the environment, problems which are common
to all manufacture of high-quality gratings by holographic techniques.

The groove profile obtained in the photoresist was studied in an electron
microscope. In Fig. 5, obtained from a grating blazed for infrared, the asymmetry achieved by the superposition of grating exposures is evident.

*Fig. 5 Electron micrograph of
the profile of a holographic
grating of 600 grooves/mm generated from two superimposed
interference exposures.*

In Fig. 6, the distribution of relative diffraction efficiency among the
various diffraction orders is plotted.

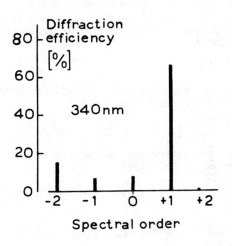

Fig. 6 *Diffraction efficiency in
the various diffraction orders*

The data have been obtained in reflection from an Al-coated grating irradiated
at normal incidence at the blaze wavelength $\lambda = 340$ nm. The ratio of the efficiencies in the +1 and -1 diffraction orders is 9.5, which is more than the
value of 6.5 calculated from the design data.

The wavelength dependence of the efficiency in the +1 order is shown in Fig. 7
for two polarization directions.

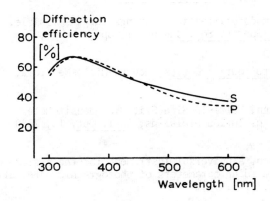

Fig. 7 *Diffraction efficiency
for P- and S- polarized light
as a function of wavelength.
Ebert angle 10°.*

A maximum relative efficiency of 67 per cent is obtained at the blaze wavelength 340 nm, with an Ebert angle of 10 degrees. This value is twice that
to be expected from a purely sinusoidal grating and it is also somewhat larger
than the value calculated from the design data for linear transfer.

SUMMARY

It has been shown that the synthesis of asymmetrical grating profiles by incoherent superposition of sinusoidal exposure distributions is possible, in spite of the extraordinary accuracy problems, by use of a reference grating and a phase control system. By this technique, blazed holographic gratings with enhanced efficiency have been produced.

ACKNOWLEDGMENT

This work has been supported by the Swedish Foundation for Optical Research and by a grant from the Swedish Board for Technical Development.

REFERENCES

[1]   I.J. Wilson, R.C. McPhedran, and M.D. Waterworth, A new high performance diffraction grating, Opt. Commun. 9, 263 (1973).

[2]   M.C. Hutley, Interference (holographic) diffraction gratings, J. Phys. E: Sci. Instrum. 9, 513 (1976).

[3]   N.K. Sheridon, Production of blazed holograms, Appl. Phys. Letters 12, 316 (1968).

[4]   M.C. Hutley, Blazed interference gratings for the ultra-violett, Optica Acta 22, 1 (1975).

[5]   J.J. Hanak and J.P. Russel. Permanent holograms in glass by RF sputter etching, RCA Review 32, 319 (1971).

[6]   Y. Aoyagi and S. Namba, Blazing of holographic grating by ion etching technique, Japan J. Appl. Phys. 15, 721 (1976).

[7]   R.C. McPhedran, I.J. Wilson and M.D. Waterworth, Profile formation in holographic diffraction gratings, Opt. Laser Technol. 5, 166 (1973).

[8]   G. Schmahl, Holographically made diffraction gratings for the visible, UV and soft X-ray region, J. Spectr. Soc. Japan 23, Suppl. No. 1, 3, (1974).

[9]   Goodman, J.W., Introduction to Fourier Optics, McGraw-Hill, New York, 1968, p. 69.

[10] S. Johansson and K. Biedermann, Multiple-Sine-Slit Microdensitometer and MTF evaluation for high resolution emulsions, Appl. Opt. 13, 2280 (1974).

[11] K. Biedermann and S. Johansson, A universal instrument for the evaluation of the MTF and other recording parameters of photographic materials J. Phys. E: Sci. Instrum. 8, 751 (1975).

[12] R.A. Bartolini, Characteristics of relief phase holograms recorded in photoresists, Appl. Opt. 13, 129 (1974).

[13] S.L. Norman and M.P. Singh, Spectral sensitivity and linearity of Shipley AZ-1350 J photoresist, Appl. Opt. 14, 818 (1975).

SESSION 14.

PATTERN RECOGNITION

Chairman: G. HAUSLER

COHERENT AND INCOHERENT AVERAGING OF ANCIENT HANDWRITTEN HEBRAÏC CHARACTERS

Jean-Marc Fournier

*Laboratoire de Physique Générale et Optique (Laboratoire associé au C.N.R.S.;
Holographie et traitement optique des signaux) Université de Franche-Comté,
25030 Besançon Cedex, France*

ABSTRACT

Averaging is carried out by patterns superposition, correlation and other
optical techniques. Paleographic studies of Hebraïc texts illustrate the
efficiency of coherent and incoherent processes.

INTRODUCTION

As an introduction to both papers by the same author a film is presented.
It concern s Optics and Paleography.
In the study of writings, it is important to apply different processings
to any text, even if several manuscripts are written by the same scriptor.
They must correspond to the general aspect of graphism in letters and
answer the questions from paleographers (1).
Through its applications, the superposition of optical images extends to
various fields in Physics (e.g. enhancement of image quality). In this paper
we essentially consider the means of superposing some components of a same
class of patterns to generate what we call "average letters". Such images
afford many qualities : they act as a picture of a series of letters, that
makes it possible to see and measure statistical data of a writing such as
average size, orientation, stability, etc... Computing of average patterns
only concerns characters that present a good regularity of penmanship and
which are sufficiently spaced with respect to each other. Note that this
concept cannot apply to cursive and unregular writings.

PATTERN RECOGNITION AND AVERAGING

Some years ago it was shown that optical processing leads to the assessment
of resemblance degrees in the morphology of letters (2). Experiments carried
out by a coherent optical correlator with matched hologram filters not only
allow to detect a given pattern but also to determine a measure of the similarity between different neighbouring characters, as recalled in figure 1 (3).
By sampling pages of a book, the concept of degree of resemblance made it possible to measure average resemblances between writings. On some cases it
allowed to separate various scribes and classify them by subjective measurements rather than by a qualitative description of writings : Fig. 2. It then
became essential to have at our disposal patterns characteristic of a type of
writing or scribe. It must be noted that averaging is automatically made in

Fig. 1. Electro-optical
approach of resemblance.

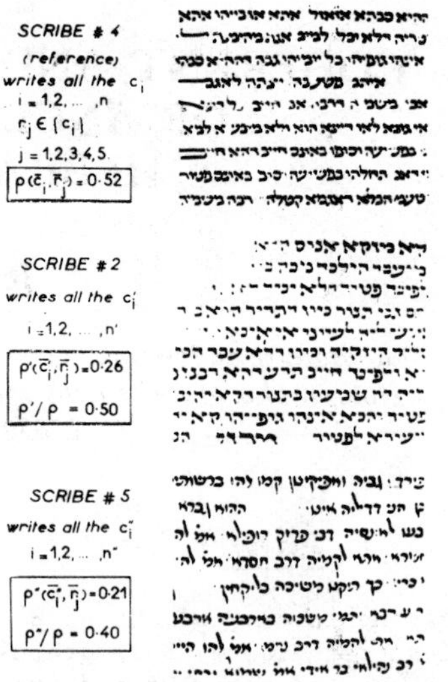

Fig. 2. Measure of resem-
blance of 3 writings.

coherent optics when pattern spectra are superimposed in the focal plane of a
lens. Unfortunately, it is impossible to obtain directly the overlapping of
input shapes at the outpout of an optical device by means of this superimposed spectra. This underlines the need for another computing method.
In the case of a text, the making of an average letter consists in extracting
a series of representations of the same character from the document (a for
example) and computing the graphism obtained by the best superposition of
these slightly different drawings (4,5). As a first stage it is necessary to
isolate the letters that are to be added. One can either extract the letters
undertest (e.g. by an optical sensor) or remove not relevant characters from
part of the text.   Actually
we just paint them white. Thus
the paleographer can see how
the morphology of a letter
changes in a text, as shown in
figure 3.

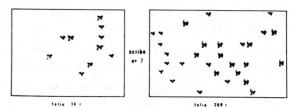

Fig. 3. Letters extraded from texts.

OPTICAL METHOD OF OVERLAPPING

We intend to have light rays travelling through the transparent input signals
and optical components, so as to construct one (or more) image where all data
are superposed. Let us consider a set of patterns to be mixed together. At
first one determines the barycenter of energy that corresponds to each pattern
-that can be achieved by means of an optical correlator-; It results in a

display of points simi-
lar to that of the pat-
terns. The correlation
function between the
two distributions (that
of the patterns and
that of the points act
as Dirac functions)
contains two terms
that represent the
average pattern of the
initial set and its
symmetrical image, as
summarized in figure 4.
We will now discuss
some properties of
different optical cor-
relation techniques

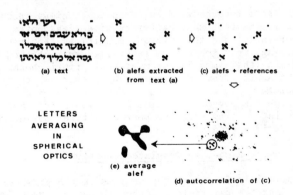

(a) text

(b) alefs extracted
from text (a)

(c) alefs + references

LETTERS
AVERAGING
IN
SPHERICAL
OPTICS

(e) average
alef

(d) autocorrelation of (c)

Fig. 4. The making of average letter.

that lead to this type of results. As is well known, it is easy to compute
correlation functions in coherent (6) as well as incoherent light (7). Because the autocorrelation function of a plane containing a set of various drawings coupled with a same set of pinholes includes the wanted cross-correlation, all techniques leading to autocorrelation functions can be used to approach pattern overlappings. Then, the main intention of the present survey
should be a comparison between these different techniques, as follows.

Coherent optical correlation

A device such as the one used in pattern recognition or measurement of a degree of resemblance does not seem to suit best because (i) it could be hard to
build a good matched-filter, (ii) pinholes used in the function to be correlated
can act as references when recording a hologram, as indicate below.

Spectrum of the Fraunhofer diffraction pattern modulus square of a function.

This is described by the diagram :

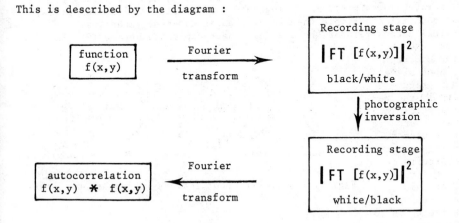

In other words, this technique corresponds to the recording of a multiple
reference-beam Fourier hologram in which spectra are modulated by a carrier

of the same frequency. This leads to a unique reconstruction of images of letters. The fact that better results (less noisy) are provided by photographic reversal and the rigorous conditions required by linear photographic recording imply to perform several experiments before acceptable results are obtained (the gamma factor must be a priori chosen and respected ; this cannot be achieved by constant weighing over all the frequency ranges of the spectra).

Although the use of real time correlators such as non-linear optical systems seems interesting (8), the poor quality of output responses and the cost of such devices prohibit them.

Spectrum of a speckle modulated spectrum

The principle is almost the same as in the previous case. It consists in recording the far field diffraction pattern of a function f(x,y) modulated by a speckle t(x,y). A Fourier transform of this recording gives the wanted autocorrelation function f(x,y) ✳ f(x,y) multiplied by a speckle pattern which is the autocorrelation of t(x,y). This process does not require special conditions of recording and is easy to use -it is only a matter of matching the granularity of the diffuser to the size of the processed patterns-.

Other techniques

We thought of considering other processes, namely using different methods of modulation transfer function measurement and interferometers. Such techniques were not suited to our problem, except in the case of incoherent optical processing mentioned below.

Incoherent averaging

There are disadvantages of incoherent optical processing techniques : the fact of working in the Gaussian region and poor resolution in the output images. Yet the shadow correlation systems had a successful development for twenty years or about in the past. We have adapted some of these techniques to our problem, because of the appeal of real-time, low-cost, and simplicity of processing. Morover for experts in Paleography who do not specialize in Optics, their approach is more simple than sophisticated coherent techniques.

The classical autocorrelators have already been explained before (9). The device presented here allows the overlapping of input signals, as shown in figure 5. It is roughly the reverse of a multipinhole array camera that can be used to multiply images for as microcircuits manufacturing for example. In this case the most important difficulty is to

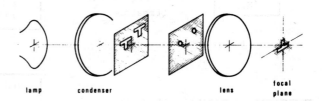

lamp   condenser   lens   focal plane

Fig. 5. Superposition of letters in incoherent light.

match the size of letters (it is common to use large letters) to the pinholes. This was studied theoretically by considering pinholes as Fresnel zone plates or holograms, and verified experimentally (10). Figure 6 shows the results of

incoherent superposition of three geometric shapes and the densitometric profile of the noiseless correlation of two slits.

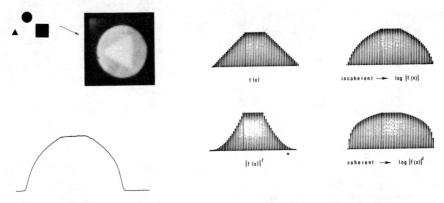

Fig. 6. Patterns superposition
and photometric curves.

Fig. 7. Coherent and
incoherent observation.

Remark.

Coherent optical correlation processes give the amplitude of the correlation
function of two functions ; As far as visual observation is concerned, one
can see the square modulus of this correlation function. On the other hand
incoherent optical techniques lead to the intensity of correlation functions :
therefore the eye (or photographic emulsion) can see the wanted information
directly. This is summarized in figure 7 : starting from the same function
f(x) -that is the cross correlation of two slits- approximately the logarithm
of the square of the function is observed in the coherent case, and the logarithm of the function itself in the incoherent one.

EXPERIMENTAL RESULTS

Hebraïc block characters are
well suited to the methods
of averaging. Some experiments using such letters
were made on the manus-
cript referenced H. 316 at
the National Library of
Paris, because its characters had been used before
in computing measurements
of degrees of resemblance.
Results shown here were obtained by Fraunhofer dif-
fraction of a speckle modulated spectrum as noted
above. Figure 8 shows the
evolutions of the average
pattern obtained by super-
imposing 2, then 3, then 4,

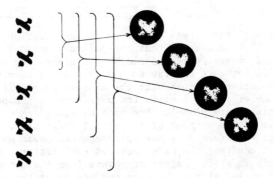

Fig. 8. Superposition of letters.

and then 5 letters drawn by the same hand.

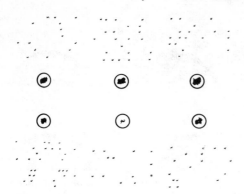

In figure 9 six pieces of
text where only letters
"aleph" remain are displayed
side by side with 6 average
letters built with these data.
The latter shows the resultant
of very fluctuant patterns
from a weak stability writing.

Fig. 9. Set of alephs and
corresponding average patterns.

Figure 10 presents samples
of average letters taken
out from different folios
that represent the writings
of four scribes. This
points out the evolution
of a script throughout the
same book and evinces that
comparisons between different writings is possible
through their average letters.

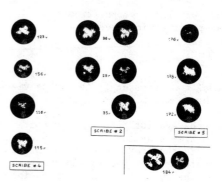

Fig. 10. Average letters.

EXTENSION OF THE TECHNIQUE AND NEW DEVELOPMENTS

Let us first stress 3 aspects of the question.
(i) We have assumed that the points coupled which each letter in order to
build average patterns can be represented by delta functions. In practice
it is not easy to draw such points which eventually act as pinholes. The best
resolution attainable in the average images is linked to the size of the
pinholes : this implies the use of reference points as small as possible.
Satisfactory pinholes are made by recording correlation peaks on a lith-type
film at the outpout of an optical correlator.
(ii) When recording multiple reference beam holograms it is necessary to
keep constant the ratio between each letter and its coupled pinhole. This
leads to shift the set of points from the set of letters by a reasonable
amount and also to balance the energy of the two sets of data.
(iii) All the letters of a writing must be used a priori to the building of
the average letter of this writing. In practice, if a set of letters contains
n times the same character, it is possible to determine theoretically the

distance between the superposition of all the n characters and the superposition of p of them only (p $<$ n). So we can compute the minimum number p which is sufficient to keep below a given distance (which can be as small as desired). The root mean square σ of a gaussian function that represent the dispersion of a writing can represent the stability of this writing. For example, according to this model, with σ = 0.2 , 9 letters are sufficient to approach the average pattern with an accuracy of 10 %, and the use of 33 letters corresponds to an accuracy of 5 % . It seems also interesting to determine the minimum number of letters necessary to represent a writing by its average letters. Unfortunately, this cannot commonly be used because we do not have an a priori concerning the writings under test.

CONCLUSION

Many optical techniques can produce superposition of letters which are characteristic of writings. The great simplicity of incoherent shadow correlation techniques make them useful for people who are not specialists in Optics while the results given by coherent methods make them more attractive for applications which require more accuracy. As pointed out in the study of similarity the concept of resemblance appears important for classification of patterns :
in average letters the parts common to all letters are present in the average character whereas the differences between the various original letters represent noise only.
An analytical knowledge of the profiles of average patterns gives important results in handwriting study. The curves that describe these profiles would enable to extract characteristics of writings such as dispersion (or stability) and to compare them to theoritical models coming from statistical laws. This can be a great help in the field of expert appraisement. Can the brand of a scribe be traced out by measuring the dispersion of a given style of writing ?

REFERENCES

1. J.-M. Fournier, Traitements optiques d'écritures,
   Rev. d'Histoire des Textes, to be publish (1976).

2. J.-Ch. Viénot, Optical inform. proces. applied to the study of hebraïc
   manuscripts, 5 th. World Jewish Cong., Jerusalem, (1969).

3. J.-M. Fournier and J-Ch. Viénot, Fourier transf. holog. used as matched
   filters in hebraïc Paleography, Israel Journ. of Techn., 9, 281 (1971).

4. J.-M. Fournier and J.-Ch. Viénot, Mesures sur des tracés de lettres au
   moyen de techn. holog., in Les Techn. de Lab. dans l'Etude des Manuscrits,
   C.N.R.S., Paris, 1972.

5. J.-Ch. Viénot, J. Duvernoy, G. Tribillon, and J.-L. Tribillon, Three
   methods of inf. assessment for opt. data proc., Appl. Opt., 12, 950 (1973).

6. A. Vander Lugt, F.B. Rotz and A. Klooster, Optical and Electro-optical
   Inf. Proc., 135, Tippet and al., M.I.T., 1965.

7. L. Bragg, Lightning calculations with light, <u>Nature</u>, 3898, 69   (1944).

8. S.H. Lee, Math. operations by opt. proc., <u>Opt. Engineer.</u>, 13, 196 (1974).

9. See for ex.: O.M. Minot, Automatic devices for recogn. of visib. two-dim.
   patterns, a surv. of the field, <u>Techn. Memo. 364</u>, U.S. Navy Electr. Lab.,
   1959.

10. B. Cretin, Correl. en éclairage incohérent ; appl. à la superposition
    d'images et à la réalisation de formes moyennes, <u>D.E.A.</u>, Besançon, 1974.

ONE-DIMENSIONAL FOURIER TRANSFORM
FOR THE INSPECTION OF PHOTOMASKS

L. U. Almi* and J. Shamir

Dept. of Electrical Engineering, Technion - Israel Institute of Technology, Haifa, Israel

ABSTRACT

The inspection of photomasks is very important for the production lines in the manufacturing of integrated circuits. A recently developed method of pattern comparison by one-dimensional optical Fourier transform is generalized and applied for the detection of defects in periodic structures. Experimental results prove the feasibility of the technique that offers a simple and real-time detection of defects.

1.   INTRODUCTION

Various methods are available for pattern recognition [1,2]. Most of these methods involve some kind of photographic steps, making real time data processing impractical. Various attempts were made to overcome this difficulty, and recently a method was described [3], where one-dimensional Fourier transform (1DFT) was applied for real-time pattern comparison.

In the present work it is shown how the above mentioned method may be used for the inspection of photomasks. For completeness, in Section 2 the method described in [3] is shortly reviewed and applied directly for the inspection of photomasks. However, photomasks are frequently produced as a repetitive pattern with a large number of identical forms. Although the method described in Section 2 is adaptable to such masks by a scanning procedure, a generalization is made in Section 3, especially directed towards the detection of defects in periodic structures.

2.   THE BASIC PROCESS

The basic process involves the comparison of two given patterns $f(x,y)$ and $h(x,y)$. The patterns are placed side by side on a transparency and an optical 1DFT is performed on the x-axis, while keeping the y-axis unchanged apart from some magnification. Thus we may define the complex amplitude transfer function of the input plane as

$$t(x,y) = f(x,y) + f(x-d,y) + g(x,y) \tag{1}$$

where

$$g(x,y) = h(x,y) - f(x-d,y) \tag{2}$$

---

*L.U. Almi is now with Israel Electro-Optical Industry, Rehovot.

is the difference between the patterns f(x,y) and h(x,y), and d is the relative displacement of the two patterns. The intensity distribution in the output (Fourier) plane may be described by

$$I(u,y) = 4\cos^2\left[\frac{\pi u d}{\lambda F}\right]|F(u,y)|^2 + |G(u,y)|^2 +$$

$$+ 2\text{Real}\left\{\left[1 + \exp\left(-2j\ \frac{\pi u d}{\lambda F}\right)\right]F(u,y)G^*(u,y)\right\} \qquad (3)$$

where F(u,y) and G(u,y) are the respective 1DFT of the functions f(x,y) and g(x,y), F is the effective focal length of the transforming system, and λ is the wavelength of the illuminating light.

The first and third terms vanish for

$$u = \frac{(2n+1)\lambda F}{2d} \qquad (4)$$

where n is an integer. Thus at these positions only light from the difference function G(u,y) may appear.

This fact was effectively utilized in [3] for the recognition of digits. In the same way very small defects in photomasks may be detected in real time. Fig. 1 shows the intensity recorded along a line parallel to the y-axis for a value of u, according to eq. (4). In (a) two perfect masks were compared, while in (b) one mask included a defect occupying about 0.1% of the area of one pattern.

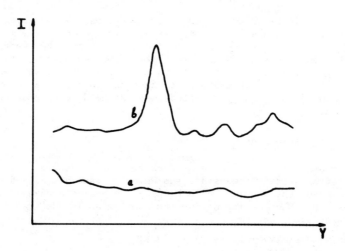

Fig. 1. Photoscan along a constant u for comparison of
two basic cells of an array as in Fig. 4.
(a) Two identical cells;   (b) one cell had a minute defect.

3. DEFECT DETECTION IN PERIODIC STRUCTURES

In the production process of integrated circuits, usually photomasks with a large number of repetitive structures are used. Since the number of rejected units strongly depends on the quality of the photomasks, a simple automatic inspection method for these masks is of great value.

A number of methods for such an inspection were suggested in the literature, some relying on the periodicity of the inspected structure [4,5]. However, these methods involve the inconvenient process of producing a photographic filter for each photomask. Although the method described in Section 2 is applicable here, it demands a time consuming scanning process. It was found that this may be avoided if the periodicity of the photomask is utilized by applying the 1DFT to the whole mask. Thus, for this case we have in the input plane of our system a transfer function of the form

$$t(x,y) = \sum_{n=o}^{N} f(x-nd,y) + g(x,y) \qquad (5)$$

where $f(x,y)$ is a column of perfect basic cells that forms a planar array, (i.e. as in Fig. 4), and $g(x,y)$ is the defect to be detected. The optical system produces the 1DFT of $t(x,y)$:

$$T(u,y) = \sum_{n=o}^{N} \exp\left(-\frac{2\pi j n d u}{\lambda F}\right) F(u,y) + G(u,y)$$

or

$$T(u,y) = \frac{\sin\frac{(N+1)du}{\lambda F}}{\sin\frac{\pi du}{\lambda F}} \exp\left(-\frac{j\pi N du}{\lambda F}\right) F(u,y) + G(u,y) \ . \qquad (6)$$

The intensity to be detected in the output plane is proportional to the squared absolute value of eq. (6), which resembles eq. (3). The difference is that instead of a sinusoidal modulation we have the modulation proportional to

$$\sin\left[\frac{\pi(N+1)du}{\lambda F}\Big/ \sin\frac{\pi du}{\lambda F}\right]^2 , \qquad (7)$$

that approaches a comb function as N increases. In such an intensity pattern it is difficult to locate the zero positions, however, there is a wide region of low intensity level if $G(u,y)$ is absent. Therefore the intensity distribution produced by a defect $g(x,y)$ can easily be detected by integrating the energy in such a dark region.

4. EXPERIMENTAL RESULTS

The inspection technique as described in the previous section was investigated by a number of measurements, where controlled defects were introduced in various periodic masks. The optical system used a collimated laser beam illuminating the photomask, and a simple combination of a spherical and a cylindrical lens that performed the required transformations. A light-sensitive cell scanned the output plane parallel to the y-axis for a constant u. The aperture of the detector was adjusted to collect the light from an appreciable

fraction of one dark band in the output plane. Fig. 2a illustrates one of the
masks, where one circle is displaced. Fig. 2b is a photograph of the output
plane, while Fig. 2c is the trace of a scan along a dark region indicating the
strong double peak corresponding to the row (a specific value of y), where
the defect appears.

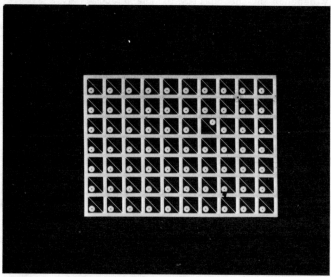

Fig. 2.   (a) A repetitive pattern with one circle misplaced.

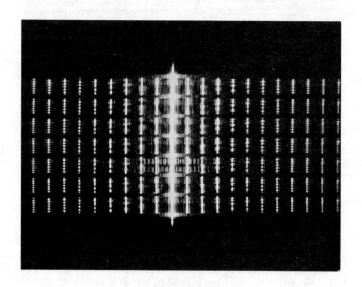

Fig. 2.   (b) 1DFT of (a).

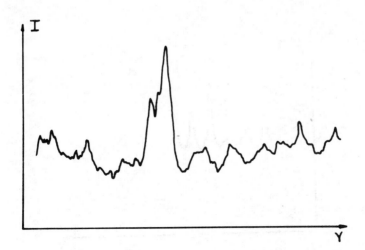

Fig. 2. (c) Scan along a dark band (vertical direction) of (b).

Fig. 3 represents another experiment, where one of the dark spots is missing. Although the defect is minute, it is detectable on the trace by a higher peak.

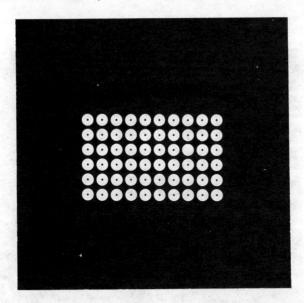

Fig. 3. As Fig. 2, the defect being a missing dot
         in the center of one of the circles.
         (a) The pattern.

Fig. 3.   As Fig. 2, the defect being a missing dot
          in the center of one of the circles.
          (b) Photoscan as in Fig. 2c.

For the final check we used a real IC photomask where we introduced a defect
(Fig. 4) that also clearly manifested itself on the trace.

Fig. 4.   As Fig. 3, but with a real photomask
          with a defect indicated by arrow.
          (a) The pattern.

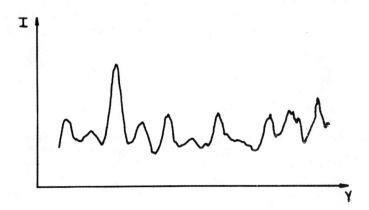

Fig. 4.   As Fig. 3, but with a real photomask
          with a defect indicated by arrow.
          (b)   Photoscan as in Fig. 2c.

5.   CONCLUSION

A simple method was described for the real time detection of defects in photomasks. The method described uses a scanning device that may be replaced by a
detector array to facilitate automatic detection.

At present it appears that for the detection of very small defects, the original comparison method of two patterns has a better SNR than the present generalization for periodic structure. However, it is reasonable to believe that
most of the defects occuring in practice are detectable by the new method for
which the ultimate detection level has still to be determined.

ACKNOWLEDGMENT

We wish to thank Mr. Y. Feinman of our faculty for performing part of the
measurements.

REFERENCES

[1]   A. Vander Lugt, Proc. IEEE, 62, 1300, 1974.

[2]   J.R. Ulmann, Opto-Electronics, 6, 319, 1974.

[3]   L.U. Almi, J. Shamir, Opt. Com. In print.

[4]   L.S. Watkins, Proc. IEEE, 57, 1634, 1969.

[5]   G. Winzer, N. Douklias, H.P. Kraft, Frequenz, 25, 220, 1971.

NEW PROGRESS IN OPTICAL WRITING APPRAISAL APPLIED TO THE "DREYFUS AFFAIR"

Jean-Marc Fournier

*Laboratoire de Physique Générale et Optique (Laboratoire associé au C.N.R.S.;
Holographie et traitement optique des signaux) Université de Franche-Comté
25030 Besançon Cedex, France*

ABSTRACT

Statistical informations dealing with manuscripts are extracted by enhancing characteristic spatial frequencies of writings. New experiments give closer results in writing appraisal of Dreyfus' files.

A film concerned with Optics and the study of writings is presented before this lecture.

INTRODUCTION

One of the perennial problems for those dealing with the study of writings is to identify the scriptor of a text. Generally the transfer from a pattern known beforehand to the drawing of a written letter is described by a complicated non linear process, but linear space-invariant optical imagery may sometimes represent a good approximation. It appears, then, as an application of the interesting property of lenses consisting in their large processing capacity. An experimental approach to the limits of the possibility of evincing the characteristic features of a scribe over specific frequency bands of the Fourier domain is achieved in some files of the Dreyfus affair.

THE CONCEPT OF FORM FACTOR

Anyone who has ever received a note from a friend knows how often it is possible to identify the sender at once from his or her hand-writing. Similarly in optical processing, recognition of the pertinent features of a handwritten sample is operated. Writing characters is drawing ; each letter can be described as made up of a *stereotype* and a disturbance. The stereotype is the very concept of the letter, rooted in our memory. The disturbance function makes it possible to step from the stereotype to the manuscript form as seen by the eyes. We call it *form factor* because it rules character making : letters can be angular or rounder, moreless inclined, supple, rough, etc...
The transfer from a given pattern to the graphism of a written letter is an a priori non-linear function. However, in many cases, linear approximation gives a good representation of it. This permits to use space-invariant optical imagery. As an example, figure 1 shows how, starting from an arbitrary stereotype of the words *" Laboratoire d'Optique Besançon"* the letters morphology is modified by correlation involving three different functions successively (1).

SEARCHING FOR INVARIANTS

As noted earlier, in spite of the many
possible varieties of drawn letters
that derives from the writer's environment at the time the text was written,
the "brand" remains, characterizing
the writer's "hand".
Instead of decomposing letters into
geometric elements by sequential analysis as in some classical expert's
reports, we analyse each form in terms
of its spatial frequencies by Fourier
transform. Our purpose being to extract
statistical data coming from the out-
lines of the letters, we endeavour to
suppress what corresponds to the geometrical coordinates of the letters in
the page. This involves to use sphero-
cylindrical optics to make the best of
the fact that writing is drawn along
parallel lines.
As nobody can exactly draw a stereotype,
the disturbance can never be exactly
deduced and therefore, nor can one
know what in the spectrum of a letter

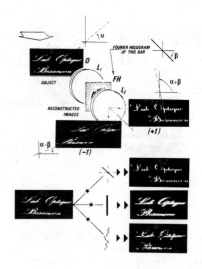

Fig. 1. Perturbation of a
model by optical correlation.

is dependent on the form factor. However, by comparing two identical letters,
two letters a, for instance, one finds the differences inputable to the disturbance only. If one considers several texts, in the same language and of sufficient length, they will reflect the same probability of each having, for the
same number of letters, as many a's, as many b's, an so on, even if their con-
tents are different. When this very important condition is fulfilled comparisons between spectra of different texts can bring out informations concerning
the form factors. In other words, when the frequency of appearance of letters
remains stable in various texts, the letters they contain can bee successfully
analysed by optical computers (2).

The method uses a sphero-
cylindrical optical sys-
tem through a two-step
process as summerized in
figure 2. At first we
record the square moduli
of spectra of the texts
undertest. These spectra
include characteristic
frequencies of the letter
morphology and frequen-
cies coming from the po-
sition of letters in the
texts. Secondly, a uni-
dimensional Fraunhofer
diffraction of these re-
cordings gives a uni-
dimensional autocorrela-
tion of the texts.

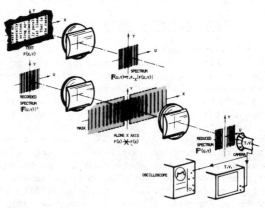

Fig. 2. Extraction of Form Factors.

By filtering off all the terms corresponding to the positions of letters in a text, spectra are observed as the results of an other diffraction. Then, characteristic frequencies appear : they come from the drawing of letters. At this stage, the position factor of the letters is removed. An electronic filtering is in cascade with that process. It enhances the features we seek for. The results are displayed on a T.V. screen. The video signal is filtered and displayed on an oscilloscope (3).
Nota : The letters in various texts have not necessary the same height. That is why a photographic operation is convenient for matching their graphisms : all the characters must have a common "average size" in order to obviate any subsequent discussion about the results.
If some frequency bands remain unchanged among the spectra of several texts written by the same person, these bands can be considered as the invariants for the individual's hand-writing. As such, they are the constants that must be brought out in making the expert analysis of a document (4).

THE DOCUMENTS OF THE DREYFUS AFFAIR

Let us briefly recall some historical events. In 1894, a cleaning lady working for the French Intelligence Service intercepts -at of a wastebasket at the German Embassy- a handwritten memo listing a number of documents pertaining to the national defence and intended for the enemy : that is the "bordereau" : Fig. 3. The investigations were initially conducted in secret within the War Ministry. A jewish officer Captain Dreyfus, was suspected, and curiously several expert's reports

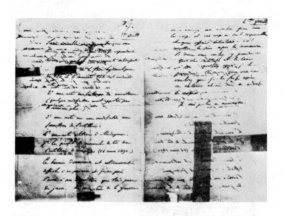

Fig. 3. The "bordereau".

Monsieur le Président de la Ré
Accusé, puis condamné,
une preuve d'écriture, pour le crime
le plus infame qu'un soldat puisse
commettre, j'ai déjà déclaré, et je
déclare encore que je n'ai pas écrit
la lettre qu'on m'impute, que je n'a

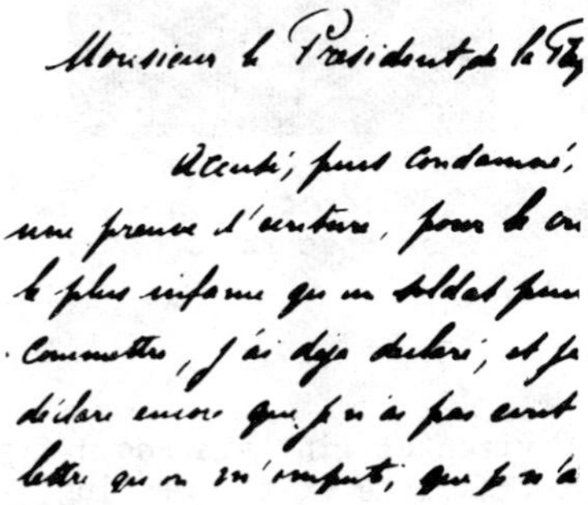

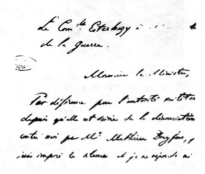

Fig. 4. Dreyfus's writing.

Le Comte Esterhazy à
de la Guerre.
Monsieur le Ministre,
Par déférence pour l'autorité militaire
depuis qu'elle est saisie de la dénonciation
contre moi par Mr Mathieu Dreyfus,
je me suis imposé le silence et je ne réponds ni

Fig. 5. Esterhazy's writing.

pointed out that the peculiarities of the letters -the "idiosyncrasies"- of
the memo were found in the writings of Dreyfus. In 1895 Alfred Dreyfus was
sentenced to jail and sent to the Devil's Island. His family and some friends
of his claimed that there had been a plot against Dreyfus ; politicians and
journalists went into polemic about the affair and quite rapidly another
French officer, Count Esterhazy, was suspected. In 1905, after a famous review trial, Dreyfus was discharged.
In the lot of documents contained in the files of the Dreyfus Affair we chose
two letters by Alfred Dreyfus : Fig. 4, sent while he was in detention : one
on the War Minister (February 14, 1895), the other to the President of the
Republic (October 5, 1895), and three pieces of correspondence from Count
Esterhazy : a letter to the War Minister (November 20, 1897) : Fig. 5, a letter to a baron (June 29, 1894), and a letter to the War Minister (September
14, 1898).

RESULTS

Fig. 6 is a representation of the spectra of 3 processed writings (Dreyfus,
bordereau, Esterhazy). Frequencies are displayed along the horizontal axis.
A classical imagery is made along the vertical direction.

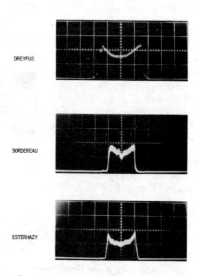

Fig. 6. 1-dimensional   Fig. 7. Densitometric analysis
  spectra of processed   of spectra.
        writings.

The photometric curves corresponding to one horizontal line are shown figure 7.
They represent a "line" taken perpendicular to the line of the writing, and
covering the entire height of the text. It is then necessary to add all such
densitometric curves in order to carry out a statistical analysis of the overall text. These sums of the analysis obtained by using television signals

are displayed on an oscilloscope as
shown in figure 8. They emphasize the
main characteristics of each writing
and allow their discrimination.
It is noted that in all cases the characteristic frequencies are situated
in the medium and low frequency bands
(5). This proves that invariant will
generally be sought for in the rough
parts of the handwriting (manner of
shaping the letter) rather than in the
minute details. This information can
successfully help experts because they
traditionally focus their attention on
the idiosyncrasies and details of the
writings they have to appraise.
It would be easy to determine the similarity between Esterhazy's writing and
that of the bordereau ; Unfortunately
a degree of resemblance which not at-
taind 100 % cannot provide indisputa-
ble proof that Esterhazy wrote the
memo. However, there is no overlapping
between the spectral form factor of the
memo and those of the Dreyfus writing
to reject the Captain as a possible
writer of the famous document. Facing

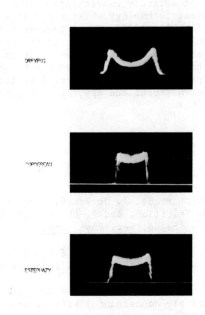

Fig. 8. Sum of photometric curves.

the assumption that the size of the
sample -that is the bordereau- is sufficient -as far as the number of letters
is concerned-, it can be said that Dreyfus has definitely not written the
bordereau.

RECENT ADVANCES

Up to now most of the results appear as statistical. It seems of interest to
know how far this kind of investigation can go in the domain of the individual.
However one shuld keep in
mind not to loose any in-
formation. To do so it is
necessary to work with the
complete geometrical field
therefore to play with
spherical optics. Experi-
ments are on progress.
Preliminary results are
illustrated in figure 9
where spectral representations are given of a text,
a part of text, and a word
written by Dreyfus and a
text, a part of text, and
the same word by Esterhazy
respectively. Series of
measurements performed on
about an hundred pictures

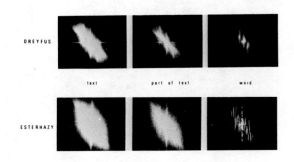

Fig. 9. Spectra of writings.

are not entirely convincing of the necessity of such a sophisticated and
complexe technology whist rapid looks at the spectra seem to give the evidence of obvious differences between the two writings (see for example that sort
of whiskers observable in the Dreyfus's that cannot be detected in the
Esterhazy's at least in the study of a few particular words).
At this point no conclusion can be made. Indeed we probably have to consider
new parameters for much more fine interpretation, for instance thickness of
the downstrokes and upstrokes which are obviously different of the two
writings.

REFERENCES

1. D. Charraut, J. Duvernoy, J.-M. Fournier, J.-P. Goedgebuer, A. Lacourt,
   G. Tribillon, and J.-Ch. Viénot, Quelques aspects récents du trait. opt.
   des images, Rev. Phys. Appl., 11, 227 (1976).

2. J. Duvernoy, Erreurs liées à la détermination des facteurs de forme en
   traitement optique de l'information, Opt. Comm., 16, 350 (1976).

3. J.-M. Fournier, Approche analogique d'une expertise en écriture : un
   exemple concernant l'affaire Dreyfus, Le Courrier du C.N.R.S., 16, 23
   (1975).

4. J.-M. Fournier and J.-Ch. Viénot, Mesures sur des tracés de lettres au
   moyen de techn. holog., in Les Techn. de Lab. dans l'Etude des Manuscrits,
   C.N.R.S., Paris, 1972.

5. J. Duvernoy and D. Charraut, Détermination de la bande semantique et de la
   bande graphique dans un texte imprimé, Opt. Comm., 14, 56 (1975).

SESSION 15.

BIO-MEDICAL APPLICATIONS

Chairman: R. WEIL

QUANTUM NOISE IN CODED APERTURE IMAGING

Francis Sanchez

Institut d'Optique, Université de Paris-Sud, 91405 Orsay, France

ABSTRACT

The signal to quantum noise ratio is calculated for general Coded Aperture Imaging. The geometrical parameters are optimized, assuming a quasi-uniform object and ideal receptors, and taking into account a parasitic radiation. The optimal potential performances are compared with that of Pinhole Imaging. The conditions for the intermediate receptor in order to achieve these potential performances are also derived.

INTRODUCTION

The technique of Shadow Holography has been applied to X-ray astronomy (1) and to γ-ray bio-medical applications (2). These applications lie in domains where the only available image forming system is pinhole imaging (P I), and where the limiting factor for quality of images is the quantum noise.

In fact, Shadow Holography is a particular case of Coded Aperture Imaging (C A I), In general the coding function is not necessarily a Fresnel Zone Plate, and hence reconstruction process is different. This field is a rapidly growing one, and it seems there is need to clarify the ideas concerning the quantum signal to noise ratio (S N R). This question has been investigated by several authors (3,4,5) for Shadow Holography taking into account only the receptor properties as constraint factors before optimization. It seems that one more fundmental way of dealing with this problem is to compute intrinsic S N R performance (i.e. the detector being in no way limiting), leading to optimization of the coding function and geometrical parameters, and finally derive the conditions the detector must fulfil in order to achieve this intrinsic performance. This is the purpose of the present paper.

Of primary importance will be the comparison between S N R performances of C A I and P I optimized under the same constraints. However, it must be kept in mind that C A I differs from P I in two respects. First, it is a two-step process, meaning that S N R and collection efficiency are not simply related. Second, it has tomographic properties, i.e. one may select one plane section of the object during the reconstruction. This last point means that C A I can be significantly compared to static P I only for plane object. As dynamic imaging systems will not be considered in the present paper, we limit our considerations to plane objects.

It is also supposed that the object is quasi-uniform, i.e. it has details of small contrast, as is the case in tumor-detection applications. We only consider the S N R value at the center of the image. The constraint can be summarized as follows : a plane quasi-uniform object of area s and total quantum radiance N_o is to be resolved by the imaging system into M cells.

It is immerged in an unlimited parasitic radiation whose relative luminance, with respect to that of the object, is q. The minimal attainable distance from the object is ℓ_m. This last parameter is, of course, of primary importance since if $\ell_m = 0,$ optimization is trivial, consisting in putting receptor and object into contact.

557

In Fig.1 the geometrical parameters to be optimized under the above constraints are represented. These are : ℓ the distance object-code, the magnification γ defined as the ratio between code-receptor and object-code distances, and α, the angular radius of co de-contour viewed from the center of the object. For P I the geometrical parameters are the same, but α and ℓ are not independent.

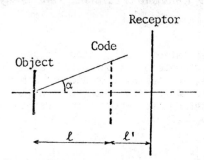

Fig. 1. Geometrical parameters

The magnification is $\gamma = \ell'/\ell$.

PRINCIPLE OF CODED APERTURE IMAGING

When quantum noise is not considered, one may denote the object by a continu-ous function $O(x,y)$. For simplicity we shall denote it simply by O.The inter-mediate receptor receives a coded image I' by a convolution relation involving the code transmission :

$$I' = O * \tau \qquad\qquad\qquad (1)$$

In the second step, this picture is decoded by a convolution operation with a decoding function :

$$I = I' * \tau' \qquad\qquad\qquad (2)$$

The overall process is a linear one with impulsional response $\tau * \tau'$. For effective decoding, this function must have a high maximum at its center, but it is not generally sufficient for correct imaging. In particular, while one has liberty to choose τ' (for instance by computional decoding). τ is necessari-ly real positive resulting in a shoulder for the response $\tau * \tau'$. This shoulder can introduce important distorsions, in particular for large objects. Fortunately the undesired terms in the development of τ may be eliminated by the techni-que of multi-coding that consists in successively recording several coded images with different codes and to deal with a linear combination of these coded images. The over-all process is still linear but the distorsions are greatly reduced, as demonstrated by Houle and Joy {6} in the case of three-fold shadow holography. It can be shown that the S N R is reduced only by a factor of about $\sqrt{3}$. In the same way, two-fold coding associated with incohe-rent reconstruction is feasible , although the overall system cannot transmit low spatial frequencies.

Another type of distorsion appears in Shadow Holography in the case of cohe-rent reconstruction (not computed), since the final optical detector is essentially non-linear, producing non-linear distorsions.

However in this case the d.c. bias is not a handicap, but rather an advantage, as it reduces non-linearities {7}. It has been shown that the over-all S N R is reduced by a factor of $\sqrt{2}$ (3).

From these considerations, it results that the S N R value may be calculated for the simple scheme of Eq.(1) and (2). The order of unity factors coming from multi-coding or non-linearity will be easily deduced in each particular case. It is also to be noticed that this simple scheme can be applied to the P I as well, the second step being merely the image readout.

S N R GENERAL EXPRESSION

The last remark indicates that one may derive a general expression valid for P I and C A I, as well as for coherent or incoherent reconstruction. Indeed, the total signal at the image central cell may always be written :

$$A = \sum_{n=1}^{n'_o} e^{j\phi_n} + \sum_{n=1}^{n'_p} e^{j\phi'_n} \tag{3}$$

where n'_o is the random number of photons that has come from the object central cell which plays a role in the reconstruction, while n'_p is the random number of photons that play a role in the reconstruction but do not come from the object central cell. Of course n'_o and n'_p are independent Poisson random variables, and are independent with the statistics of the ϕ_n. By defining the signal as the mean value of the first term of A, and the noise as the r.m.s. of A, one obtains

$$S N R = \frac{N'_o}{\sqrt{N'_t}} < e^{j\phi} > \tag{4}$$

where $N'_o = \langle n'_o \rangle$, $N'_t = \langle n'_o \rangle + \langle n'_p \rangle$. For incoherent reconstitution $\langle e^{j\phi} \rangle = 1$. For coherent reconstitution with Fresnel zone plate $\langle e^{j\phi} \rangle = 1/2$ (sinusoidal zone) or $1/\pi$ (binary zone). In the following we shall drop the term $\langle e^{j\phi} \rangle$, as well as the coefficient in $N'_o N'_t{}^{-1/2}$ that depends on code transmission τ and τ' (which is of order of unity for binary functions). We now concentrate our attention on the geometrical parameters. Denoting by δs, 0, C, D the respective contours of central object cell, object, code and decoding diaphragm (located on receptor plane), one obtains

$$N'_o = \frac{N_o}{\pi s} \min |G(\delta_s, C); G(\delta_s, D)|$$

$$N'_t = \frac{N_o}{\pi s} |G(0, C, D) + q\, G(\infty, C, D)|$$

where $G(0,C,D)$ is the geometric extent sustained by object, code and decoding contours, and $G(\infty,C,D)$ is the same quantity when dimensions of the object increase indefinitly.

If, for a fixed C, D is increased from the zero value, N_o' increases more rapidly than $\sqrt{N_t'}$, until D reaches the extent C' (projection of code contour on receptor plane from the central object point). For a greater extent of D, N_o' is stationnary, but N_t' still increases. It results that optimal choice of D is D $\equiv$ C'. So one obtains

$$S\,N\,R = \left[\frac{N_o\,\pi\,s\,\sin^4\alpha}{M^2\,[G(0,C,C') + qG(\infty,C,C')]}\right]^{1/2} \tag{5}$$

PINHOLE IMAGING

For the P I system, α is determined by the desired resolution

$$\alpha = \sqrt{\frac{s}{\pi\ell^2 M}}\,\frac{\gamma}{1+\gamma} \qquad (\alpha<<1)$$

and $\qquad G(0,C,C') \equiv G(\infty,C,C') = \pi\,\alpha^2\,\frac{s}{M}$

So the S N R for P I, denoted $S\,N\,R_{p.i}$ is

$$S\,N\,R_{p.i} = \frac{\gamma}{1+\gamma}\left[\frac{N_o\,s}{M^2\pi\ell^2\,(1+q)}\right]^{1/2} \tag{6}$$

The optimization of geometrical parameters leads to a choice for ℓ its minimum value ℓ_m and for γ a value $>>1$, so the optimal S N R denoted $(S\,N\,R)_{p.i}$ is

$$(S\,N\,R)_{p.i} = \left[\frac{N_o}{M^2 p(1+q)}\right]^{1/2} \tag{7}$$

with $\qquad p = \dfrac{\pi\ell_m^2}{s}$ $\tag{8}$

The condition for (7) to be valid is pM $>>$ 1.

CODED APERTURE IMAGING S N R

For C A I one must distinguish between large object region (p<<1) and small object region (p>>1). In either cases calculation of the geometrical extent G can be made only at the limits $\gamma<<1$ or $\gamma>>1$.

Large Object Case (p<<1)

$\underline{\gamma<<1.}$ In this case $G(0,C,C') = G(\infty,C,C') = (\pi\ell\,\mathrm{tg}\alpha)^2$ and

$$S N R = \left[\frac{N_o s}{M^2 \pi \ell^2 (1+q)} \right]^{1/2} \sin\alpha \cos\alpha \tag{9}$$

So optimal values for ℓ and α are ℓ_m and $\frac{\pi}{4}$. S N R has then half the value of $(S N R)_{p.i}$.

$\underline{\gamma \gg 1}$. Then $G(0,C,C') = G(\infty,C,C') = (\pi\ell \sin\alpha \, tg\alpha)^2$ and S N R is proportional to $\cos\alpha$. Optimal value is $\alpha \to 0$ but this solution is not distinct from that of the P I system.

<u>Small Object Case (p≫1)</u>

$\underline{\gamma \ll 1}$. One has $G(0,C,C') = \pi s \sin^2\alpha$ and $G(\infty,C,C') = (\pi \ell \, tg\alpha)^2$, so

$$S N R = \left[\frac{N_o \sin^2\alpha}{M^2 \left[1+q \, \frac{\pi\ell^2}{s} \, \frac{1}{\cos^2\alpha}\right]} \right]^{1/2} \tag{10}$$

Optimal value for ℓ are α are ℓ_m and Arc $tg(1+1/pq)^{-1/4}$, and

$$opt \, S N R = \frac{\sqrt{N_o}}{M} \frac{1}{\sqrt{pq} + \sqrt{1+pq}} \tag{11}$$

$\underline{\gamma \gg 1}$. One has $G(0,C,C') = \pi s \sin^2\alpha$ and $G(\infty,C,C') = (\pi \ell \sin\alpha \, tg\alpha)^2$, so :

$$S N R = \left[\frac{N_o \sin^2\alpha}{M^2 \left[1+q \, \frac{\pi\ell^2}{s} \, tg^2\alpha\right]} \right]^{1/2} \tag{12}$$

Optimal value for ℓ and α are ℓ_m and Arc $tg(pq)^{1/4}$, and

$$opt \, S N R = \frac{\sqrt{N_o}}{M} \frac{1}{1+\sqrt{pq}} \tag{13}$$

All these results are summarized in Table 1. It shows that the advantage in geometrical collection efficiency of C A I over that of P I, is accompanied by an advantage in S N R only for small objects and weak parasitic field. In other cases the optimized S N Rs are of the same order of magnitude.

CONDITIONS ON INTERMEDIATE RECEPTOR

In the preceding, it has been supposed that intermediate receptor was ideal. Now, we are looking for the conditions it must fulfil in order to realize the ultimate performance calculated above. The receptor parameters of interest are ε quantum efficiency for object radiation, ε' quantum efficiency for parasitic radiation, S the area, P the number of resolved cells, Q the receptor noise equivalent number of counts. Concerning the influence of the receptor noise it must be noted that it is not, as is also true for the quantum radiation noise, directly related to the geometrical collection efficiency.

The final detector will be always assumed to be ideal, since at the reconstruction step, time is unlimited.

The influence of quantum efficiencies ϵ and ϵ' in S N R value (Eq.5) is to replace N_o by ϵN_o and q by $q\frac{\epsilon'}{\epsilon}$. Of course the optimal choice is $\epsilon = 1$ and $\epsilon' = 0$, but this trivial elimination of parasitic field will not be considered to be achievable.

The minimum condition for S is derived by writing that the projection of the code on the detector plane from the extreme point of object must lie within the detector area :

$$S_{min} = \gamma^2 \, s\left[1 + \frac{1+\gamma}{\gamma} \, \sqrt{p} \, tg\alpha\right]^2 \tag{14}$$

The minimum condition for P is derived by writing that the projection on detector of one object cell from the center of the code must have an area inferior to that of the receptor resolution cell :

$$P_{min} = \frac{M \, S}{\gamma^2 s} \tag{15}$$

The maximal tolerable value Q_M of Q is obtained by writing that the equivalent number of receptor noise counts on the domain C' is equal to the collection number N_t'. Introducing the geometric efficiency ratio $\eta = N_t'/N'_{tPI}$, one obtains

$$Q_M = N_o(1+q)\eta\frac{S}{s} \frac{1}{p^2M^2(1+\gamma)^2 \, tg^2\alpha} \tag{16}$$

The applications of Eqs.(14), (15) and (16) are given in the Table. Concerning specific noise problems at the reconstruction stage, these will be facilitated for high counting levels, i.e. for high values of η. It is seen from the Table that η is of order $(pM)^2$ for the cases for which $\alpha_{opt} = \pi/4$ and of order pM^2 in the other cases. In this respect, C A I is always preferable to P I since $pM \gg 1$ (if $pM \ll 1$ one-step "contact" recording is better).

CONCLUSION

The geometrical parameters in the coding stage step have been shown to determine the overall quantum S N R value of a general C A I process. The optimal configuration has been defined for different cases, assuming the intermediate receptor to be ideal. The optimized S N R values for C A I and P I are of comparable performance except for the case $p \gg 1$, $pq \ll 1$ (i.e. small object and weak parasitic field) for which C A I gives a better S N R value by a factor $\sqrt{p}$, whatever the number of resolution points may be. Conditions on the intermediate receptor are also computed. These conditions are very different for P I and C A I. As a matter of example, let us take a specific case in Nuclear Medicine, as a human liver for which $\ell_m \simeq 5$ cm, $p \approx 4$, $q \approx 1$, and let us compare two different systems.

First, in an Anger camera ($S = 25 \times 25$ cm^2, $P \simeq 600$) associated with P I, second a reinforced photographic film ($S = 35 \times 35$ cm^2, $P \simeq 10^5$) associated with C A I. This second system is preferable when high-resolution is necessary ($M > 600$). Experiments are conducted in our laboratory in this domain.

BIBLIOGRAPHIE

{1} L. Mertz and N.O. Young, Fresnel Transformation of Images. Proc. Int Conf. Opt. Inst., London, 305, (1961).

{2} H.H. Barrett, Fresnel Zone Plate Imaging in Nuclear Medecine. J. Nucl. Med., 13, 382, (1972).

{3} H.H. Barrett and G.D. de Meester, Quantum Noise in Fresnel Zone Plate Imaging. Technical Memory, Raytheon Company (1973).

{4} A.G. Lindgren, D.K. Guha and J.E. Spence, A Noise Analysis of Fresnel Zone Plate Imaging Systems. International Optical Computing Conference, Zurich, 54, (1974).

{5} M.L.G. Joy and S. Houle, The Potential Performance of Off-axis Fresnel Zone Plate Gamma Imaging Systems on Arbitrary Objects.

{6} S. Houle and M.L.G. Joy, On Axis Multiple Fresnel Zone-Plate Coded Apertures. Report n° 16 of Inst. of Biomed. Eng., University of Toronto, 1975.

{7} J. Fonroget, Y. Belvaux et S. Lowenthal, Fonction de Transfert de Modulation d'un Système de Gammagraphie Holographique. Opt. Comm.,15, 76 (1975).

F. Sanchez

	p	pq	γ	α_{opt}	ρ_0	n	S_{min}	P_{min}	$\dfrac{Q_{max}}{N_0(1+q)}$
P I	$\ll 1$		$\gg 1$	$[pM]^{-1/2}$	1	1	$\gg s$	M	$[pM]^{-1}$
			$\ll 1$	$\pi/4$	$1/2$	$(pM)^2$	ps	$\gg Mp$	p
C A I	$\gg 1$	$\ll 1$	$\gg 1$	$\mathrm{Arc\ tg}[pq]^{-1/4}$	$\sqrt{p}$	pM^2	$\gg s\left[\dfrac{p}{q}\right]^{1/2}$	$M\left[\dfrac{p}{q}\right]^{1/2}$	1
							$s\left[\dfrac{p}{q}\right]^{1/2}$	$\gg M\left[\dfrac{p}{q}\right]^{1/2}$	
		$\gg 1$	$\gg 1$	$[pq]^{-1/4}$	$\left[1+\dfrac{1}{q}\right]^{1/2}$	$\dfrac{pM^2}{1+q}$	$\gg s\left[\dfrac{p}{q}\right]^{1/2}$	$M\left[\dfrac{p}{q}\right]^{1/2}$	$\dfrac{1}{1+q}$
			$\ll 1$	$\pi/4$	$\dfrac{1}{2}\left(1+\dfrac{1}{q}\right)^{1/2}$	$\dfrac{(pM)^2}{1+1/q}$	ps	$\gg Mp$	$\dfrac{p}{1+1/q}$

Table 1. Geometrical optimization and corresponding S N R values (ρ_0) with respect to the S N R of the P I. n : collection efficiency ratio. The three columns on the right indicate the conditions for the intermediate receptor. Q_{max} is the tolerable equivalent number of the receptor noise counts on the whole receptor surface.

IMAGE ENHANCEMENT FOR OBJECTS EMBEDDED IN SCATTERING MEDIA

Z. Weinberger* and E. Lange**

*Jerusalem College of Technology, Jerusalem, Israel
**National Physical Laboratory of Israel, Jerusalem, Israel

ABSTRACT

Images of objects embedded in scattering medium may be enhanced by spatial filtering in a coherent optical system. Spatial filtering is proposed to filter the optical image of a transilluminated breast for the diagnosis of mammarian growths.

INTRODUCTION

In 1929 M.Cutler proposed the transillumination of the human breast as a diagnostic tool for mammary growths.[1] Skilled analysis of the shadow patterns formed by transillumination allow malignant and benign tumors to be distinguished.

In contrast to X-ray analysis, the optical examination does not destroy normally healthy cells. Consequently the breast may be reexamined as often as medically necessary without risk of harm. However, only a few experts have learned to analyze the subtleties of the shadow pattern for tumor analysis. The difficulty in analysis is due to scattering properties of the fatty breast tissue which considerably diffuses the shadow pattern.

E.L. O'Neill has shown both theoretically and experimentally how spatial filtering may be applied to filter additive noise from a signal when the spatial frequencies of the two signals are sufficiently different.[2] In this study we propose a similar optical system for filtering the tumor signal from the light scattered by the breast tissues.

Figure 1 shows how much an optical system would be applied for an examination.

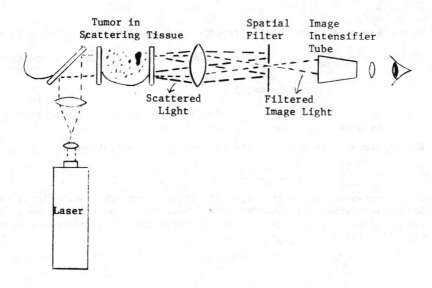

As a preliminary test of our proposal we have studied the enhancement of objects immersed in highly scattering milk water solutions. The scattering properties of milk water solutions are ideal for relating scattering theory to image enhancement theory, but are not particularly applicable to the scattering properties of living breast tissue. The reason for this is that scattering in milk solution is caused by micron size particles. The scattering of fatty tissue may be due to refractive index changes which take place over a spatial extent similar in size to the tumorous growth. In the former, use of simple pinhole filtering suffices to separate the signal from the scattered light. For scattering by tissue, pinhole filtering alone converts scattered phase differences into intensity differences which appear as speckle in the image plane. However, a small amplitude vibration of the tissue will suppress the speckle pattern and allow the direct light to be viewed.

Visibility of Objects Embedded in Scattering Media

Visibility, which is a psychophysical quantity, has been given the physical measure:

$$V = \frac{Bmax - Bmin}{Bmax + Bmin} \tag{1}$$

where Bmax and Bmin are the maximum and minimum brightness of the object field.

If we consider an object in a scattering medium, then part of the direct image bearing light will be scattered and appear uniformly as background radiation in both the maxima and minima of the object field.

Thus for an object in a scattering medium:

$$V' = \frac{B'max - B'min}{B'max + B'min + 2A} \tag{2}$$

Where B' is the brightness of the direct image bearing light and A is the uniform brightness scattered over the object field.   A is proportioned to B-B'.

In our experiments our model "tumor" was a metal screw so that Bmin vanished.

Writing $\beta = \dfrac{B'max}{B\ max}$ and for Bmin = 0 we may write in place of equation (2) :

$$V' = \frac{\beta}{\beta + 2\alpha(1-\beta)} \tag{3}$$

The proportionality constant α is the ratio of the light scattered onto the image area to the total scattered light.

For dilute solutions α is independent of concentration.

If

$$\beta_1 = \frac{B'(1)}{B} \tag{4}$$

is the ratio of the image brightness after scattering to the image brightness before scattering for some "standard" scatterer concentration-path length product c,d,   , then for any other concentration-path length product c d we will obtain:

$$\beta_n = \beta_1^n \tag{5}$$

where

$$\frac{c\,d}{c'd'} = n \qquad (6)$$

Equations (5) and (6) express the Lambert-Beer law for scattering in dilute solutions.

In our experiments the path length in the scattering medium was approximately 75 mm and our "standard" concentration was 0.2 parts of milk per thousand parts water.

From our measurements of the "standard" solution we found that

$$\beta_1 = 0.63$$

Substituting for β into equation (3) we obtain

$$V = \frac{0.63^n}{0.63^n + 2\alpha\,[1-0.63^n]} \qquad (7)$$

The experimental arrangement used is similar to that illustrated in Figure I, where the laser source was a HeNe laser emitting 6328 $\overset{o}{A}$ radiation and in place of the breast we had a rectangular container with the milk-water solution.

At the image plane photographic film was substituted for the image intensifier tube.
Without enhancement direct measurements of α gave

$$\alpha \approx 0.07$$

The curve labelled "without filter" in Figure 2 gives V' as a function of milk drop concentration as calculated with equation 7.

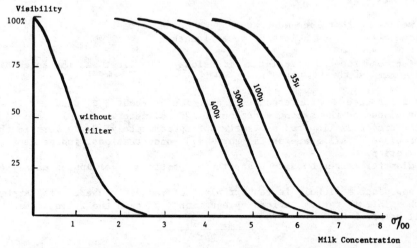

Figure 2 –Visibility of object immersed in milk-water
            solutions as a function of milk concentration
            with and without fourier plane filtering.

Figure 3 shows a photograph of the pattern observed for a 2.5‰ concentration
For a 3.5‰ concentration the screw is no longer visible .

Figure 3   Photograph of screw as observed in optical system
without fourier plane filtering.
c= $2.5\%_{oo}$ d = 75mm

Image Enhancement

In a coherent optical system a great deal of the information of an object of
dimension D will, in the fourier plane, be contained within a region of angular size P where

$$P \overset{\sim}{\sim} \frac{2\lambda}{D} \tag{8}$$

where λ is the wavelength of the coherent radiation.

If f is the focal length of the lens than the extent of this region is given
by:

$$2r = fP = \frac{2\lambda f}{D} \tag{9}$$

where we assume that lens aberrations can be neglected.
Experimentally we used a lens of focal length:
f= 62mm
Our object had two characteristic dimensions. The first was the screw diameter $D_1 \sim 3mm$ and the second was the screw pitch $D_2 \sim 0.64mm$.

In the first case r   corresponds to a region of about 30μ   diameter in the
fourier plane and the second a region of 120μ diameter.
The scattered radiation had a diameter of approximately 2r = 18mm in the
fourier plane. This diameter is, for small concentrations, independent of
concentration.
[This diameter corresponds to a scattering particle D_s of approximately 4μ
diameter]
If we now place a pinhole in fourier plane of radius $\bar{r}$ where $\bar{r}$ is larger
than the minimum radius defined by equation (7) then the filtered visibility will be given by

$$V'' = \frac{\beta}{\beta + 2\left(\frac{\bar{r}}{r_s}\right)^2 \alpha (1-\beta)} \tag{10}$$

where $(\frac{\bar{r}}{r})^2$ is the ratio of scattered light transmitted through the filter
to the scattered light transmitted without the filter.
Equation 10 is shown in figure 2 for various vilters and for $2r_s \approx$ 18mm

Figure 4 shows the relation of the filter diameter required for an object dimension according to equation (9) as calculated for the prameters of our experiment (λ = 0.63 x 10^{-3} mm, f=62mm)

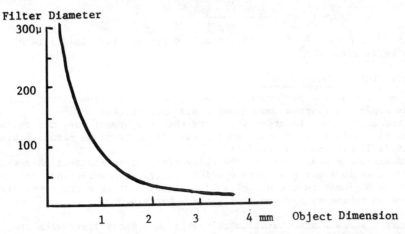

Figures 5,6, and 7 show the enhancement of the screw for a 75 mm path of water with a 3.3% concentration. Without enhancement the object is liminaly visible.

Figure 5   Enhanced image of screw. Filter diameter   400 μ .
Thread and screw clearly visible.

   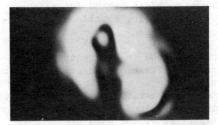

Fig.6 Enhanced image of screw,filter diameter   100 μ – screw thread no longer visible

Fig 7 Enhanced image of screw,filter diameter   35μ screw image is significantly distorted

It is important to observe that we can combine equations (9) and (10) to obtain the minimal object size $\overline{D}$ which may be recovered if a minimal visibility is required.   From (9) and (10) we obtain:

$$\left(\frac{\overline{D}}{D_s}\right)^2 \geq \frac{2\alpha\,(1-\beta)V''}{\beta\,(1-V'')} \tag{11}$$

Thus α, β and D_s for a breast will determine the minimal tumor size which can be discovered.

Scattering by Fatty Tissue

We have not as yet performed scattering experiment from living tissue. We have however performed some experiments with chicken fat.

In the first place in order to insure that the breast does not refract the incoming radiation, we must confine the tissue by plane parallel plates. This is illustrated in Figure I.

Also since the scattering by the fatty tissue also scatters phases we discover a serious speckle pattern which results when a randomly scattered pattern is incident on the pinhole. The pinhole act as a schlieren filter to yield an intensity pattern from a phase pattern.

Figure 8 shows a screw visible through a thin fatty layer with the speckle pattern. As the pinhole decreases in size the speckle will increase in size. For barely visible objects the speckle severely impairs recognition of the image. Since the speckle is due to phase scattering, the speckle pattern moves about rapidly in small movements of the tissue. This allows us to smooth the speckle by a small amplitude vibration of the tissue. This is illustrated in Figure 9.

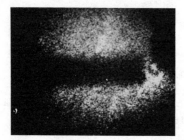

Figure 8 Object behind fat tissue   Figure 9 Smoothing of speckle pattern
        showing speckle due to fil-   of Figure 8 by vibrating the
        tering   tissue

Breast tissue is highly transparent to visible light, but not so the blood vessels. The blood vessels become more transparent in the near infra-red region of the spectrum [3] so that it is probable that near infra-red radiation will be required for the analysis. If so then an image conversion tube will be required to convert the infra-red light to visible light for visual inspection by the examiner.

The preliminary results of this study are encouraging, but further studies of the in-vivo spectral scattering properties of the breast are required.

REFERENCES

1. M.Cutler   Transillumination as an aid in the diagnosis of
              breast lesions. Surgery, Gynocology and Obstetrics
              <u>48</u> 721-729 (1929).

2. E.L. O'Neill Spatial Filtering in Optics. IRE Transactions on
              Information Theory <u>2</u> 56-65 (1956).

3. E. Loewinger, A. Gordon, A. Weinreb, J. Gross - Analysis of a
              method micromethod for transmission oximetry of whole
              blood. J. of Applied Physiology <u>19</u> 1179 (1964)

A REAL–TIME OPTICAL PROCESSOR
FOR PATTERN RECOGNITION OF
BIOLOGICAL SPECIMENS

Silverio P. Almeida*, James Kim-Tzong Eu* and Peichung F. Lai,*
John Cairns Jr.,** and Kenneth L. Dickson**

*Department of Physics, ** Department of Biology and Center for Environmental Studies,
Virginia Polytechnic Institute and State University, Blacksburg, Virginia 24061

ABSTRACT

A real-time optical processor (Ref 1, 2, 3) is described and results are presented concerning the automatic identification and counting of biological
specimens called diatoms (algae). The processor makes use of matched spatial
filters (Fourier transform holograms) for its identification of the diatoms.
Some of the matched filters have been averaged so as to account for orientation and sizing problems which occur with the diatoms. The input used for the
processor is in the form of 35mm photographs of the specimens taken by a phase
contrast microscope.

The input 35mm film strips are automatically positioned to each frame by an
on-line PDP-11-40 computer. Once the input is in position the operator types
on the Decwriter the coordinates of the desired spatial filter; this filter is
then moved into position via stepper motors under computer control. The autocorrelation is then performed and the output dots of light are received by a
vidicon camera, digitized and stored into the computer core and onto magnetic
tape for later processing/analysis. At the same time, a second vidicon camera
receives the input image and displays it on a second monitor along side the
monitor containing the output autocorrelation dots.

The immediate application studied, diatom pattern recognition, can be used to
monitor water pollution.

INTRODUCTION

Management of aquatic ecosystems requires a clear understanding of the goals
to be achieved, current information on the status of the ecosystem, and the
motivation and technology to take action. A major determinant of the effectiveness and efficiency of ecological quality control is the lag time in the
feedback of biological information. If the lag time is too great, the control
measures may repeatedly overshoot or undershoot the desired goal. The bio-
monitoring system described in this paper is based on the use of a coherent
optical spatial filtering system to identify diatoms rapidly. Ideally a system of this type could function within a computer based network of "in-stream"
monitoring stations measuring deviations from the "normal" or reference community of diatoms. The system is based on the filters designed to select diatoms
of one particular structure from a mixture of diatoms of varying structure

(i.e., to "identify" the type of diatom which matches that particular filter). A mechanical system fitted to a reference "library" of spatial filters of many different species of diatoms could then rapidly scan a sample consisting of a diatom community and greatly reduce the time required for sample analysis.

The diatoms constitute a group of unicellular algae which have been found to reflect complex shifts in water quality (Ref 4, 5, 6). Both changes in the kinds of diatom species present and in the species-abundance relationships in a community can provide valuable information in the assessment of the effects of water pollutants.

What follows is a description of the prototype optical processor used to pattern recognize diatoms. The methods employed for matched spatial filtering, the on-line PDP-11-40 computer hardware/software system are briefly presented as well as some recent results.

OPTICAL PATTERN RECOGNITION SYSTEM

Diatom Patterns

The taxonomy of the diatoms is based primarily on the recognition of certain patterns of physical form which are apparent in the cleared silicon dioxide frustule of the cell. These patterns are visually matched against some reference descriptor (usually a taxonomic key) by the taxonomist in order to identify the diatom.

The present method of identification and classification of diatoms requires a highly trained person and is very time-consuming. To classify hundreds of diatoms in thousands of samples rapidly is not feasible. Hence, some form of objective classification which is automated or semi-automated would be very useful to study the water quality changes as a function of diatom species-abundance distributions.

Diatoms possess unique features which lend them as good candidates for spatial filters. They are fairly regular in their patterns within a given species; they have a hard silicon dioxide skeleton, the species structures possess varying degrees of spatial frequencies. Also, detailed studies of diatoms and their classification have been made and provide one with a library reference set for comparisons. Some studies of diatom orientations and size variations have also been made.

Spatial Matched Filtering of Diatoms

Our purpose is to automatically recognize the different species of diatoms and to count the number of each species presented in the unknown input by optical methods. Optical pattern recognition is essentially a measure of similarity between the unknown input pattern and the known reference pattern (Ref 7). This operation can be described by a correlation function $V(x)$, where

$$V(x) = \int U_n(x') U_m^*(x'-x) \, dx' \tag{1}$$

U_n and U_m are unknown and known reference pattern respectively.

The correlation coefficient, that is

$$V(x) = \int U_n(x')U_m^*(x'-x)dx' \qquad \text{at } x = 0 \qquad (2)$$

is particularly important because this will give the correlation peak when $U_n(x')$ and $U_m(x')$ have high similarity.

The desired correlation operation can be implemented in a few ways (Ref 8, 9, 10). We have chosen to use Vander Lugt's method. One advantage of this method is that once the filter bank is constructed, the matched filtering can be done in real-time. The optical matched filtering process is well documented in (Ref 8). We will only briefly describe the way we implement the matched filtering operation in our particular system of diatom recognition and counting.

The matched spatial filter of an object function $U(x)$ is defined as $\tilde{U}^*(\nu)$, where $\tilde{U}^*(\nu)$ is the Fourier transform of the function $U^*(-x)$, where * means complex conjugate, and we have used one dimensional notation for simplicity.

In our system we construct a filter plate which contains a bank of n reference matched filters: $\tilde{U}_1^*(\nu)$, $\tilde{U}_2^*(\nu)$.....$\tilde{U}_n^*(\nu)$, each can be timeshared. (n=100 for the present system). By matching a particular filter $\tilde{U}_m^*(\nu)$ with an unknown input $U_i(x)$, which may contain one or more of the n possible diatoms, one obtains immediately behind the filter plate the wave field

$$U(\nu) = \tilde{U}_i(\nu)\tilde{U}_m^*(\nu) \qquad (3)$$

where $\tilde{U}_i(\nu) = \int U_i(x)\exp[-2\pi i\nu x]dx$ and $\tilde{U}_m^*(\nu)$ is the matched filter being used.

When one or more $U_m(x)$ are present in the input, we will obtain one or more autocorrelation peaks at the output plane. These peaks are distributed corresponding to respective location of the diatoms in the input plane. The output function $V(x)$ when a single $U_m(x)$ is present in the input is given by

$$V(x) = \int \tilde{U}_m(\nu)\tilde{U}_m^*(\nu)\exp[-2\pi i\nu x]d\nu$$

$$= \int U_m(x')U_m^*(x'-x)dx' \qquad (4)$$

namely, the autocorrelation function.

The critical positioning requirement of the filter inherent to Vander Lugt's system has to be solved. We have empirically calculated the tolerable displacement of the filter from its optimum position (Ref 3). This displacement $\Delta\xi$ is related to the system parameters λ, f, and Δx by the relation

$$\Delta\xi \leq \frac{f}{4\Delta x} \qquad (5)$$

where λ is the wavelength of the laser beam, f the focal length of the Fourier transform lens and Δx the object width at the input. The parameters used in our system gives $\Delta\xi \leq 2.7$ μm.

We have devised a system which can address any one of the n component filters on the filter plate to within 2.5 μm of its optimum position. The addressing of filter is under the control of a PDP-11-40 computer.

There are several factors that will degrade the system performances; namely, (a) size variations among the same species of diatoms, (b) orientation differences and (c) variations in depth of focus. To minimize these problems we constructed a special kind of matched filter which we call an "incoherently averaged filter".

We use N patterns of diatoms with different size variations or different orientations or different depths of focus as filter-generator input, one pattern at a time. The exposure of each component pattern is then overlayed to give rise to a composite matched filter with amplitude transmittance proportional to I (assuming linear exposure) where I is given by

$$I = N + \sum_{n=1}^{N} |\tilde{U}_n(\nu)|^2 + \sum_{n=1}^{N} \tilde{U}_n^*(\nu)\exp[-2\pi i\nu_0 x] + \sum_{n=1}^{N} \tilde{U}_n^*(\nu)\exp[2\pi i\nu_0 x] \quad (6)$$

where N is a constant equal to the number of patterns used in the input and ν_0 is the carrier frequency of the reference beam used in constructing the filter. The third term in the right hand side $\sum_{n=1}^{N} \tilde{U}_n^*(\nu)\exp[-2\pi i\nu_0 x]$ is the desired composite filter with carrier frequency of ν_0. The composite filter thus obtained can take into account variations in size, depth of focus and orientation differences.

ON-LINE COMPUTER HARDWARE/SOFTWARE CONFIGURATIONS

The optical processor described in the previous section is on-line to a PDP-11-40 computer with 16k of memory. The computer together with the hardware interface (MPX-Buffer unit) and software serves the following functions: it advances the input 35mm film drive, positions the spatial filter stage to a desired (X, Y coordinate) filter, receives the autocorrelation output from the digitizer, statistical analysis, plots and displays the data. It also provides storage of a data base library for individual filters. A brief description of the hardware/software is presented here. Shown in Fig. 1 is a schematic of the optical processor. Further descriptions and details are cited in (Ref 11).

A. Hardware Components

The commercially available components used were slightly modified for the application described. Figure 2 shows the block diagram of the system hardware:

(a). Two sets of B&W video cameras and monitors are used for displaying of initial unknown input to be identified and the autocorrelation dots.

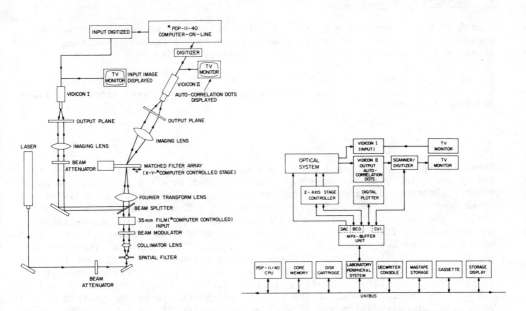

Fig. 1. Schematic view of the overall optical processor.

Fig. 2. Block diagram of the system hardware.

(b). A 6-bit Colorado video scan converter/digitizer (CVI-260) is used for data acquisition and display purposes.

(c). An Aerotech stage controller is used for positioning the filter stage into the desired location.

(d). A high speed laboratory peripheral system DEC/LPS-11 for the digital I/O interface.

(e). A disk pack DEC/RK-05 is for real-time monitoring.

(f). A magtape DEC/TU-10 is for bulk storage.

(g). A cassette DEC/TU-60 is used as the back-up storage of the software support programs.

(h). A HP-7210A digital plotter is for plotting the data.

(i). Either a DEC-writer or a Teleray display can be used as console terminal.

(j). A special designed electronic circuits called MPX-Buffer unit is for channel multiplexing, interfacing and timing purposes.

(k). A Tektronix-613 storage scope is used as a display output.

B. Software Components

A real-time disk operating system (RT-11) acts as the system supervisor and
provides: the linkage of system programs, support programs, execution of programs, all input-output operations and commands all peripherals.

The system programs included under RT-11 are the: peripheral interchange program, macro assembler, linker, text-editor, online debugger, librarian, BASIC
and FORTRAN translators.

EXPERIMENTAL RESULTS

Shown in Fig. 3 is the input used for the matched filter study presented in
this paper. The diatoms shown have been arranged in a circular pattern on a
microscope specimen slide by a diatom taxonomist using a micromanipulator. A
phase contrast positive photograph was made on a 35mm slide which was then
used as the input to the processor. This slide enables us to demonstrate a
matched spatial filter "averaged" in order to select out different orientations
of the same species. The species chosen is the elliptical one (inner set of
eight). A composite of four of these species was used to select out the total
of eight. Figure 4 shows the eight autocorrelation dots obtained from the
single composite filter. It is noted that sufficient symmetry within individual diatoms exists as well as the reflective symmetries in the pattern to provide good autocorrelation signals. The intensities of the dots can be seen to
vary, depending on which diatoms were used for the composite filter. Nevertheless, the cross-correlations among the other diatom species was below the
threshold for the accepted eight signals.

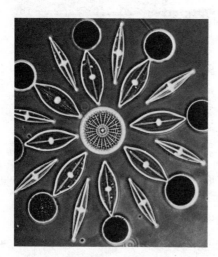

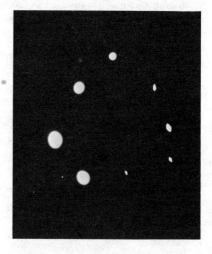

Fig. 3. Diatoms used as
input to the processor.

Fig. 4. Autocorrelation signals obtained
with composite matched spatial filter
for the elliptical diatoms (inner eight).

CONCLUSION

The results presented in the previous section are, thus far, representative of the matched spatial filters made for diatom identification. The averaged composite method discussed in this paper for orientation differences of diatoms has also been applied to depth of focus and size variations (Ref 11). The optical processor, thus far, is being used to continually test new filters.

ACKNOWLEDGMENTS

The authors wish to thank Dr. Charles W. Reimer and Mr. J. P. Slocomb, for the biological specimens used to provide the input for this paper.
This research was supported in part by the National Science Foundation/RANN Division, U. S. Army Medical Research and Development Command and the Environmental Protection Agency.

REFERENCES

1. J. Cairns, Jr., K. L. Dickson, G. R. Lanza, S. P. Almeida, and D. Del Balzo, Coherent optical spatial filtering of diatoms in water pollution monitoring, Arch. Mikrobiol. 83, 141 (1972).

2. J. Cairns, Jr., K. L. Dickson, J. P. Slocomb, S. P. Almeida, J. K. T. Eu, C. Y. C. Liu and H. F. Smith, Microcosm pollution monitoring, Trace Subs. in Environmental Health, D. Hemphill, Ed. 8, 223 (1972), University of Missouri, Columbia.

3. S. P. Almeida and J. K. T. Eu, Water pollution monitoring using matched spatial filters, Applied Optics, 15, No. 2, 510 (1976).

4. R. Patrick, M. H. Hohn and J. H. Wallace, A new method for determining the pattern of diatom flora, Not. Nat., No. 259, (1954).

5. R. Patrick, N. A. Roberts and B. Davis, The effect of changes in pH on the structure of diatom communities, Not. Nat., No. 416, (1968).

6. R. Patrick, The effects of increasing light and temperature on the structure of diatom communities, Limnology and Oceanography, 16: 12 (1971).

7. Goodman, J. W., Introduction to Fourier Optics, McGraw-Hill, New York, 1968.

8. A. Vander Lugt, Signal detection by complex spatial filtering, Trans. IEEE IT, 10, 139 (1964).

9. J. D. Armitage and A. W. Lohmann, Character recognition by incoherent spatial filtering, Applied Optics, 4, 461 (1965).

10. C. S. Weaver, S. D. Ramsey, J. W. Goodman and A. M. Rosie, The optical convolution of time functions, Applied Optics, 9, No. 7, 1672 (1970).

11. S. P. Almeida, J. K. T. Eu, P. F. Lai, J. Cairns, Jr., K. L. Dickson, An interactive hybrid optical/digital processor for matched filtering of biological specimens, Proc. Int'l. Optical Computing Conf., Capri, Italy (1976).

AN OPTICAL DIGITAL APPROACH TO THE PATTERN RECOGNITION OF COAL-WORKERS' PNEUMOCONIOSIS*

Henry Stark and David Lee

Department of Engineering and Applied Science, Yale University, New Haven, Conn. 06520

ABSTRACT

An optical-digital system for the pattern recognition of coal-workers' pneumoconiosis is described. The system consists of a coherent-optical configuration to generate the Fourier irradiance of the radiographic image and a digital computer to process the data. The interface is a TV camera. The technique was tested in a pilot study using sixty-four chest films of normal and abnormal lungs. Five different feature selection algorithms were tested. We found that the Hotelling trace criterion and the Sammon-Foley algorithm, together with a k-nearest neighbor classifier, furnished excellent normal/abnormal results.

The advantage of using a coherent optical system to generate the Fourier irradiance is that high-resolution two-dimensional Fourier transforms are obtained instantaneously, without digital computation. The combined optical-digital technique furnished 91% correct classification with a missed detection rate of less than three percent. The results attest to the feasibility of this approach to the medically significant problem of CWP classification.

INTRODUCTION

In this paper we present the results of a pilot study dealing with machine-aided classification of coal-workers' pneumoconiosis (CWP) in which an optical-digital computer (ODC) was used. Computers of this type typically use a coherent optical system (COS) for preprocessing and a digital computer for final processing. The COS in our configuration was used for generating two-dimensional spatial Fourier transforms while data-vector generation, feature selection, and classification were done with a digital computer.

Optical-digital computers, also called hybrid computers, are in increasing use in image processing. A recent special issue of the IEEE Transactions on Computers [1] discusses a number of such systems and their applications to various problems.

CWP is a disease that is endemic to the mining profession: NIOSH data [2] show that approximately thirteen percent of active miners and as many as twenty-eight percent of retired miners are afflicted with CWP. Recent

*Work supported by NSF grant GK 38308. This paper is for the ICO Conference proceedings only.

government legislation now gives each coal worker the right to obtain a chest X-ray at regular (<5 years) intervals for detecting the onset or progression of CWP. There are 185 or so film readers actively screening for this disease. The levels of compensation furnished to miners depend in part on the diagnosed severity of CWP. To deal with both the problems of interreader variation and anticipated overburdening of the present, manual, techniques, it would be highly desirable to develop machine techniques for mass screening of susceptible groups. Since the large majority of X-ray film apparently do not show signs of CWP, the machine techniques should be effective in screening out definite normals.

Kruger [2] has attempted a computer-aided diagnosis of CWP using both all-digital and optical-digital techniques. His results were encouraging; the normal/abnormal classification accuracy was not less than 88 percent while the comparable rate for physicians ranged from 83.0 to 97.9 percent. The false-normal (missed detection) computer rate never exceeded four percent while the comparable physician rate ranged from one to seven percent.

The medical importance of the CWP problem has prompted additional recent research in this area [3], [4], [5]. The results point to the potential feasibility of using machine-assisted techniques for classifying CWP.

The results of our own study were also encouraging. We concentrated more on the feature-selection aspect of the problem and indeed found that feature selection, was of central importance. We did not have to make any Gaussian assumptions about feature statistics and used no parametric techniques. Films were classified in normal/abnormal classes and only a single k-nearest neighbor (k-NN) classifier was used. We found that the k-NN classifier performed very well when used in conjunction with suitable feature selection algorithms.

AN OPTICAL-DIGITAL SYSTEM

Figure 1 shows the ODC used for this project. The coherent light source is a 15 milliwatt helium-neon laser operating at a wavelength of 6328A. Lens L1 furnishes a collimated beams and lens L2 generates the Fourier transform of the input data in its focal plane. A photographically reduced portion of the X-ray film is inserted in a phase-matching liquid to negate unrelated phase errors due to emulsion relief. The phase errors, if left uncorrected, can submerge the meaningful CWP-related data in the Fourier irradiance. A simple but excellent discussion of this problem is furnished by Goodman [6]. The film sample in the phase-matching liquid is a transparency of 35 millimeter high-contrast film. The whole film assembly is placed against lens L2 to minimize vignetting [7]. A microscope objective images the Fourier irradiance onto a ground glass plate and a spatial filter (not shown in Fig. 1) removes the zero-order aperture-diffracted light to protect the TV camera and allow the full dynamic range of the data to be recorded. The zero-order light is proportional to the square of the average amplitude transmittance of the film, and since the latter is largely independent of the manifestations of CWP, its removal did not bother us in any way. The video signal from the TV camera is displayed on a monitor and simultaneously transmitted to a digitizing system. Figures 2 and 3 show a normal and abnormal radiograph respectively and Figs. 4 and 5 show the TV monitor display of the Fourier irradiances of a normal and abnormal case respectively. The brightness distributions in Figs. 4 and 5 are proportional to the magnitude-square of the Fourier

transform of the input data. The distance from the origin is proportional to
spatial frequency. The horizontal or u spatial frequency is measured along
the abscissa; the vertical or v spatial frequency is measured along the

ordinate. The radial spatial frequency $w = \sqrt{u^2 + v^2}$ is measured along a
radial direction from the origin. Since these spatial frequencies are associated with the reduced scene, their absolute value is not meaningful. The
vertical waveform shows the video level along the scan line in the center.
The other two white lines on the left are reference lines for digitizing purposes. The Fourier irradiance associated with the abnormal film (Fig. 5)
shows more scattered light at higher spatial frequencies.

The digitizing system consists of a bandwidth compressor (Colorado Video Inc.
model 260) which samples a single field point per horizontal scan line. By
sampling only one point per line or 525 points per frame, the system does not
need expensive A/D converters and the data outflow rate is commensurate with
the rate at which the computer can accept data. A general purpose META-4
computer is used for all digital processing. Signal traffic between the bandwidth compressor and the META-4 is handled by the controller. A more detailed
description of the electronic system is given elsewhere [8].

Once the data is in digitized form, the computer organizes the data into
vectors, one vector for each film sample. The organization of the Fourier
irradiance into vectors is done with the aid of a computer-generated mask
consisting of 16 wedges and 16 semi-annular regions (Fig. 6).

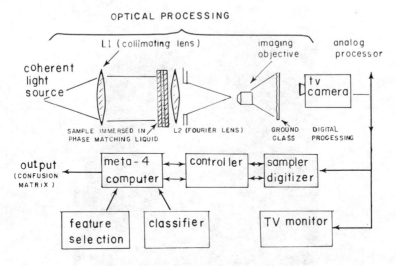

Fig. 1. An optical-digital system for the pattern recognition of CWP.

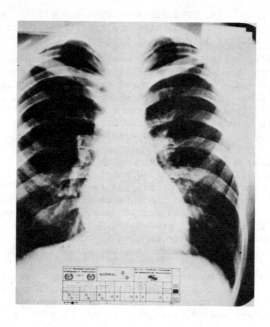

Fig. 2. A normal radiogram of the chest.

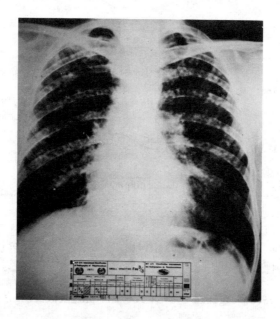

Fig. 3. An abnormal radiogram of the chest (classification, r(m) 3/3).

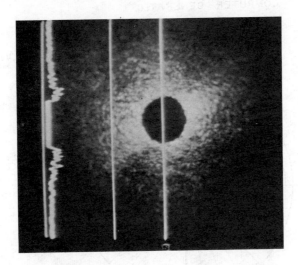

Fig. 4. Monitor display of the Fourier irradiance generated
by a normal X-ray film.

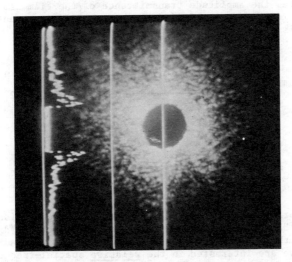

Fig. 5. Monitor display of the Fourier irradiance generated by
an abnormal X-ray film.

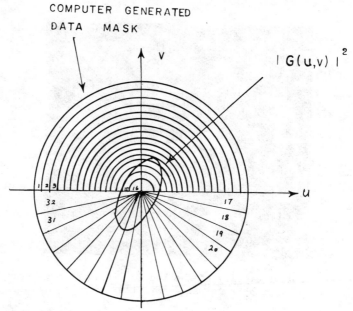

Fig. 6. The 32 sectors of integration generated by **the** computer to
produce the raw-datum vector $\underline{X}$ for a film sample.

If $g(x, y)$ denotes the amplitude transmittance of the film sample and $G(u, v)$
denotes the Fourier transform of $g(x, y)$ then the ith component of the raw
datum vector $\underline{X}$ is simply

$$x_i = \frac{\iint\limits_{S_i} |G(u, v)|^2 \, du \, dv}{\int\limits_{-\infty}^{\infty}\!\!\int |G(u, v)|^2 \, du \, dv}, \qquad i = 1, \ldots, 32 \qquad (1)$$

where

$$G(u, v) = \int\limits_{-\infty}^{\infty}\!\!\int g(x, y) e^{-j2\pi(ux+vy)} \, dx \, dy, \qquad (2)$$

and S_i is the surface element of the ith sector or wedge region. The elements
$\{S_i\}i = 1, \ldots, 32$, are disjoint, and when assembled into a "disk", they cover
that portion of the Fourier plane over which significant irradiance is observed. Since we are interested in the relative spatial-frequency distribution of power, we normalize so that $\sum\limits_{i=1}^{32} x_i = 1$. This is achieved by divi-

ding by the denominator in Eq. (1) and helps to eliminate factors that lead
to errors, such as average gray-level variations from film-to-film, variations in laser light during the course of the experiment, etc.. The

computation of the vector $\underline{X} = (x_1, \ldots, x_{32})^T$ is very rapid since $|G(u, v)|^2$ is computed essentially at the speed of light. If there are a total of M films then our data base will consist of M 32-component vectors. The components obtained from the semi-annular ring integrations give information about the rotationally symmetric portion of the spectrum. The components obtained from the wedge integrations give information about the angularly distributed portion of the spectrum.

The underlying reason for using the Fourier irradiance as the basis for a signature for CWP as manifest in radiograms, is the fact that the texture of abnormal X-ray film is different from the normal film [9]. The main radiographic CWP manifestation is a profusion of opacities in the affected zones which add to or mask the normal lung vascularities. The presence of these radiopacities alters the visual impression of texture and, at least so it is argued, the qualitative changes in texture can be quantitatively translated through the Fourier irradiance spectrum.

The set of vectors $\{\underline{X}\}$ represents the preprocessed data base. When all of these vectors have been computed, the problem of diagnosing CWP can be modelled as a classical two-stage pattern recognition problem involving feature selection and classification. We discuss these next.

FEATURE SELECTION AND CLASSIFICATION

The aim of feature selection is to reduce the dimensionality of the raw-datum vectors while finding a set of coordinates in which the different classes nicely separate. The reduction in dimensionality is quite important for at least two reasons: first, it leads to fewer computations and hence saves money and time and second, it helps to avoid the pitfalls associated with the "curse of dimensionality". Meisel [10] suggests that to avoid apparently significant separations of identical classes, the ratio of the number of samples per class to the dimensionality of the pattern space should not be less than three.

Five well known feature selection/data reduction algorithms were considered: projection on one dimension; the Karhunen-Loeve transform; the Fukunaga-Koontz transform; the Foley-Sammon transform; and the Hotelling scatter criterion. The latter two seemed well suited for our problem and we felt beforehand that they might furnish better results than the others. A brief description of these algorithms is given below.

Projection on One-Dimension (POD)

In the POD algorithm, a good component is one for which the ratio of the difference of class means to combined class scatter is large. The quality factor for the $k\underline{^{th}}$ coordinate is given by

$$Q_k = \frac{\mu_{1k} - \mu_{2k}}{\frac{1}{N_1} \sum_{j=1}^{N_1} |x_{1jk} - \mu_{1k}| + \frac{1}{N_2} \sum_{j=1}^{N_2} |x_{2jk} - \mu_{2k}|} \tag{3}$$

where μ_{ik} is the kth component of the sample mean μ_i, N_i is the number of samples in class i and x_{ijk} is the kth component of the jth sample of the ith class (i = 1, 2). The Q's are computed for all N components in the original space and ranked so that $Q_{p(1)} \geq Q_{p(2)} \geq \cdots \geq Q_{p(L)} \geq \cdots$. The coordinates associated with the L largest Q's are chosen as the reduced L-dimensional feature space (L << N). Since the POD algorithm involves no rotation of the space and makes no use of the covariability of the data, its power as a feature selector is impeded.

Karhunen-Loeve Transform

The K-L transform is widely discussed in the literature [11] because of its optimal properties in fitting and representing data. However it is usually not optimum with respect to discriminating data and its use in feature selection must be considered very cautiously.

Fukunaga-Koontz Transform [12]

The F-K transform is an attempt to improve the performance of the K-L transform by insuring that the eigenvectors which best fit one class are poorest for the other class. In this way Fukunaga and Koontz hope to increase the likelihood that the transformed space will enable separability between the classes. The major problem with the F-K transform is that it does not exploit the information contained in the difference of the mean vectors. For this reason it did not furnish good results in our case.

Foley-Sammon Transform [13]

The F-S transform is specifically aimed at improving discrimination ability and therefore differs in a conceptual way from the K-L transform which is optimal with respect to data fitting. The optimality criterion for the F-S transform is the maximization, as a function of a vector $\underline{d}$, of the ratio of projected class mean differences to the sum of the projected within-class covariance along $\underline{d}$ (the modified Fisher ratio). The modified Fisher ratio is given by

$$R(\underline{d}) = \frac{[\underline{d}^T(\mu_1 - \mu_2)]^2}{\underline{d}^T(C_1 + C_2)\underline{d}} \tag{4}$$

where $\underline{d}$ is an N-dimensional column vector on which the data are projected, μ_i is the sample mean of the ith clasa and C_i is the sample covariance matrix of the ith class (i = 1, 2).

The F-S transform, specifically designed for the separation of classes, is a useful technique when there is a significant difference in the class means and the underlying distributions are unimodal.

Hotelling Trace Criterion [14]

In this approach the criterion for class separability is the number J given by

$$J = \text{tr } S_2^{-1} S_1, \tag{5}$$

where

$$S_1 = \sum_{i=1}^{2} P_i (\mu_i - \mu_0)(\mu_i - \mu_0)^T, \tag{6}$$

$$S_2 = \sum_{i=1}^{2} P_i C_i, \tag{7}$$

$$\mu_0 \equiv P_1 \mu_1 + P_2 \mu_2, \tag{8}$$

C_i is the covariance of the ith class, P_i is the a priori probability of the ith class, and all the other symbols have been defined earlier. The matrix S_1 is a measure of the between-class scatter while S_2 is a measure of the within-class scatter. To find the eigenvalues and eigenvectors of $S_2^{-1} S_1$, S_1 and S_2 must be diagonalized simultaneously. In practice this may require some non-trivial computations. The trace-criterion algorithm gave excellent results.

For classification, we used an ordinary unweighted k-nearest neighbor algorithm. Because the dimensionality of the feature vectors and the number of samples in the data base were small, we used no preprocessing for the k-nearest neighbor computations. For two classes, the k-nearest neighbor rule can be defined as follows. Let X_1, X_2,...., X_n be a sequence of n independent sample points of known classification and let X be the point (i.e. our datum vector) to be classified. Among X_1, X_2, ..., X_n, let X_1^*, X_2^*, ..., X_k^* be the k-nearest neighbors of X, as determined by an appropriate distance measure. Then X is assigned the classification associated with the majority of its k-nearest neighbors. As a distance measure, the generalized Euclidean distance is commonly used. The near-optimum (Bayes) performance and the relative simplicity of the k-nearest neighbor algorithm makes it a very attractive option when a classifier is required.

The computer classification procedure consisted of removing one sample point from the data set for subsequent use as a test point. The remaining samples serve as the prototypes. After classification, the test sample is returned to the sample pool and a new test point is chosen. This procedure is repeated until every point in the sample has been classified.

It should be noted that while samples play the dual role of test and prototype, at no time are a set of prototypes used to find separation planes which are then used to separate the prototypes, as is sometimes the case in linear discriminant analysis. In our use of the k-nearest neighbor classifier, a test sample never influences the decision rule while it is being classified. For this reason, the k-NN procedure is free from the bias that is associated with linear discriminant analysis, when training and test samples are the same. Our method is a variation of the "leave-one-out" method discussed by K. S. Fu [15]. It is very useful when sample sizes are small.

CLASSIFICATION OF RADIOGRAPHS AND DATA BASE

The 1971 ILO U/C International classification of radiographs of pneumoconioses uses two figures for the profusion category and a letter to indicate the type of profusion. Other symbols for very large opacities and other significant disease manifestations are also in use [4]. However for our purposes, the profusion category (no. of opacities) and opacity size were of greatest interest. The profusion category is indicated by two numbers separated by a slash. The higher the number, the more advanced the disease. The first number is the actual category; the number after the slash indicates that this alternative was seriously considered. Thus a category 2/1 indicates a film in category 2 but category 1 was seriously considered.

Our data set consisted of sixty-four samples. Of these twenty-seven were judged normal (0/, 0/0, 0/1) and thirty-seven were judged abnormal (2/2, 2/3, 3/2, 3/3). Both rounded (p, q, r) and irregular (s, t, u) opacities were present. The area of examination roughly corresponds to a single lung zone, i.e., a third of a lung lobe.

The films were made available by Dr. William Crawford, formerly of the Department of Radiology at Yale Medical School. About twenty of the films were international standards and furnished by the ILO. The remaining films were furnished by Dr. Alan H. Purdy of NIOSH. They were independently classified by five board-certified readers, including Dr. Crawford. The final classification was assigned according to majority rule.

We found that the feature selection algorithms were critically important. Our results with the POD algorithm, the Karhunen-Loeve transform and the Fukunaga-Koontz transform were quite poor as the correct classification percentage hovered around seventy percent. The missed detection (false normal) rate was around ten percent and it was not clear what was the effective reduction in dimensionality. These poor results were not overly surprising in view of the properties of these feature selection algorithms.

On the other hand, the results obtained with the Foley-Sammon transform and the Hotelling trace criterion were very encouraging. For both the F-S transform and the trace criterion algorithm, the correct classification rate was near ninety percent. In both cases, the reduction in dimensionality was essentially from 32 to 1. This suggests the efficient exploitation of inherently strong correlations in the Fourier irradiance data. The results were consistant vis-a-vis the value of k in the k-nearest neighbor rule, i.e., for $3 \leq k \leq 10$, the classification performance remained the same.

The results of the classification algorithm when the F-S transform was used for feature selection is summarized in the following confusion matrix:

true class

	Normal	Abnormal
Normal assigned class	22	2
Abnormal	5	35

Of thirty-seven abnormals, only two were missed giving a missed detection rate of 5.4 percent. The overall correct classification rate was 89 percent.

When the Hotelling trace criterion was used, the results were:

	true class	
	Normal	Abnormal
Normal	22	1
assigned class Abnormal	5	36

Only one abnormal out of thirty-seven was missed giving a missed detection rate of only 2.7 percent. The overall correct identification rate was 90.7 percent.

CONCLUSIONS

We have attempted to use an optical-digital approach to the problem of normal/abnormal classification of CWP. Although our data base was smaller than we would have liked, the results of our study, especially when taken together with Kruger's results, suggest that the ODC technique is a promising method for the machine classification of CWP films into normal/abnormal classes. Our data base did not include any 1/0 and 1/1 cases. But 1/0 or 0/1 data is so borderline that they represent a source of confusion even to radiologists (hence the 1/0, 0/1 classification). In general, borderline data rated 1/- can be conservatively classified in a screening procedure by biasing the k-nearest neighbor decision rule (or, for that matter, any other decision rule that is used).

It is appropriate to point out that our coherent optical system employed standard components and we did not use either precision grade lenses or specifically designed, so-called, Fourier optics. No careful precautions were taken with respect to positioning the film samples in the optical system. Once the feature-selection transformation matrix is determined, the only significant computation in processing a CWP film is in applying the k-nearest neighbor algorithm. A special purpose dedicated computer could be used for this task.

REFERENCES

(1) IEEE Trans. Comp. (Special Issue on Optical Computing), vol. C-24, pp. 337-456, (1975).

(2) R. P. Kruger, W. B. Thompson and A. F. Turner, Computer diagnosis of pneumoconiosis, IEEE Trans. Sys. Man and Cybernetics, SMC-4, 40, (1974).

(3) E. L. Hall, R. P. Kruger and A. F. Turner, An optical-digital system for automatic processing of chest X-rays, Optical Engineering, June 1974.

(4) E. L. Hall, W. O. Crawford, and F. E. Roberts, Computer classification of pneumoconiosis, IEEE Trans. Biomed. Eng., BME-22, 518, (1975).

(5)   J. R. Jagoe and K. A. Paton, Measurement of pneumoconiosis by computer, IEEE Trans. Comput., vol. C-25, 95, (1976).

(6)   J. W. Goodman, Introduction to Fourier Optics, McGraw-Hill, New York 1968, pp. 154-155.

(7)   M. Born and E. Wolf, Principles of Optics, Perganon Press, New York, 1965, p. 188.

(8)   H. Stark, An optical digital computer for parallel processing of images, IEEE Trans. Comput., C-24, 340, (1975).

(9)   R. N. Sutton and E. L. Hall, Texture measures for automatic classification of pulmonary disease, IEEE Trans. Comput., C-21, 667, (1972).

(10) W. B. Meisel, Computer-Oriented Approaches to Pattern Recognition, Academic Press, New York, 1972, pp. 12-15.

(11) Y. T. Chien and K. S. Fu, On the generalized Karhunen-Loéve expansion, IEEE Trans. Inform. Theory, IT-13, 518, (1967).

(12) K. Fukunaga and W. L. G. Koontz, An application of the Karhunen-Loéve expansion to feature selection and ordering, IEEE Trans. Comput., C-19, 311, (1970).

(13) D. H. Foley and J. W. Sammon, Jr., An optimal set of discriminant vectors, IEEE Trans. Comput., 281, (1975).

(14) K. Fukunaga, Introduction to Statistical Pattern Recognition, Academic Press, New York, 1972, pp. 260-264.

(15) K. S. Fu, Pattern recognition in remote sensing of the earth's resources, IEEE Trans. Geosci. Electron., GE-14, 10 (1976).

HOLOGRAPHIC ANALYSIS OF TYMPANIC MEMBRANE VIBRATIONS IN HUMAN TEMPORAL BONE PREPARATIONS USING A DOUBLE PULSED RUBY LASER SYSTEM

Gert von Bally

Hals-Nasen-Ohrenklinik der Westf. Wilhelms - Universität, Kardinal-von-Galen-Ring 10, D-4400 Münster, W. Germany

SUMMARY

Double pulsed holography was used to study the vibrations of tympanic membranes in human temporal bone preparations. A displacement pattern characteristically depending on amplitude, frequency, and phase of the eliciting sound pressure could be found in middle ears lacking of pathological changes. Characteristic deviations from this normal vibratory pattern could be detected in cases of pathological alterations of the mechanical properties of the middle ear. This may provide the possibility to determine the location of disorders in cases of conduction deafness without opening the tympanic cavity. It could also be shown that due to its phase sensitivity double pulsed holography is in principle capable of detecting unsymmetric vibrations and phase opposition of different sections of the human tympanic membrane. This may be important for further investigations on the transfer function of the middle ear.

INTRODUCTION

The function of the human middle ear, located in the temporal bone, is that of a mechanical impedance matching transformer. The vibrations of the tympanic membrane elicited by sound waves are transmitted by the ossicular chain to the fluid system of the inner ear (Fig. 1.). The ossicular chain, consisting of three small bones - malleus, incus, and stapes -, acts like a leverage system. The tympanic membrane vibrations are transformed into movements of the stapes footplate of less volume displacement but greater force. Among others there are two reasons to study the vibration pattern of human tympanic membranes:

The first reason refers to the possibility of locating disorders in the middle ear without removing the tympanic membrane. In cases of conduction deafness caused by pathological changes in the middle ear the known audiometric and x-ray diagnostic techniques often fail to locate exactly the cause for these malfunctions. For that purpose so far the middle ear has to be opened. As pathological changes of the mechanical properties have an influence on the vibratory pattern of the tympanic membrane a vibration analysis may provide the possibility of a differential diagnosis of these dysfunctions without opening the tympanic cavity (1).

The second reason for these investigations is concerned in the study of the influence of tympanic membrane vibrations on the

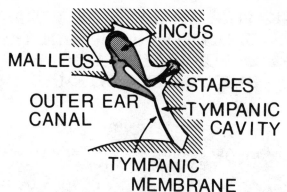

Fig. 1. Cross section of the human middle ear
(muscles and ligaments not included)

transfer function of the middle ear. Especially the volume displacement of the tympanic membrane is of interest because of its importance to the evaluation of the impedance of the middle ear. For
both reasons the vibratory pattern has to be determined. The investigations described below should prove whether double pulsed holography is in principle a suitable method for a high resoluting,
threedimensional, non-contact analysis of the vibrations of human
tympanic membranes.

EXPERIMENTAL ARRANGEMENT

Controlled by a Pockels cell a double pulsed ruby laser emits two
light pulses of 35ns halfwidth. The interval between both pulses
can be varied between 100 µs and 900 µs. Because of this short exposure time an in vivo application of double pulsed holography is
possible (2). Figure 2 shows the principle of the optical setup.

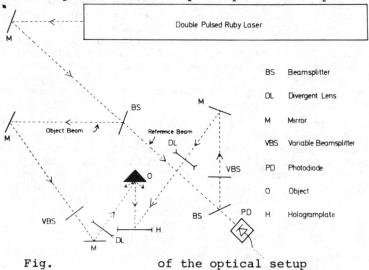

Fig.   of the optical setup

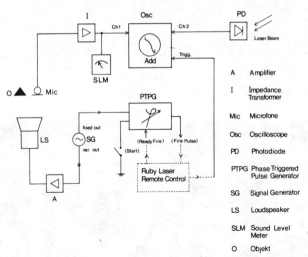

Fig. 3. Principle of the laser pulse release control

As usual in holographic arrangements the laser beam is splitted into an object and a reference beam. The laser pulses are controlled
by means of a fast photodiode and a storage oscilloscope. Soft
coated variable beamsplitters can be used for intensity matching
of the object and reference beam because the optical energy density
at the location of the object - i.e. the tympanic membrane - could
be proved not to be greater than $0,25\text{mJ/cm}^2$. Therefore an in vivo
application seems to be harmless. Vibrations of the tympanic membrane are generated in the free sound field of a loudspeaker
(Fig. 3.). The test signal triggers a phase triggered pulse generator which provides a control pulse (fire pulse). This fire pulse
can be delayed continously with respect to the phase of the test
signal. The laser pulses are released and the storage oscilloscope
is triggered by the fire pulse. The acoustic signal - i.e. the
sound pressure - is recorded by a probe microphone positioned right
at the rim of the temporal bone preparation while the optical signal - i.e. the laser pulses - is picked up by the photodiode. The
addition of both,pulses and sound pressure oscillation, is shown on
the oscilloscope screen for a visual control of the phase relation
between both signals (see Fig. 6.). A sound level meter connected
also to the microphone output measures the actual sound pressure
level (SPL).

<u>PREPARATION</u>

So far 20 human temporal bone preparations were used for basic investigations. The outer ear canal was resected. A small hole was
drilled into the tympanic cavity near to the incudo-mallar joint to
allow pressure equalization. To improve its reflectivity the normally translucent tympanic membrane was coated with a thin layer of
bronce powder (3). In order to obtain an equal scattering of the
laser light from all parts of the funnel shaped tympanic membrane
an additional coating of magnesium oxide was necessary (4).

RESULTS

Normal Vibration Pattern
In order to classify alterations of the vibratory pattern of the
human tympanic membrane for diagnostic purposes a standard has to
be confirmed. Such a normal vibration mode could be demonstrated
in general agreement with other authors (5), (6).

      110dB   105dB   100dB   95dB
                    2,0kHz; 270°/450°

Fig. 4. Vibration pattern of a left tympanic membrane at various
sound pressure levels and a frequency of 2kHz. The given phase
specifications (270°/450°) indicate the release points of the
laser pulses in relation to the sound pressure oscillation.

As an example for this normal vibratory pattern Fig. 4. presents
that of a left tympanic membrane at various sound pressure levels
and a frequency of 2kHz. The laser pulses were released at the peak
points of the sound pressure oscillation (270°/450°) as specified
in Fig. 4. This typical vibratory pattern shows a larger displacement above the manubrium - i.e. that part of the malleus which is
fixed to the tympanic membrane (see Fig. 1.) - and less vibration
amplitude beneath it. The manubrium itself - here running from the
upper left to the lower right in each interferogram - vibrates considerably less than the adjacent parts of the tympanic membrane. A
vibration of pars flaccida - i.e. a small part of the tympanic membrane around the upper end of the manubrium with different structure and tension - is also detected in most specimen. This can be
seen in the upper left corner of each interferogram in Fig. 4.

Without pathological alterations of the mechanical properties of
the middle ear vibratory patterns different in principle to that
shown in Fig. 4. could only be detected at higher frequencies.
Above 2kHz the displacement pattern becomes more irregular and
breaks up as demonstrated in Fig. 5.

Because of its phase sensitivity double pulsed holography is capable of investigating the phase relation between the membrane vibration and the corresponding stimulus oscillation as demonstrated in
Fig. 6. The interferograms in Fig. 6. show the vibration pattern
of the same left tympanic membrane as in Fig. 4. and 5. at 2kHz and
100dB SPL. The corresponding phase relation between the laser
pulses and the sound pressure oscillation can be seen on the photographs of the oscilloscope screen on the right side of each inter-

1,5kHz   2,0kHz   3,0kHz
                          105dB; 90°/270°

Fig. 5. Vibration pattern of a left tympanic membrane at various
        frequencies (same specimen as in Fig. 4.).

ferogram. This phase relation is varied in steps of 45° while the
interval between both pulses is kept constant to half a period in
respect to the acoustic oscillation. Most interference fringes –
demonstrating the greatest displacement – are detected when both
laser pulses are released in the peak points of the sound pressure
oscillation. Both pulses released in the zero crossing points of
the acoustic oscillation do not cause interference fringes. On the
assumption of a sinusoidal membrane vibration it can be concluded
that sound pressure and tympanic membrane vibrate in phase.

In Fig. 7. interferograms of the same specimen but at a frequency
of 3kHz are shown. Comparison of interferograms in the same line
demonstrates a repetition of the same vibration pattern after a
180°-phase shift of both consecutive laser pulses in respect to the
acoustic stimulus. With less phase shift, however, the fringe pattern changes substantially (c.f. interferograms e.g. in the same
column). Some parts of the tympanic membrane are displaced even
when the laser pulses are released in two adjacent zero crossing
points of the sound vibration, thus showing that at 3kHz not all
parts of the tympanic membrane vibrate in phase with the eliciting
sound pressure.

Comparison of interferograms taken at the same sound pressure level,
frequency, and identical or opposite releasing phases of the laser
pulses within and between Fig. 4., 5., 6., and 7. demonstrates that
the fringe number is reproducable with an accuracy of one fringe.

Influence of Changes in the Mechanical Properties of the Middle Ear
Alterations of the mechanical properties of the middle ear cause
characteristic deviations from the demonstrated normal vibratory
pattern of the tympanic membrane. As an example Fig. 8. presents
interferograms of a left tympanic membrane at 2kHz and various
sound pressure levels. In the upper line the normal vibratory pattern can be seen, while in the lower line the fringe pattern of the
same tympanic membrane after excision of the incus is shown. In
contrast to the normal vibration pattern here the displacement of
the tympanic membrane beneath the manubrium (lower left fringe
system) is equal or even greater than above it.

Figure 9. shows the vibration pattern of a right tympanic membrane
with a lesion at its lower right rim. At 1kHz a normal (in respect

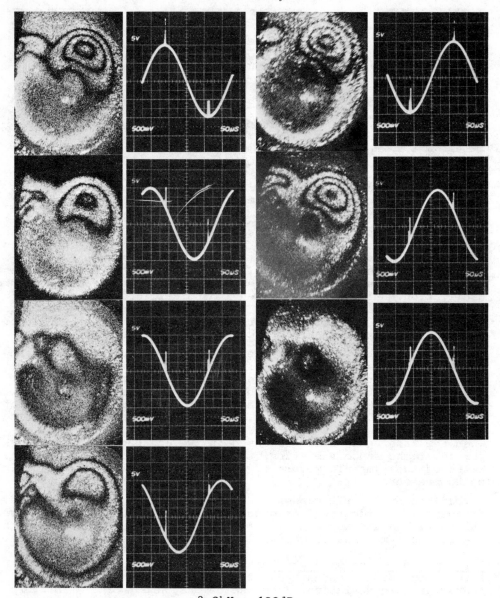

2,0kHz; 100dB

Fig. 6. Vibration pattern of a left tympanic membrane, laser pulses released at various phases of the sound pressure oscillation as indicated on the photographs of the oscilloscope screen.

to 115dB SPL) upper centre of vibration can be seen, while the corresponding lower one is missing, depending on the changed impedance of the different parts of the tympanic membrane. A frequency of 2kHz causes nearly no displacement even at this high sound pressure level.

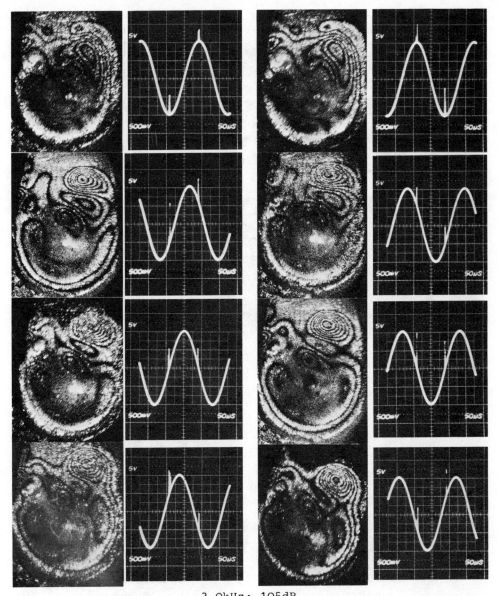

3,0kHz; 105dB

Fig. 7. Vibration pattern of a left tympanic membrane (same specimen as in Fig. 6. but at different frequency and sound pressure
level), laser pulses released at various phases of the sound pressure oscillation.

The vibration pattern of a right tympanic membrane in the presence
of a serous otitis media is demonstrated in Fig. 10. It is shown
that the fringe pattern is broken up into several sections even at

frequencies below 2kHz caused by the fluid in the tympanic cavity.

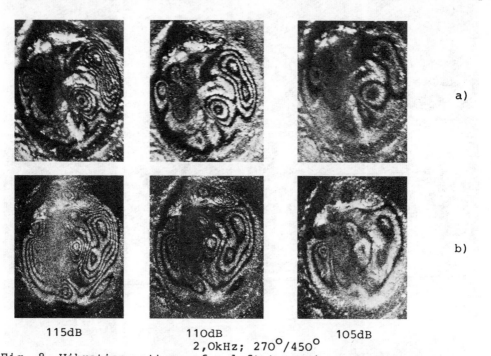

a)

b)

115dB   110dB   105dB
                  2,0kHz; 270°/450°

Fig. 8. Vibration pattern of a left tympanic membrane at various
sound pressure levels a) before and b) after excision of the incus.

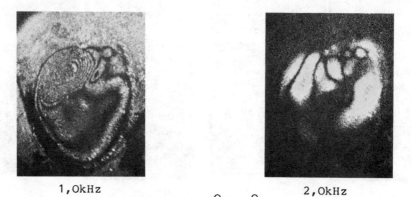

1,0kHz   2,0kHz
                  115dB; 90°/270°

Fig. 9. Vibration pattern of a perforated right tympanic membrane
(lesion at the lower right rim) at two different frequencies.

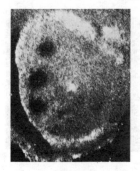

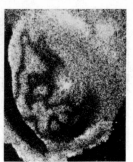

  1,0kHz   1,5kHz   2,0kHz
                       105dB; 270°/450°
Fig. 10. Vibration pattern of a right tympanic membrane in
         presence of a serous otitis media at various frequencies.

Phase Opposition and Unsymmetric Vibration
Figure 11. presents the vibration pattern of a left tympanic membrane at various sound pressure levels and a frequency of 1kHz
after excision of the incus and resection of the malleus above the
manubrium. Three sections of the tympanic membrane can be distinguished separated by bright nodal lines , thus indicating a phase
opposition of adjacent areas. This is of interest because so far
for the evaluation of the impedance of the tympanic membrane from
its volume displacement uniform phase of the vibration of the entire membrane was assumed,   especially at low frequencies (5).

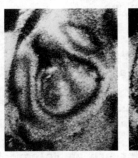

   110dB   105dB   100dB   95dB
                       1,0kHz; 270°/450°
Fig. 11. Examples for nodal lines separating sections vibrating in
phase opposition, vibration pattern of a left tympanic membrane at
various sound pressure levels after excision of the incus and resection of the malleus above the manubrium.

The vibration pattern of a right tympanic membrane perforated at
its lower rim is demonstrated in Fig. 12. Both interferograms of
Fig. 12. were taken at 1kHz and 115dB SPL. As shown by the corresponding photographs of the oscilloscope screen the laser pulses
were shifted by 180° in respect to the phase of the sound pressure
oscillation. From these photographs it can be gathered that the
difference in sound pressure between two consecutive laser pulses

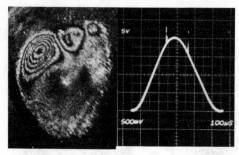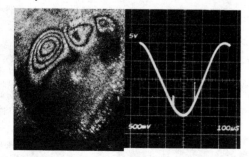

1,0kHz; 115dB

Fig. 12. Example for unsymmetric vibration of a right tympanic
membrane perforated at the lower rim.

is equal for both interferograms. The fringe number in the lowest
left section of the membrane vibration,however, indicates a significant displacement difference in both cases, thus demonstrating
an unsymmetric vibration.

It should be pointed out that so far in the absence of pathological
changes in the mechanical properties of the middle ear no phase
opposition or unsymmetry in the vibrations of human tympanic membranes could be detected. Double pulsed holography, however, - contrary to other holographic techniques (7) - turned out to be capable in principle of investigating such phenomena in the vibration
pattern of tympanic membranes as demonstrated by these experiments.

REFERENCES

(1) G. von Bally, Biomed.Techn. 21, 123 (1976)

(2) P.R. Wedendal and H.I. Bjelkhagen, Appl.Opt. 13, 2481 (1974)

(3) S.M. Khanna and J. Tonndorf, J.Acoust.Soc.Am. 51, 1904 (1972)

(4) A.L. Dancer, R.B. Franke, P. Smigielski, F. Albe, and H. Fagot,
    J.Acoust.Soc.Am. 58, 223 (1975)

(5) J. Tonndorf and S.M. Khanna, J.Acoust.Soc.Am. 52, 1221 (1972)

(6) W. Fritze, K. Burian und O. Schwomma, Proc. Symposium 1976,
    SFB88, Conference on Holography in Medicine, Münster (1976)

(7) C. Sieger, Proc. Symposium 1976, SFB88, Conference on Holo-
    graphy in Medicine, Münster (1976)

This work was supported by grants of the Deutsche Forschungs-
Gemeinschaft within the Special Research Area 88/B3.

SESSION 16.

MEASURING METHODS I

Chairman: L. LEVI

REAL TIME SPECTRAL RECOGNITION BY MATCHED FILTERING IN INTERFERENTIAL SPECTROMETRY

André Maréchal

Institut d'Optique Orsay, France

Modern spectrometry offers rich possibilities by convenient use either of interferometric devices (Fourier transform spectrometry) or of selective modulation (grid spectrometry). Progresses in resolution and luminosity have been spectacular in the last decade, and new domains of high resolution and low luminosity have been explored. In those cases of extreme performances, the instruments are generally highly sophisticated, delicate, and expensive. Nevertheless, Fourier transform spectroscopy is now developing for practical applications : we have examined the possibilities of interferometric devices to routine spectral analysis in chemistry, biology, pollution detection etc... and are now aware of the interesting characteristics of those mountings by the fact that they are luminous, flexible and very simple. They need no computer and are very suitable for low resolutions. We shall describe first the basic principle, and later focus on the various possibilities resulting from the direct access to the interferogram and the application of the mathematical properties of the Fourier transform : Fourier derivation, Fourier correlation with a reference spectrum, Fourier correlation of derivatives etc... All those possibilities result from techniques of matched filtering for spectral recognition, obviously similar to analog techniques for pattern recognition.

A) BASIC PRINCIPLE OF THE SIMS

The "Spectromètre Interférentiel à Modulation Sélective"(SIMS) combines the use of interferences and selective modulation. Its principle has first been pointed out by R. Prat (1), and G. Fortunato (2) has shown that it is the only way to obtain a high "étendue" and consequently a high optical signal.

The leading idea of the device is
- the production of very luminous interference fringes in a plane.
- the analysis of those fringes by a moving grid and the production of a photoelectric signal obtained by synchronous detection.

1 - The Interferometric Device

In order to obtain luminous fringes with an extended source, it is necessary
to manage in order that the position of the fringes should not depend upon
the position of the emitting point on the source. Fortunato has shown that
the only solution is the Prat mounting comprising Fig. 1.
- a doubling device producing two laterally shifted images of the source,
- a converging lens.
The fringes located in the focal plane of the lens, and produced by various
coherent coupled images of the points of the source, do not move when the
point source moves ; as a conséquence, the flux can be important by the fact
that there is no fundamental limitation on the solid angle of the beams.

S.I.M.S.

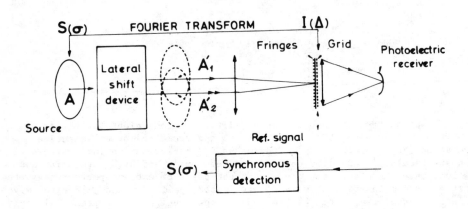

Fig. 1.

2 - Selective Modulation

If now we put a periodic grid in the fringes and move it, the outgoing flux
is modulated only for the wavelength for which the fringe separation is matched to the grid spacing : it is then possible to modulate selectively one
wavelength (or small spectral region) and use synchronous detection in order
to produce the spectral signal. The modulated flux will be proportional to
the source luminance for the wavelength for which fringe separation, $i = \lambda/\alpha$
(where α is the angle between the two interfering rays), is matched to the
grid period. Changing the wavelength can easily be performed by acting on the
angle which depends upon the lateral shift produced by the interferometer : we
use now a very simple polarizing interferometer invented by G. Nomarski
Fig. 2. and based on the properties of two Wollaston prisms located between
polariser and analyser. The adjustment of the distance between the two prisms
allows the variation of the wavelength.

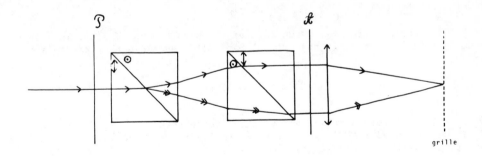

Interféromètre utilisant deux prismes de WOLLASTON pour obtenir un dédoublement variable.

Fig. 2.

3 - Luminosity

The high gain in "étendue", with respect to classical devices, leads to appreciable advantages : as an example, the fluorescence spectrum of anthracene has
been obtained on a bench mounting down to concentrations of 10 ppB. Nevertheless, we have to take into account the increase of noise due to the unmodulated flux, and a detailed discussion done by Fortunato has led to various encouraging conclusions.

4 - Resolution

The response of the apparatus to monochromatic light (laser) has shown to be
in agreement with theoretical predictions : we perform the analog Fourier analysis of a sample of N fringes, and the resolution is equal to N and does not
depend directly on the luminosity : if we increase the number of fringes, we
increase the resolution at the same time, and in fact, luminosity and resolution do no depend upon the same parameters and are practically independent.

B) OPTICAL DATA PROCESSING ON THE SIMS

It turns out that the SIMS is a luminous spectrometer, but another feature can
also be significant : we have in the focal plane of the lens the interferogram
of the source, i.e. the Fourier transform of the spectrum. It is then possible
to operate on this interferogram in order to obtain various signals representing linear operations on the spectrum. In other words, we have at our disposal the Fourier transform and we can take advantage of this situation.

1 - The Spectrum

If $S(\sigma)$ is the spectrum (as a function of the wave number σ) the interferogram is $I(\Delta)$, where Δ is the optical path difference and they are related by

$$i(\Delta) = I(\Delta) - I_0 = \int S(\sigma) \cos 2\pi\sigma\Delta\alpha\sigma$$

where $i(\Delta)$ is the variable part of the interferogram and I_0 the average of I. If we only need $S(\sigma)$, we operate an analog Fourier analysis of $i(\Delta)$ by using a movable periodic grid.
The response function is normally a sinc function, but procedures of apodisation can be applied by using a convenient smoothing screen on the interferogram.

2 - Derivation

If, instead of wishing to obtain the spectrum $S(\sigma)$, we prefer to obtain directly one of its derivatives, we can use the general properties of the Fourier transform ; as an example, the F. T. of $S'(\sigma)$ is proportionnal to $\Delta I(\Delta)$, which means that, in order to obtain a signal representing S', we should multiply the interferogram by Δ, what is very simple : we put in the interferogram a "two triangles" mask, and replace the ordinary periodic grid by a grid made up of two zones : for $\Delta > 0$ and $\Delta < 0$, the black and white bars are interchanged in order to multiply the signal by a negative factor for $\Delta < 0_2$. Fig. 3. represents the result obtained on the transmission spectrum of NO^2, where the carves $S(\sigma)$ and $S'(\sigma)$ are represented.

Fig. 3.

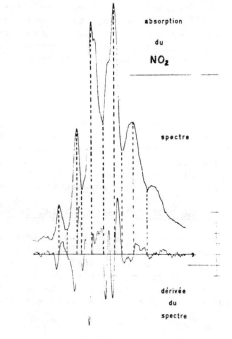

absorption du NO₂

spectre

dérivée du spectre

We should notice that happily the Fourier derivation does not operate the
derivation of the noise which is also the case of synchronous detection
methods for obtaining the derivative).

3 - Correlation

In order to recognize a reference spectrum S (σ), it is useful to correlate
the observed spectrum S (σ) with the reference S_R. This means that we should
multiply the interferograms i (Δ) and I_R (Δ), which is possible if we put in
the interferogram plane a screen representing I_R (σ), i.e. the interferogram
of the reference spectrum. If we move the interferogram, the signal obtained
is a linear combination of the signals produced by the various wavelengths and
we act at the same time on the various elements composing the spectrum.

4 - Correlation of Derivatives

Spectral signals S or S_R are always positive. This means that even if S and
S_R are totally different, they can be both non-zero in some spectral domain
and the correlation will be positive, which is misleading. On the other hand,
derivatives are positive or negative and it seems to be safer to correlate the
derivatives rather thant the spectra ; in order to perform this correlation,
we have to multiply Δ i(Δ) by Δ i_R (Δ) which means that we have to use a Δ^2
filter on the interferogram i_R. It is noticeable that this operation tends to
eliminate the central part of the interferogram, which brings no useful information, and use the parts of the interferogram that correspond to an appreciable value of Δ.

5 - The Case of Quasi Periodic Spectra

Spectra of interesting molecules (SO^2, NO^2) can be quasi periodic Fig. 4.

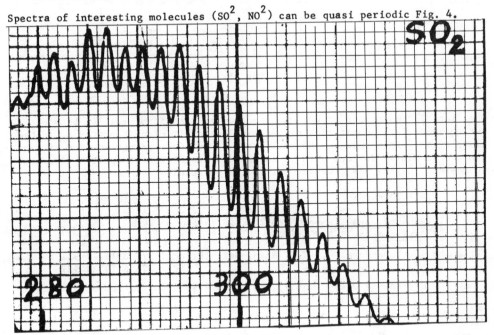

Fig. 4.

and this has an immediate consequence on the interferogram Fig. 5. , which
has a noticeable contrast (fringes reappearing) for a given value of Δ . In
those conditions, we can select by a convenient mask this value of Δ in the interferogram and base the correlation on 2 conditions :
- existence of a critical value of Δ where the fringes have an appreciable
  contrast,
- correlation of spectra.
It seems that this procedure is safer than the simple correlation of spectra
through multiplication of interferograms.

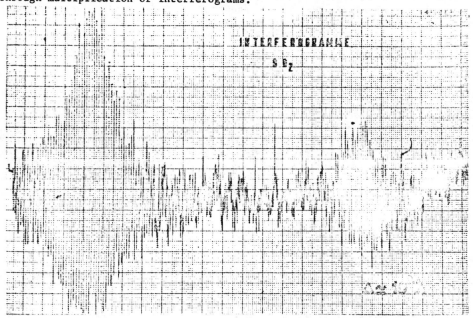

Fig. 5.

CONCLUSION

Once more in the domain of optics, the Fourier transform relation between
two characteristic quantities allows simple and efficient operations. In the
present case, it should be obviously very useful for the realization of simple and sensitive devices used in automatic recognition of spectra : we hope
to present, in the near future, more detailed results.

BIBLIOGRAPHY

Prat, R., Japan J. Appl. Phys. Suppl. 1,448 (1965)
          Optica Acta 13,2,73 (1965)

Fortunato G. et Maréchal A., C.R. Acad. Sci. (France) 274 B(1972) 931
                                       and   276 B(1973) 527

Fortunato G., Thesis Orsay June 76, to be published in the Nouvelle Revue
          d'Optique.

LIQUID CRYSTAL DISPLAY FOR REAL-TIME
MULTICHANNEL OPTICAL SPECTRAL ANALYSIS

Philippe Peltié, Guy Labrunie, Daniel Chagnot

LETI/CENG, Avenue des Martyrs, 85X, 38041 Grenoble Cédex, France

ABSTRACT

A nematic liquid crystal multichannel electrical to optical converter (ETOC) has been achieved and tested. Such a device, with
its processing and addressing electronics, is able to display simultaneously 18 electric signals of 64 analog samples each on a
coherent optical set-up yielding their spectral analysis distributions by one-dimensional Fourier transform.

The sampling frequency can be chosen between 25 Hz and 25 KHz,
and the overall spectral accuracy reaches 4 to 6% for periodic
signals. Such a real-time device can find applications in the
field of electric signals continuous analysis and survey, such
as in mechanics (engine vibrations), medicine (patients survey by
means of their electroencephalogram), noises analysis and recognition, ...

INTRODUCTION

Research in Optical Data Processing has reached during these last
years an interesting state of the art because of the discovery
and development of new materials and devices for optical-to-optical and electrical-to-optical converters. In that way, the various optical informations processors have become more and more powerful and rapid, which is most important : in fact, the two main
qualities of optics in this field, the reasons why studies are
still going on are the huge capacities -owing to the two dimen-
sions- and the intrinsic real time of optical processing.

Nematic liquid crystal optical-to-optical (Ref 1) and electrical-
to-optical (Ref 2, 3) converters have been shown to exhibit very
attractive properties due to their high optical and electro-optical qualities. This paper presents a further development of these
nematic liquid crystal (N.L.C.) displays applied to optical spectral analysis of several electrical signals by optical Fourier
Transform (O.F.T.).

This kind of application presents two interesting aspects :

- The optical processing -O.F.T.- is the most simple one can imagine and the most reliable one can perform in coherent optics
(Fig. 1).

P. Peltié, G. Labrunie, D. Chagnot

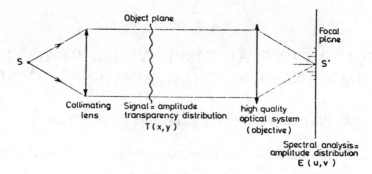

$$E(u,v) = \iint T(x,y) \ exp-2\pi i (ux+vy) \ dxdy$$

Fig. 1. Optical Fourier Transform (principle)

- Real-time spectral analysis is very attractive for examining and surveying such signals as mechanical, medical, acoustical and other electrical ones. In fact, several types of industrial users would have to be interested by our results.

OPERATING PRINCIPLE OF THE DEVICE

The device is schematically shown Fig.2.

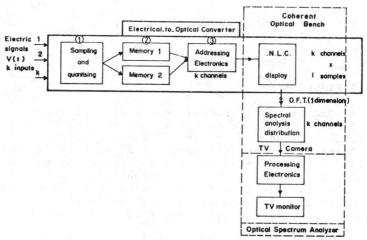

Fig. 2. Schematics of the apparatus

It is made of :

1. An optical spectrum analyzer achieved by MATRA-Division Optique (Ref 4), mainly consisting in a compact coherent optical bench allowing to perform several one-dimensional O.F.T. (with a diffracting objective, followed by a cylindrical lens). The multichannel spectral analysis is detected by a TV camera set and displayed on a TV monitor via a processing electronics permitting to superimpose on the TV monitor several adjustable spectral distributions I(u).

2. A N.L.C. multichannel ETOC (electrical-to-optical converter), consisting itself in a processing and operating electronics plus the N.L.C. display which is placed in the object plane of the optical bench. The electronics has k electric signals V(t) inputs (in this particular device, 18 inputs) and consists mainly in :

a) A sampling and quantising block 1 : 64 samples defined on 16 analog levels (4 bits) are taken in each channel (incoming signal)

b) Two buffer memories -block 2- allowing to work in real-time and to fit the operating conditions of blocks 1 and 3. The content of the first memory can be displayed on the optical bench while the second one is filled with the entering signals, and vice-versa, thus permitting true real-time operation without losing information.

c) An operating or addressing electronics -block 3- displaying the memory content onto the N.L.C. display, with k channels of 1 samples (here : 18x64). Thus, the operating principle of the device can be schematically represented as shown Fig. 3 : an input (a) electric signal V(t) is sampled with an adjustable frequency f_s and quantised on 16 analog levels (b) ; the resulting information $V_s(t)$ is displayed via an amplitude transparency function t(x) proportional to $V_s(t)$, with samples of width a and period p(c). By O.F.T., the spectral amplitude distribution is obtained in the focal plane (d), which gives by quadratic detection -eye, photodetector, TV camera- the spectral intensity distribution (e) or power spectrum I(u) of V(t).

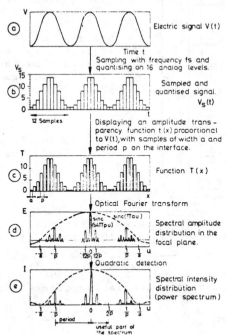

Fig. 3. Spectral analysis with the N.L.C. ETOC

MULTIPLEXED ANALOG ADDRESSING

The multiplexed binary addressing principle we have been using for several years has been widely described (Ref 2). Let us just recall that we use the so-called electro-optic "field effect" achieved under a.c. electric field in N.L.C. with an homeotropic orientation. We have experimentally checked (Ref 5) that in these conditions, the dynamic optical birefringence $\Delta n(t) = n(t) - n_0$ of the N.L.C. submitted to a voltage step of r.m.s. value V and null mean value is :

$$\Delta n(V,t) = \Delta n_{oo} \cdot \exp(AV^2 t) \qquad \text{for } \Delta n \ll n_e - n_0 \qquad (1)$$

n_0, n_e, n : ordinary, extraordinary and variable N.L.C. index.
Δn_{oo} : residual birefringence due to thermal fluctuations.
A : constant depending on the N.L.C. physical constants and cell thickness.

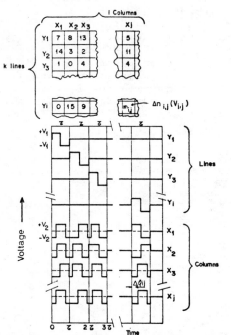

Now, using our multiplexed analog addressing (Fig. 4) where X-Y displays are sequentially operated, one line at a time, with pulses of r.m.s. value V_1 on the lines, V_2 on the columns and common width τ, the informations are presented simultaneously for the whole line on all the columns by means of variable electric phase shifts of columns pulses versus the line pulse. If V_1 and V_2 are chosen great enough, the whole display can be sequentially addressed in a time shorter than the intrinsic N.L.C. dynamic storage time (typically 10 to 100 ms).

Fig. 4. Multiplexed analog addressing.

If $\Delta\psi_{i,j}$ is the phase shift between the pulses applied to line i and column j, the r.m.s. value V in formula (1) is given by :

$$V^2_{i,j} = (V_1 - V_2)^2 + 4 \frac{\Delta\psi_{i,j}}{\pi} V_1 V_2 \quad \text{with} \quad \Delta\psi_{i,j} \in [0, \pi]. \qquad (2)$$

And taking into account the cumulative effects due to the k lines (Ref 2) :

$$\Delta n_{i,j} = \Delta n_{oo} \exp. A. \left[(V_1 - V_2)^2 + 4 \frac{\Delta\psi_{i,j}}{\pi} V_1 V_2 + (k-1)V_2^2\right]\tau. \qquad (3)$$

Now, once the display is placed between crossed polarizers oriented at 45° from the N.L.C. main axes, the amplitude transparency
of sample (i,j) is :

$$t(i,j) = t_o \cdot \sin(\frac{\pi e}{\lambda} \cdot \Delta n_{i,j})$$

With e : N.L.C. thickness
     λ : monochromatic wavelength

   $t_o \sim 0.85$-0.9 (losses are mainly due to reflections)

We have theoretically and experimentally determined (Fig. 5) the
curves $t(\Delta\psi)$ in the following conditions :

.$V_1/V_2 = \sqrt{k}$ which has been shown to lead to the best values for
the contrast C = t(180°)/ t(0°) for a given complexity k.

. The liquid crystal is M.B.B.A. (p-Methoxybenzilidene- p-n- buty-
laniline) under thickness e~8µm.

. The wavelength : λ= 6328 Å (He-Ne laser).

. The theoretical curve is for a single frame ; the experimental
ones are for 10 and 25 frames/s.

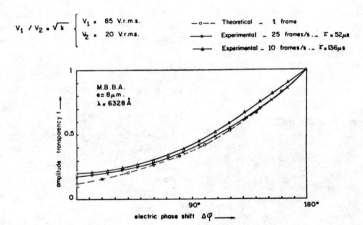

Fig. 5. Electro-optic analog addressing : Typical behavior

Hence, with these calibrating curves, the quantised analog level
$m_{i,j}$ (m $\in$ [0,15]) of sample (i,j) is represented by an amplitude
transparency t(i,j)-proportional to the corresponding value of
$V_S(t)$- obtained with $\Delta\psi_{i,j}$.

PROCESSING AND ADDRESSING ELECTRONICS

18 signals can be processed and displayed simultaneously. Their
levels have to be adjusted to 0-1 Volt before a single sampling
clock can pick up the first sample successively on each channel,
then the second... The sampling frequency can be chosen between
25 Hz and 25 Khz per channel.

Then (Fig. 6), each analog sample is converted on 4 digits, taking into account the non-linear correction corresponding to the
curve $t(\Delta\Psi)$, and registered in one of the memories. By order of
the "writing" clock, completely independent of the "reading" clock,

the N.L.C. display can be addressed as soon as the memory is fil-led up.

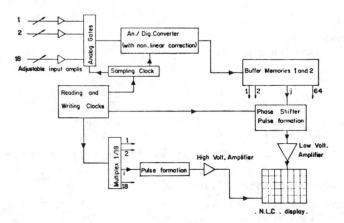

Fig. 6. Processing and addressing electronics

A phase shifter and a pulse former deliver the operating pulses on each column via a low voltage (V_2) amplifier, while the line pulses are provided by a 1/18 multiplex, and 18 pulse formers and high voltage (V_1) amplifiers.

Figure 7 shows this electronics, where one can notice :

.2 commercially available low vol-tage (lower left) and high voltage (lower right) supplies.

. 18 electric signal inputs (cen-ter)

.16 printed circuits of 4 pulse formers, phase shifters and low voltage amplifiers each (upper left).

.The printed circuit of 18 pulse formers and high voltage amplifiers (upper right).

. The input amplifiers and various clocks and modes of operation con-trols (upper right).

Fig. 7. Photography of the ope-rating electronics (see text for details).

The pulses modulating frequency can be chosen between 625 Hz and 90 KHz, thus allowing pulsewidths τ = 11µs to 1.6 ms.

The writing process can be triggered from a single frame up to 70 frames/s.

V_1 can be varied in the range 50-150 V.r.m.s. and V_2 in the range 10-40 V.r.m.s., with always $V_1/V_2 = \sqrt{k}$ = 4.24 for keeping on the

best contrast values,

N.L.C. DISPLAY

Figure 8 shows a typical N.L.C. display. It is made of two floated glass plates, exhibiting good enough flatness and attractive physico-chemical surface properties. Each plate is carefully cleaned, and a Sn doped In_2O_3 transparent conductive ($3.10^{-4}\Omega.cm$ or 30 $\Omega/\square$ for a thickness of 0.1 μm) layer is deposited in a diode sputtering chamber. This electrode is engraved by chemical etching or ion milling with conventional photolithographic techniques. The first plate is etched in 18 rows 450 μm wide and 50μm spaced with their electric connections, the second one in 64 columns 80 μm wide and 20 μm spaced with their connections.

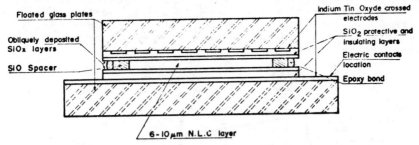

Fig. 8. Typical N.L.C. display

Each electrode is then covered with a SiO_2 chemically protective and electrically insulating layer about 0.2μm thick, and an obliquely deposited SiO_x (1<x<2) coating permitting to fix the direction of the variable optical index n in the N.L.C. layer. At last, a 6-10 μm thick SiO spacer is deposited through a thin metallic mask on one of the plates, the two free surfaces are processed with a surfactant inducing a good homeotropic orientation in the liquid crystals, and the display is hermetically sealed with epoxy.

Afterwards, the cell is filled up in vacuum with M.B.B.A. by means of a tiny hole previously drilled in one plate or spared in the spacer.

Figure 9 shows a photography of the display assembled in a mounting allowing a careless handling.

Fig. 9. Photography of the N.L.C. display in its mounting.

This mounting has two main functions :

- Industrial electric connectors (where the addressing electronics will be connected) are electrically coupled inside to the In_2O_3 rows and columns by means of printed circuits pressed on the glass plates.

- The display, which would be fragile otherwise, is thus mechanically, chemically and even somewhat thermally insulated.

OPTICAL SPECTRUM ANALYZER

This device, schematically shown Fig. 10, is a Fourier transform coherent optical bench consisting in a low power He-Ne laser, a spatial filter, a high quality diffracting objective and a cylindrical lens imaging the channels of the object plane (N.L.C. display) in the spectral plane, sometimes improperly called focal plane. The N.L.C. display is placed in the image principal plane of the diffracting objective, between crossed polarizers, and the spectra can be detected :

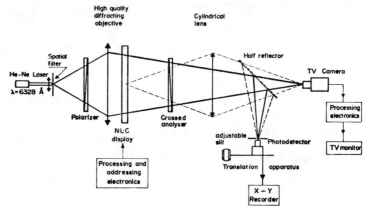

Fig. 10. Optical spectrum analyzer. Schematics.

- either by a TV camera set displaying the spectra and, by means of a processing electronics, their spectral intensity distributions on a TV monitor (Fig. 11).

- or by a photodetector linearly translated with its adjustable slit and connected to a X-Y recorder for fixed signals spectra.

The samples and channels spatial periods (p=100µm and q=500µm) have been chosen to fit the geometrical characteristics of the device : the distance object-spectral plane D=800 mm, and the cylindrical lens magnification $\simeq -1$ lead to a distance of $\frac{\lambda D}{p} \sim 5$ mm between peaks 0 and 1/p in the spectral plane. Hence, the useful part in the spectral plane measures 2.5x9 mm^2, dimensions easily detected by the 0.5" TV camera tube (useful dimensions $\sim 7 \times 9$ mm^2)

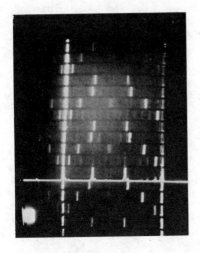

Fig. 11. Photography of the TV monitor during a typical multichannel spectral analysis.

RESULTS

The spectra and their spectral analysis distributions I(u) (Fig.3
and 11) are the result of several physical contributions. Mathematically, the transparency function can be written :

$$T(x) = t(x) \ x \left[\sum_{k=-\infty}^{+\infty} rect(\frac{x}{a}) \circledast \delta(x-kp) \right] x \ rect \ (\frac{x}{64p})$$

Where t(x) is the quantised transparency (see §2) :

$$rect \ (\frac{x}{a}) \begin{cases} =1 & for \\ =0 & for \end{cases} \quad \begin{matrix} -a/2 < x < +a/2 \\ x \notin [-a/2, +a/2] \end{matrix}$$

$\sum_{-\infty}^{+\infty} \delta(x-kp)$ is the Dirac function repeated with a p spacing.

$\circledast$ is the convolution product.

x is the scalar product

The Fourier transform of T(x) is :

$$G(u) = (64 \ ap).g(u) \ \circledast \left[\sum_{k=-\infty}^{+\infty} \delta(u-k/p) x \ sinc(\pi au) \right] \ \circledast \ sinc \ (64\pi pu)$$

Where sinc $(\pi au) = \dfrac{sin(\pi au)}{\pi au}$

And we detect : $I(u) = G(u) \ x \ G^*(u) = |G(u)|^2$

Physically, this means that g(u) which we are interested in is repeated with a period 1/p (due to the geometrical sampling period
p in the object plane), each peak being enlarged by sinc (64 πpu)
due to the finite width of the pupil, and the whole spectrum is
multiplied by sinc (π au) corresponding to the diffraction of one
sample.

This is clearly shown Fig. 3.

Taking into account all the types of possible errors (statistic,
filtering, quantising and analog to digital conversion, detection)
the accuracy of the spectral analysis can be estimated between 4
and 6% for periodic signals.

With the optical spectrum analyzer shown Fig. 10, we have observed real-time multichannel spectral analyses (Fig. 11)of periodic and aperiodic signals. We can also observe "fixed signals spectra", when displaying always the content of the same memory ; by this way, the spectral intensity distributions I(u) of various signals have been recorded. Here are the most significant results.

a) The sinus function, the spectrum of which includes a single frequency (with its secondary peaks) :

$$f(t) = \sin (2\pi.\nu.t)$$

b) The "square function" (Fig. 12), defined by :

$$f(t) = \text{rect} (\frac{2t}{T}) \otimes \sum_{-\infty}^{+\infty} \delta(t-kT)$$

Its spectrum presents the main peaks for frequencies ν (intensity 1), 3ν (intensity $\frac{1}{9}$), 5ν (intensity 1/25),...

c) The amplitude modulated (A.M.) sinusoïdal function $f(t) = (a+b.\sin 2\pi ft).\sin 2\pi\nu t$ (Fig.13) exhibits a spectrum with 3 characteristic peaks at frequencies ν, $(\nu-f)$, $(\nu+f)$ and f.

d) The frequency modulated (F.M.) sinusoïdal function $f(t) = \sin 2\pi(\nu t+m \sin 2\pi ft)$ gives a spectrum with m characteristic lines f spaced on each side of frequency ν; the spectral bandwidth is $[\nu-\Delta F, \nu+\Delta F]$, where $\Delta F = m.f$ is the modulation frequency excursion. Figure 14 shows the spectral distribution for m=1.

e) Up to now, following the Shannon theorem, we have chosen $f_s>2\nu$ for the sake of a correct quantisation. This leads theoretically to a frequency resolution $\delta\nu = f_s/64$, which we have experimentally checked by measuring the peaks halfwidth in the spectra. Thus, once f_s is chosen, the lowest frequency that can be detected is $f_s/64$, and the highest is $f_s/2$, according to Shannon. And if we have, for example, an A.M. sinusoïdal function : $f(t) = (a+b \sin 2\pi ft) \sin 2\pi\nu t$ with $\nu/f>32$ (Fig. 15), it is necessary to take $f_s<64f$, hence $f_s< 2\nu$ to detect and resolve in the spectrum the $(\nu-f)$ and $(\nu+f)$ peaks. In this example (Fig.15), the peaks of frequencies 48, 50 and 52 Hz corresponding to the first order spectrum are placed near the 1/p (61 Hz) frequency, whereas the peaks near the 0 frequency (on the left) belong to the folded second order spectrum.

f) Particularly attractive and conclusive is the spectral distribution of pseudo-random noises, such as a binary coded noise generated by a N stages shift register with a F frequency clock. The autocorrelation function of such a noise is a triangle of base width $2\Theta = 2/F$, repeated with a period $T = (2^N - 1)\Theta$. This is why we actually observe (Fig. 16 where :

N = 4 and F = 1 KHz, hence 2Θ= 2 ms and T = 15 ms)

a spectrum made of a series of peaks separated by 1000/T = 66,6Hz.

g) Figure 17 shows a typical example of application of the device to an electroencephalogram spectral analysis, including the "α line", the intensity of which is characteristic of the patient's condition.

In this application, one can easily imagine a medical survey of

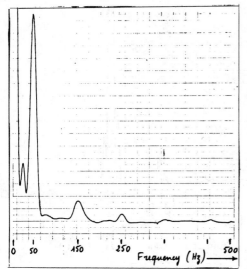

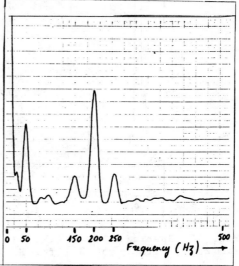

Fig. 12. Spectrum of the
"square function" :

$$rect\ (\frac{2t}{T}) \circledast \sum_{-\infty}^{+\infty} \delta(t-kT)\ with\ 1/T=$$

50 Hz and f_s = 1000 Hz.

Fig. 13. Spectrum of an A.M.
sinus function (a+bsin 2πft).
sin $2\pi\nu$t with :ν=200Hz,f=50Hz
and f_s = 1000 Hz.

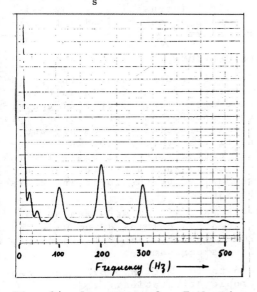

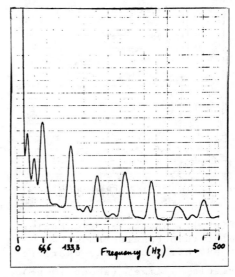

Fig. 14. Spectrum of a F.M. sin.
function sin $2\pi(\nu t+m.sin\ 2\pi ft)$
with ν=200Hz, f=100Hz, m=1 and
f_s = 1020 Hz.

Fig. 16. Spectrum of a binary
coded noise (pseudo-random noise) generated by a N stages
shift register with a F frequency clock. Here, N=4, F=1000 Hz
and f_s = 1000 Hz .

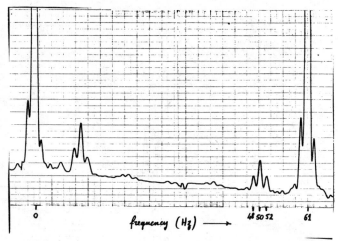

frequency (Hz) ⟶

Fig. 15. Spectrum of an A.M. sinus function (a+b sin 2πft)
sin 2πνt, with ν =50 Hz, f=2 Hz and f_s = 61 Hz.

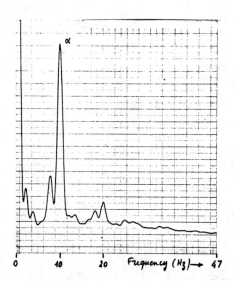

Frequency (Hz) ⟶

Fig. 17. Spectrum of a typical
electroencephalogram with
f_s =94 Hz. Notice the characteristic "α line" (about 10 Hz).

many patients simultaneously by
real-time measurement of their
"α line" intensities, and possibly with an automatically triggered alarm.

CONCLUSION

This paper demonstrates the possibilities of nematic liquid crystals electrical-to-optical converters applied to such field as optical information processing, and
particularly real-time multichannel optical spectral analysis. The
device we have presented can be
summarized by the following characteristics :

- 18 rows 450 µm wide and 50 µm
spaced, 64 columns 80 µm wide and
20 µm spaced, and 16 transparency
levels per sample.

- 6 to 10 µm of M.B.B.A. in homeotropic orientation.

- Typical addressing voltages are
85 Vr.m.s. on the lines and 20 Vr.
m.s. on the columns providing a total displaying time of 0.9 ms. at
25 frames/s and 3.6ms for a single
frame.

- Sampling frequency f_s=25 Hz to 25 KHz, giving an actual frequency resolution of $f_s/64$.

- Single frame to 70 frames/s.

- Spectral analysis accuracy = 4 to 6% for periodic signals.

- Optical transparency in ON state = 85 to 90%.

Hopefully, we believe that such a device can find industrial applications, mainly in the fields of medical or mechanical analysis and survey.

ACKNOWLEDGEMENTS

Many thanks are due to our colleagues A.M. CARTIER, J.Y. BESCOND, J. LIZET and M. SARO for their participation in conceiving and realizing the device.

We also gratefully acknowledge J. POULEAU of Société Matra-Division optique for kindly lending its optical spectrum analyzer.

This work has been partly supported by Direction des Recherches et Moyens d'Essais in France under contract n°73/803.

REFERENCES

(1) J. Grinberg et al, Photoactivated birefringence liquid-crystal light valve for color symbology display, IEEE Trans. on Electron Dev. 22,775 (1975)

(2) G. Labrunie, J. Robert and J.Borel, A 128x128 electro-optical interface for real-time data processing, Rev. Phys.Appl. 10, 143 (1975).

(3) G. Labrunie, Analog electro-optical interface for spectral analysis, Rev. Phys. Appl. 10, 213 (1975).

(4) D. Enard et J. Pouleau, Analyseur spectral-Application à l'analyse des clichés de photographies aériennes. Journées d'études du CNES sur le traitement des images (13-15 décembre 1973) à Paris. Page 173 du recueil des exposés publié par le CNES (Département des Affaires Universitaires - 129 Rue de l'Université - 75327 Paris Cédex 07).

(5) G. Labrunie and J. Robert, Transient behavior of the electrically controlled birefringence in a nematic liquid crystal, J. Appl. Phys. 44, 4869 (1973)

PARAMETERS FOR A HOLOGRAM SPECTROGRAPH

Zvi Weinberger

Jerusalem College of Technology, Jerusalem, Israel

ABSTRACT

An ordinary grating spectrograph may be converted to a hologram fourier transform spectrograph by replacing the diffraction grating with a diffracting
screen consisting of two identical randomly structured patterns placed side-
by-side.

In addition to the usual parameters which limit the properties of a grating
spectrometer, the resolving power of the hologram spectrograph is limited by
the mean "grain" size of the diffracting screen.

INTRODUCTION

The line and band spectra of atomic and molecular systems form the empirical
basis for our present understanding of atomic and molecular structure .[1]
These spectra, as formed by the usual spectroscopic tools, give a direct insight into the discrete energy states allowed for an atom or molecule. The
description of an atomic or molecular system in terms of its energy levels
has been called the "Schrödinger picture" of the system. [2]

Recently R.G. Gordon and others have devoted attention to the study of the
time development of molecular systems,[2] or to the "Heisenberg picture" of
quantum mechanics. If we focus our attention to the internal structure of the
band spectra formed by the usual spectroscopic methods, then we interpret this
structure as the fourier transform of the time correlation function of the
system.[2,3]

R.G. Gordon [2] and more recently V.G. Cooper [3] have used digital means to
accomplish the fourier transform of ordinary spectral data in order to study
the time development of molecular systems.

The fourier analysis of optical transitions may be accomplished directly by
means of fourier transform spectrometers. The usual methods of fourier spectroscopy are based on the motion of a mirror of the Michelson Interferometer.

The fourier analysis can also be performed spatially by separating the light
emenating from the transition into two beams and then recombining the radiation to form an interference pattern. This method, first described by Stroke
and Funkhouser is called hologram spectroscopy. [5]

The system of Stroke and Funkhouser is based on the fringes of equal thickness
formed by a Michelson Interferometer with relatively inclined mirrors. Modifications of the interferometric system used in obtaining fringes have been

described by Yoshara and Kitedo and by Doki and Suzuski. [6,7] The latter
authors have discussed, in part, the resolving power and light gathering
power of the hologram spectrometer.

In this study we will consider the properties of an ordinary grating spectrograph converted to function as a hologram spectrograph. This conversion is
accomplished by replacing the diffraction grating by a diffracting screen
consisting of two identical randomly structured patterns placed side by side.
An advantage of the proposed spectrograph (illustrated in Figure 1) is its
inherent stability and ruggedness. The diffracting screen may be displaced
laterally and longitudinally in the optical path without affecting the properties of the spectrograph. This may be contrasted to the critical alignment
required for fourier spectroscopy based on a two-beam interferometer. Another
advantage of the proposed system is that it permits the use of a line source
as opposed to the small circular source required for use with the two beam
interferometer-spectrograph.

We will discuss in detail the resolving power, light gathering power and the
short-wave attenuation of the hologram spectrometer. These spectrometric
parameters are dependent upon the separation between the identical random patterns, their structure, the source size and the size and resolution of the
recording film.

II. Description of the Spectrograph

The hologram spectrograph (illustrated in Figure 1) consists of a line source
of spectral distribution S (σ) and of angular width 2Θ with respect to the
collimator, the diffracting screen and a transform lens. The diffracting
screen consists of two spatially adjacent patterns of transmission t(x,y)and
t(x-d,y) respectively, where d is the separation between the patterns.

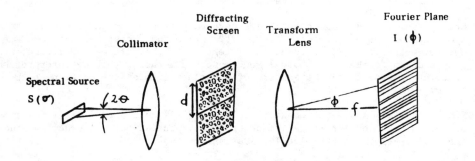

FIG. 1   Spectrograph for recording the Fourier Transform of S (σ)

t(x,y) is a rapidly varying function of the coordinates and can be experimentally realized by photographing the speckle pattern formed by coherently illuminating a ground glass screen (8). For an incoherent line source placed parallel to the y direction we can expect that the intensity of the pattern dif-

fracted by the screen in the y direction will be relatively structureless.
For this reason we will not concern ourselves for the remainder of this discussion with the structure of the screen, nor with the diffraction pattern
of the screen, in the y direction.

At the focal plane of the transform lens we have for the power distribution
in the Fraunhofer diffraction pattern of the screen for light of wave number
σ and incident at angle θ' to the screen: (8,9)

$$I'(\sigma,\alpha) = 2\,S(\sigma,\theta')\,(1 - \cos 2\pi\sigma\alpha d)\,\big|T(\sigma,\alpha)\big|^2 \tag{1}$$

$$\alpha = \sin\theta' - \sin\phi$$

ϕ = diffraction angle

$T(\sigma,\alpha)$= Fraunhofer pattern of $t(x,\theta')$

The power distribution at the focal length of the transform lens will be given
by the integral of equation (1) with respect to σ and θ'.

We will assume that T (σ,α) varies slowly with respect to wave-number and
angle of incidence when compared to the harmonic term of equation (1).   For
this assumption we can write for the integral of equation.(1)

$$I(\phi) = \int_{-\theta}^{\theta} d\theta' \int_{0}^{\infty} d\sigma\, I'(\sigma,\alpha)$$

$$= 2\left\{ \big|\overline{T_1(\phi)}\big|^2 \int_{-\theta}^{\theta} d\theta' \int_{0}^{\infty} d\sigma\, S(\sigma,\theta')\right.$$

$$\left. - \big|\overline{T_2(\phi)}\big|^2 \int_{-\theta}^{\theta} d\theta' \int_{0}^{\infty} d\sigma\, S(\sigma,\theta')\,\cos(2\pi\sigma\alpha d)\right\}$$

$$\tag{2}$$

$\big|\overline{T_1(\phi)}\big|^2$ and $\big|\overline{T_2(\phi)}\big|^2$ are mean values for $\big|T(\phi)\big|^2$ in
the respective integrals.   These mean values will be different for the first
and second integrals of the right hand members of equation (2) and are also
dependent on the nature of S(σ).

We can integrate equation (2) with respect to θ' if we assume that S(σ)
is independent of θ' and allow the small angle approximation for $\sin\theta'$.
Alternatively,if we assume that S(σ,θ') = S(σ) $\cos\theta'$ then equation (2)
can be integrated directly.   In either case we obtain:

$$I(\phi) = C\big|\overline{T_1}\big|^2 - 4\big|\overline{T_2}\big|^2 \sin\theta \int_{0}^{\infty} S(\sigma)\,\operatorname{sinc}(2\sigma d\sin\theta)\cdot \tag{3}$$
$$\cos(2\pi\sigma d\sin\phi)\,d\sigma$$

Film with linear response will record this pattern. The spectral sensitivity
of the film and also the spectral absorption characteristics of a sample which
may be introduced between the collimator and the diffracting screen are to be
included in S (σ) for the remainder of the analysis.

The second term of equation (3) is the fourier transform of

$$S(\sigma) \, \text{sinc} \, (2\sigma d \sin\theta)$$

It is this function recorded in the ϕ plane which is the experimental purpose of the hologram spectrometer . The time parameter τ of the optical transition is given by:

$$\tau = \frac{d}{c} \sin\phi$$

where c is the velocity of light.

In order to discuss the resolving power and light gathering power and short wave cut-off of the spectrometer it will be useful to return to the frequency domain in order to utilize the methods developed by Jacquinot for the description of these parameters. [10]

Let us then consider the recovery of S(σ) from equation (3).

III. The recovery of S (σ)

The first limitation of the spectrograph is that it is not S(σ) which is recovered through a fourier transform of I(ϕ) but rather :

$$F(\sigma) = S(\sigma) \, \text{sinc} \, (2\sigma d \sin\theta)$$

As the sinc function vanishes for :

$$2\sigma d \sin\theta = 1 \quad,$$

then only for wavelengths :

$$\lambda > 2 \ d \sin\theta \tag{4}$$

can S(σ) be reliably recovered. This means that the short-wave cut-off of the spectrograph is at least equal to   the product of the separation between the patterns and the angular width of the light source. We will see shortly that the finite resolution of the film also implies a cut-off wavelength.

If the focal length of the transform lens of the spectrograph is f   , then we have

$$I(\phi) = I\left(\frac{x}{f}\right)$$

where the coordinate x is measured from the center of the focal plane of the transform lens. (The transform lens must obey the sine condition as closely as possible. [11])

Let us perform the inverse fourier transform optically by placing the transparency in a collimated coherent light beam of wave number s   as illustrated in figure 2. If the amplitude of the collimated light is distributed as L(x) then the diffraction from the transparency will be given by:

$$K(\psi) = \int_{-\infty}^{\infty} I\left(\frac{x}{f}\right) L(x) \, \exp\left(2\pi i s x \sin\psi\right) dx \tag{5}$$

where ψ is the angle of diffraction from the transparency. The finite width of the transparency is also included in L (x) which vanishes outside the area of the transparency.

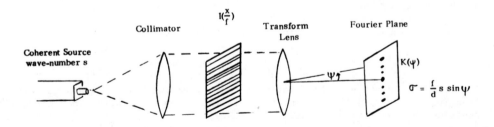

FIG. 2   Optical system for the recovery of S (σ)

It follows from fourier theory that equation (5) may be written as:

$$K(\psi) = \mathcal{F}(I) \otimes \mathcal{F}(L) \tag{6}$$

where $\otimes$ denotes the convolution operation.

For simplicity we will assume that L is uniform across the transparency of side a. We then have:

$$\mathcal{F}(L) = \int_{-a/2}^{a/2} exp\,(2\pi i\,s x\,\sin\psi)\,dx$$

$$= \frac{a}{2}\,sinc\,(as\,\sin\psi) \tag{7}$$

we will interpret this result after we evaluate $\mathcal{F}$ (I).

For I ($\frac{x}{f}$) given by equation (3) we have:

$$\mathcal{F}(I) = c_1 \int_{-\infty}^{\infty} |\overline{T_1}|^2 exp\,(2\pi i\,xs\,\sin\psi)\,dx$$

$$+ 4 \int_{-\infty}^{\infty} |\overline{T_2}|^2 \int_{0}^{\sigma} F(\sigma)\,\cos(2\pi\sigma\frac{x}{f}d)\,exp\,(2\pi i\,xs\,\sin\psi)\,dx\,d\sigma \tag{8}$$

The first term on the right hand side of the above relation is the fourier transform of $|\overline{T_1}|^2$ and is restricted to a small extension about the origin in the ψ plane. From elementary considerations it follows that a measure of this extent is given by:

$$\Delta\psi \sim \frac{\overline{\sigma}}{s}\frac{b}{f} \tag{9}$$

where b is a measure of the microstructure of the screen (a mean "radius" of the grain size) and $\overline{\sigma}$ a representative wave number of the original spectral source.

The second term of equation (8) is equal to:

$$4\ \widetilde{\mathcal{F}}(|\overline{T}_2|^2)\ \otimes \int_0^\infty \int_{-\infty}^\infty F(\sigma)\ \cos(2\pi\sigma\tfrac{x}{f}d)\ \cos(2\pi s\, x\, \sin\psi)\,dx \quad (10)$$

where $\mathcal{F}(|\overline{T}_2|^2)$ has the spatial extent given approximately by equation (9).

The second member of the convolution is, by fourier theory, equal to:

$$\frac{fa}{d}\ F\left(\frac{f}{d}\ s\ \sin\psi\right) \quad\quad\quad (11)$$

In words: if we neglect the effects of convolving F with $\mathcal{F}(\overline{T})$ and the sinc relation (equation (7)) then the pattern diffracted by the transparency corresponds to the spectral distribution $F(\sigma)$.

The correspondence between the wavenumber and the angle of diffraction is given by:

$$\sin\psi = \frac{\sigma\,d}{s\,f} \quad\quad\quad (12)$$

IV. The resolving power of the spectrograph.

Substituting from equations (7), (10) and (11) into equation (6) we have, outside the region about the origin in the ψ plane:

$$K(\psi)=F\left(\frac{f}{d}\,s\,\sin\psi\right)\otimes \mathrm{sinc}\,(as\,\sin\psi)\otimes\mathcal{F}(|\overline{T}_2|^2) \quad (13)$$

This convolution of F with the sinc function, with its dependence on the size of the recording film, and with $\mathcal{F}(|\overline{T}_2|^2)$ which is dependent on the microstructure of the diffracting screen limits the resolution capabilities of the hologram spectrometer.
The half-width of the sinc function is given by:

$$\Delta\psi_1 = \sin\psi = \frac{1}{a\,s}$$

The half-width of $\mathcal{F}(|\overline{T}_2|^2)$ may be approximated by equation (9) as:

$$\Delta\psi_2 \approx \frac{\overline{\sigma}\,b}{s\,f}$$

The half width of the convolution of the sinc function with $\Delta\psi_2$ is given by

$$\Delta\psi^2 \approx \Delta\psi_1^2 + \Delta\psi_2^2 = \left(\frac{1}{as}\right)^2 + \left(\frac{\sigma\,b}{s\,f}\right)^2 \quad (14)$$

The half-width resulting from this convolution implies that the spread in wave-numbers measured at a particular diffracted angle will be given by:

$$\Delta\sigma = \Delta\left(\frac{f}{d}\,s\,\sin\psi\right)\approx\frac{f}{d}\,s\,\Delta\psi = \frac{1}{d}\left\{\left(\frac{f}{a}\right)^2 + \sigma^2 b^2\right\}^{1/2}$$

The resolving power of the spectrograph is given by:

$$R=\frac{\sigma}{\Delta\sigma} = \frac{d}{\left\{\left(\frac{\lambda f}{a}\right)^2 + b^2\right\}^{1/2}} \quad (15)$$

$$= \frac{d}{b} \quad \text{for } b \gg \frac{\lambda f}{a} \quad (16a)$$

$$= \frac{d\partial}{\lambda f} \qquad \text{for } b \ll \frac{\lambda f}{\partial} \tag{16b}$$

assuming as representative numerical value that: $\frac{f}{\partial} \sim 2$ and $b \sim 5 \mu$, we observe that in the ultra-violet and visible regions of the spectrum equation (16a) defines the resolving power. Equation (16b) is similar in value to the resolving power of a diffraction grating of size 2d. in the infra-red region of the spectrum, we must use the more exact relation (15).

V. The dependence of the cut-off wavelength on film resolution.

In recording the pattern I ($\frac{x}{f}$) on film we have made two assumptions. The first is the complete linearity of the recording process and the second that the resolution of the film is infinite. The linearity property of the film is of extreme importance if we wish to reconstruct S (σ) from I ($\frac{x}{f}$) as described in the previous section. If, however, we are satisfied with $I(\frac{x}{f})$ then the problem of film linearity may be secondary.

Due to the grain structure of the film, the higher spatial frequencies recorded on the film are recorded with decreasing visibility. If we assume that until a specified spatial frequency the recorded visibility is unity and afterwards the resolution drops sharply to zero, then the problem of film resolution becomes equivalent to the finite interval readings made in Fourier Transform spectroscopy based on the Michelson Interferometer. [4]

Following the treatment of Vanasse and Sakai [4[for the Michelson Interferometer F.T.S. we observe that the highest value of $\frac{\sigma d}{f}$ in the cosine term of equation (10) which can be recorded on the film is equal to ν_o where ν_o is the number of cycles per centimeter which the film is capable of resolving. This gives for the cut off wave number

$$\sigma_m < \frac{f \nu_o}{d}$$

or

$$\lambda_c > \frac{d}{f \nu_c} \tag{17}$$

(centimeter units for all quantities)

The above relation should be used in conjunction with equation (4) for the cut-off wavelength due to the finite angular size of the source. For maximum throughput   value of the cut-off wavelengths of the two relations should be equal. Equating the two relations we obtain:

$$\sin \theta = \frac{1}{2 f \nu_o} \tag{18}$$

VI. The light gathering power of the spectrograph

The throughput of the spectrograph will be proportional to the angle made by the source with the collimator and the size of the diffracting screen. The size of the diffracting screen is not limited to twice the separation between the patterns. The patterns can overlap each other and still satisfy the transmission requirement:

$$t_o (x,y) = t (x,y) + t (x-d,y)$$

used for the derivation of equation (1).

Such screens can be made by double exposure photographs of speckle patterns with the film moved a distance d between exposures. These screens have been described by Burch and Tokarski[8]and Debrus et al. [9]

We observe that throughput can be increased at the expense of increasing the short-wave cut -off length. The best compromise is probably that given by equation (18).

VII SUMMARY

The important arguments and results advanced in this proposal may be summarized as follows:

1. The applications of hologram spectroscopy should be in those areas where the fourier transform of the spectrum is more pertinent thatn the spectrum presentation. Cooper has shown this to hold for scattering problems [3]. Physically the fourier transform of the spectrum is the auto-correlation of the temporal behavior of the source.

2. The resolving power of the spectrograph is given by

$$R = \frac{\sigma}{\Delta\sigma} = \frac{d}{\left\{ \left(\frac{\lambda f}{a} \right)^2 + b^2 \right\}^{1/2}}$$

3. The short-wave cut-off wavelength is dependent on the angular size of the source as:

$$\lambda_c = 2d \sin\theta$$

and on the film resolution measured in cycles per centimeter as

$$\lambda_c > \frac{d}{f \nu_0}$$

This implies that, if the film resolution limits the spectrograph, then, for maximum throughput:

$$\sin\theta = \frac{1}{2 f \nu_0}$$

REFERENCES

1. G. Hertzberg - Atomic Spectra and Atomic Structure, Dover, New York (1944)

2. R.G. Gordon- Molecular Motion in Infra-Red and Raman Spectra, J. Chem Phys 43 1307 (1965)

3. V.G. Cooper - Analysis of Fabry-Perot Interferograms by Means of Their Fourier Transforms - Applied Optics 10 525 (1971)

4. G. A. Vanasse and H. Sakai   - Fourier Spectroscopy - Prog. inOptics 6 271 Ed.   E. Wolf. North Holland, Amsterdam 1967

5. G.W.Stroke and A.T. Funkhouser - Fourier Transform Spectroscopy Using Holographic Imaging Without Computing and with Stationary Interferometers Physics Letters 16 272 (1965)

6. K. Yoshihara and A. Kitede   Holographic Spectra Using a Triangle Path Interferometer   Japan J. Appl. Phys. 6 116 (1967)

7. T. Doke and T. Suzuki - Attainment of High Resolution Fourier Transform Spectroscopy - Applied Optics 10 1137 (1971)

8. J.M. Burch and J.M.J. Takarski - Production of Multiple Beam Fringes from Photographic Scatterers - Optics Acta $\underline{15}$ 101 (1968)

9. S. Dehrus, M. Francon, S. Mallick, M. May and M.L. Rubin - Interference and Diffraction Phenomena produced by a New and very Simple Method. Applied Optics $\underline{8}$ 1157 (1969)

10. P.J. Jacquinot - Luminosity of Spectrometers - J.Opt.Soc. Am. $\underline{44}$ 761 (1954)

11. K.Von Bieren - Lens Design for Optical Fourier Transform Systems Applied Optics $\underline{10}$ 2739 (1971)

MEASUREMENT OF OPTICAL TRANSFER FUNCTIONS AND CORRECTION OF IMAGES IN SPECTROSCOPIC SYSTEMS

Hitoshi Kanamori and Katuyuki Kozima

Kyoto Technical University, Matsugasaki, Kyoto 606, Japan

ABSTRACT

The O.T.F. of a monochromator was determined by measuring responses of light beams having sinusoidal spectra, which are produced using a Fabry-Perot interferometer. Unresolvable adjacent peaks in observed spectra were split by the correction using O.T.F.s.

INTRODUCTION

In our previous articles, optical transfer functions (O.T.F.s) of a plane-grating monochromator have been determined from the Fourier transformation of line profiles (1,2). By this method, error in determining line profiles and in computing their Fourier transformation are probably creep in.

The direct measurement of O.T.F.s can be performed by measuring the responses of sinusoidal inputs having frequencies varied in a wide range. In order to measure the O.T.F.s of spectroscopic systems, we present a light source which emits sinusoidal spectra.

Examples of spectra corrected by O.T.F.s are shown.

LIGHT SOURCE EMITTING SINUSOIDAL SPECTRA

When quasi-monochromatic rays of wavelength λ pass through a Fabry-Perot interferometer at an emergent angle θ, the separation of adjacent peaks Δ in the spectrum of transmitted rays (see Fig. 1(a)) is given by

$$\Delta = \lambda^2/2nt\cos\theta, \qquad (1)$$

where, t is the separation of the plates of the interferometer and n is the refractive index of the medium between the plates.

The ratio of Δ to the half intensity width of the peaks δ is called the finesse N, i.e.,

$$N = \Delta/\delta, \qquad (2)$$

635

which is composed of:
the finesse due to the reflectivity R of the two plates,

$$N_R = \pi\sqrt{R}/(1 - R),$$ (3)

the finesse due to the roughness λ/M of the plates,

$$N_D = M/2,$$ (4)

and the finesse due to the field angle α,

$$N_F = 4\lambda/t\alpha^2.$$ (5)

The total finesse N is approximately given by

$$1/N^2 \cong 1/N_R^2 + 1/N_D^2 + 1/N_F^2 .$$ (6)

If we adjust the field angle α to make

$$N \cong 2,$$ (7)

the spectrum of the transmitted rays become approximately sinusoidal with the frequency $1/\Delta$ as shown in Fig.1(b). If N_F is dominant, then the field angle should be

$$\alpha \cong \sqrt{2\lambda/t}.$$ (8)

We have set a light source shown in Fig.2. In order to get quasi-monochromatic rays, light from a tungsten-iodine lamp is filtered by an interference filter. The rays are emitted to a Fabry-Perot interferometer in normal direction to the plates.

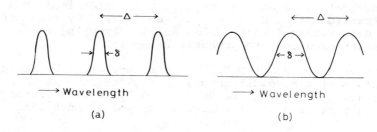

(a)   (b)

Fig.1   Spectra after transmitting through a Fabry-Perot interferometer.

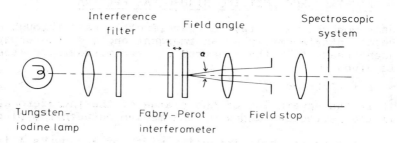

Fig.2   The light source emitting sinusoidal spectra.

Fixed parameters of our light source are,

$$\lambda = 546 \text{ nm} , \quad n = 1 , \quad \theta = 0 , \quad R = 0.95 , \quad \text{and } M = 30.$$

Therefore, for example, if we choose

t = 0.1 mm, then Δ = 1.49 nm and α = 0.104 radian,

or if   t = 0.03 mm, then Δ = 4.97 nm and α = 0.191 radian

We have adjusted the frequencies $1/\Delta$ in a wide range by varying the separation of plates t, and the finesse by the field angle α. The shapes and amplitudes of sinusoidal spectra were calibrated by a plane-grating spectrometer whose resolution limit was good enough as 0.01 nm.

Typical spectra for t = 0.1 and 0.03 mm are reproduced in Fig.3. Peak heights are normalized. Frequencies measured were a litte bit higher than those calculated. Proper field angles were much less than those calculated, because of poorness of parallelism of beams and poorness of focussing of lenses.

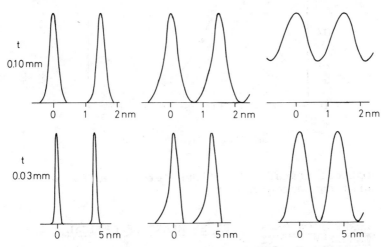

Fig.3   Spectra emitted from light source: wave lengths (abscissas) are measured from left peaks.

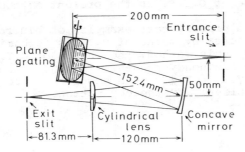

Fig.4   The side-view of a monochromator tested.

OPTICAL TRANSFER FUNCTION DETERMINED FOR A
PLANE-GRATING MONOCHROMATOR

With the light source above mentioned, we can determine modulus
of O.T.F.s. of spectroscopic systems, by measuring the ratio of
the amplitude of output-sinusoidal spectrum to that of input-
spectrum of the systems, for each frequencies.

Figure 4 shows the side view of a simple plane-grating monochromator whose resolution limit is poor as 0.4 nm. Its modulus of
O.T.F. measured at 546 nm is plotted in Fig.5.

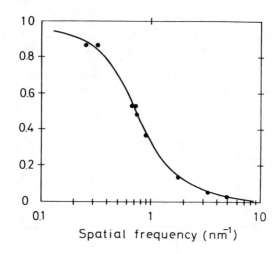

Fig.5   Optical transfer functions determined for a plane-
         grating monochromator.

CORRECTION OF SPECTRA BY MEANS OF OPTICAL
TRANSFER FUNCTIONS

In correcting images by means of an O.T.F., the information of
phase shift is necessary. We have not yet established how to
determine the phase of O.T.F. in the present method.

In Figs.6 and 7, we present here examples of blurred spectra and
their corrected spectra by means of O.T.F.s determined from line
profiles (1,2). By this method, the phases of O.T.F.s have been
determined in wide frequency ranges. In Fig.6, the Na-D doublet
is resolved clearly. In Fig.7, the shoulder corresponding to
578.97 nm appears clearly by the correction.

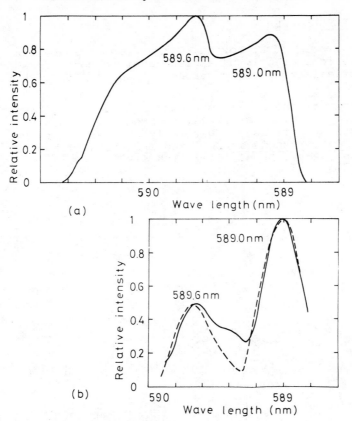

Fig.6   Profiles of Na-D doublet:
        (a) uncorrected,
        (b) corrected by an experimental O.T.F.(solid curve)
            and by a theoretical O.T.F.(broken curve); both
        O.T.F.s were derived from the Fourier transformation
        of line profiles.

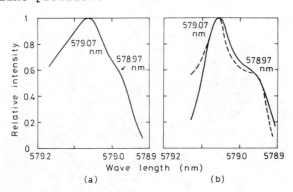

Fig.7   Profiles of two lines of Hg 6P-6D radiation:
        Legend as in Fig.6.

REFERENCES

(1) H. Kanamori and K. Kozima, Correction of a spectral image
    formed incoherently by a plane-grating monochromator by means
    of optical transfer functions, Japan. J. appl. Phys. 13, 1685
    (1974).

(2) H. Kanamori and K. Kozima, Correction of a spectral image
    formed by a plane-grating monochromator by means of optical
    transfer functions——a partially coherent case, Proc. ICO
    Conf. Opt. Methods in Sci. and Ind. Meas. Tokyo 1974, Japan.
    J. appl. Phys. Suppl. 14-1, 199 (1975).

SESSION 17.

MEASURING METHODS II

Chairman: J. SHAMIR

APPLICATION OF HOLOGRAPHY IN CAVITATION AND FLOW RESEARCH

H. J. Raterink* and J. H. J. Van der Meulen**

*Institute of Applied Physics TNO-TH, Stieltjesweg 1, Delft, The Netherlands
**Netherlands Ship Model Basin, Haagsteeg 2, Wageningen, The Netherlands

ABSTRACT

"In-line" Fraunhofer holography has been successfully applied to obtain three-dimensional information on boundary layer flow and cavitation phenomena on two axisymmetric bodies located in the test section of a high speed recirculating water tunnel. The boundary layer flow was visualized by injecting a 2 percent sodium chlorite solution from the nose of the bodies. With the applied method the influence of polymer additives was also investigated.

INTRODUCTION

Cavitation is defined as a group of phenomena which are associated with the occurrence of vapour bubbles or cavities in liquids, in particular their appearance and growth from nuclei by pressure reduction, their disappearance and the physical, chemical and biological effects thereby produced. In flow systems and propulsive devices cavitation can cause erosion, noise and vibrations. These effects are caused by the high temperatures and pressures arising from the spherical collapse of vapour bubbles or by the formation of microjets during the non-spherical collapse of vapour cavities. Usually, model tests are made to predict or avoid the difficulties related to cavitation. An important parameter to characterize cavitation is the cavitation number σ, defined by

$$\sigma = \frac{P_o - P_v}{\frac{1}{2} \rho V_o^2} ,$$

where P_o denotes the static pressure, and V_o the uniform flow velocity of the liquid at a great distance from the body. P_v is the vapour pressure of the liquid and ρ its density. A large value of σ corresponds to a non-cavitating flow and a small value to a cavity flow. The transition between these flow conditions is called cavitation inception.

The proper scaling of model test results to the prototype is obscured by the complexity of the factors involved. Acosta (1) emphasized the need for a thorough understanding of the basic fluid

643

mechanics of the liquid flow surrounding the bodies in which cavitation takes place. This statement was based on an earlier
study by Arakeri and Acosta (2) in which the boundary layer flow
on a hemispherical nosed body was visualized by the use of the
schlieren method. Cavitation inception could be correlated with
the occurrence of laminar flow separation, a phenomenon not known
until then. In spite of the fact that the schlieren method was
successfully applied to visualize the boundary layer flow, the
need for a more advanced method, to obtain three-dimensional and
more detailed information, still existed.

"In-line" holography has been applied earlier to measure the
nuclei spectrum of water in a cavitation tunnel. (Ref. 3, 4). The
method has also been introduced by the Institute of Applied
Physics TNO-TH at the Netherlands Ship Model Basin. Holograms
were made in a section upstream of the test section at various
test conditions. In contradiction with the results of other experiments (Ref. 3, 4), hardly any bubbles of detectable sizes
(5 μm or larger) were found to be present in the flow. Bader (5)
used the method of "off-axis" double pulse holography to study
bubble fields generated by acoustic waves. By switching the angle
between the reference beam and the object beam from $-30°$ to $20°$
during the short time lag between the two exposures, the dynamics of the bubbles could be investigated.

The main advantages of the present method of "in-line" holography
are: (1) the required holographic system is simple, (2) three-
dimensional information on the object is stored in the hologram,
which permits detailed analysis afterwards, using the reconstructed image, and (3) high speed flows can be investigated by applying a pulse laser. Also time information can become available
if the laser pulse contains a number of pulses, separated in
time. In the following chapters the applied holographic set-up
will be described and some results will be shown of the boundary
layer flow and cavitation phenomena on two axisymmetric bodies.

DESCRIPTION OF EXPERIMENTAL SET-UP

The experiments were made in the high speed water tunnel of the
Netherlands Ship Model Basin. This tunnel is provided with a 50
mm square test section with rounded corners; the maximum speed in
the test section is 30 m/s. A schematic diagram of the applied
optical system is shown in Fig. 1. A Korad K-1QH ruby laser, provided with a KDP-crystal for frequency doubling ($\lambda = 0.347$ μm) is
used as light source. The pulse duration is 25 nanoseconds and
the maximum energy 4 mJ in the TEM_{OO} mode. Two or three pulses,
with pulse separations of 50 or 100 μsec. can be generated in one
laser pulse. A telescopic system (L_2 and L_3) is used to obtain a
laser beam with a diameter of 30 mm. The laser beam is reflected
to the test section by a mirror; the power of the reflected laser
beam depends on the reflection coefficient of this mirror. Two
optical glass windows are mounted in the walls of the plexiglass
test section. The stainless steel model, having a diameter of
10 mm, is mounted in the test section in such a way that the nose
is illuminated by the laser beam over a length of about 20 mm.

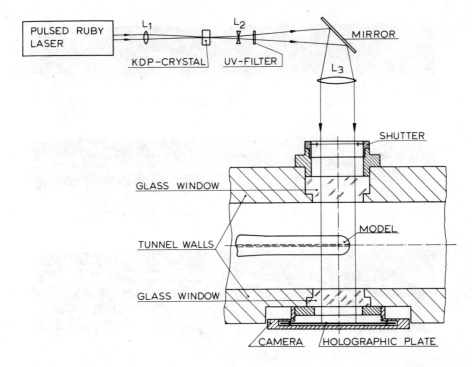

Fig. 1. Schematic diagram of optical system for making holograms
        of cavitation or flow phenomena in test section of tunnel.

Besides the contour of the model, also information concerning
bubbles or particles in the fluid flow between the windows is
stored in the hologram. Agfa-Gevaert Scientia Plates 8E56 and
8E75, with a resolution up to 3000 lines/mm were used as recording material. The optical path z, equivalent to that in air,
between the center of the test section and the holographic plate,
is 34.6 mm. It means that the condition for "in-line" Fraunhofer
holography: $z \gg d^2/\lambda$, in which d is the diameter of the objects
to be recorded, is fulfilled for $d < 15$ μm, which implies that
the disturbance of the twin image is minimal. The minimum diameter of the object (bubble, cavity) to be reconstructed is determined by the resolution power of the applied holographic
plate and by Abbe's condition that the first order of the diffraction pattern at least must be recorded. Due to the high resolution of the applied plates, Abbe's condition will determine
mainly the minimal size of the object to be recorded. With the
described set-up, bubbles with a diameter down to about 5 μm in
the center region of the test section could be recorded. The
allowed maximal speed V_{max} of the bubbles is determined by the
condition that the displacement of the bubble with diameter d,
perpendicular to the optical axis, must not exceed d/4 during the
exposure time. For a bubble (or particle) with a diameter of 5 μm
and an exposure time of 25 nsec., V_{max} is about 50 m/s. For the
reconstruction of the holograms a c.w. He-Ne laser was used. This

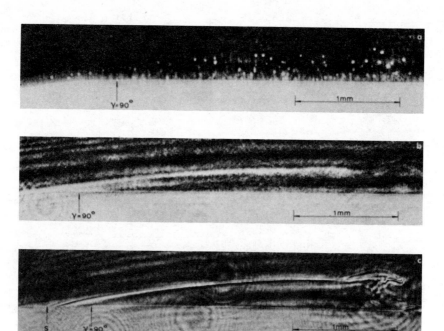

Fig. 2. Photographs showing laminar separation bubble and subsequent transition to turbulence on hemispherical nose. The
flow is from left to right (V_O = 5 m/s) and is visualized
by (a) injected particles, (b) heated water and (c) 2
percent NaCl solution.

means that all distances in the reconstruction, parallel to the
laser beam are reduced with respect to the corresponding distances
of the original object. The reconstructed image was observed with
a microscope with a magnification between 40x and 200x.

FLUID FLOW OBSERVATIONS

A new technique had to be developed for the purpose of boundary
layer flow visualization. Originally, small latex particles were
injected from a hole located at the stagnation point of a hemispherical nosed model (Fig. 2a). This method proved to be rather
complicated. A much simpler method consists of making holograms
when injecting a fluid with a slightly different index of refraction from the fluid in the tunnel. Injection of heated water gave
good results (Fig. 2b), but injection of a 2 percent NaCl solution gave excellent results (Fig. 2c). An amount equivalent to a
displacement thickness of 2-3 μm proved to be sufficient. In the
reconstructed image plane interference is observed between the
transmitted laser beam and laser light reflected and diffracted

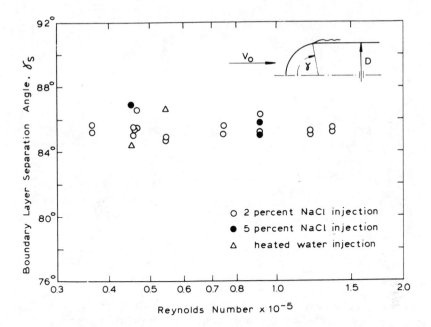

Fig. 3. Boundary layer separation angle, γ_S, as a function of
Reynolds number for hemispherical nose.

by the model and the injected NaCl layer. Although the method
does not provide data on the boundary layer thickness itself, it
does provide information on its laminar or turbulent character
and on the location of flow separation or transition to turbulence. From Fig. 2c it is found that the hemispherical nose exhibits a laminar separated flow region (bubble), which corresponds with the earlier observations made by Arakeri (6) and Arakeri and Acosta (2). Further downstream, transition to turbulence
occurs and reattachment of the shear layer is observed. In the
transition region the flow is still visualized by the NaCl, but
when the turbulence becomes more developed, mixing of the NaCl
prevents any further observations. The position of separation on
the hemispherical nose is given in Fig. 3. In this figure the
angular position of laminar boundary layer separation, γ_S, is
plotted against the Reynolds number Re (Re = $V_O D/\nu$, where D is
the diameter of the model and ν the kinematic viscosity). It is
found that γ_S does not depend on Re, which is consistent with
theoretical predictions; its mean value is 85.5°. Results of the
length to height ratio L/H of the separated bubbles are plotted
in Fig. 4. The mean value of L/H is 10.6. The data obtained by
injecting 2 percent NaCl, 5 percent NaCl or heated water match
quite well. This may be regarded as proof that the injection itself does not affect the flow at the location of separation or
transition.

The influence of polymer additives on the boundary layer flow

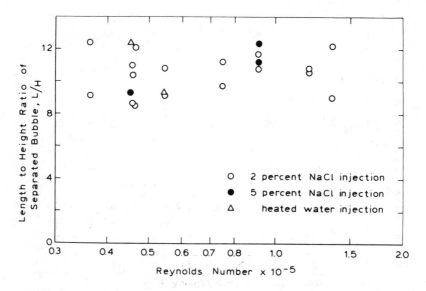

Fig. 4. Length to height ratio of separated bubble, L/H, as a
function of Reynolds number for hemispherical nose.

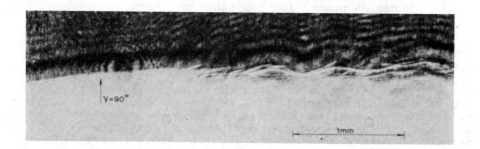

Fig. 5. Photograph showing boundary layer flow about hemispherical
nose. Injection of a 2 percent NaCl + 500 ppm Polyox WSR-
301 solution. The flow is from left to right (V_O= 4 m/s).

about the hemispherical nose was investigated by the injection of
a 500 ppm (parts per million by weight) Polyox WSR-301 solution
from the nose of the model. To visualize the flow, sodium chlorite was added to the solution. Polyox WSR-301 is one of the
grades of poly (ethylene oxide) which have a molecular weight of
several millions. These polymers are widely known for their
ability of turbulent-flow friction reduction. Fig. 5 shows a
photograph of the boundary layer flow on the hemispherical nose,
when Polyox WSR-301 is injected. It is found that the polymer
additive suppresses the occurrence of laminar boundary layer separation. This phenomenon has not been observed before.

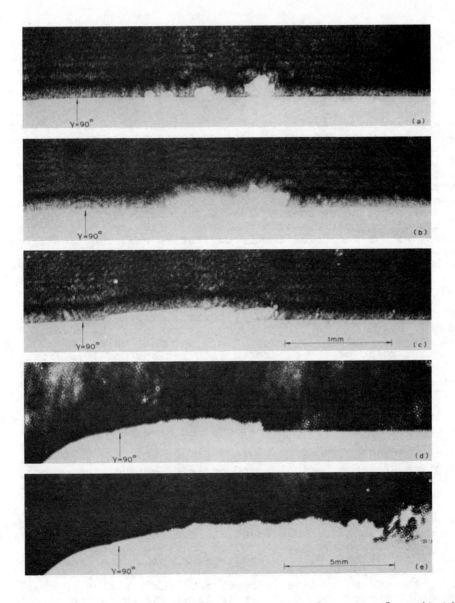

Fig. 6. Photographs showing progressive development of cavitation
on hemispherical nose. The flow is from left to right
$(V_O = 13 \text{ m/s})$. (a) $\sigma = 0.60$; (b) $\sigma = 0.59$; (c) $\sigma = 0.56$;
(d) $\sigma = 0.47$; (e) $\sigma = 0.39$.

CAVITATION STUDIES

The appearance of cavitation on the hemispherical nose is closely
related to the occurrence of laminar boundary layer separation.
Arakeri (6) showed that cavitation bubbles are first observed at
the location of transition and subsequent reattachment of the
separated shear layer, where strong vortices occur. This type of
cavitation is usually called bubble cavitation. An example is
shown in Fig. 6a. The larger bubbles at the location of transition are preceded by smaller ones which, according to Arakeri (6),
are travelling upstream with the reverse flow in the separated
region. With a reduction in the cavitation number σ, the larger
bubbles create a single cavity as shown in Fig. 6b. With a
further reduction in σ, the cavity is filling the separated region and a smooth attached cavity is observed (Fig. 6c). This
type of cavitation is usually called sheet cavitation. When σ is
further reduced, the length and the height of the cavity extend,
but the first part of the cavity remains smooth (Fig. 6 d, e).
The mean value of the length to height ratio of the developed
cavities was found to be 10.2. Since the mean value of the length
to height ratio of the separated flow regions was 10.6 (Fig. 4),
it may be concluded that the shape of the developed cavity appearing on the hemispherical nose is strongly governed by the
shape of the separated flow region.

The influence of polymer additives on cavitation is a rather new
phenomenon. Since 1970, several studies have been reported in
which mainly the phenomenon itself is described. Ellis, Waugh
and Ting (7) used a blowdown water tunnel to measure cavitation
inception on hemispherical nosed bodies. The addition of polymers
to the tunnel water caused a substantial reduction of the incipient cavitation number. Van der Meulen (8) investigated the influence on cavitation by injecting polymers from the nose of the
bodies. A considerable reduction of both the incipient and desi-
nent cavitation number was found. An explanation of these effects
could, as yet, not be given. In the previous chapter it was
shown (Fig. 5) that polymer additives suppress the occurrence of
laminar boundary layer separation on the hemispherical nose.
Hence, the strong vortices occurring at the position of transition and reattachment of the shear layer and being the principal mechanism for cavitation inception (Ref. 9) are suppressed
and thus cavitation will start at a much lower cavitation number.

The cavitation studies were extended by investigating a second,
blunt-nosed axisymmetric body. With the flow visualization method
it was found that (in accordance with theoretical predictions)
this body did not exhibit laminar boundary layer separation. The
appearance of cavitation on the blunt nose can best be described

Fig. 7. Photograph of triple exposure hologram showing three
        stages of cavity growth near nose of blunt model. The
        time separations are 50 μs and 100 μs respectively. The
        flow is from right to left (V_o = 10 m/s; σ = 0.31).

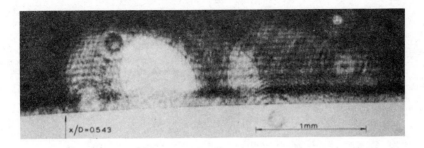

Fig. 8. Photograph of triple exposure hologram showing three
        stages of travelling bubble along blunt nose. The time
        separations are 50 μsec. and 100 μsec. respectively.
        The flow is from left to right (V_o = 10 m/s; σ = 0.31).

as "travelling bubble cavitation". The justification for this definition originates from the photographs presented in Fig. 7 and
Fig. 8. These photographs were taken from one hologram, where
three pulses were generated by the ruby laser. The separation
between the first and second pulse was 50 μsec., the separation
between the second and third pulse 100 μsec. Fig. 7 shows the
growth of the cavity near the nose of the model. The cavity
grows and, at the same time, travels along the surface with a
velocity slightly below that of the surrounding fluid. When the
cavity reaches a certain height, its shape becomes more like an
attached bubble, as shown in Fig. 8. The attached bubble hardly
grows, travels along the surface and, finally, collapses.

Further details of the fluid flow and cavitation phenomena on
the hemispherical and blunt nose are presented by Van der
Meulen (10).

ACKNOWLEDGEMENT

The authors acknowledge the support by the National Defense Organization of the Netherlands and the Royal Netherlands Navy.

REFERENCES

1. Acosta, A.J., "Cavitation and fluid machinery", Conference on Cavitation, Paper No. C267/74, Edinburgh, Scotland, September 1974.

2. Arakeri, V.H. and A.J. Acosta, "Viscous effects in the inception of cavitation on axisymmetric bodies", Journal of Fluids Engineering, ASME, Vol. 95, December 1973, pp. 519-527.

3. Feldberg, L.A. and K.T. Shlemenson, "The holographic study of cavitation nuclei", Proceedings IUTAM Symposium, Leningrad, June 1971, NAUKA Publishing House, Moscow 1973, pp. 239-242.

4. Peterson, F.B., "Hydrodynamic cavitation and some considerations of the influence of free-gas content", Ninth Symposium on Naval Hydrodynamics, Paris, August 1972, Proceedings Vol. 2, 1975, pp. 1131-1186.

5. Bader, F.E., "Kurzzeitholografische Untersuchungen von Kavitationsblasenfeldern, Ph.D. dissertation, Göttingen, 1973.

6. Arakeri, V.H., "Viscous effects in inception and development of cavitation on axisymmetric bodies", Report No. ENG. 183-1, California Institute of Technology, January 1973.

7. Ellis, A.T., J.G. Waugh and R.Y. Ting, "Cavitation suppression and stress effects in high-speed flows of water with dilute macromolecule additives", Journal of Basic Engineering, ASME, Vol. 92, September 1970, pp. 405-410.

8. Van der Meulen, J.H.J., "The influence of polymer injection on cavitation", Conference on Cavitation, Paper No. C149/74, Edinburgh, Scotland, September 1974.

9. Arakeri, V.H., "A note on the transition observations on an axisymmetric body and some related fluctuating wall pressure measurements", Journal of Fluids Engineering, ASME, Vol. 97, March 1975, pp. 82-86.

10. Van der Meulen, J.H.J., "A holographic study of cavitation on axisymmetric bodies and the influence of polymer additives", Ph.D. dissertation, Enschede, 1976.

ANGULAR DISTORTIONS IN REFLECTING SCANNING SYSTEMS

M. Nagler, G. Neumann and J. Zioni

Rehovoth Instruments Ltd., Rehovoth, Israel

INTRODUCTION

Many imaging systems, information retrieval systems, and laser scanning systems make use of optical-mechanical scanners composed of two or more reflecting elements, which determine the output direction of an input ray (Ref. 1,2). Inherent in such scanning mechanisms are distortions in the scanned field since the latter is a non linear function of the scanning angle. The exact shape and magnitude of the distorted field is a function of the orientation of the rotation axis of each mirror, relative to the direction of the input beam, and the position of the scanned field plane.

In this paper we describe the derivation of the general case where the two axes of rotation are in a random position, and reduce it to an explicit formula for three specific practical cases.

GENERAL SOLUTION

In this section we describe the solution for the direction of the output ray, given an input beam entering a two mirror system, rotating about two random axes.

Following the notation of R.E. Hopkins (3) we have for the reflection matrix of one mirror:

$$\left[r \right] = \begin{bmatrix} (1-2L^2) & -2L\,M & -2L\,N \\ -2L\,M & (1-2M^2) & -2M\,N \\ -2L\,N & -2M\,N & (1-2N^2) \end{bmatrix} \tag{1}$$

where L, M, N are the direction cosines of the normal to the reflecting plane, (pointing opposite to the reflecting plane).

The reflecting surfaces used usually rotate about one axis and it will be useful to specify the rotation about that axis using a single variable angle.

Since the general direction of a vector in space can be achieved by starting from a given vector and rotating by the Euler Angles (4) we will use these angles to describe our system.

Let us start for instance with a mirror whose normal in its reference system is given by (0,-1,0). We first rotate the normal by an angle X about the Z axis, and obtain a new coordinate system termed by (X',Y',Z'). We next rotate the normal by ψ about the Y' axis. The new X'' axis will now be the new axis of rotation. The mirror will rotate by a variable angle ϕ about that axis (Fig. 1).

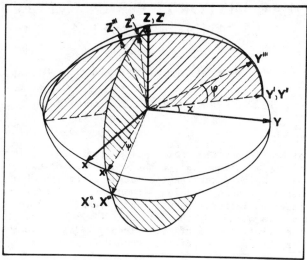

Fig. 1

In mathematical form we obtain for the normal, as seen from the reference system:

$$\overrightarrow{M}_{1ref} = \begin{bmatrix} \cos\chi & -\sin\chi & 0 \\ \sin\chi & \cos\chi & 0 \\ 0 & 0 & 1 \end{bmatrix} \times \begin{bmatrix} \cos\psi & 0 & -\sin\psi \\ 0 & 1 & 0 \\ \sin\psi & 0 & \cos\psi \end{bmatrix} \times \begin{bmatrix} 1 & 0 & 0 \\ 0 & \cos\phi & -\sin\phi \\ 0 & \sin\phi & \cos\phi \end{bmatrix} \times$$

$$\times \begin{bmatrix} 0 \\ -1 \\ 0 \end{bmatrix} = \begin{bmatrix} \cos\chi \ \sin\phi \ \sin\psi + \sin\chi \ \cos\phi \\ \sin\chi \ \sin\phi \ \sin\psi - \cos\chi \ \cos\phi \\ -\sin\phi \ \cos\psi \end{bmatrix} \tag{2}$$

$\overrightarrow{M}_{1ref}$ is the normal as seen from the reference coordinate system. As stated, the angles χ and ψ define the axis of rotation relative to the reference system and ϕ is the variable angle of rotation about that axis.

In a similar way we define the normal to the second mirror first by rotating by δ about the X axis, next by ξ about the Y'$_1$ axis and finally by the variable angle θ about the Z''$_1$ axis. In mathematical form:

$$\overrightarrow{M}_{2ref} = \begin{bmatrix} 1 & 0 & 0 \\ 0 & \cos\delta & -\sin\delta \\ 0 & \sin\delta & \cos\delta \end{bmatrix} \times \begin{bmatrix} \cos\xi & 0 & -\sin\xi \\ 0 & 1 & 0 \\ \sin\xi & 0 & \cos\xi \end{bmatrix} \times \begin{bmatrix} \cos\theta & -\sin\theta & 0 \\ \sin\theta & \cos\theta & 0 \\ 0 & 0 & 1 \end{bmatrix}$$

$$x \begin{bmatrix} 0 \\ 1 \\ 0 \end{bmatrix} = \begin{bmatrix} -\cos\xi \sin\theta \\ \cos\delta\cos\theta + \sin\delta \sin\xi \sin\theta \\ \sin\delta \cos\theta - \cos\delta \sin\xi \sin\theta \end{bmatrix} \qquad (3)$$

The normal to the mirror in (3) has the opposite sign from the one in (2) as the two mirrors are facing each other.

By inserting the direction cosines obtained in (2) and (3) into (1) we obtain two reflection matrices, which then have to be multiplied in the proper order to give the transfer function of the scanning system.

Rather than solving explicitly the tedious general case, we will now describe solutions for three specific cases. As we are interested only in angles and not in the cartesian coordinate of the output we will look at the output as displayed at the focal plane of an ideal lens. In this way the output beam is represented by the principal ray, and the three cases to be analyzed can be easily compared.

Case 1: $\psi = \delta = \xi = 0$

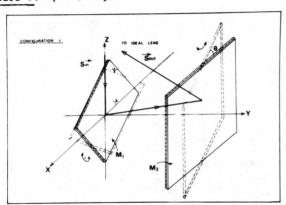

Fig. 2

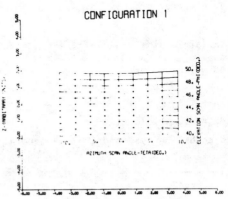

CONFIGURATION 1

Fig. 3

In this case, by inserting the angles ψ, δ, ξ, χ, θ and ϕ into (2) and (3) respectively we get:

$$\vec{M}_{1ref} = \begin{bmatrix} \sin\chi \, \cos\phi \\ -\cos\chi \, \cos\phi \\ -\sin\phi \end{bmatrix} \tag{4a}$$

$$\vec{M}_{2ref} = \begin{bmatrix} -\sin\theta \\ \cos\theta \\ 0 \end{bmatrix} \tag{4b}$$

and by substituting for the direction cosines in (1) we get

$$\begin{bmatrix} r_1 \end{bmatrix} = \begin{bmatrix} 1-2\sin^2\chi \, \cos^2\phi & \sin2\chi\cos^2\phi & \sin\chi \, \sin2\phi \\ \sin2\chi \, \cos^2\phi & 1-2\cos^2\chi\cos^2\phi & -\cos\chi \, \sin2\phi \\ \sin\chi \, \sin2\phi & -\cos\chi \, \sin2\phi & \cos2\phi \end{bmatrix} \tag{5a}$$

$$\begin{bmatrix} r_2 \end{bmatrix} = \begin{bmatrix} \cos2\theta & \sin2\theta & 0 \\ \sin2\theta & -\cos2\theta & 0 \\ 0 & 0 & 1 \end{bmatrix} \tag{5b}$$

Let the input ray be defined by:

$$\vec{S}_{in} = \begin{bmatrix} 0 \\ 0 \\ -1 \end{bmatrix} \tag{6}$$

then:

$$\vec{S}_{out} = \begin{bmatrix} r_2 \end{bmatrix} \cdot \begin{bmatrix} r_1 \end{bmatrix} \cdot \vec{S}_{in} = \begin{bmatrix} \sin2\phi \, \sin(2\theta-\chi) \\ -\sin2\phi \, \cos(2\theta-\chi) \\ -\cos2\phi \end{bmatrix} \tag{7}$$

The intercept of this ray with the plane Y = f is given by:

$$x/(\sin2\phi \, \sin(2\theta-\chi)) = -f/(-\sin2\phi \, \cos(2\theta-\chi)) = z/(-\cos2\phi) \tag{8}$$

from which we obtain:

$$x = f \, \text{tg} \, (2\theta-\chi) \tag{9a}$$

$$z = -f \, \text{ctg} \, 2\phi/\cos(2\theta-\chi) \tag{9b}$$

and combining (9a) and (9b) we obtain:

$$\frac{z^2}{f^2 \text{ctg}^2 2\phi} - \frac{x^2}{f^2} = 1 \tag{10}$$

For a constant ϕ this equation describes a hyperbola, whereas for a constant θ we obtain a straight line x = const.

The scan of an entire field was simulated by a computer program on an IBM 370. The angle θ varied from -10. to 10. deg. around 0. and ϕ- from +40 to +50 deg. around 45. The result is presented in Fig. 3.

<u>Case 2</u>: $\delta = \pi/4$, $\psi = \chi = \xi = 0$ (Fig. 4).

From (2) (3) we obtain:

$$\overrightarrow{M_1} = \begin{bmatrix} 0 \\ -\cos\phi \\ -\sin\phi \end{bmatrix} \tag{11a}$$

$$\overrightarrow{M_2} = \begin{bmatrix} -\sin\theta \\ \cos\delta\cos\theta \\ \sin\delta\cos\theta \end{bmatrix} \tag{11b}$$

hence from (1)

$$[r_1] = \begin{bmatrix} 1 & 0 & 0 \\ 0 & -\cos2\phi & -\sin2\phi \\ 0 & -\sin2\phi & \cos2\phi \end{bmatrix} \tag{12a}$$

$$[r_2] = \begin{bmatrix} \cos2\theta & \cos\delta\sin2\theta & \sin\delta\sin2\theta \\ \cos\delta\sin2\theta & 1-2\cos^2\delta\cos^2\theta & -\cos^2\theta\sin2\delta \\ \sin\delta\sin2\theta-\cos^2\theta\sin2\delta & 1-2\sin^2\delta\cos^2\theta \end{bmatrix} \tag{12b}$$

We again take

$$\overrightarrow{S}_{in} = \begin{bmatrix} 0 \\ 0 \\ -1 \end{bmatrix} \tag{13}$$

and obtain

$$\overrightarrow{S}_{out} = \begin{bmatrix} \cos\delta \; \sin2\theta \; \sin2\phi -\sin\delta \; \sin2\theta \; \cos2\phi \\ \sin2\phi-2\cos^2\delta\cos^2\theta\sin2\phi + \cos^2\theta\sin2\delta\cos2\phi \\ -\cos^2\theta \; \sin2\delta \; \sin2\phi -\cos2\phi + 2\sin^2\delta\cos^2\theta \; \cos2\phi \end{bmatrix} \tag{14}$$

which, upon substituting $\delta = \pi/4$ becomes

$$\vec{S}_{out} = \begin{bmatrix} \sin2\theta \ \sin(2\phi- \pi/4) \\ \sin2\phi \ \sin^2\theta + \cos2\phi \ \cos^2\theta \\ -(\sin2\phi\cos^2\theta + \cos2\phi\sin^2\theta) \end{bmatrix} \qquad (15)$$

We define

$$\cos2\phi + \sin2\phi = A \qquad (16a)$$

$$\cos2\phi - \sin2\phi = B \qquad (16b)$$

hence (15) can be written as:

$$\vec{S}_{out} = \begin{bmatrix} \sin2\theta \ \sin(2\phi-\pi/4) \\ 1/2(A+B\cos2\theta) \\ -1/2(A-B\cos2\theta) \end{bmatrix} \qquad (16)$$

This time the ideal lens is positioned such that its focal plane is at $z = -f$. By the same steps as before we obtain:

$$x = -f\sqrt{2} \ B\sin(2\theta)/(A-B \ \cos2\theta) \qquad (17a)$$

$$y = -f(A+B\cos2\theta)/(A-B\cos2\theta) \qquad (17b)$$

The graph describing these curves is presented for the same θ and ϕ as before in Fig. 5. We see that the field distortion is rather large and unsymmetric about the (0,0) point.

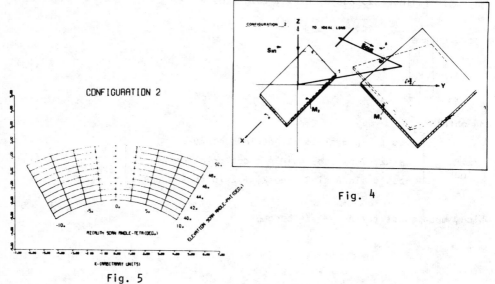

CONFIGURATION 2

Fig. 4

Fig. 5

Configuration 3: $\delta = \pi/4$, $\psi = \chi = \xi = 0$ (Fig. 6).
In this case as seen we preserve the angles of configuration 2, but this time
the input ray is given by

$$\vec{S}_{in} = \begin{bmatrix} 0 \\ 1 \\ 0 \end{bmatrix}$$

along the +y direction and being reflected first by M_2.

The previous matrices are multiplied in reversed order, and the focal plane
of the imaging mirror is at $y = f$. We obtain:

$$\vec{S}_{out} = \begin{bmatrix} \dfrac{\sqrt{2}}{2} \sin2\theta \\ -\cos2\phi\sin^2\theta + \sin2\phi\cos^2\theta \\ -\sin2\phi\sin^2\theta - \cos2\phi \cos^2\theta \end{bmatrix} \qquad (19)$$

and

$$x = \sqrt{2} \, f \, \sin2\theta \, / \left[2 \cdot (\sin2\phi\cos^2\theta - \cos2\phi\sin^2\theta) \right] \qquad (20a)$$

$$z = -f \, (\sin2\phi\sin^2\theta + \cos2\phi\cos^2\theta)/(\sin2\phi\cos^2\theta - \cos2\phi\sin^2\theta) \qquad (20b)$$

The curves described by 20a, b are presented in Fig. 7. The amount of distortion in this configuration is larger than in configuration 1 but smaller
than in 2.

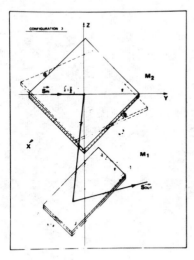

Fig. 6

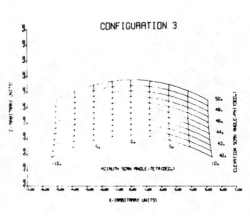

Fig. 7

CONCLUSION

We see that in developing a system which makes use of two reflecting surfaces to scan or display a field, distortions are always present but can be minimized. The amount of distortion allowed may be decided by an overall system performance criteria. We hope that these calculations will help other people in deciding on their scanning arrangement.

REFERENCES

1) R.F. Leftwich and D.W. Fisher, Macro and micro thermography, Proc. Tech. Prog., Electro-optical System Design Conf., N.Y. City Sept. 1972 p. 237.

2) P.J. Brosens and E.P. Gruda, "Applications of galvanometers to laser scanning", Electro Optical System Design Mag., (Apr. 1974) p. 32.

3) R.E. Hopkins, Military Standardization Handbook - Optical Design, Sec. 13, Defense Supply Ag. Wash. (1962).

4) H. Goldstein, Classical Mechanics, p. 107, Addison-Wesley, Cambridge, 1953.

LASER SCANNERS FOR DEFECT DETECTION

P. Ziekman* and J. W. Klumper**

*SKF Engineering and Research Centre, Nieuwegein, Netherlands
**Technisch Physische Dienst TNO-TH, Delft, Netherlands

ABSTRACT

Detection of defects on machined metal surfaces by means of laser scanning techniques is discussed. As a study of the mechanism of defect detection in this way, a diffraction pattern analysis has been made. In order to see if this theory could lead to sensitizing techniques, some experiments have been carried out using objects machined in one direction and exhibiting small cracks. Preliminary results show a good agreement with the trends as predicted from the theory and look rather promising.

An example of optical signal processing by means of sampled measurements of the spatial distribution of the intensity in the field diffracted by a machined metal surface is a recently developed laser scanner for cylinder bore inspection. Apart from the capability of the system to detect porosity defects, machining defects are also detected.

INTRODUCTION

Detection of defects in transparent and on non-transparent objects by means of a laser scanner is well known in the field of non-destructive testing. With this technique, defect detection is accomplished whenever the amplitude of the output signal of a photon detector, which receives light reflected or transmitted by the object, exceeds certain threshold values. The significance of the use of a laser is that, due to the coherence properties of its light, the wavefront can be focussed to a diffraction limited point. If this D.L. point is used as the scanning spot on smooth surfaces the highest sensitivity attainable for defect detection by means of optical techniques is obtained. An example of such a laser scanner, which detects fine scratches on optical surfaces, has been reported by Baker (1).

In many cases, industrial demands also involve surfaces which are rough compared with optical surfaces to be inspected with a high sensitivity. However, the output of the photon detector is then usually a signal which varies rapidly in amplitude. This means that a small defect will not always contribute to a significant change in the output signal, and thus will not be detected by such a simple technique.

Similarly, it is usually impossible to detect machining defects on engineering components using this technique, since machining defects change the surface

661

662 P. Ziekman, J. W. Klumper

statistics but cannot be observed by point measurements alone.

It may be concluded, therefore, that defect detection on rough surfaces with a laser scanner by means of amplitude measurement of the output signal of a photon detector placed in the direct beam reflected from or transmitted by the surface is not possible in a number of practical cases.

DIFFRACTION PATTERN ANALYSIS

In order to study the mechanism of rough surface laser scanning with respect to defect detection, a diffraction pattern analysis has been made.

As a model to describe a class of rough surfaces we chose a one-dimensional facet-surface, diagrammed in Fig. 1. As a type of defect we considered small cracks.

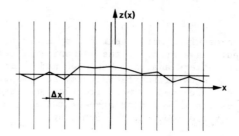

Fig. 1. The one-dimensional facet-surface

The model is built up of a number of subsequent straight line facets having random slopes. All facets have an equal projection length Δx, defined as the projection of a facet on the x axis, as can be seen from Fig. 1. A restriction is made to surfaces having large facet projection lengths compared with the wavelength λ. The one-dimensional surface height function $z(x)$ can thus be expressed as:

$$z(x) = \sum_k z_k(x) \, \text{rect}\{\frac{x}{\Delta x} - (k + \tfrac{1}{2})\}$$ (1)

where $z_k(x)$ describes a straight line coinciding with the k-th facet and the rect-function is defined as:

$$\text{rect}(x) = 1 \text{ if } -\tfrac{1}{2} < x \leq \tfrac{1}{2}$$
$$\text{rect}(x) = 0 \text{ otherwise.}$$

A reference line may be defined in the usual way as a centre line such that

$$\int_L z(x) \, dx = 0$$

where $\int_L$ denotes integration over a sufficiently large part of this line.

It is believed that the choice of this particular model provides a means of adequately describing diffraction phenomena which occur when a laser beam is scanned over a number of technically interesting metal surfaces.

We start our analysis by causing a collimated laser beam of diameter Δx, the facet projection length, to scan the surface perpendicular to its centre line. In practice, this situation can be approximated by inserting a positive lens

into the light path between laser and test object, allowing for an appropriate
misfocussing. A diagram of this object illumination method is shown in Fig. 2,
which also shows the signal detection plane being the back focal plane of the
above-mentioned lens.

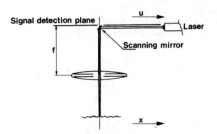

Fig. 2. Object illumination and signal detection
plane in basic scanning configuration.

At any point in the signal detection plane, a signal rapidly varying in amplitude will be received as a result of the diffraction patterns of the individual
facets that follow each other.

For the k-th facet the diffracted light is advanced or retarded in phase by an
amount of $(4\pi/\lambda) z_k(x)$, so that the corresponding complex amplitude $\psi_k(\nu)$ in the
diffraction pattern is obtained by Fourier transforming of the complex amplitude
in the object plane (2):

$$\psi_k(\nu) = \int_{-\infty}^{+\infty} \text{rect}\{\frac{x}{\Delta x} - (k + \tfrac{1}{2})\} \exp\{-\frac{4\pi i}{\lambda} z_k(x)\} \exp(-2\pi i x\nu)\, dx \qquad (2)$$

where the reduced co-ordinate $\nu = u/\lambda f$ has been used and unit illumination
amplitude has been assumed. The symbols u, λ, and f are for the co-ordinate in
the detection plane, the wavelength of the light and the focal length of the
lens respectively.

Substituting $m_k x + q_k$, the equation of a straight line having a slope m_k, for
$z_k(x)$, the above integral can be evaluated. The intensity $I_k(\nu)$ diffracted by
the k-th facet can then be calculated by multiplying $\psi_k(\nu)$ with its complex
conjugate, yielding

$$I(\nu,m) = (\Delta x)^2 \text{ sinc}^2 \{\Delta x(\nu + \frac{2m}{\lambda})\} \qquad (3)$$

where we have dropped the index k and have $I(\nu,m)$ dependent on the variable m
also. $I(\nu,m)$ now represents the intensity in the detection plane at a position
defined by the reduced co-ordinate ν, obtained by diffraction from a facet
with slope m. The sinc-function is defined as:

$$\text{sinc}(x) = \frac{\sin(\pi x)}{\pi x}$$

Since $I(\nu,m)$, upon scanning, is considered noise the mean square value of
$I(\nu,m)$ is the mean square value of the noise:

$$\sigma_N^2(\nu) = \int_{-\infty}^{+\infty} \{I(\nu,m) - <I(\nu,m)>\}^2 w(m)\, dm \qquad (4)$$

where w(m) is the probability density function of m and $<I(\nu,m)>$ is the expectation value of $I(\nu,m)$. The probability density function w(m) can be determined e.g. by a technique proposed in a recent paper by Tanner (3). If w(m) is known the integral for $\sigma_N^2(\nu)$ can be evaluated numerically.

However, a direct qualitative trend can be inferred from this analysis. Figure 3 shows three different plots for $I(\nu,m)$ as defined by eq. (3).

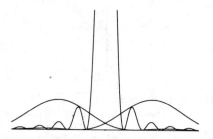

Fig. 3. Comparison of diffraction signals from a "normal"
roughness facet, having facet projection length Δx
and zero slope, and from "crack" facets having
facet projection lengths smaller than Δx and non-
zero slopes.

The highly peaked one in the middle symbolizes $I(\nu,m)$ as obtained from a "standard" facet, having a projection length Δx. The value m = 0 has been used to make the drawing. Other values of m within the restriction of w(m) could have been used too and would have resulted in the same plot but shifted over a distance of $|2m/\lambda|$. The two other plots symbolize $I(\nu,m)$ as obtained from a crack. A crack is characterized by facets having slopes outside the population defined by w(m) of a "normal" surface and by a facet projection length which is small compared with the "normal" facet projection length, Δx. Therefore the central peak of $I(\nu,m)$ produced by a crack is smaller, is shifted outside the shifting range for the "normal" peaks and spreads out in space more than the comparatively narrow peak of the "normal" function $I(\nu,m)$.

From this it may be concluded that a minimum signal to noise ratio can be expected in a detection area around $\nu = 0$ and that, depending on the specific nature of w(m), an optimum defect detectability is obtained at a certain off-axis detection area. Similar results have been found by Reich and Coleman (4).

EXPERIMENT

Experiments have been carried out using a scanner set-up as shown in Fig. 4. As a test object a fine polished steel plate containing an artificially made fatigue crack has been used.

The light of a low power CW He-Ne laser is expanded by a microscope objective O, a spatial filter SF and a positive lens L_1 into a collimated beam which is reflected from a scanning mirror M ("galvanoscanner") having its centre of rotation coinciding with the front focus of a positive lens L_2. The latter focusses the beam onto the test surface. The intensity of the light diffracted by the metal surface is measured by a silicon photodiode-amplifier combination D which is placed in the back focal plane of a positive lens L_3.

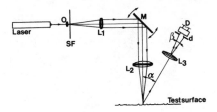

Fig. 4. Scanner set-up used for experiments

The entrance pupil of this pickup system, specified by an aperture angle β, is controlled by an iris diaphragm d in front of the detector D. The set-up has been arranged so as to allow for a variable direction of detection (α in Fig. 4).

The half width diameter of the Gaussian scanning spot in our experiments was approximately 25 μm, which is large compared with the crack width.

The first experiment concerned the measurement of the signal arising from the surface roughness alone. Using a true rms meter, the rms value σ_N of the light intensity was measured as a function of α, which is equivalent to ν in the previous section. The results are shown in Fig. 5 for three different aperture angles. The experimental data for this figure have been obtained by averaging the results of a number of measurements at different positions on the test sample.

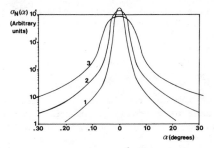

Fig. 5. Experimentally determined graphs of $\sigma_N(\alpha)$, obtained from
        a fine polished steel surface.
        Curve 1: aperture angle 2°
        Curve 2: aperture angle 5°
        Curve 3: aperture angle 10°

In a second experiment the signal amplitude produced by the crack was measured, again as a function of α. The measured values $V'_{cr}(\alpha)$, obtained in this way were corrected by subtraction of the mean value of the surface roughness signal from them, thus yielding a "true" $V_{cr}(\alpha)$.

Signal to noise ratio graphs, produced by plots of $V_{cr}(\alpha)/\sigma_N(\alpha)$ as a function of α, are shown in Figs. 6 and 7 for different crack geometries and different aperture angles respectively.

These experimental results are the first of a series of experiments which all relate to the detection of small defects on rough surfaces. These results bear a preliminary character.

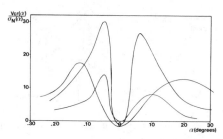

Fig. 6. Experimentally determined graphs of signal to noise ratio
$V_{cr}(\alpha)/\sigma_N(\alpha)$, at three different scanning positions on
the crack and a detector opening of 5°.

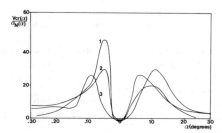

Fig. 7. Experimentally determined graphs of signal to noise ratio
$V_{cr}(\alpha)/\sigma_N(\alpha)$, at one crack position and at three different
detector openings.
Curve 1:   aperture angle 2°
Curve 2:   aperture angle 5°
Curve 3:   aperture angle 10°

However, a few conclusions may be drawn from these results at this stage.
An agreement with the trends as predicted from the theoretical diffraction
pattern analysis has been obtained, so that - at least for this type of defect
and surface - a minimum signal to noise ratio is reached at positions near
$\alpha = 0$ ($\nu = 0$) and maximum values are found at certain off-axis positions. A
suggestion for a technical realization of a laser scanner for detection of this
type of defect on this type of surface involves a pick-up system comprising a
lens-spatial filter combination, where the spatial filter consists of a central
stop and a diaphragm.

LASER SCANNER FOR CYLINDER BORE INSPECTION

The laser scanner for cylinder bore inspection, primarily but by no means
exclusively intended for inspection of automotive cylinder bores, is a joint
development of the SKF Engineering & Research Centre and the Cranfield Unit
for Precision Engineering * (5).

---

* Cranfield, Bedford, England.

A prototype of the system has recently been completed. The main innovative aspect* of it is that it features a combination of two separate techniques of electro-optical detection and signal processing. This enables the system to detect and to discriminate the two types of defects which may occur:

- major defects, such as the blow holes and areas of porosity shown in Fig. 8, which are characterized by a gross change of reflectivity of the surface over the area concerned;

- machining defects, such as automotive cylinder bores not having a honed pattern of the proper roughness and orientation, which leads to incorrect lubrication and a too high oil consumption. Defects of this type ("retained boring marks") are also shown in Fig. 8.

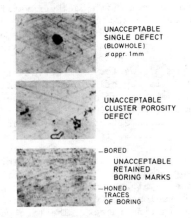

Fig. 8. Examples of typical cylinder bore defects.

Inspection for both types of defects is carried out simultaneously by the arrangement shown in Fig. 9.

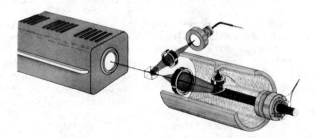

Fig. 9. Schematic diagram of laser scanner for cylinder bore inspection

Light derived from a low power CW He-Ne laser illuminates - via a beam splitter, two telescopically arranged positive lenses and a 45° oriented mirror in between them - a spot on the cylinder bore surface. The 45° mirror and the positive

---

*Patents applied for

lens closest to the cylinder bore surface perform a circular scan while the whole system makes an axial movement which is locked to the circular movement of mirror and lens. This generates a helical scan covering the whole cylinder bore surface.

Detection of "major" defects is accomplished by amplitude measurement of the signal obtained from a silicon photodiode, which receives light reflected from the cylinder bore after passage of the 45° mirror-telescopic system combination in return and is reflected off the beam splitter, as shown in Fig. 9. As is apparent from the description above the cylinder bore is scanned by the unexpanded laser beam. The simplest way to do this would have been by means of the 45° mirror alone. But again, since we are dealing with rough surface scanning, this would have resulted in a "$\nu = 0$ detection", as meant in the previous section on diffraction pattern analysis. This, in turn, would have meant a poor signal to noise ratio, a fact which has been verified experimentally. These experiments have revealed that, by scanning the unexpanded beam without the additional optical system shown in Fig. 9, it was not possible to detect all unacceptable single and cluster defects. Increasing the sensitivity by means of focussing the laser beam onto the cylinder bore surface was undesirable not only since a bigger spot allows for a relatively coarse pitch of the helical scan but primarily because this way of object illumination was very suitable for detection of machining defects, to be discussed below. Therefore, by illuminating the cylinder bore with the relatively coarse spot of the unexpanded laser beam the noise contribution effect due to the surface roughness could be reduced, while a sufficient increase in sensitivity could be obtained by introduction of above described optical pick-up system. This part of the system is equipped with an automatic accept/reject decision making digital electronic signal processor. What this processor basically does is make a weighted measurement of the total amount of defect area in the cylinder bore. Weighted means that defects belonging to an area where the defect density is high are supplied with a weighting factor. The total amount of defect area is further evaluated in such a way that, if in a certain area the acceptability threshold has not been exceeded and a sufficiently large area of acceptable surface follows, the system is automatically reset so that the evaluation begins from this part of the cylinder on. All these respective comparator levels, threshold values and the weighting factor are according to users'/customers' demands and are selected by simple thumbwheel settings.

Detection of machining defects by means of the electro-optical system described above, turned out to be impossible since not all types of possibly occuring machining defects give rise to a gross change in reflectivity of the cylinder bore surface. In order to analyse experimentally the diffraction patterns obtained from a properly machined cylinder bore surface and such surfaces exhibiting machining defects, a simple experiment was set up. This is shown in Figs. 10 and 11, where pieces of cut cylinder bores - a properly machined one and one with machining defects respectively - were put on an optical bench and illuminated with an unexpanded laser beam so that the diffraction patterns could be observed on a screen some distance away from them. As can be seen from these figures the information about the surface texture is not in the overall reflectivity of the surface concerned, but in the spatial distribution of the reflected light intensity. In the case of an automotive cylinder bore (used throughout our experiments) which is properly machined, the diffraction patterns consist of very distinct intensity lobes, exhibiting an "X"-geometry due to a set of honing marks present on the cylinder bore. See Fig. 10. The angle between the arms of this "X" indicates the angle between the honing marks. In the equipment sketched in Fig. 9, the diffraction patterns developed by scanning are sampled at the entrances of a number of optical fibres arranged around the lens closest to the cylinder bore surface. The exits of these fibres are re-arranged and grouped together where they deliver their output signals as a number of pulses to a number of photon detectors arranged on a stationary ring. This part of the system is also equipped with an automatic accept/reject decision making digital electronic signal processor. This processor makes a count of the number of missing pulses due to improper machining and performs basic functions similar to those described under the processor for major defects.

Fig. 10. Diffraction pattern obtained from cylinder bore exhibiting proper machining.

Fig. 11. Diffraction pattern obtained from cylinder bore with incorrect honing.

CONCLUSION

Detection of two types of defects on rough metal surfaces - small cracks and machining defects - by means of a laser scanner has been discussed. Preliminary results concerning small crack detection have shown that - at least in one particular case of scanning on a rough surface - the laser scanning technique can be made sensitive enough.

An example of a successful development of sensitized defect detection on a rough surface has been reported as the laser scanner for cylinder bore inspection. Moreover, this system incorporates an optical signal processing technique for detection of machining defects. Further developments, for instance by extending the theory of diffraction pattern analysis, in the field of laser scanning for defect detection may be expected.

REFERENCES

(1) L. R. Baker, Some scanning techniques used in automatic inspection, paper presented at the Conference on the Engineering Uses of Coherent Optics, University of Strathclyde (1975).

(2) Goodman, J. W., Introduction to Fourier Optics, McGraw-Hill, New York, 1968.

(3) L. H. Tanner, The use of laser light in the study of metal surfaces, Optics and Laser Technology 8, 113 (1976).

(4) F. Reich and W. J. Coleman, High-speed surface flaw detection, Optical Engineering 15, 48 (1976).

(5) P. A. McKeown, P. Cooke and W. P. N. Bailey, The application of optics to the quality control of automotive components, Proceedings SPIE 60, 77 (1975).

SESSION 18.

NON-OPTICAL HOLOGRAPHY

Chairman: J. POLITCH

DESIGN OF GLOW DISCHARGE PLASMA LAMPS FOR ECONOMIC REAL-TIME MICROWAVE HOLOGRAPHY

N. S. Kopeika

Department of Electrical Engineering, Ben-Gurion University of the Negev, Beer Sheva, Israel.

ABSTRACT

Advantages of glow discharge lamps for real-time microwave holography and optimum design for such applications, using both the subnormal and abnormal glow, are described. In particular, internal signal amplification by the subnormal glow discharge is derived for the first time.

INTRODUCTION

Real-time microwave or millimeter wave holographic detection is difficult to implement because of problems stemming from the large recording aperture required for good image quality (1). Large area detectors such as nematic liquid crystals obviate the slow mechanical scanning used with single 'point' detectors but require large incident power densities (2), are limited by ambient thermal drift (3), and have slow response times (4). The last is disadvantageous for imaging moving objects or for monitor display utilizing normal frame rates, as might be desired in concealed weapon imaging (5), for example. Quasi real-time schemes involving circular scanning with a single point detector have been suggested (6), but the projected scan time of one second is still too slow for many moving objects as well as for ordinary frame rates. Photochromic recording processes are much slower (7). Utilization of thermographic television to record the thermal image on a microwave absorber (8) is limited by the high price of an infrared imaging system. In addition, although the sensitivity of the last scheme is 1 - 2 orders of magnitude better (8) than that of any nonscanned mapping technique described above, it is still orders of magnitude less than that available using single 'point' detectors in scanning techniques. Mechanical scanning can be obviated by the use of real-time electronic scanning techniques involving arrays of sensitive detectors. However, since image resolution is dependent upon the quantity of detectors in the array, the detector price becomes a significant item in determining the economic feasibilities of such schemes. Microwave diode detectors and detector mounts in such quantities are prohibitively expensive.

Commercial glow discharge lamps, whose individual price is ordinarily less than a dollar, have been shown to exhibit excellent sensitivity, as compared to the much more expensive diode detectors, to both millimeter wave (9)-(11) and microwave (12) radiation in both video (10) - (12) and synchronous (9) detection modes. While $50\mu W.cm^{-2}$ microwave irradiance levels are required in the infrared thermography scheme described above in order to be detected, irradiance levels as low as a $nW.cm^{-2}$ at 10 GHz (12) or less than a $\mu W.cm^{-2}$ at 70 GHz (10), (11) can be detected by these lamps over bandwidths as large as 10 kHz even in video detection schemes without external antennas. Sensitivity in synchronous detection schemes is even orders of magnitudes better (9). Radiation levels even orders of magnitude lower can easily be detected using small external antennas and noise bandwidth reduction in both types of detection schemes. Thus, the microwave power levels and radiation exposure required in microwave holographic systems can be considerably reduced, as well as over-all cost, by implementing electronic scanning techniques involving arrays of glow discharge lamps to record the hologram in real-time. The capability of these devices to detect, in principle, two simultaneous signals linearly (9), and thus their suitability for holography (1) have been demonstrated, and commensurate recording, processing, and display schemes utilizing them and noncoherent to coherent image converters for real-time image reconstruction have been proposed (1).

The proposed schemes involve either parallel or sequential integration, as in Fig. 1.

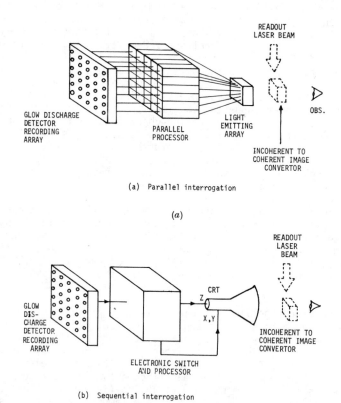

(a) Parallel interrogation

(*a*)

(b) Sequential interrogation

Fig. 1.   Proposed real-time microwave hologram processing and display.

In the first case a light-emitting array of incoherent miniature incandescent or neon lamps whose light intensity is linearly proportional to the current in them could follow a glow tube detector recording array with the detected signal of each lamp in the recording array controlling the intensity of the lamp corresponding to it in the light-emitting array. An incoherent light pattern corresponding to the hologram would then appear on the face of the light-emitting array. An incoherent image converter could then prepare the hologram for reconstruction via a laser beam.

The principle of sequential interrogation involves rapid electronic scanning of the output of the detectors in the hologram-recording array with a single processing network, as in Fig. 1(b). Since this unit must handle the entire hologram data at normal frame rates, the very large bandwidth of this processing system would result in much more noise than that of the parallel interrogation scheme. The advantage of the sequential interrogation scheme, however, is that a single display device such as a cathode-ray tube can be employed to display the hologram information for subsequent analogue image retrieval. In this scheme the output of the electronic processor is used to intensity modulate the cathode-ray tube (1).

Real-time reconstructed image quality has been exhibited by incoherent-to-coherent image converters with image quality comparable to that obtained directly from original photographic transparency records of microwave holograms (13).

Other advantages of glow discharge detectors arrays as compared to diode detector arrays aside from the economic one include less sensitivity to temperature changes (14), greater dynamic range and electronic ruggedness (10), the ability to detect sudden increases in radiation levels without being damaged (10) and an almost isotropic directivity which allows them to detect radiation depolarized as a result of reflection by the hologram object. In addition they can be used in environments such as the Van Allen belt, nuclear reactors, or space systems subject to intense ionizing radiation fields where many other types of sensors cannot operate (15).

GLOW DISCHARGE DETECTION OF MICROWAVE SIGNALS

Lamp operation as a detector is quite simple. The gas is broken down with a dc source, as in Fig. 2.

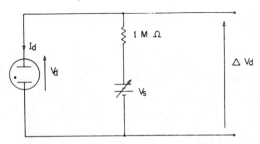

Fig. 2.   Glow discharge detector biasing circuit for subnormal glow mode.

The mechanism of detection by dc glow discharge plasmas of microwave radiation involves electron kinetic energy enhancement and thus ionization rate
enhancement by the microwave electric field (16) and/or losses of signal
current to diffusion processes (17). The former increases the discharge
current, while the latter decreases it. Small dc changes generated by the
detected microwave signal are difficult to sense, even in bridge circuits
(15), because of voltage drift problems. Therefore, most effective use is
made of glow discharge detectors for radiation detection if the signal to be
detected is modulated, in which case detector sensitivity is limited by ac
noise (10), (18), (20). When the discharge current is very low, on the order
of microamperes, the lamp is characterized by a negative dynamic resistance
when it is biased to the subnormal glow, as indicated in Fig. 3. Here,
losses of signal current to diffusion processes are quite significant. The
dc glow and its high field (19), (20), are very localized and envelope only
a very small portion of the electrode length and electrode separation. Thus,
the plasma volume where the dc field is weak is much larger than the volume
where it is strong. This allows electrons generated by the microwave signal
ionization rate enhancement to diffuse away from negative space charge regions to the tube walls. As such, these electrons are then not available to
ionize gas atoms through energy provided primarily by the dc field and thus
to internally amplify the signal.

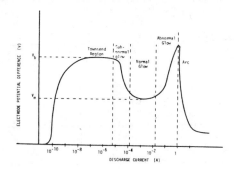

Fig. 3.   Typical voltage-current characteristic of a gas discharge. V_b and
V_m are breakdown and maintaining voltages, respectively.

Such diffusion current losses can be minimized if the average electron energy
provided by the dc field is about 69% of that required for ionizing collision
purposes (20). With such dc biasing, the microwave signal average electron
velocity enhancement, given in the steady state by

$$\Delta \vec{v} = \frac{q}{m}\left(\frac{\nu - j\omega}{\nu^2+\omega^2}\right) \vec{E} \tag{1}$$

is minimized. In Eq.(1), q and m are electron charge and mass, respectively,
ν is the average frequency per electron of elastic collisions with neutral
gas atoms, and ω is the radian frequency of the microwave signal electric
field $\vec{E}$. Minimizing the average electron microwave signal velocity enhancement minimizes the diffusion coefficient enhancement ΔD which is proportional
to Δv. On the other hand, the microwave signal ionization rate enhancement
is independent of the dc electron energy and, for ω much greater than the
plasma frequency ω_p, is given by (16), (20),

$$\Delta \nu_i = \frac{q\, \eta_o \nu}{m\, V_i(\nu^2 + \omega^2)}\, P_D \tag{2}$$

in the steady-state, where η_o is the wave impedance of free space, V_i is the ionization potential in volts of the gas mixture, and P_D is the incident microwave signal power density. The detection current is

$$\Delta I_D = qVn\Delta \nu_1 G$$

$$= \frac{q^2 dn\eta_o \nu\, G}{mV_i(\nu^2 + \omega^2)}\, P_s \tag{3}$$

where n is the electron number density, V is the plasma volume, d is the distance through the discharge traversed by the received microwave signal power P_s, and G is the internal signal amplification. From Eq. (3) the detector is seen to be a square-law device. The direct proportionality of detector response with incident microwave signal power has been verified experimentally up to incident power density levels on the order of $3W\cdot cm^{-2}$ with argon glow lamps (21).

Internal Gain

The internal gain G stems from ionizing collisions with neutral atoms involving electrons generated from the microwave signal ionization rate enhancement. In the abnormal glow, where the load resistor of Fig. 2 is on the order of a kilohm (9), (12), (19), secondary electrons resulting from such collisions involving signal electrons increase the signal current by a factor of approximately 10^6 in ordinary commercial indicator lamps (20). In the subnormal glow, positive ions generated from both the initial microwave signal ionization rate enhancement and secondary collisions involving such electrons resulting from the strong dc field bombard the cathode and produce secondary emission electrons which amplify the signal considerably further. The initial signal electron multiplication through ionizing collisions can be derived from the continuity equation for a dc discharge perturbed by a small incident ac field (17). The change in drift current is (17)

$$\Delta I_D \propto \mu[\Delta E_{dc}\frac{dn}{dx} + E_{dc}\frac{d\Delta n}{dx}] = (\Delta \nu_i)\, n + \nu_i(\Delta n)$$

$$- (\Delta D)\nabla^2 n - D\nabla^2(\Delta n) \tag{4}$$

where μ is electron mobility and ν_i is the ionization rate provided by the dc electric field E_{dc} in the x-direction. This equation actually is a sum of two equations:

$$\mu\Delta E_{dc}\frac{dn}{dx} = (\Delta \nu_i)n - (\Delta D)\nabla^2 n \tag{5a}$$

$$\mu E_{dc}\frac{d\Delta n}{dx} = \nu_i(\Delta n) - D\nabla^2(\Delta n). \tag{5b}$$

The second equation describes the signal electron multiplication through ionizing collisions. Since μE_{dc} equals the electron drift velocity v, if diffusion losses in Eq. (5b) are neglected for the sake of simplicity, then over the incremental path dx the signal electron number density increases at the rate

$$\frac{d\Delta n_e}{\Delta n_e} = \frac{\nu_i}{v}\, dx. \tag{6}$$

In these ionizing collisions involving signal electrons the detected signal current is also increased at the same rate by the addition of positive ions. Thus, in terms of total charge carriers,

$$\frac{d\Delta n_s}{\Delta n_s} = \frac{2\nu_i}{v}\, dx \;, \tag{7}$$

or, over an interval x', signal charge carriers increase at the rate

$$\Delta n_s = n_s e^{\,2\frac{\nu_i}{v}x'} \;. \tag{8}$$

The effect of diffusion losses in Eq. (5b) is to diminish the coefficient of x' in the exponent of Eq. (8) and thus the internal signal amplification. Ionization collisions by signal positive ions are neglected here because the energy $\overline{E}_I$ of positive ions of mass $\underline{M}$ compared to that $\overline{E}_e$ of electrons for a given electric field is very low, i.e., (20)

$$\frac{\overline{E}_I}{\overline{E}_e} = \frac{m}{M}\;. \tag{9}$$

If the whole discharge is irradiated by the EM signal, Eq. (8) must be integrated to include initial enhanced ionization electrons over the entire inter-electrode space. Assume that radiation uniformly absorbed in the discharge provides $n\Delta\nu_i$ electrons per meter^3 per second. As there are $n\Delta\nu_i dx$ initial signal electrons per meter^2 per second produced by the microwave signal ionization rate enhancement, the electron population per unit cross-sectional receiver area over the electrode separation $\underline{b}$ is increased at the rate

$$b\nu_i\Delta n = n\Delta\nu_i \int_0^b e^{\frac{2\nu_i x}{v}}\, dx$$

$$= \frac{n\Delta\nu_i v}{2\nu_i}\,(e^{\,2\frac{\nu_i b}{v}} - 1) \tag{10}$$

or

$$\Delta n = \frac{n\Delta\nu_i}{2\nu_i t_d}\,(e^{\,2\nu_i t_d} - 1) \tag{11}$$

where $t_d = b/v$ is the average electron transit time from cathode to anode. In the abnormal glow $\nu_i t_d \gg 1$ and the signal current gain is

$$G_{abg} = \frac{(\Delta n)\nu_i}{n\Delta V_i} \approx \frac{e^{\,2\nu_i t_d}}{2\nu_i t_d} \tag{12}$$

In the subnormal glow secondary electron emission as a result of signal positive ion bombardment of the cathode is a significant source of signal gain.

One half of the signal current carrier increase consists of positive ions. Each positive ion produces δ secondary electrons upon its return to the cathode, where δ is the secondary electron coefficient and depends on the ratio of E_{dc} to gas pressure, the cathode material, and the gas mixture itself. Thus, $\delta(\Delta n)/2$ secondary electrons are emitted by the cathode. Each secondary electron generates an electron signal current gain of $e^{2\nu_i t_d}$ at the anode. The number of positive ions produced per initial secondary electron is equal to the number of electrons produced minus the original electron, or $\frac{e^{2\nu_i t_d}}{2} - 1$. These positive ions then generate $\delta(\frac{e^{2\nu_i t_d}}{2} - 1)$ secondary electrons. Thus, at the beginning of the second cycle $\delta^2 \frac{\Delta n}{2} (\frac{e^{2\nu_i t_d}}{2} - 1)$ secondary electrons are emitted from the cathode. The number of ion pairs produced in the gas is $\delta^2 \frac{\Delta n}{2} (\frac{e^{2\nu_i t_d}}{2} - 1)^2$ and the number of electrons reaching the anode, consisting of those emitted from the cathode and those they produce in the gas through ionizing collisions on the way to the anode, is $\delta^2 \frac{\Delta n}{2} (\frac{e^{2\nu_i t_d}}{2} - 1) \frac{e^{2\nu_i t_d}}{2}$. This multiplication takes place an infinite number of times. The electron number density reaching the anode as a result of secondary emission is thus

$$\Delta n_{se} = \frac{\Delta n}{2} (1 + \delta\frac{e^{2\nu_i t_d}}{2} + \delta^2(\frac{e^{2\nu_i t_d}}{2} - 1) \frac{e^{2\nu_i t_d}}{2}$$
$$+ \delta^3(\frac{e^{2\nu_i t_d}}{2} - 1)^2 \frac{e^{2\nu_i t_d}}{2} + \cdots\cdots$$

$$= \frac{\Delta n (1 + \delta)}{2 - \delta(\frac{e^{2\nu_i t_d}}{2} - 2)} . \tag{13}$$

The positive ion number density reaching the cathode is increased to

$$\Delta n_{spi} = \frac{\Delta n}{2} [1 + \delta(\frac{e^{2\nu_i t_d}}{2} - 1) + \delta^2(\frac{e^{2\nu_i t_d}}{2} -1)^2 + \cdots$$

$$= \frac{\Delta n}{2 - \delta(e^{2\nu_i t_d} - 2)} . \tag{14}$$

Hence, for the subnormal glow, summing eqs. (13) and (14),

$$G_{subg} = \frac{(2 + \delta) G_{abn}}{2 - \delta(e^{2\nu_i t_d} - 2)} . \tag{15}$$

In the abnormal glow, much of the secondary emission signal energy from the cathode is lost through elastic collisions or trapped near the cathode by the negative space charge sheath surrounding it. In this case, Eq. (15) reverts to Eq. (12) because although secondary emission does take place, it is hardly felt at the anode. Thus δ effectively approaches zero. For a fixed modulation frequency the subnormal glow response peaks sharply at the discharge current where the denominator of Eq. (15) approaches zero, corresponding to a secondary emission coefficient equal to $2/[\exp (2\nu itd) - 2]$. As the modulation frequency changes so too does the subnormal glow equivalent impedance, which, to both signal and noise, behaves as a tuned filter whose center frequency is (18), (19).

$$f_o = \frac{\mu_i}{b} \left(\frac{T}{273} \right) \left(\frac{E_{dc}}{p} \right) \tag{16}$$

where μ_i is positive ion mobility at 1 torr pressure and 273° K temperature, T is gas temperature in degrees Kelvin, and p is gas pressure in torr. As the device impedance changes so too does the electric field and with it the number of ionizing collisions an electron undergoes per transit and the secondary emission coefficient δ. Thus, as explained in detail elsewhere (18), (19), (22), the subnormal glow discharge acts as a variable video frequency tuned-filter detector with video bandwidth also proportional to f_o in Eq. (16).

Although the subnormal glow internal signal amplification is much larger than the abnormal glow internal signal amplification, as shown by Eq. (15), for microwave signals, the signal-to-noise ratios in both cases are approximately the same provided the subnormal glow discharge current is such that Eq. (15) is maximized (19). This is so because of the large cathode secondary emission noise characteristic of the subnormal glow (18), (19), which is amplified similarly by the discharge. This noise diminishes greatly with increasing discharge current because of increasing negative space charge effects surrounding the cathode (20) which diminish cathode shot noise as well (23).

The abnormal glow video frequency spectral response is flat out to a cut-off frequency determined by detector risetime (19). In common commercial indicator lamps the abnormal glow risetime is on the order of a microsecond.

Microwave Spectral Response

Since $\mu \Delta E_{dc} = \Delta v$, Eq. (5a) reduces on the left hand side to $\Delta \frac{dn}{dt}$, which describes electron number density gain as a result of enhanced ionization and loss as a result of diffusion to the tube envelope walls. At low current carrier densities the diffusion is of the free electron type and D is proportional to $\underline{v}$, thus making ΔD proportional to Δv in Eq. (1). The microwave spectral response in Eq. (3) depends on $v \cdot (\omega^2 + v^2)^{-1}$. Since the actual diffusion current loss in the subnormal glow is proportional to ReΔv, which has the same spectral dependence, but different amplitude, the subnormal glow microwave spectral response is flat (20). For $n \gtrsim 10^{14}$ m^{-3}, as in the abnormal glow, forces exerted by positive ions greatly reduce the loss of electrons by diffusions to the walls. In this case the diffusion is ambipolar. As the ambipolar diffusion coefficient is determined primarily by the positive ions, whose velocity and collision frequencies are much less than that of electrons, the microwave abnormal glow spectral response is determined primarily by $v \cdot (v^2 + \omega^2)^{-1}$ (20), (24). In this case peak response occurs when the gas pressure and dc bias are such that $v = \omega$. If the electrode length approaches one half the carrier frequency wavelength, the response in the abnormal glow can be improved further as a result of such antenna properties (24).

Stability

Stability problems arise in the subnormal glow because of the negative dynamic resistance. Stability requires of the load resistor R_L and the discharge current I_D are (19), (22):

$$|R(I_D)| \leq R_L \tag{17}$$

$$|R(I_D)| \leq \frac{L(I_D)}{R_L \, C(I_D)} \tag{18}$$

where $R(I_D)$, $L(I_D)$, and $C(I_D)$ are the discharge dynamic resistance, inductance, and capacitance respectively. The subnormal glow is characterized by a very small negative glow of volume less than a millimeter^3 at the cathode. Fluctuations in the glow itself if it changes location on the cathode can change the tube current-voltage characteristic (22), and with it the dc and dynamic resistances and equivalent reactances of the discharge and thus the tube electronic parameters. Such variations in subnormal glow discharge characteristics can be largely overcome by using several lamps in series in the circuit of Fig. 2. An additional advantage of using several tubes in series in the subnormal glow is that with such biasing the video bandwidth of the detection system is narrowed further and acts as a tuned filter with an even higher 'Q' than otherwise. The internal detector noise is limited in this way, thus decreasing the detection system noise equivalent power (NEP). If N glow lamps are used in series, the equivalent impedance of the combination is

$$Z_n(s) = \frac{N}{C(I_D)} \, \frac{[\, s + \frac{R(I_D)}{L(I_D)} \,]}{[s^2 + \frac{R(I_D)}{L(I_D)} s + \frac{1}{L(I_D) \, C(I_D)}]} \, . \tag{19}$$

When these are paralleled with a load resistor R_L the equivalent impedance is

$$Z(s) = \frac{N}{C(I_D)} \, \frac{s + \frac{R(I_D)}{L(I_D)}}{s^2 + 2 \, \alpha s + \omega_0^2} \tag{20}$$

which describes a video frequency tuned filter of bandwidth

$$2\alpha = \frac{N}{R_L \, C(I_D)} + \frac{R(I_D)}{L(I_D)} \tag{21}$$

at center frequency

$$\omega_0 = [\frac{1}{L(I_D) \, C(I_D)} \, (1 + \frac{NR(I_D)}{R_L}) \,]^{\frac{1}{2}} \tag{22}$$

To achieve subnormal glow stability R_L should be very large, on the order of a megohm. Necessary but not sufficient conditions for stability are that Eqs. (21) and (22) remain positive despite $R(I_D)$ being negative (22). The discharge inductance and capacitance decrease and increase, respectively,

with increasing discharge current(19). The effect of increasing $\underline{N}$ in Eq. (21)
is to allow the discharge to be stable at lower values of discharge current
where α, which is determined by the variance of the positive ion energy
distribution, decreases. At very low values of discharge current, where the
number of ionization and excitation collisions is very small, the ion energy
distribution is Druyvesteynian rather than Maxwellian (25). The variance of
the former is less than that of the latter. Thus, by using several lamps in
series in the subnormal glow variation in lamp characteristics as a function
of time can be largly overcome (22) and detector noise bandwidth can be
limited, thus improving sensitivity considerably. Indeed, experiments performed here using two 5AH neon indicator lamps in series biased to the subnormal glow to detect X-band (3 cm wavelength) radiation indicate a five-fold
rather than two-fold, improvement, which stems not only from the increased
response but also from the considerably decreased noise bandwidth. The
slightly higher prices involved with using more lamps are offset by the much
lower power dissipation. Power dissipation in the abnormal glow is about 3/4
W per lamp. In the subnormal glow, it is about a milliwatt per lamp (19)
primarily because of the lower discharge current as seen in Fig. 3. Risetime
requirements for the subnormal glow can be met by designing lamps for a
correspondingly high center video frequency according to Eq. (16).

CONCLUSIONS

Sensitive and economic real-time detection of microwave holograms, using
parallel processing schemes as in Fig. 1, are feasible using glow discharge
biased to either the subnormal or abnormal glow for maximum response.

The subnormal glow microwave spectral response is flat. The video frequency
spectral response is like that of a variable frequency tuned filter with
center frequency varying with discharge current which changes the dc electric
field in Eq. (16). Stability requirements for one or several lamps in series
in the circuit of Fig. 2 are described. Advantages of using several lamps
in series for the subnormal glow discharge are significant reduction in
fluctuations of lamp characteristics and noise bandwidth. The higher price
required for the greater number of lamps needed for series combinations is
offset by very low power dissipation per lamp. Signal gain in the subnormal
glow discharge is derived and is found to be quite high in accordance with
experimental measurements (19). The high signal gain is offset, however, by
correspondingly high secondary electron emission noise whose variance and
spectrum are characterized by the positive ion energy distribution. Diffusion
losses are minimized when average electron energy is about 69% of that required to ionize a gas atom.

The abnormal glow microwave spectral response behaves according to the
expression $\nu \cdot (\nu^2 + \omega^2)^{-1}$ and is maximum when $\nu = \omega$. The video frequency-
spectral response is flat out to a maximum frequency determined by tube rise-
time limitations, typically on the order of a microsecond for commercial
indicator lamps. Maximum initial, unamplified, response for both the normal
and subnormal glow discharges depends on increasing the preionization level
as described by the electron population in Eqs. (3) and (11).

Advantages of using glow discharges, as compared to diode detectors, in the
parallel processing scheme described here, and of such a technique as
compared to non-parallel processing schemes, have been described.

REFERENCES

(1)   N. S. Kopeika,
        Int. J. Electr., 38, 609 (1975).

(2)   C. F. Augustine and W. E. Kock,
        Proc. IEEE, 57, 354 (1969).

(3)   C. F. Augustine, et-al,
        Proc. IEEE, 57, 1333 (1969).

(4)   B. J. Lechner, F. J. Marlower, E. O. Nester, and J. Tults,
        Proc. IEEE, 54, 1566 (1971).

(5)   N. H. Farhat and W. B. Guard,
        Proc. IEEE, 59, 1383 (1971).

(6)   N. H. Farhat,
        J. Franklin Inst., 296, 393 (1973).

(7)   K. Iizuka,
        Appl. Optics, 12, 147 (1973).

(8)   L. G. Gregoris and K. Iizuka,
        Applied Optics, 14, 1487 (1975).

(9)   N. H. Farhat and N. S. Kopeika,
        Proc. IEEE, 60, 759 (1972).

(10)   N. S. Kopeika and N. H. Farhat,
        IEEE Trans. on Electr. Dev., 22, 534 (1975).

(11)   N. S. Kopeika and N. H. Farhat,
        ibid, 22, 540 (1975).

(12)   N. S. Kopeika, B. Galore, D. Stempler, and Y. Heimenrath,
        IEEE Trans. on Microwave Theory and Tech., 23, 843 (1975).

(13)   M. Wu and N. H. Farhat,
        Proc. IEEE, 63, 1254 (1975).

(14)   F. A. Benson and G. Mayo,
        J. Sci. Instru., 31, 118 (1954).

(15)   N. S. Kopeika and A. P. Kushelevsky,
        Proc. IEEE, 64, 369, (1976).

(16)   N. H. Farhat,
        Proc. IEEE, 62, 279 (1974).

(17)   N. S. Kopeika,
        Proc. IEEE, 63, 981 (1975); 1737 (1975).

(18)   N. S. Kopeika,
        Int. J. Electr., 39, 209 (1975).

(19)   N. S. Kopeika,
        IEEE Trans. on Plasma Science, 4, 51 (1976).

(20)   N. S. Kopeika,
        (to be published).

(21)   N. H. Farhat,
        Proc. IEEE, 52, 1053 (1964).

(22)   N. S. Kopeika, S. Glazer, and R. Katz,
        Int. J. Electr., 40, 481 (1976).

(23)   Davenport, W. B. Jr. and Root, W. L.,(1958),
        An Introduction to the Theory of Random Signals and Noise,
        McGraw-Hill, New York.

(24)   N. S. Kopeika, D. Even-Zur, and I Fishbein,
        Proc. IEEE, 64, 382 (1976).

(25)   Von Engel, A., (1965), Ionized Gases,
        Clarendon Press, London.

IMAGING THROUGH ABERRATING MEDIA BY MEANS OF PHASE-MODULATED REFERENCE BEAM

G. Papi, V. Russo, S. Sottini

Istituto di Ricerca sulle Onde Elettromagnetiche del C. N. R., Firenze, Italy

ABSTRACT

Imaging through random aberrating media can be of interest not only in optics but also in microwaves and in acoustics. It has been found that holography offers some unique advantages for this problem and a microwave holographic technique has been already proposed. With this technique the corrected image is obtained from a hologram that is the product of the aberrated hologram with another, where the object is substituted by a point source, compensating the aberration. This method is particularly convenient in the case of time invariant media because it allows the recording of the compensating hologram in advance, while it seems to have some disadvantages for time-variant media. Here we describe a new method easier than the previous one in many cases, particularly for time variant media. The method consists of recording an <u>in line</u> microwave hologram of the object through the aberrating medium. The reference source, located near the object, works also as compensating source. The off-set angle is simulated by a suitable phase modulation of the reference beam. In this way the object and reference beams pass through the same portions of the perturbing medium and, at the same time, a suitable off-set angle can be obtained in the optical reconstruction.

Preliminary tests have been carried out that, although not completely satisfactory, seem to confirm the feasibility of the method.

INTRODUCTION

Holography offers some unique advantages for imaging through random aberrating media. This problem has received considerable attention in the past years and different holographic techniques have been investigated in the optical region (Refs 1 to 6).

More recently imaging through aberrations has been studied also for application in microwave and in acoustics, and a microwave holographic technique has been proposed which accomplishes the imaging by multiplying two microwave holograms. The first hologram is obtained with the object wave aberrated by the random medium; the other one, recorded with a point source in the place of the object, produces the compensating wavefront. This method overcomes the difficulties that would arise by using, at microwaves, the techniques already tested in optics (Ref 7). It is particularly convenient in the case of time invariant aberrating media while it shows some disadvantages for time variant media.

After a presentation of this method, here we describe a new technique which seems to be preferable to the previous one in many cases, particularly for time-variant media. The compensation is obtained by passing both the reference and the object beam through the aberrating medium analogously to the well known optical technique (Ref 1), but here the microwave hologram is recorded in line while the off-set angle necessary for the optical reconstruction is simulated by a suitable phase modulation of the reference beam. An usual reference wave cannot be used at microwaves where large off-set angles are necessary (Ref 8), while the object and the reference beam must pass through highly correlated portions of the perturbing medium (Ref 1;7).

COMPENSATION OF THE ABERRATION BY MEANS OF MULTIPLICATION OF TWO HOLOGRAMS

The imaging through random aberrating media is usually accomplished by compensating the aberrated object wavefront with a phase perturbation conjugate of the aberration (Ref 9). With the multiplication technique it is achieved, making the product of two microwave holograms. As sketched in Fig. 1, the first hologram is recorded with the object wave aberrated by the random medium; the second one, where the object is substituted by a point source, produces the compensating wavefront. For the optical wavereconstruction, the product of these two holograms must be recorded onto an optical transparency. It can be easily shown that the transparency gives rise to the corrected image when it is illuminated by laser light.

For a good correction it is necessary that the rays from the object and the rays from the compensating point source received at any one point on the holographic aperture, have passed through identical (i.e. highly correlated) portions of the perturbing medium. It can be expressed in a quantitative form introducing the coherence width Δ, that is the average linear dimension over which the wavefront emerging from the aberrating medium shows a phase perturbation within $\lambda/8$. With the notations of Fig. 1, it can be easily shown (Ref 1) that the <u>coherence width</u> must satisfy to the condition:

$$\Lambda \geq d/D.\ell/2 \qquad\qquad (1)$$

This condition does not introduce any limitation in the reference wave and, in particular, the off-set angle can be done as large as necessary for a good reconstruction. It is very important at microwaves where the recording off-set angle is usually strongly reduced in the optical reconstruction as a consequence of the much smaller wavelength. In particular this technique, as shown

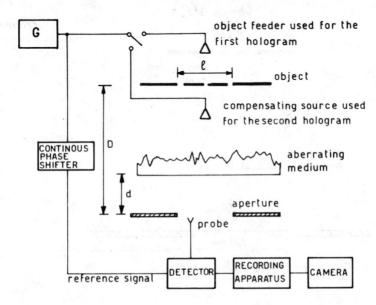

Fig. 1. Set up for recording the two microwave holograms used in the multiplication technique.

in Fig. 1, allows that the off-axis reference can be obtained by a suitable signal internal to the detecting apparatus (Refs 8, 10) It avoids the possibility that the aberrating medium disturbs the reference wave.

This microwave holographic technique has been tested at 10 GHz with the arrangement shown in Fig. 2. The object was a transparency constituted by three slits each 5 cm wide, separated by 16 cm. The aberrating medium was simulated by two longitudinal sections of cylinders of a paraffin compound ($n \simeq 1.8$), whose dimensions and curvature (50 cm width, 1.5 cm maximum thickness) were chosen in such a way to perturb as much as possible the object wavefront without going far from condition (1) ($\Delta \simeq 8.3$ cm, while $d/D.\ell/2 \simeq 8.5$ cm). For laboratory tests a real reference beam was used that reached the holographic aperture without being affected by the aberrating medium. The microwave holograms were recorded on photographic plates by using the procedure sketched in Fig. 3 and described in detail in a previous work (Ref 7).

In order to have the product of the two microwave holograms, they have been recorded in successive steps on transparencies at CRT

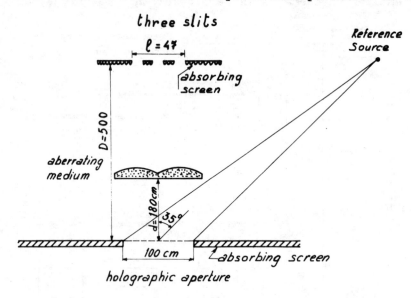

Fig. 2. Arrangement used in the tests of the multiplication technique

camera and then they have been superimposed to obtain, by contact print, a third transparency representing their product. The new hologram, reduced in size and illuminated by laser light, has given the corrected image of the object shown in Fig. 4c. For comparison, Fig. 4a shows the image of the same object obtained from a microwave hologram in absence of aberrations and Fig. 4b the object image from an usual hologram recorded in presence of the aberrating medium.

Although some noise comes in sight near to the image of Fig. 4c, by taking into account that the experiments were carried out at the limit of condition (1), the results can be considered satis-

factory. This procedure can be used only in the case of time
invariant aberrating media. This limitation can be overcome by
accomplishing the multiplication directly at microwaves. The two
microwave holograms are simultaneously detected point by point.
Then the **signals** can be separated because the sources used for
making the holograms work at two slightly different frequencies.
Experiments have been carried out using the same object and experimental disposition of the previous case. The result is shown
in Fig. 4d. The quality of the corrected image is comparable with
that of Fig. 4c.

COMPENSATION OF THE ABERRATION BY MEANS OF PHASE-
-MODULATED REFERENCE BEAM

The technique above described shows a rather low efficiency of
the corrected image because it is diffracted by a transparency

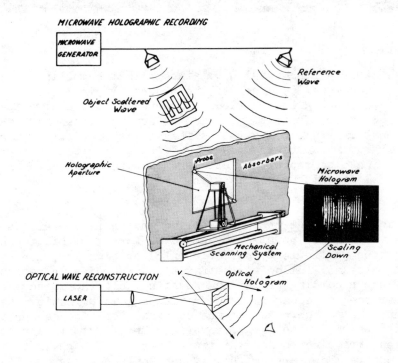

Fig. 3. Sketch of the microwave holographic recording
and of the optical wave reconstruction.

a   c

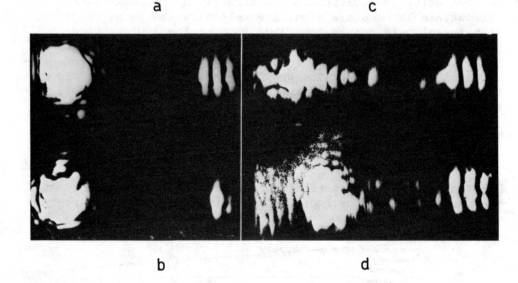

b   d

Fig. 4. Experimental results obtained with the multiplication
        technique. a) Image of the object obtained from a micro-
        wave hologram recorded in absence of the aberrating me-
        dium. b) Image of the same object in presence of the
        aberration. c) Corrected image of the object from a holo-
        gram recorded in presence of the aberration. d) Corrected
        image of the object obtained with the procedure of the
        electronic multiplication.

which is the product of two holograms and therefore diffracts
more waves than an usual hologram does. Moreover this technique
seems to be particularly suitable only for time invariant aberrations because, following the first procedure above reported,it
allows the recording of the compensating hologram in advance and
no coherence is requested between the wave illuminating the object and the source used for the compensation. The second procedure can be used also for time variant media but it requests
different frequencies for the two holograms. Though this difference can be made very slight, it can influence the quality of
the corrected image in the case of thick aberrating media. Moreover a special apparatus for the electronic multiplication is requested. For these reasons a new technique is here described
which often seems to show advantages with respect to the previous
one. The method consists of recording an <u>in line</u> microwave

hologram of the object through the aberrating medium. A real reference source must be used, located near the center of the object
(Fig. 5). This reference works also as compensating source,

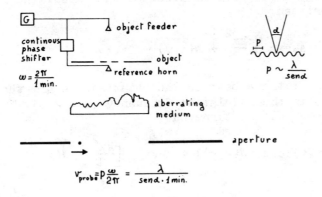

Fig. 5. Set up for the holographic imaging through
        perturbing media by means of the phase mo-
        dulated reference beam.

analogously to the optical technique already mentioned (Ref. 1).
However, here, the off-set angle is not real but simulated by a
suitable phase modulation in time of the reference beam. In this
way the object and the reference beams pass through highly correlated portions of the perturbing medium and, at the same time, an
off-set angle can be obtained large enough for a good optical wave
reconstruction. This technique, used also to put the reference signal directly into the detector (Ref 10), needs that the hologram
is detected by means of a scanning system.

Referring to Fig. 5, let us consider, for the sake of simplicity,

an x scanning, that is a system which scans the holographic aperture line by line moving in the x direction only. If an x,y scanning is performed point by point, it is supposed that the velocity v_y, in the y direction is much greater than that v_x in the
x direction. Assuming, for simplicity, that the aberrating medium
is sufficiently close to the holographic aperture so that intensity variations are not significant, its amplitude transmittance
can be represented by: $\exp[jW(x,y)]$ and the object wave reaching
the aperture turns out to be:

$$\underline{a}(x,y) \cdot \exp[,W(x,y)]$$

$\underline{a}(x,y)$ represents the object wave in absence of aberrations. Analogously, with usual approximations the time modulated reference
wave can be written:

$$A \cdot \exp\ j\left[\frac{kx^2}{2D} + W(x,y) - 2\pi\nu t\right]$$

where D is the distance between the point source and the aperture,
$k = 2\pi/\lambda$ and ν is the modulation frequency ($\omega = 2\pi\nu$). Therefore
the interference pattern turns out:

$$I(x,y) = \left|\underline{a}(x,y)\ \exp[jW(x,y)] + A\ \exp\ j\left[\frac{kx^2}{2D} + W(x,y) - 2\pi\nu t\right]\right|^2$$

$$(2)$$

Along a line of coordinate x, the detection occurs at time t =
$= x/v_x$ therefore (2) can be written also:

$$I(x,y) = \left|\underline{a}(x,y)\ \exp[jW(x,y)] + A\ \exp\ j\left[\frac{kx^2}{2D} + W(x,y) - 2\pi\ \frac{\nu}{v_x}\ x\right]\right|^2 =$$

$$= |\underline{a}|^2 + A^2 + \underline{a}^*(x,y)A\ \exp\left[j\left(\frac{kx^2}{2D} - 2\pi\ \frac{\nu}{v_x}\ x\right)\right] +$$

$$+ \underline{a}(x,y)\ A\ \exp\left[-\ j\left(\frac{kx^2}{2D} - 2\pi\ \frac{\nu}{v_x}\ x\right)\right]$$

$$(3)$$

Thus the hologram results unaffected by the presence of the
aberrating medium, on the other hand it can be easily seen that
the phase of the spherical reference wave: $kx^2/2f - 2\pi\nu x/v_x$ is
equivalent to that of an analogous wave with an off-axis angle
α given by

$$\sin\ \alpha/\lambda = \nu/v_x \qquad\qquad (4)$$

Therefore the rotation of the phase shifter of the reference
source must be related to the velocity v_x of the detector to give
the wanted value of the simulated off-axis α necessary for a
good optical reconstruction.

As already mentioned, the arrangement of the reference source at
the center of the object, minimizes the limitations of the object
field requested for a good correction. The coherence width must
still satisfy condition (1). It can be easily understood observing that the position of the compensating source with respect
to the object is the same used in the multiplication method but
now the compensating source works also as reference and therefore condition (1) introduces now a limitation in the choice of
the reference wave.

Also this technique has been tested at 10 GHz. Apart from the
scanning system, the recording set up was the same used for the
tests reported in the previous section, but only 1-dimensional
holograms could now be detected. It depended on the scanning lines of our recording apparatus that approach arcs of circles
(Ref 8) and, as a consequence, cannot be easily synchronized with
the phase shifter. Therefore, for simplicity, the holograms were
obtained by moving the point probe only on a line parallel to
the x axis, while the other dimension was artificially introduced
by means of a rapid excursion in the y direction of the spot of
the oscilloscope whose intensity was modulated by the RF signal
from the probe. The CRT camera recorded the microwave holograms
so obtained, which were later reduced in size for the optical
reconstruction.

The arrangement is shown in Fig.5, the object transparency and
the aberrating structure were the same used for the experiments
already described. The same geometry was also maintained so that
the two members of (1) had yet the values: $\Lambda \sim 8.3$ cm and
$d/D.\ell/2 \sim 8.5$ cm. The shape of the object transparency as well
as that of the aberrating structure seemed to be suitable to
minimize the inconvenience of the only 1-dimensional scanning.
The period of the rotation of the phase shifter $T = 1/\nu$ was
1', and the velocity v_x of the probe ~ 5.3 cm/min. As a consequence the simulated off-set angle was $\sim 35°$.

The experimental results are shown in Fig. 6. In particular
Fig. 6a and b are analog respectively to Fig. 4a and b. The only
difference is that these images have been obtained from only
1-dimensional holograms with the procedure described. These images must be compared with that shown in Fig. 6c. The hologram
was recorded in line in presence of the aberration that was

corrected with the technique of the phase modulated reference
beam.

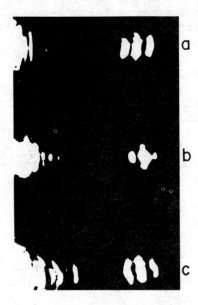

Fig. 6. Experimental results obtained with the technique of the
        in line phase modulated reference beam. a) Image from a
        hologram recorded in absence of aberrations. b) Image
        from a hologram of the same object in presence of the
        aberration. c) Corrected image of the object.

The results of this preliminary test have been deteriorate by the
impossibility of using usual 2-dimensional holograms and by the
recording system that introduced a valuable noise whose effects
are evident in the irregularities or the images of Fig. 6 in the
vertical direction. Nevertheless it seems that the compensation,
although not completely satisfactory, has been accomplished.

CONCLUSIONS

A new holographic technique has been proposed to image through
aberrating media at microwaves. It consists of recording an in
line microwave hologram through the aberration using the reference as compensating wave. The off-set angle necessary for the optical reconstruction is simulated by a suitable phase modulation
of the reference beam. This technique has been compared with that,
already suggested and tested, of the multiplication of two holograms.

The new method seems to be easier than the previous one in many

cases, particularly for time-variant media.It could be applied,
for instance to the holographic analysis of portions of ionosphe-
re (Refs 11,12) or to plasma diagnosis avoiding the disturbance
of the containers. Other interesting applications of this technique can be suggested in acoustic to biomedical applications
(Ref 13), to the mapping of subsurface cavity, to nuclear reactor
surveillance and to undersea viewing (Ref 14). It is to be noted
that the phase modulated reference wave could be convenient to
record usual acoustical or microwave holograms with the in line
geometry, when the technique of internal reference signal cannot
be used.

REFERENCES

(1) J.W.Goodman, W.H.Huntley Jr., D.W.Jackson and M.Lehmann,
     Wavefront reconstruction imaging through random media,
     Appl.Phys.Lett. 8, 311, (1966)
(2) E.N.Leith and J.Upatnieks, Holographic imaging through dif-
     fusing media, JOSA 56, 4, 523, (1966)
(3) K.A.Stetson, Holographic fog penetration, JOSA 57, 8, 1060,
     (1967)
(4) J.D.Gaskill, Imaging through a randomly inhomogeneous medium
     by wavefront reconstruction, JOSA, 58, 5, (1968)
(5) H.Kogelnik and K.S.Pennington, Holographic imaging through
     a random medium, JOSA 58, 2, 273, (1968)
(6) O.Bryngdahl and A.Lohmann, Holographic penetration of an inho-
     mogeneous medium, JOSA 59, 9, 1245 (1969)
(7) G.Papi, V.Russo and S.Sottini, Imaging by microwave holography
     through aberrating media, IEEE Trans on AP (to be publ.Nov.76)
(8) G.Papi, V.Russo and S.Sottini, Microwave holographic interfe-
     rometry, IEEE Trans on AP-19, 6, 740 (1971)
(9) J.W.Goodman, Introduction to Fourier optics, McGraw Hill,
     San Francisco, 261-268, 1968
(10) G.A.Deschamps, Some Remarks on radio frequency holography,
     Proc.IEEE, 570, (April 1967)
(11) G.L.Rogers, A new method of analysing ionospheric movement
     records, Nature 4509, 613 (1956)
(12) Schmidt G., Oksman J., Tauriainen A., Geophysical interpreta-
     tion of the results of the holographic method, Proc. of the
     Beacon Sat. Inv. of the Ionosphere Struct. and ATS-6 Data
     1, 137, Izmiran, Moscow, 1975
(13) F.L.Thurstone, Acoustical imaging of biological tissue, in
     Acoustical Holography, ed. by Metherell and Larmore, II,
     265, Plenum Press N.Y. 1970
(14) Hildebrand B.P. and Brenden B.B., An Introduction to Acoustical
     Holography, Cap.8, Plenum Press N.Y., 1972

HOLOGRAPHIC DETECTION OF ACOUSTIC RADIATION SOURCES

Sadayuki Ueha, Shigeru Oshima and Jumpei Tsujiuchi

Imaging Science and Engineering Laboratory, Tokyo Institute of Technology
4259 Nagatsuta, Midori-ku, Yokohama 227, Japan

ABSTRACT

The computer reconstruction of a partially coherent acoustic
extended source, as well as coherent and incoherent sources, can
be made, as an extension of holographic techniques, from interferograms using the propagation law of mutual intensity, and is
useful for detection of acoustic radiation sources. Using a
motor-cycle engine and a tweeter array as objects, reconstruction
experiments are carried out together with the theoretical
approach.

INTRODUCTION

Acoustic noise pollution by machine tools, traffic vehicles and
airplanes is a serious problem, and mapping or locating acoustic
radiation sources with high resolution is essential in the
reduction of these noises.

For spatially coherent acoustic sources, holographic techniques
have been successfully used [1] [2]. For incoherent or partially
coherent sources, however, conventional imaging methods used in
optics can hardly be applicable, because imaging elements like
lenses or Fresnel zone plates are difficult to provide. Only a
two-dimensional phased array, which is a kind of passive sonar,
can be used, but the apparatus is complicated and expensive.

Recently we have proposed an imaging method using the propagation
law of mutual intensity as an extension of the holographic
technique, which is applicable not only for coherent sources but
also for both partially coherent and incoherent sources[3]. In
this method radiation sources can be reconstructed from an
interferogram which is recorded by measuring the cross-correlation of signals at every point in an observation plane with a
signal at a fixed point chosen as the reference point. Using a
motor-cycle engine and a tweeter array as partially coherent and
incoherent sources respectively, computer reconstruction
experiments from their interferograms are carried out.

697

THEORY

If the mutual intensity of two points, x_1 and x_2, in the object plane, as shown in Fig. 1, is $J_0(x_1,x_2)$ the mutual intensity distribution $J(u_1, u_2)$ of two points, u_1 and u_2, in an observation plane is given by

$$J(u_1,u_2) = \iint_{\Sigma\Sigma} J_0(x_1,x_2)\frac{\exp[ik(S_1- S_2)]}{S_1S_2} dx_1dx_2 \qquad (1)$$

according to the propagation law of mutual intensity [4], where S_1 is the distance between x_1 and u_1, S_2 that of x_2 and u_2, and k the propagation constant, and the inclination factors are assumed to be unity.

Using the Fresnel approximation;

$$S_1 = R + \frac{(u_1 - x_2)^2}{2R}$$

$$S_2 = R + \frac{(u_2 - x_2)^2}{2R} \qquad (2)$$

eq.(1) is rewritten as

$$J(u_1,u_2) = \iint_{\Sigma\Sigma} J_0(x_1,x_2)\exp[ik\frac{(u_1-x_1)^2-(u_2-x_2)^2}{2R}] dx_1dx_2 \qquad (3)$$

apart from a constant factor, where R is the distance between the object plane Σ and the observation plane Ω. In practice the mutual intensity $J(u_1,u_{20})$ in the plane Ω can be obtained by cross-correlating signals at every point u_1 with a signal at a fixed point u_{20} chosen as a reference. The real part of the mutual intensity $Re[J(u_1,u_{20})]$ may be called an interferogram, or a hologram in special cases [5].

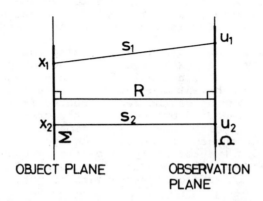

OBJECT PLANE   OBSERVATION   Fig. 1
                      PLANE

As the equation

$$\mathrm{Re}\{J(u_1,u_{20})\} = \frac{1}{2}\{J(u_1,u_{20}) + J^*(u_1,u_{20})\} \qquad (4)$$

holds and the second term of eq.(4), as well as in Gabor type holography, yields only a conjugate image of the object, we shall confine ourselves only to the first term. Then the inverse Fresnel transformation of $J(u_1,u_{20})$ gives

$$J_0'(x : u_{20}) = \int_\Omega J(u_1,u_{20}) \, \exp[-ik \, \frac{(x - u_1)^2}{2R}]du_1$$

$$= \exp[-\frac{ikx^2}{2R}] \iiint_{\Sigma\Sigma\Omega} J_0(x_1,x_2)\exp[\frac{ik}{2R}\{x_1^2-(x_2-u_{20})^2\}]$$

$$\times\exp[\frac{ik}{R}(x-x_1)u_1]du_1dx_1dx_2 . \qquad (5)$$

If the plane Ω is so large that the relation

$$\int_\Omega \exp[\frac{ik}{R}(x-x_1)u_1] \, du_1 = \delta(x-x_1) \qquad (6)$$

is satisfied, where $\delta(x)$ is the delta function, eq.(5) is reduced to

$$J_0'(x:u_{20}) = \int_\Sigma J_0(x , x_2) \, \exp[-\frac{ik}{2R}(x_2 -u_{20})^2]dx_2 . \qquad (7)$$

This calculation corresponds to the reconstruction process of the object from the hologram, or more generally from the inter-ferogram.

For the case where the intensity of the source element of the object does not change with respect to time and only the phase changes, the mutual intensity $J_0(x_1,x_2)$ becomes

$$J_0(x_1,x_2) = <A(x_1)A(x_2)\exp\{i[\Phi(x_1,t) - \Phi(x_2,t)]\}>$$

$$= A(x_1)A(x_2)\gamma(x_1,x_2), \qquad (8)$$

where $A(x)$ and $\Phi(x,t)$ are the amplitude and the phase of the radiation at a point x in the object, $\gamma(x_1,x_2)$ is the complex degree of coherence, and $<\ >$ denotes average with respect to time. Using eq.(8), eq.(7) becomes

$$J_0'(x:u_{20}) = A(x)\int_\Sigma A(x_2)\gamma(x ,x_2)\exp\{- \frac{ik}{2R}(x_2-u_{20})^2 \}dx_2 \qquad (9)$$

and if the complex degree of coherence $\gamma(x_1, x_2)$ depends only on the distance $x_1 - x_2$, we have finally

$$J_0'(x:u_{20}) = A(x)\int_\Sigma A(x_2)\gamma(x,x_2)\exp[-\frac{ik}{2R}(x_2-u_{20})^2]dx_2$$
$$= A(x)A'(x:u_{20}), \tag{10}$$

where

$$A'(x:u_{20}) = \int_\Sigma A(x_2)\gamma(x,x_2)\exp[-\frac{ik}{2R}(x_2-u_{20})^2]dx_2 \tag{11}$$

means the disturbance which would arise at the reference point u_{20} from the source whose amplitude distribution is modulated by the complex degree of coherence around a point x in question.

In the general case, if we suppose, for simplicity, that the complex degree of coherence is expressed by

$$\gamma(x, x_2) = \exp[-(x - x_2)^2/a^2], \tag{12}$$

where a is the radius of a circular area within which the object will be estimated as spatially coherent, i.e. the area of coherence. Then, $A'(x:u_{20})$ will be the disturbance at u_{20} from the area of coherence around a point x.

Two extreme special cases, that is, $a\rightarrow\infty$ and $a\rightarrow0$ are considered. The former corresponds to the case of a spatially coherent source, and

$$\gamma(x,x_2) = 1 \tag{13}$$

then we have

$$A'(x:u_{20}) = \int_\Sigma A(x_2)\exp[-\frac{ik}{2R}(x_2-u_{20})^2]dx_2$$

$$= \text{constant} \tag{14}$$

and finally from eqs. (10) and (14)

$$J_0'(x : u_{20}) \propto A(x); \tag{15}$$

this means that the calculated image by eq. (10) corresponds to the amplitude distribution of the object, and the processing results in ordinary holography.

The latter case represents a spatially incoherent source where

$$\gamma(x,x_2) = \delta(x - x_2) \tag{16}$$

then

$$A'(x:u_{20}) = \int_{\Sigma} A(x_2)\delta(x-x_2)\exp[-\frac{ik}{2R}(x_2-u_{20})^2]dx_2$$

$$= A(x)\exp[-\frac{ik}{2R}(x-u_{20})^2]. \tag{17}$$

This means that the modulus of $A'(x:u_{20})$ is proportional to the amplitude $A(x)$, and $J_0'(x:u_{20})$ gives the intensity distribution of the object. This is the case of van Cittert-Zernike's theorem, and the absolute value of the calculated image by eq. (10) shows the intensity distribution of the object.

EXPERIMENT

Setup of experiment

The experimental setup is almost the same as that of scanned acoustical holography except for signal processing. A schematic diagram of the system is shown in Fig.2. A microphone M_1 which receives the object signal scans in an observation plane whose dimensions are 63 cm in both the X and Y directions and another microphone M_2 which receives the reference signal is spatially fixed, as mentioned in the previous section.

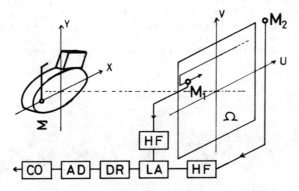

Fig.2. Schematic diagram for making a hologram, HF; heterodyne filter, LA; lock-in amplifier, DR; data recorder, AD; A/D converter, Co.; computer.

The distance between the observation plane and the object is R.
The microphone M_1 is moved by a pulse motor and stopped at every
sampling point where the signal is picked up. To produce an
interferogram the output signals of microphones M_1 and M_2 are put
into a lock-in amplifier as the reference and the object signals
repectively. The interferogram signal, which is a product of the
reference and object signals, is time averaged at every sampling
point to reduce temporal fluctuations. This is also carried out
by the lock-in amplifier with an appropriate averaging time. The
output signal of the lock-in amplifier is recorded on magnetic
tape using a data recorder together with the position signals for
the microphone M_1.

The reconstruction of the image from the interferogram is carried
out by using the discrete Fresnel transformation by a computer
and the reconstructed image is then displayed by a line printer
using an overprinting technique [6]. The frequency used for
producing the interferogram is selected by means of a tunable
band-pass filter in the lock-in amplifier. Two narrow band-pass
filters or heterodyne filters can also be used both for the
reference and object signals if their frequency spectra are
extended.

Experimental result

As a partially coherent source a motor-cycle engine is employed.
The motor-cycle engine whose vertical and horizontal dimensions
are about 35 cm and 34 cm respectively, cooled by a blower, is
placed 55 cm from the observation plane.
The spectrum of the noise at a revolution of 5000 rpm is shown
in Fig. 3. Since the spectrum is rather similar to white-noise,
a narrow bandwidth of 100 Hz and a large averaging time of 3
seconds are used.

An interferogram at a frequency of 4.2 KHz is recorded using a
sampling interval of 3 cm. The resolution of the image reconstructed is 7.5 cm according to Rayleigh's criterion and the
number of sampling poins is 484. The interferogram and the
reconstructed image together with the shape of the engine are
shown in Fig.4.

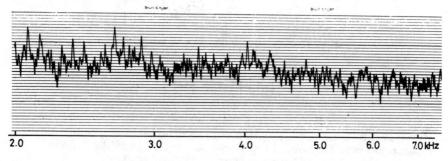

Fig. 3   Spectrum of a motor-cycle engine.

At this frequency, the cylinder-head where combustion takes place vibrates most strongly as expected, as well as in the neighborhood of the kick-starter. Two outside white-portions below the body correspond to the stays which support the engine. The central white-portion just below the body is probably due to echoes between the body and the floor.

As another object a tweeter array forming the letter H is employed. The tweeters are driven at a frequency of about 10 KHz with separate oscillators at slightly different frequencies, and this object can be regarded as an incoherent source. The diameter of the element tweeter is 3 cm and the distance R is 100 cm in this case. The reconstructed intensity, shown in Fig.5, is slightly different from tweeter to tweeter although all the tweeters are driven with sinusoidal signals of equal amplitude. This difference is probably accounted for by misalignment and/or differences in the conversion efficiencies of the tweeters.

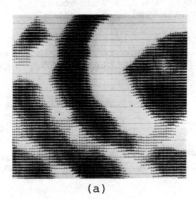
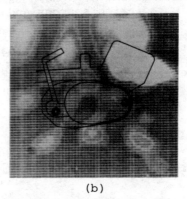

(a)   (b)

Fig. 4   Experiment on a motor-cycle engine at a frequency of 4.2 KHz.
(a) Interferogram,   (b) Reconstructed image.

Fig.5. Reconstructed image of a tweeter array whose shape is the letter H.

CONCLUSION

Using the propagation law of mutual intensity a holographic imaging method applicable for partially coherent objects is proposed and it can be reduced to principles used in holography and in radio astronomy [7] if the degree of coherence is extended to either extreme. The interferogram can be obtained by using almost the same technique as used for scanned acoustical holography, and its reconstruction can be carried out by computer as well as by optical transformation.

Using a lock-in amplifier interferograms of a motor-cycle engine and a tweeter array are recorded at certain frequencies to obtain good reconstructed images. If another frequency is used to record an interferogram, mapping of sources which vibrate just at that frequency can be performed. This method is applicable to any other machine or apparatus and is useful for locating acoustic radiation sources and for analyzing the vibration of machines.

The authors wish to express their gratitude to Prof. Umezawa, Mr. H. Hojo and Mr. J. I. Hill of Tokyo Institute of Technology for their cooperations.

REFERENCES

[1] S.Ueha, M. Fujinami, K. Umezawa and J. Tsujiuchi, Mapping of noise-like sound sorces with acoustical holography, Appl. Optics 14, 1478 (1976).

[2] S. Ueha, M. Fujinami, K. Umezawa and J. Tsujiuchi, Imaging of acoustic radiation sources with acoustical holography, Opt. Acta 23, 107 (1976).

[3] S. Ueha, S. Oshima and J. Tsujiuchi, Image reconstruction by using the propagation law of mutual intensity, Opt. Commun., to be published.

[4] Born, M. and Wolf, E., Principles of Optics, Pergamon Press, Oxford, 1975.

[5] O. Bryngdahl and A. W. Lohmann, Interferograms are image holograms, J. Opt. Soc. Am. 58, 141 (1968).

[6] I. D. G. Macleod, Pictorial output with a line-printer, I.E.E.E. Trans. Comput., 19, 160 (1970).

[7] Bracewell, R. N., Radio astronomy techniques, Handbuch der Physik (ed, S. Flugge, Springer) 54, 42 (1962).

PRELIMINARY CONSIDERATIONS OF MICROWAVE HOLOGRAPHIC IMAGING OF WATER WAVES

E. L. Rope, G. Tricoles, On-Ching Yue

*General Dynamics Electronics Division, P. O. Box 81127,
San Diego, California 92138*

ABSTRACT

A new approach is given to the diagnostics of microwave scattering by water waves. The method is microwave holographic imaging of water waves in a laboratory tank. Both mechanically and wind-generated waves were imaged. Although preliminary the results demonstrate feasibility.

INTRODUCTION

The scattering of microwaves by the sea depends on many factors, including the shape of the surface and the frequency and polarization of the incident wave (Ref 1). Analytical connections between properties of the scattered field and the surface are difficult to establish because the surface is complex and imprecisely known, especially for growing waves. The difficulties have stimulated diagnostic measurements with radars (Ref 2) and laboratory measurements with wave tanks (Refs 3,4). Photographs of the sea have been utilized to obtain surface spectra (Ref 5).

This paper presents a new approach to the diagnostics of microwave scattering by water waves. The method is microwave holographic imaging of water waves in a laboratory tank. Both mechanically and wind generated waves were considered. Two techniques were utilized. In one, holograms were formed by interference between forward-scattered microwaves and unreflected waves. The images were produced by laser light reconstruction. The other technique also used a bistatic arrangement, but the scattered field was measured by a down-looking antenna. Images were computed.

The method of this paper differs from earlier work. The method uses images, as do the optical methods; however, the images are formed with microwaves, so the scattering is not scaled. Of course, the image resolution is low because of the long wavelength. Earlier microwave experiments measured radar cross section, significant for radar systems, or Doppler spectra, related to water wave kinematics such as orbital velocities. The purpose of imaging is to locate scattering regions and determine their temporal behavior.

705

HOLOGRAPHIC IMAGING BY FORWARD SCATTER

This section describes microwave holographic images of mechanically generated water waves for oblique microwave incidence and
forward scattering in approximately specular directions from the
mean water surface.

Hologram Formation

Holograms were formed by illuminating a water wave tank with
16.50 GHz microwaves. Figure 1 shows the arrangement. The tank
was 5 m long and 1 m wide; water depth was 28 cm. Water waves
were generated by periodically moving a wedge vertically with a
motor and eccentric cam at one end of the tank. The other end
had a sloping, perforated baffle to reduce water wave reflections.
Figure 2 shows a side view of water waves, with length and speeds
approximately 33 cm and 68 cm/sec respectively. Wavelength and
speed were varied by adjusting the motor and cam. Figure 3 shows
the microwave apparatus. The transmitting antenna was an electromagnetic horn, aperture 3.8 by 5 cm; it was placed at the same
end of the tank as the water wave source. Polarization was
vertical. An array of 16 receiving antennas, each with a diode
detector, was at the other end of the tank. The array detected
the fringe pattern formed by interference between the field
scattered by the water surface and that proceeding directly from
the transmitter to the receiving array. Two values of antenna
spacing were used in the array, either 2.54 or 5.08 cm. A set of
16 light bulbs, one for each antenna and detector, displayed the
fringe pattern. Each bulb was connected to the microwave detector through an amplifier. The array of lamps was supported by
the carriage of a rectangular recorder that was placed in a small
darkened chamber and photographed with a Polaroid camera. The
carriage was displaced linearly with time to record the fringe
pattern as a function of time and therefore water wave motion.
The film and amplifiers were non-linear so the brightness was
proportional to microwave intensity for only a 6 dB range. The
display was essentially binary. Figure 4 shows a photograph of
the fringe patterns for an exposure approximately one second.
Narrow vertical sections corresponding to 1/56 second were
utilized as one dimensional holograms. Photographic reduction
produced 1/500 scale optical holograms.

Optical Wavefront Reconstruction

Images were produced by illuminating the 1/500 scale holograms
with a 6328 Å beam directly from a Spectra Physics 112 laser.
Images were recorded by a Polaroid camera.

A preliminary experiment established the scale of the images.
The object was a metallic hemicylinder on a sheet of absorbing
material placed over the water tank. The hemicylinder was moved
to a sequence of positions, near the center of the propagation
path, to increase the angle between the direct, reference wave
and the reflected, object wave; see Fig. 5. The off-axis position of the images increased linearly with displacement.

AE01   Fig. 1.

Wave tank and antennas for intensity holograms; the
antenna in the foreground transmitted, and the receiving
array is in the background. The water wave source was
below the transmitting antenna.

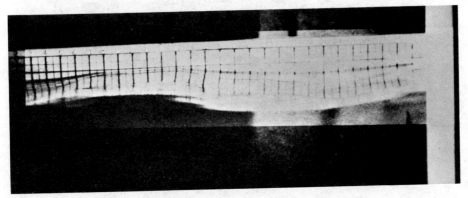

AE02   Fig. 2.

Mechanically generated water waves.

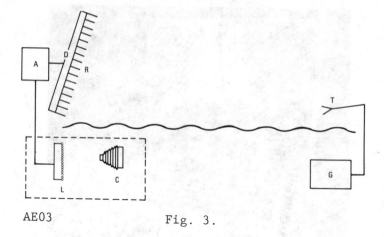

AE03   Fig. 3.

Apparatus for forming intensity holograms. G is a
continuous wave signal generator; T a transmitting
antenna; R a receiving array of 16 antennas; D represents detectors, and A amplifiers; L is an array
of 16 miniature lamps; C is a camera.

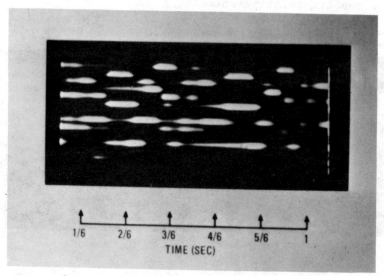

AE04   Fig. 4.

Time exposure photograph of fringe patterns formed by
reflection of 16.50 GHz waves from mechanically driven
waves; picture width corresponds to a one second temporal interval; arrows indicate positions of a narrow
slit used to form one-dimensional holograms.

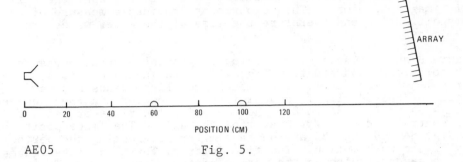

POSITION (CM)

AE05   Fig. 5.

Arrangement for imaging hemicylinder; initial position
was 60 in.; final position 120 in.

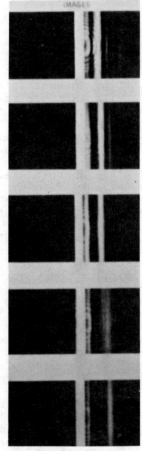

AE06   Fig. 6.

Images of metallic hemicylinder 8.9 cm diameter for a
sequence of positions; array length was 38 cm.

Figure 6 shows a sequence of images, and Fig. 7 graphs the off-
axis image positions. This experiment provided a measure of
demagnification and therefore the size of the water waves from
the images.

Figure 8 shows images of mechanically generated water waves for
six holograms produced from Fig. 4 for the indicated times. The
array length was 76 cm. The images are bright bands and have
little of the detail of optical images. However, in the sequence
of images in Fig. 8A to 8D, the images on the right or left of
the central order shifts away from the axis. The image positions
in Fig. 8A and Fig. 8D are quite similar, suggesting the periodicity of the wave motion. The time interval for these five images
is 0.5 second, which is the period of the water waves. The
sequence also shows images as split bands. Since the waves did
not break, but were continuous, this result suggests that the
microwave reflections may arise from more than one part of the
water wave.

IMAGING WIND GENERATED WAVES

This section describes images of wind generated waves for microwaves incident at small grazing angles. A bistatic arrangement
was used with the transmitting antenna at one end of the tank,
and the receiving antenna with axis vertical near the center.
Reconstructions were computed.

In the experimental arrangement, Fig. 9, microwaves of frequency
16.5 GHz, wavelength 1.82 cm, were radiated by a vertically
polarized horn antenna aperture 3.2 cm by 5 cm. Incidence was
grazing with the grazing angle approximately 3°. The receiving
antenna was a small horn aperture 2 cm by 3.3 cm. This aperture
gave better image resolution than a larger antenna. The aperture
was parallel to the mean water surface so the broad main lobe was
directed downward. The receiving antenna was supported 15 cm
above the water surface by a boom attached to a vertical track
that permitted height adjustments. The water waves were generated by wind from a motor-driven, squirrel cage blower. A tunnel
and baffles smoothed the airflow somewhat. Even with this
precaution, the wave profiles varied across the tank, as can be
seen in Fig. 10. The water waves have two sets of wavelengths.
The longer waves have wavelengths between 12 and 15 cm, and the
capillaries have wavelengths between 1 and 2 cm. Microwave
phase and intensity were measured with a network analyzer that
utilized a reference field directed by a waveguide path. The
receiving antenna in Fig. 9 was covered with absorbing material
to attenuate direct radiation. Both antennas were stationary;
however, an aperture was synthesized while the water waves moved
beneath the receiving antenna. Data were digitized from an
oscilloscope display. Data were recorded for a short interval
that correspond to the time estimated for water waves to advance
15 cm, roughly the wavelength of the longer water waves. Although
the interval was precisely controlled, the relation to the period
is only an estimate because of nonuniformity of the water waves.

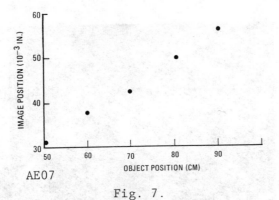

AE07

Fig. 7.

Distances of images in Fig. 6 from zero order.

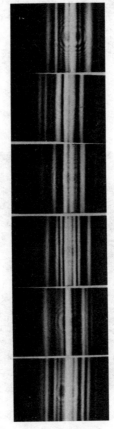

AE08   Fig. 8.

Optical reconstructions of mechanically generated
water waves; images correspond to instants denoted
in Fig. 4.

AE09   Fig. 9.

Wind-generated water waves as seen from transmitter.

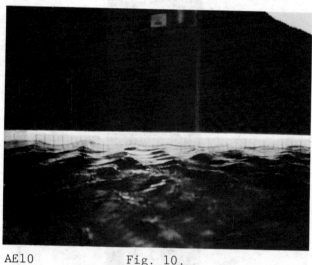

AE10   Fig. 10.

Side view of wind-generated waves.

Theory of Image Computation

Images were computed digitally from the measured phase and intensity data with algorithm based on the angular spectrum concept (Ref 6). This algorithm has been described in Ref. 7. Briefly let h(x,y,o) be the measured field distribution with the mean water surface at -z. For fixed z the angular spectrum is the two-dimensional Fourier transform:

$$H(f,g,z) = \int\int_{-\infty}^{\infty} h(x,y,z) \exp[-j\,2\pi\,(fx + gy)]\,dx\,dy. \tag{1}$$

The inverse is

$$h(x,y,z) = \int\int_{-\infty}^{\infty} H(f,g,z) \exp[j\,2\pi\,(fx + gy)]\,dx\,dy. \tag{2}$$

For homogeneous, isotropic regions, with λ the wavelength,

$$H(f,g,z) = H(f,g,o)\,P(f,g,z,\lambda) \tag{3}$$

where

$$P(f,g,z\,\lambda) = \exp j\,2\pi\,(\lambda^{-2} - f^2 - g^2)^{1/2} \tag{4}$$

From Equations 1 through 4

$$h(x,y,z) = \int\int_{-\infty}^{\infty} H(f,g,o)\,P(f,g,z,\lambda)$$

$$\exp\ j[(2\pi\,(fx + gy)]\,dx\,dy \tag{5}$$

$$= \int\int_{-\infty}^{\infty} h(u,v,o)\,p(x-u,\ y-v,\ z,\ \lambda)\,du\,dv. \tag{6}$$

Equation 6 shows h(x,y,z) can be determined from h(x,y,o). Computations were done with Equation 5 by Fourier transforming h(x,y,o) to obtain H(f,g,o), multiplying by P(f,g,z,λ), and inverse transforming.

Computed Images of Wind Driven Waves

As mentioned above, images were formed from data measured during the time interval required for water waves to travel 15 cm. The data were sampled; 50 samples were utilized with spacing corresponding to 0.28 cm. The desk calculator used for computations had capacity for 128 input data values so field of view is 36 cm.

Figure 11 shows a reconstructed image; the quantity is $|h|$ from Equation 6. The image requires interpretation. It is arbitrarily normalized to the largest computed value. The largest maxima have horizontal spacings corresponding to the periods of the capillary waves. Figure 11 suggests that the scattering in the upwind direction for grazing incidence arises from the crests of the capillary waves. Their spacing varies from 1¾ to 2 cm. This brackets the resolution of 1.82 cm expected for a 15 cm aperture at 15 cm distance when only the object is scanned. In Fig. 11 only one group of maxima is present although a second group at a spacing of 15 cm might be expected. A second group is missing apparently because the synthetic aperture has restricted size and reflections are received from only a portion of the illuminating surface. This interpretation follows from some auxiliary measurements and images.

To interpret images like those in Figure 11 we imaged deterministic objects. The first example was a pair of metallic cylinders. This object was imaged for two methods of collecting data. In one, the object and illuminating antenna were moved together past a fixed receiving antenna. The image is shown in Fig. 12; the illuminator was on the right side of the figure. The pair of cylinders was also imaged by moving it below the receiving antenna with the illuminator also stationary. Images were blurred. The explanation was apparent on examining the computed spectrum of the input data. The spectrum was shifted. The shift results from the linear phase shift caused by the increasing path length between object and illuminator. When the spectrum was centered on the spatial frequency axis, the resulting image was identical to that of Fig. 12.

The second example utilized a set of 11 aluminum wedges that simulated water waves. The wedges had base 12.7 cm and height 2.54 cm. The bases were coplanar as in Fig. 13. Data were recorded by scanning the object and illuminator. The center of the spectrum was displaced by -5/18 wavelength from the origin. The spectrum was shifted 5/18 wavelength and the image of Fig. 13 was obtained. The positions of the maxima coincide closely with those of the vertices of illuminated wedges. Note that only 3 maxima are distinctly produced in a region of approximately 38 cm although the field of view is 81 cm. The reason is that data were collected only for an aperture of 45 cm. This second example strongly suggests the scattering arises from the vertices. It seems reasonable to consider the crests of the capillaries as the sources imaged in Figure 11.

ACKNOWLEDGEMENTS

Helpful discussions were held with Denzil Shillwell and Dr. Jack Wright of the U.S. Naval Research Laboratory, Professor Walter Munk of the University of California, San Diego, and James Willis of the U.S. Naval Air Systems Command.

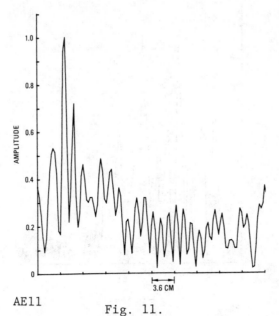

AE11

Fig. 11.

Images of wind-driven water waves with 18° spectrum shift; amplitude normalization is arbitrary to the largest computed value.

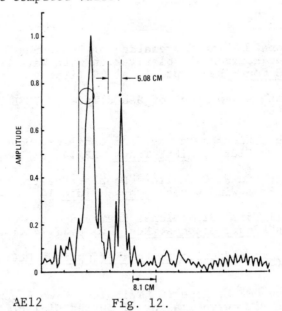

AE12   Fig. 12.

Images of two cylinders; object was moved and the antennas were stationary; spectrum was shifted -120° for the calculation.

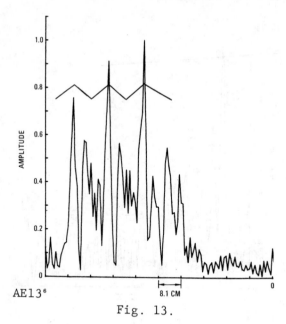

Fig. 13.

Images for a set of metallic wedges; illuminator
and object were scanned.

REFERENCES

1. F. C. Macdonald, "The Correlation of Radar Sea Clutter on
    Vertical and Horizontal Polarization with Wave Height and
    Slope," IRE Conv. Rec., pt. 1, p. 29 (1956)

2. J. W. Wright "A New Model of Sea Clutter," IEEE Trans., AP-16
    p. 217 (1968)

3. J. W. Wright "Backscattering from Capillary Waves with Appli-
    cation to Sea Clutter" IEEE Trans. AP-14, p. 749 (1966)

4. J. W. Wright and W. C. Keller "Doppler Spectra in Microwave
    Scattering From Wind Waves", Phys. Fluid 14, p. 466 (197).

5. D. Stillwell, Jr. "Directional Energy Spectra of the Sea From
    Photographs" Jour. Geophys. Res. 74, p. 1974 (1969)

6. J. Goodman, Introduction to Fourier Optics, McGraw-Hill, New
    York (1968)

7. On-Ching Yue, E. L. Rope, G. Tricoles, "Two Reconstruction
    Methods for Microwave Imaging of Buried Dielectric Anoma-
    lies", IEEE Trans. C-24, p. 381 (1975)

AUTHOR INDEX

717

SUBJECT INDEX

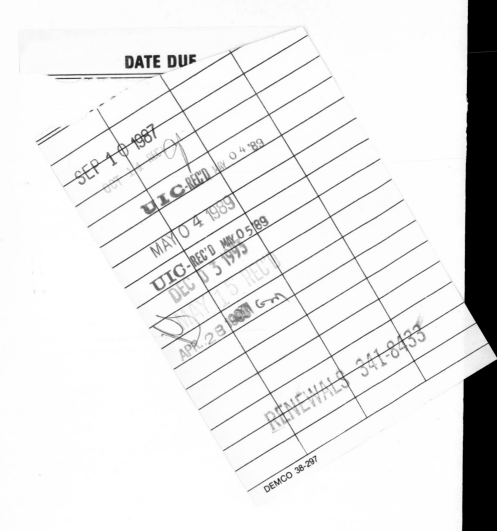